Textbook of Petrology

VOLUME ONE

PETROLOGY OF THE IGNEOUS ROCKS

Textbook of Petrology

VOLUME TWO

PETROLOGY OF THE SEDIMENTARY ROCKS

by J. T. Greensmith & the late F. H. Hatch
& R. H. Rastall

Fifth Edition

PETROLOGY OF THE
IGNEOUS ROCKS

THE LATE

F. H. HATCH

O.B.E., PH.D., M.I.C.E.

Past-President Inst. Mining and Metallurgy
Past-President Geol. Soc. of South Africa

A. K. WELLS

D.SC.

Formerly Reader in Geology
University of London, King's College

AND

M. K. WELLS

D.SC.

Reader in Geology
University of London, University College

REWRITTEN THIRTEENTH EDITION

THOMAS MURBY & CO
40 MUSEUM STREET LONDON W.C.1

First Edition, January 1891
Second Edition, October 1892
Third Edition, January 1903
Fourth Edition, September 1905
Fifth Edition, January 1909
Sixth Edition, June 1910
Seventh Edition, January 1914
Eighth Edition, February 1926
Ninth Edition, January 1937
Tenth Edition, entirely revised, 1949
Eleventh Edition, 1952
Twelfth Edition, completely revised, 1961
Thirteenth Edition, completely revised, 1972

ISBN 0 04 552008 9 hardback
0 04 552009 7 limp

George Allen & Unwin Ltd
are the proprietors of
Thomas Murby & Co

PRINTED IN GREAT BRITAIN
BY PHOTOLITHOGRAPHY
UNWIN BROTHERS LIMITED
WOKING AND LONDON

PREFACE

THE first edition of this book, a slender volume of 128 pages written by the late Dr F. H. Hatch, O.B.E., was published in 1891. In 1926 one of us (A.K.W.) joined Dr Hatch in a complete revision. Ten years later, after the death of Dr Hatch, more drastic alterations were made in the scheme of classification of the igneous rocks, mode of occurrence being replaced by grain-size or degree of crystallinity. In the tenth edition another of the main props of former classifications—silica percentage—was discarded in favour of the mineral contents of the rocks. As might be expected, the subject matter and illustrations of the original book have through successive editions been completely replaced. Although the authors feel justified in claiming that the work is now entirely their own, it is considered expedient to retain the sub-title 'Hatch and Wells', which has become familiar through usage by several generations of students.

In order to save space for the inclusion of new material, we have omitted some of the crystallographic and other descriptive data from the chapters dealing with mineralogy. The descriptions of minerals are especially tailored for use in the practical study of igneous rocks, and as such form an integral part of the whole book without which we believe much of the value of the petrographical and petrological parts of the text would be lost. Experience suggests that students will continue to find it helpful to have integrated accounts of the mineralogy and petrology under one cover, despite the fact that authoritative text-books of mineralogy are available, to which reference may be made for a more systematic and complete treatment of the subject.

Since the last edition was published in 1961, fundamental advances have been made in understanding the nature of the major structual features of the Earth's crust, and of the rôle of igneous activity in their development. The significance for petrology of modern concepts in global geophysics and geology is recognized by the inclusion of a new chapter on the geological setting of igneous activity.

The vast increase in quantitative data on igneous rocks from all parts of the world, together with improvements in experimental techniques, have brought about major changes in thinking about many of the fundamental problems in petrology. This is perhaps most noticeable in regard to the nature and origin of magmas. The degree of interest currently shown in the origin of basalts and basaltic magmas—now known to be derived by partial melting of the upper mantle—rivals that shown a decade or so previously in the origin of granites. The chapter dealing with the distribution and origin of basalts has therefore been re-written. Similarly, discussions of the origins of some of the other rock-groups, notably andesites and trachytes, have been brought up to date and widened in scope.

Although these discussions of the distribution and origin of the various rock-groups constitute some of the most radical changes that have had to be made for this edition, we have tried to incorporate them in such a way that they do not divert attention from what we regard as the fundamental objective of the book: to demonstrate the principles of igneous petrology as far as possible through the evidence of the rocks. Petrogenetic theories are often based to a large extent on data that are relatively remote from the experience of most students. Teaching experience has convinced us that despite the natural attraction that new and changing theories hold for students, the fundamental need is knowledge of the rocks themselves. In this book, therefore, emphasis is placed upon petrographic aspects of petrology: features of the mineral composition and texture which the reader will be able to verify by his own observation. Our own knowledge of rocks has grown significantly as a direct result of preparing the petrographic illustrations. We hope that these drawings, accompanied by their detailed legends, will act as sign posts, encouraging readers to make their own discoveries by similar means.

An important part of petrography concerns the precise naming and classification of rocks. Unfortunately, at the present time, widely varying schemes of classification are in use in different parts of the world, with the result that many aspects of petrological research are seriously hampered. We therefore welcome the efforts currently being made by a sub-commission of the International Geological Congress to formulate a scheme of classification with the intention that it might be accepted universally. In fact, the broad structure of our own scheme, evolved over a period of many years, conforms fairly closely with the proposals of the Commission that have been presented so far. We have therefore slightly modified our classification, particularly in regard to the quartz-bearing rocks, in order to make it compatible with the new proposals. These matters and others, particularly the significance of chemical composition in determining the mineral-content of a rock, are discussed fully in an enlarged and virtually new chapter on classification.

It is a pleasure to record our thanks for the generous help we have received from friends and colleagues during the preparation of this edition.

A. K. WELLS
M. K. WELLS

CONTENTS

Part IV: Igneous Activity in the British Isles

GENERAL WORKS OF REFERENCE

R. Balk, *Structural Behaviour of Igneous Rocks*, Geol. Soc. of America, Memoir 5, 1937.

T. F. W. Barth, *Theoretical Petrology*, New York, 1952.

N. L. Bowen, *The Evolution of the Igneous Rocks*, Princeton, 1928.
The Later Stages of the Evolution of the Igneous Rocks, Journ. Geol. xxiii (1915), Supplement, 91 pp.

W. L. Bragg, *The Atomic Structure of Minerals*, 1937.

C. A. Cotton, *Volcanoes as Landscape Forms*, Wellington, N.Z., 1944.

R. A. Daly, *Igneous Rocks and their Origin*, New York, 1914.
Igneous Rocks and the Depths of the Earth, New York, 1933.

W. A. Deer, R. A. Howie and J. Zussman, *Rock-forming Minerals*, in five volumes, London; vol. 1, 1962; vol. 2, 1963; vol. 3, 1962; vol. 4, 1963; vol. 5, 1962.

A. Harker, *The Natural History of Igneous Rocks*, 1909.

A. Holmes, *Petrographic Methods and Calculations*, 1921.

A. Johanssen, *Descriptive Petrography of the Igneous Rocks*, in four volumes, Chicago; vol. i, 1931; vol. ii, 1932; vol. iii, 1937; vol. iv, 1938.

P. F. Kerr, *Optical Mineralogy*, 3rd edn., 1959.

A. Rittmann, *Volcanoes and their Activity*, translated from the Second German Edition by E. A. Vincent, New York, 1962.

S. J. Shand, *Eruptive Rocks*, 3rd edn., 1947.

H. G. Smith, *Minerals and the Microscope*, 4th edn., completely revised by M. K. Wells, London, 1956.

F. J. Turner and J. Verhoogen, *Igneous and Metamorphic Petrology*, 2nd edn., 1960.

L. R. Wager and G. M. Brown, *Layered Igneous Rocks*, Edinburgh, 1968.

through which the magma penetrated. Changes of this nature involving the addition of material of magmatic origin are said to be metasomatic and **metasomatism** in favourable circumstances may result in the formation of rocks closely resembling those formed by crystallization of a magma. Lava before consolidation *is* magma; but magma is not necessarily wholly liquid. It is a matter again of direct observation that magma often rises to the surface bringing with it (as an essential part of itself) well-formed crystals suspended in liquid. Thus the lava erupted by Vesuvius is sometimes charged with crystals of leucite, while the commonest type of lava, basalt, frequently contains phenocrysts, of olivine and augite. In certain circumstances it is inferred that some bodies of magma consist very largely of concentrates of such crystals, lubricated by a minimum amount of liquid, and therefore possessing only limited mobility.

The rocks termed igneous include lavas of a wide range of compositions; but these are closely associated with other rock bodies which consolidated below ground-level and which therefore cannot be observed in the process of formation. It is inferred, however, from the facts of mineral composition and textural quality, that these rocks also originated from the same material as the associated lavas. The argument may be developed by working from the known to the unknown: thus basalt forming lava flows is demonstrably magmatic. A widespread type of basalt consists essentially of the minerals olivine, augite and plagioclase. The last two minerals are often involved in a distinctive intergrowth known as ophitic texture. This texture and mineral association are found also in rocks occurring as minor intrusive sheets (dykes and sills). Thin sheets of either kind may consist of basalt identical with that forming the lava flows; but the thicker ones cooled more slowly and a coarser grain was consequently developed. These medium-grained rocks of basaltic composition are termed dolerites; but although they are given a distinctive name on account of their different grain-size and mode of occurrence, it is obvious that they were formed from the same material as the lava basalts. Finally, large rock masses (major intrusions) composed of the same minerals, often showing the telltale ophitic texture, may be exposed by deep erosion of the roots of a volcanic region. These coarse-grained rocks of basaltic composition are distinguished as gabbros. They grade into dolerites just as the latter grade into basalts, and there can be no doubt about their close genetic relationship. The three rock-types basalt, dolerite and gabbro are essentially the same in chemical composition; but, although they are not identical, such differences in texture and in the details of mineral composition as may be noted are readily understandable as arising during cooling in different environments. These three rock-types doubtless had a common origin and are hence termed comagmatic (R. A. Daly) or consanguineous (A. Harker). The latter term does not stress the magmatic nature of the rocks, but it does emphasize a common origin—they are 'of the same blood'. Many other rocks differing widely from basalts in composition are linked in precisely the same manner as basalt, dolerite and gabbro. The close genetic relationship is expressed by saying that they belong to the same family or 'clan'—in this particular case the gabbro clan.

A body of magma when intruded into the Earth's crust brings with it much heat which is ultimately dissipated by conduction into the wall- and roof-rocks causing changes in their composition and structures. Such reconstruction is termed **thermal metamorphism**, and rocks so affected fall in the metamorphic, as distinct from the igneous, category. Usually distinction between the magmatic rocks causing the metamorphism, and the metamorphic rocks resulting from it are clear and unequivocal: the metamorphic effects are restricted to a zone surrounding the intrusion and this is often shown on geological maps. The changes are well known and understood. In the deeper regions of the crust the conditions are far different, however: confining pressure and temperatures are both very much higher—the latter must be near-magmatic and much of the material is near-magmatic too. Here the boundaries between igneous and metamorphic become blurred, and it becomes increasingly difficult to distinguish between cause and effect. If subsequently the rocks from such a region are exposed at the surface it may very well be extremely difficult to interpret what is seen. Rocks which have crystallized from a melt will be closely associated with others which were not far removed from the magmatic condition. Rocks of this type include some which are very distinctive in the mass: they are often banded or streaky and are evidently admixtures of two different classes of material. Veins and streaks of the one member, usually identifiable as an igneous rock-type—often a kind of granite—were frozen-in while penetrating or soaking into the other member, consisting of a mineralogically and texturally different rock-type. Such rocks are termed migmatites and the soaking process is **migmatization**.

Naturally petrologists are interested in the origins of the rocks they study—that is, in **petrogenesis,** but excluding the lavas, we can never *know* with the same degree of certainty how a rock was made, as we can know and understand the facts of its mineral composition and texture. Therefore in this book we stress the observable facts and study the rocks as they are without becoming too deeply involved, we hope, in the more hypothetical problems of ultimate origins.

The omnibus term 'igneous rocks' is admittedly not ideal; but it is widely used, and in spite of its shortcomings it is well understood. A new name could doubtless be devised which did not stress connection with fire. S. J. Shand preferred to use the term 'eruptive rocks' for those we call 'igneous'; but surely an eruption is a breaking-out at the surface: only the extrusive igneous rocks (lavas) are really eruptive. In addition to the latter we are concerned also with the equivalent rocks which were formed in—not on—the crust of the Earth, and which are collectively termed 'intrusive', though this word also is liable to convey a wrong impression. Careful study of field relations in many cases indicates that gabbros, granites, etc., reached the positions they now occupy in the crust by *displacing* pre-existing 'country-rock'. These are genuinely intrusive; but in many other cases such evidence is lacking, and field relations suggest, if they do not prove, that the rocks in question originated in place, therefore by *replacement* rather than displacement. This applies more particularly to the rocks rather loosely termed 'granites', concerning which H. H. Read once said: 'There are granites and granites,' implying among other things that some granites originate in one way, but others in different ways. The important thing to realize is that in spite of different modes of origin, the rocks themselves are granites, lithologically. We believe strongly that a given rock-name should apply to a given mineral assemblage, irrespective of its origin. In this respect origin is of secondary importance compared with the petrographic (lithological) character of a rock. It is upon the latter that systematic classification and rock-nomenclature are based, and anything out of the ordinary (so far as origin is concerned) can be indicated by adding a qualifier to the rock-name. Thus a granite may originate by direct crystallization from a melt; but during a phase of granitization a closely similar (though not necessarily identical) granite may be formed by recrystallization of lava or volcanic 'ash' of the right composition in the root region underlying a volcano; while it is well within the realms of possibility that gaseous and fluid substances of magmatic origin rising in advance of the magma itself may so alter certain pre-existing rocks as to form new mineral associations stable under new conditions and again closely resembling, and in some instances indistinguishable from, magmatic granite.

Readers of petrological literature will quickly discover that conflicting opinions on problems of petrogenesis are strongly held, and often forcefully expressed in a manner that appears to be authoritative. Thus one learns from one source that 'all granites are magmatic'; but from another that 'no granites are magmatic'. Actually these are matters of personal opinion and they really do not matter greatly. There is no room for difference of opinion on what granite is, and what it is made of. These are matters of fact which the student can ascertain and check for himself by direct observation.

PART I

THE IGNEOUS ROCK-FORMING MINERALS

CHAPTER I

THE MAFIC MINERALS

(1) Introduction: Classification

THE rock-forming minerals may be variously classified, according to the part played by them in the structure and composition of rocks. A useful division is into primary and secondary, the former being further divisible into essential and accessory. An **essential** mineral is one whose presence is implied in the definition of the rock. An **accessory** mineral is one whose presence or absence does not sensibly affect the character of the rock. Thus, quartz, feldspar and mica are essential constituents of most granites, while zircon, sphene and apatite are accessories. It should be realized that an accessory mineral in one rock may be an important essential in another. For example, although quartz is an accessory in some gabbros, it is the characteristic mineral of granites, and no rock free from quartz can be termed granite. The accessory minerals are sparsely distributed; but although only one or two may be seen in a hand-specimen, others come to light when thin sections are examined. To make a complete list, samples of the rock must be crushed and the heavier minerals separated by liquids of high specific gravity. This method has been successfully applied in the study of a number of igneous rocks, and has resulted in a considerable advance in our knowledge of the distribution of the more uncommon constituents.[1] Many of these accessory minerals are among the most stable and most durable components of igneous rocks. They thus persist after the essential minerals have been destroyed by weathering, and by their presence add distinctiveness to the sands and other sediments formed by the degradation of igneous rocks. Their detailed study is now an important branch of petrology (see Vol. II of this work).

Minerals are **secondary** when they have resulted from the alteration or reconstruction of the primary minerals. In altered rocks both essential and accessory minerals may be replaced by secondary ones.

It should be noted that a mineral may be primary in one rock, but secondary in another; thus primary quartz occurs in granite,

[1] See, for example, Rastall and Wilcockson, *Quart. J. Geol. Soc.*, **71** (1915), 592.

while secondary quartz is liberated as a result of alteration of several species of rock-forming minerals.

In the following account, the primary essential minerals of igneous rocks are described first in as much detail as the writers consider necessary for the proper understanding of the igneous rocks in relation to their mineral composition. Particular attention is therefore paid to the internal structure of these minerals, as it is this which controls variation arising from atomic substitution within each 'family' of minerals.

The more important secondary minerals and the accessories also receive attention, but their treatment is necessarily selective.

As it is not our intention to attempt to compete with standard textbooks on mineralogy, much detail concerning the physical properties of the minerals has been omitted: such detail, if required, can be readily obtained from one of the standard works of reference.

(2) Considerations concerning Atomic Structure

In these chapters the description of most groups of the rock-forming minerals is based on their atomic structure, as revealed by X-ray analysis. The actual technique involved in the latter, and the interpretation of the experimental data are matters for specialists; but the results of their researches are most illuminating to the student of mineralogy, particularly as regards the chemical relationships between allied mineral species and the interdependence between physical characters and the intimate internal atomic structure. We have of necessity drawn largely upon W. L. Bragg's invaluable account of the atomic structure of minerals. One great difficulty arises in connection with using the results of X-ray analysis: it is far from easy to make a useful illustration representing an intricate atomic design in three dimensions. Those which we have introduced must be regarded only as diagrams—they are only first approximations to the truth; but as a teaching device they have their place, if for no other reason than that they help to make sense out of the otherwise meaningless string of symbols which represents the chemical composition of any one of these minerals.

The component atoms or groups of atoms are held together by electrical forces of attraction between oppositely charged bodies of minute size. Those carrying a positive charge are termed **cations;** those with a negative charge are **anions.** For the sake of simplicity we may speak of the size of an atom, meaning the atomic radius, which is the distance, measured in Ångström units, at which mutual repulsion sets in when two atoms approach one another.

In the essential minerals of igneous rocks we are dealing largely with silicates, *i.e.* combinations of silicon (Si) with oxygen (O), these being associated with cations in variety. Those commonly occurring include the following, in order of increasing size, the atomic radius being stated in brackets in each case:

Si (0·39)	Na (0·98)
Al (0·57)	Ca (1·06)
Mg (0·78)	K (1·33)
Fe^{2+} (0·83)	

The anions are much less varied. Among them oxygen (O, 1·32) plays the dominant role, but in certain minerals hydroxyl (OH, 1·32) or fluorine (F, 1·33) may replace it to a limited extent. Now all these anions are large compared with the cations; and we may regard the mineral structures as consisting of closely packed anions (chiefly oxygen), with the small cations tucked into the interstices between them. With some of the more complex minerals the number of kinds of atoms may be large, and the atomic design may be exceedingly intricate; but, just as an imposing modern skyscraper is built up of relatively insignificant blocks of building material, so these involved atomic structures may be resolved into simple units of structure. For our purposes only two such units need be recognized. In the first, four oxygen atoms are closely packed round a silicon atom, giving an SiO_4-group. The oxygens lie at the corners of a tetrahedron, and the silicon is of such a size that it fits snugly into

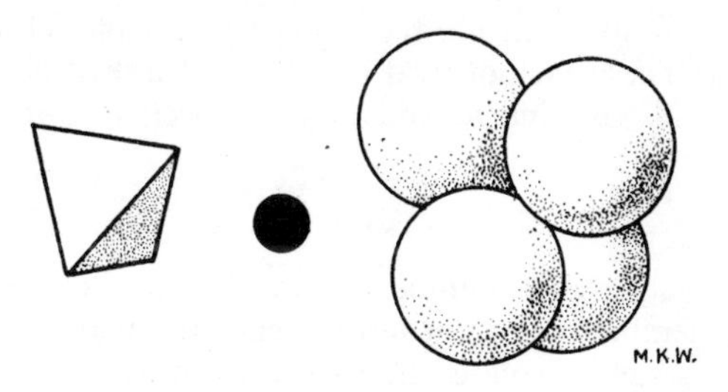

FIG. 1

Tetrahedron and SiO_4-group.

The small silicon atom is hidden between the four large oxygen atoms, each of which has its centre at a corner of the tetrahedron. The black spot shows a Si-atom to same scale as O-atoms.

the interspace between them, as shown in Fig. 1. The second fundamental unit of structure comprises six oxygens in close contact, lying, as it were, at the corners of an octahedron. There is space between them for a cation larger than a silicon, and atoms of magnesium, aluminium or iron commonly occur in such 'six-fold coordination' with the anions surrounding them. Each and every cation shares its charge, whether of one, two, three or four units, with all the anions by which it is immediately surrounded. Thus an Al^{3+} in the centre of an octahedral group of O's, has a positive charge of three units to share among the surrounding six O's, and therefore contributes a charge of $+\frac{1}{2}$ to each. On the other hand, a silicon (tetravalent with a total charge of +4), in an SiO_4-tetrahedron, contributes +1 to each of the surrounding oxygens. This is **'Pauling's principle'.** In both cases cited, as oxygen is divalent, that is, has a charge of 2 units, it follows that there must be a residual electrical charge on these units of structure. But the whole edifice is compounded of such units, packed together, with interspaces available for the introduction of as many cations as are necessary to balance out this residual negative charge. There must be no residual charge of the kind we have visualized, in a stable mineral. This necessity of balancing the total positive against the total negative charge provides a check on the accuracy of a formula representing the composition of a mineral.

In many rock specimens the component mineral grains are either dark or light in colour, and this simple fact is the basis for dividing the essential rock-forming minerals into two categories 'light' and 'dark', or more scientifically, felsic and mafic respectively. The mafic group comprises the ferro-magnesian silicates: the olivines, pyroxenes, amphiboles and micas, and these are considered in that order, which is that of increasing structural complexity, in the account which follows. In the following chapter the felsic minerals are described. These include two 'families' of anhydrous alumino-silicates, the feldspars and feldspathoids respectively, and the silica group.

OLIVINES

This group comprises a number of important rock-forming silicates. In all members of the group the essential plan of the atomic structure is the same: isolated SiO_4-tetrahedra are packed together in lines parallel to the crystal axes. In any such line parallel to an axis, they all point in the same direction (Fig. 2); but in alternate lines the tetrahedra point to the left and right, as shown in the spaced diagram of Fig. 3. The individual tetrahedra are joined one to another through the cations. These are arranged in six-fold co-ordination with the oxygen atoms, which, as the diagram shows, belong to different adjacent tetrahedra.

Thus the unit of structure is the single SiO_4-tetrahedron. Against the negative charge of 8 units supplied by the oxygens, the Si^{4+} offers a positive charge of 4 units; therefore on each tetrahedron a negative residual charge of 4 units remains to be balanced out by the addition of the requisite cations. In the several members of the olivine group the latter include Mg^{2+}, Fe^{2+}, and much less commonly

Mn^{2+} and Ca^{2+}. If all the necessary cations are Mg^{2+}, the formula becomes Mg_2SiO_4, which corresponds to the natural mineral **forsterite**. If iron occurs exclusively, the corresponding mineral is

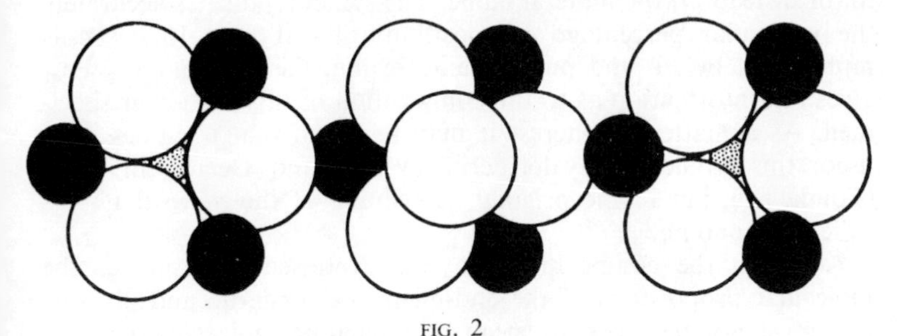

FIG. 2

The atomic structure of olivine.

A small portion of a single row of SiO_4-tetrahedra with the associated cations, Mg and Fe, shown in black, silicons stippled. This row is parallel to the b-axis.

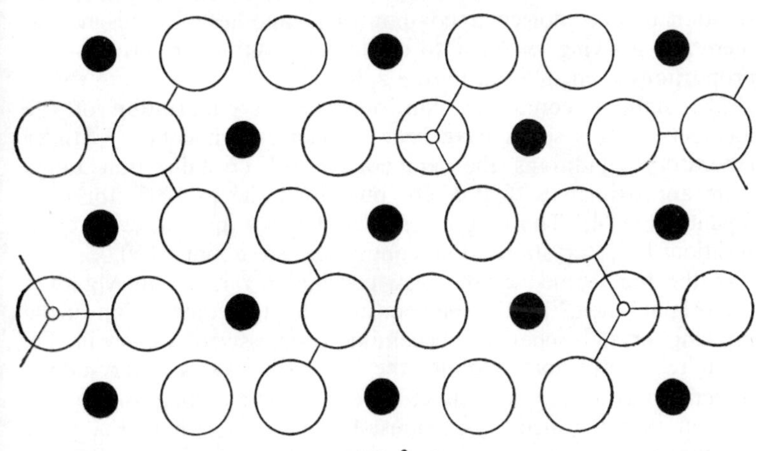

FIG. 3

The atomic structure of olivine represented formally.

Large circles—oxygens; small circles—silicons; black—magnesiums and/or irons. Two rows of SiO_4-tetrahedra are shown with their oxygens in planes parallel to (100). Each tetrahedron is associated with three metallic cations (Mg^{2+}, Fe^{2+}): these threes are alternately behind (top row, left), and in front of, the tetrahedrons as shown in Fig. 2.

fayalite, represented by Fe_2SiO_4. These two orthosilicates, as they are called, are the end-members of a continuously variable series, in which the ratio of Mg to Fe varies from 100 : 0 to 0 : 100. Such a series may be subdivided arbitrarily into a number of mineral species, with agreed ranges of composition. Unfortunately, in this case, agreement has not yet been reached as to how many divisions should be erected, but we show two recent suggestions in Fig. 4. To avoid misconception it is necessary to use a symbol, in addition to, or instead of, the mineral name. The symbol indicates accurately the molecular percentage composition: thus if pure forsterite is represented by Fo and pure fayalite by Fa, the symbol Fo_{52} Fa_{48} gives full information as to the composition of this particular specimen. As a matter of interest it may be noted that if names were used, this would be hyalosiderite (Wager and Deer), chrysolite (Tomkeieff), but in the original description of the mineral it was called hortonolite.[1]

Although the composition may be expressed in terms of the molecular proportions of the end-members, forsterite and fayalite, this must not be taken to mean that separate molecules occur in the mineral. The silicon-oxygen framework of the atomic structure runs continuously through the crystal, with the Mg^{2+} and Fe^{2+} ions evenly and statistically distributed throughout. It is best to think in terms of ionic substitution in the structure rather than solid solution of one molecular component in another: for this reason there is a growing tendency to quote compositions in terms of the proportions of metallic cations, *e.g.* $Mg_{52}Fe_{48}$.

The stability conditions controlling the crystallization of the members of this series have been studied experimentally.[2] Under laboratory conditions the temperatures of crystallization range from approximately 1890° C for pure forsterite to 1205° for pure fayalite (Fig. 4). The diagram shows that for an olivine of composition Fo_{50}, crystallization commences at about 1650° C, and that the first-formed crystals are considerably richer in Mg than the original melt.[3] As the temperature falls, the composition of the growing crystals changes, becoming progressively richer in Fe. With relatively rapid cooling, the successive layers of crystalline material may survive as distinct zones of different composition; the core in such a zoned crystal must be relatively rich in Fo, while the outermost zones must be correspondingly rich in Fa. A difference of the order of 20 per cent Fo has been noted.[4]

Reasoning on similar lines, early-formed phenocrysts in a lava should be (and are) richer in Fo than the smaller crystals in the groundmass of the lava. Further, among the latter those of largest

[1] Wagner, P. A., *Platinum Deposits and Mines of South Africa* (1929), p. 55.

[2] Bowen, N. L., and Schairer, J. F., The system MgO-FeO–SiO_2, *Amer. J. Sci.*, **29** (1935), 151.

[3] Draw a vertical line from 50 to cut the *liquidus* curve, then a horizontal line through this point shows the temperature at which crystallization commences; while the point where the horizontal line cuts the *solidus* curve indicates the composition of the first-formed crystals.

[4] Tomkeieff, S. I., Zoned olivines and their petrogenetic significance, *Min. Mag.*, **25** (1939), 229.

size, which presumably started growth early and at a relatively high temperature, are more richly magnesian than smaller ones, of later, lower-temperature formation.

The discovery of these facts concerning the variation of composition in successive crops of olivine crystals, and even within large

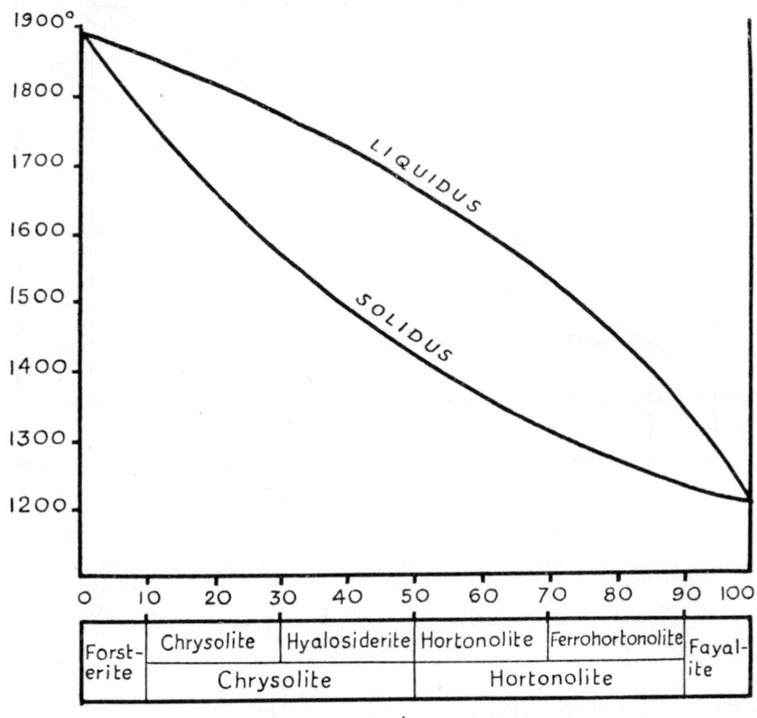

FIG. 4

Diagram showing the cooling curve for the forsterite-fayalite series.

The subdivisions of the olivine group as suggested by Wager, L. R. and Deer, W. A., *Amer. Mineralogist*, **24** (1939), 25; and Tomkeieff, S. I., *Min. Mag.*, **25** (1939), 229 are shown above and below respectively.

single olivine crystals, is only possible if means exist for accurately determining the compositions of the specimens under consideration. Obviously careful chemical analysis will give the desired information; but there are certain less costly ways available to the petrologist. In any continuously variable series like the olivines, the physical, including the optical, properties vary systematically with the composition. Much useful data has accumulated from the study of isolated olivines, and has been used to construct curves showing these variations (Fig. 5). Two sets of measurements have proved valuable in this connection: (1) the size of the optic axial angle, 2V (measured in the mineral), or 2H (measured with the Universal stage); and (2) the refractive indices, more especially the mean

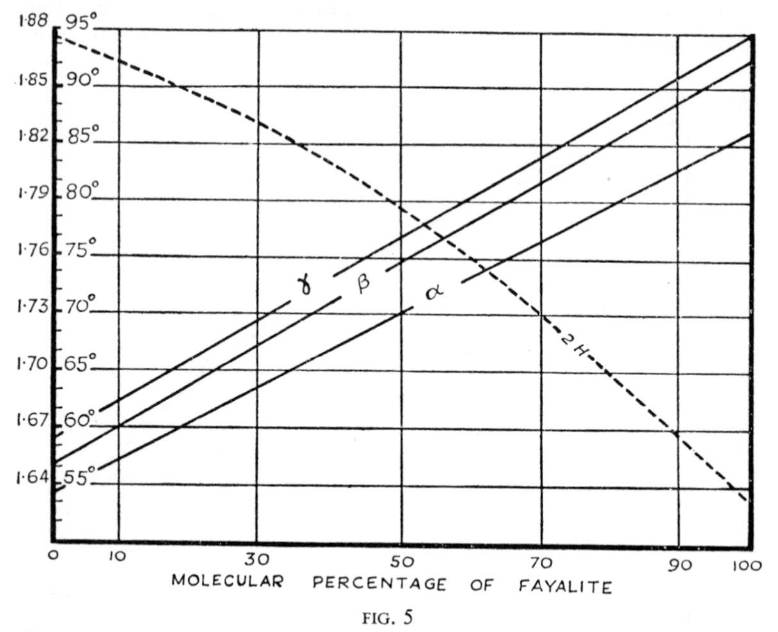

FIG. 5

Diagram showing variation in the size of the apparent optic axial angle 2H, and refractive indices, in the olivine group. *Data from Bowen and Schairer*, 1935, *Wager and Deer*, 1939 *and Tomkeieff*, 1939. 2H is the optic axial angle measured on the Universal Stage between hemispheres of refractive index 1·649.

index β measured with a refractometer. Unfortunately the actual measurement of these properties is a matter for the specialist: they involve methods and apparatus not normally available to students. It is not practicable to determine the composition of an olivine by any of the simple methods, such as extinction angle, used in other groups of minerals. While therefore one must be satisfied merely to identify the mineral generically, as it were, it is possible to go a long way towards specific identification in the light of the following facts concerning the distribution of the different members of the group.

Bowen and Schairer have shown experimentally that Mg-rich olivines—forsterite, chrysolite and hortonolite are unstable in the presence of free silica. Therefore under normal conditions these members of the group are not found in association with quartz: no granitic rock can contain Mg-rich olivines. But olivine rich in Fe can exist in the presence of free silica, though even here there is a limitation: the temperature must be relatively low to bring it within the temperature-range of rhyolitic magma. This is merely another way of stating that the only kind of olivine which can occur in a 'granitic' association of minerals is nearly pure fayalite. The latter has now been recorded in a number of quartz-bearing rock-types including pitchstones from Arran, Scotland, obsidian from the Yellowstone Park, quartz-porphyries and rhyolites from Nigeria, etc., but it still remains a relative rarity. The olivines, excluding forsterite and fayalite, are typically components of the silica-poor (Basic and Ultrabasic) igneous rocks. The amount is greatest in certain ultrabasic rocks, significantly termed olivinites and peridotites,[1] and is only little less important in certain Basic igneous rocks such as olivine-gabbros, troctolites, and olivine-basalts. Some of the last-named contain so-called 'olivine-nodules', but these are really small pieces of peridotites that were brought up by, and incorporated in, the magma during its uprise towards the surface. Well-known examples occur in the Carboniferous basalts in Derbyshire. Now in all these rocks there is a preponderance of Mg^{2+} over Fe^{2+}, and with few exceptions the general rule may be enunciated that the more basic the rock, the more richly magnesian the olivine; the more siliceous the rock, the more ferriferous the olivine.

The end-member—**forsterite**—stands in a category by itself: it is a characteristic product of thermal metamorphism.

Relatively Mg-rich common olivines possess three properties which help to distinguish them from most of the other rock-forming silicates: (1) their low SiO_2-content (pure forsterite contains only 42 per cent by weight, compared with 60 per cent in the corresponding magnesian pyroxene, enstatite); (2) their high temperature of crystallization and hence their tendency to be precipitated among the earliest minerals from a Basic magma; and (3) their high density (about 3·3 for common olivine). Because of this combination of properties, olivine plays an important role in processes of magmatic differentiation involving separation of early crystal-fractions, which in the case of olivine, tend to sink in the magma body. Taking all these characters into consideration, it is not surprising that olivine is a major component of the mantle.

General Characters of the Group

All members of the olivine group crystallize in the Orthorhombic system; but although olivine is a common mineral, well-formed crystals are rare in ordinary mineral collections. Olivine sands, formed at some points on the coasts of the oceanic volcanic islands, such as Hawaii, consist largely of singularly perfect, though small, crystals, concentrated naturally out of the olivine-basalts which

[1] From 'peridot', an alternative name for olivine of gem stone quality.

form the coastline. Many of these small phenocrysts are tabular, with the pinacoid (100) well developed, combined with the vertical prism (110) and dome (101). Other characteristic crystal habits are illustrated in Fig. 6. Forsteritic olivines are 'growth-sensitive': they readily develop different shapes in different environments.

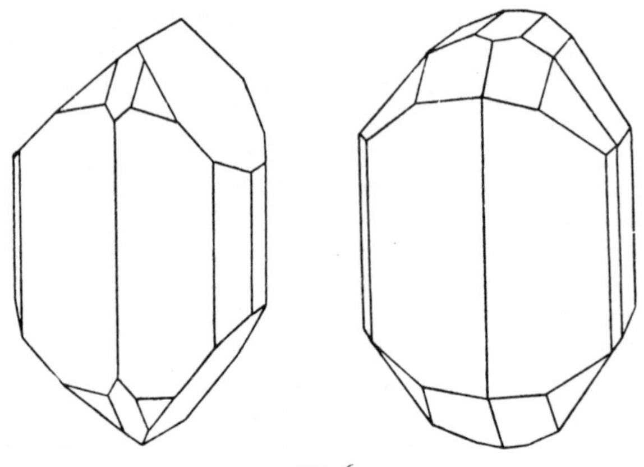

FIG. 6
Crystals of olivine.

Combination of two vertical prisms, side pinacoid, brachyprism (Okl), macroprism (hOl) (sometimes called domes), and bipyramid.
Crystal on right from St John, Red Sea, shows basal pinacoid in addition, also two brachyprisms and two bipyramids.

On account of the dominance of the tabular habit noted above, in thin sections olivine tends to show six-sided cross-sections of characteristic appearance (Fig. 116). In all members of the group the refractive indices are high, and strong surface relief, combined with complete absence of colour, the dominance of arcuate fractures over ill-developed cleavages,[1] and the strong birefringence, serve to render olivine easy to identify. The birefringence of an olivine of known composition may be read off the curve in Fig. 5; that of fayalite is very high. All olivines are prone to alteration in a distinctive manner, and in a representative collection of sections from olivine-basalts, all stages from incipient alteration to complete pseudomorphism may be studied. Commonly fibrous antigorite and/or green chrysotile appear, first along the fractures, then they spread through the body of the crystal. Less ubiquitous secondary minerals formed from olivine include bowlingite and iddingsite, much more strongly coloured 'serpentines', yellow to reddish brown in thin section. In other rocks the olivines have been

[1] Cleavages develop parallel to the pinacoids (010) and (001) in iron-rich olivines, notably in fayalite.

converted into ferruginous pseudomorphs, particularly in basalts weathered under aeolian conditions. In lamprophyres, and occasionally in other rocks, the pseudomorphs may consist largely of carbonate, with or without some form of silica such as opal, chalcedony, or quartz-mosaic. Probably the most carefully studied pseudomorphs after olivine are those occurring in the Markle-type basalts in Scotland. They consist of penninitic chlorite, about 66 per cent, haematite, 32 per cent and 2 per cent of quartz.[1]

PYROXENES

The pyroxenes constitute one of the most important groups of rock-forming silicates. Compared with the olivines, they contain a higher proportion of silica to the bases present, and are thus *metasilicates*. In different members of the group the elements iron, magnesium, calcium and sodium are present in widely varying proportions, together with smaller amounts of aluminium, manganese, and titanium.

Atomic Structure and Chemical Relationships

The study of the X-ray structure of the pyroxenes has shown that the fundamental SiO_4-tetrahedra are linked together vertically into

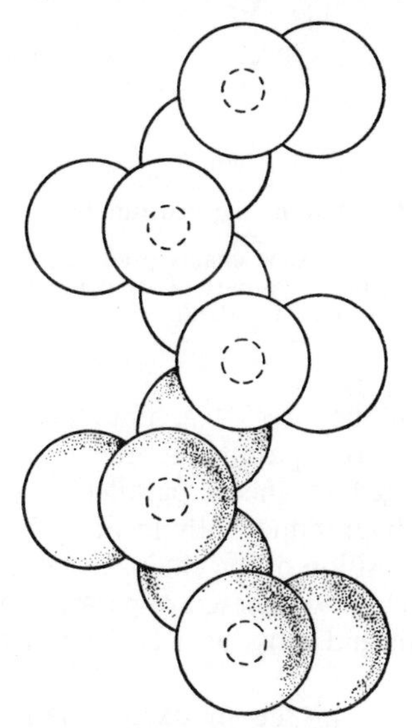

FIG. 7

Part of a chain of SiO_4-tetrahedra, as in pyroxenes. The unit of pattern contains (Si_2O_6) shown shaded. Si-atoms shown by broken circles; large circles are O-atoms.

[1] Smith, W. W., *Min. Mag.*, **32** (1959), 324.

chains, each tetrahedron sharing oxygens with those immediately above and below in the chain. The individual chains are joined together through the medium of the cations Ca^{2+}, Mg^{2+}, Fe^{2+}, etc.,

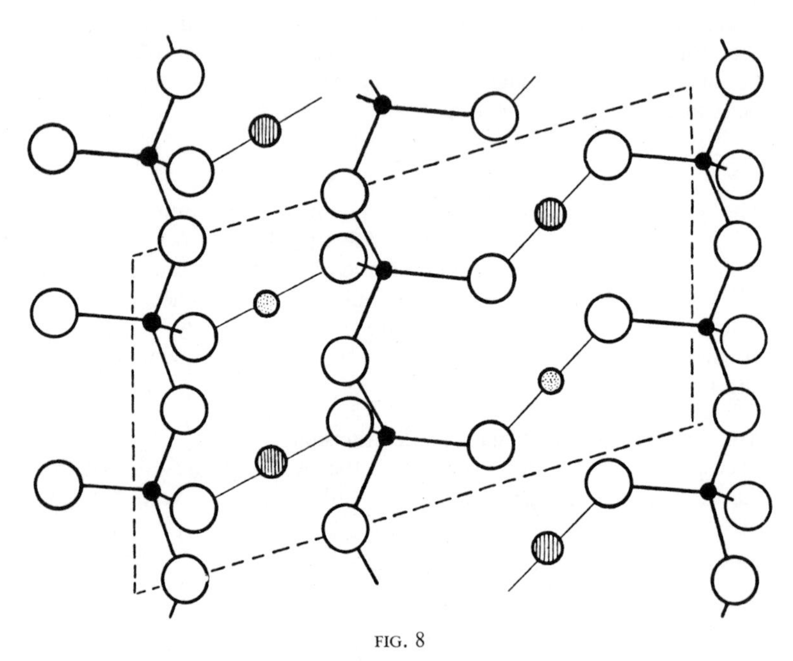

FIG. 8

Diagram showing the structure of diopside.

Portions of three parallel pyroxene chains projected on the plane (010). The unit cell is outlined, but only part of its contents is shown. Si^{4+} black; O^{2-} plain; Mg^{2+} stippled; Ca^{2+} ruled.

which are linked to the 'free' (*i.e.* not shared) oxygens. In all pyroxenes the chains run parallel to the vertical crystallographic axis, and are arranged in sheets parallel to (100). The essential features are shown diagrammatically in Figs. 7 and 8.

The form and disposition of the chains determines the positions of the cleavage planes (Fig. 9). These are parallel to the prism faces of the pyroxene crystal, and cross at a characteristic angle of 87° (or 93°).

The unit of pattern in the pyroxene chain contains Si_2O_6 on which there is a residual electrostatic charge of −4. This may be balanced out by adding two divalent cations, which may be Mg, Fe or Ca in different cases, giving $Mg_2Si_2O_6$, the formula of the mineral enstatite (En), $Fe_2Si_2O_6$, ferrosilite (Fs), or $Ca_2Si_2O_6$, wollastonite

(Wo).[1] Essentially, and ignoring certain minor constituents, all pyroxenes may be expressed as combinations of these end-members, the proportions of which may be indicated by points on a triangle

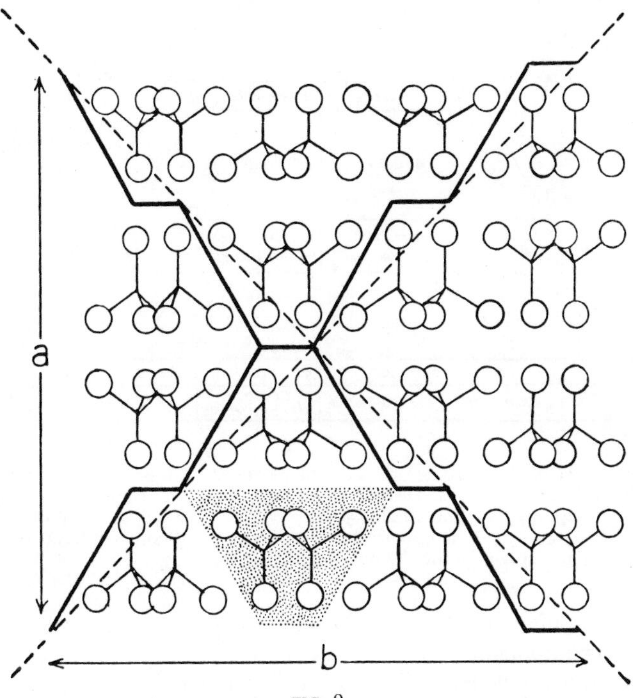

FIG. 9

Diagram showing the relationship between the atomic structure and cleavage of pyroxenes. The linked Si-O chains are shown in plan and the position occupied by one chain with its cations (not shown) is stippled. The planes of weakness are shown by heavy lines, and the resulting cleavage directions by broken lines: *a* and *b* axes shown; *c* is perpendicular to the paper.

whose apices are occupied by Wo, En and Fs as shown in Fig. 10. Virtually only the lower half of the triangle is occupied by actual pyroxenes: in other words the Wo-component in pyroxenes does not exceed 50 per cent. Points representing pyroxene compositions are not evenly distributed over the half triangle: a well-marked maximum concentration lies between Wo_{40} and Wo_{50} and another lies between Wo_{15} and the base line, Wo = 0.

[1] Although calcium metasilicate, $CaSiO_3$, here represented by Wo, is an essential component of the more complex pyroxenes, the mineral wollastonite, with the same formula, is not a pyroxene, as it has a different atomic structure. It is not a normal constituent of igneous rocks, but is characteristic of the thermal alteration of limestones. As wollastonite is like the pyroxenes, though not isomorphous with them, it has been termed a 'pyroxenoid'.

The base of the triangle embraces all possible proportions of En to Fs, thus covering the compositional range of the important group of orthopyroxenes, and the chemically identical monoclinic equivalents. The naturally-occurring members of the enstatite-ferrosilite

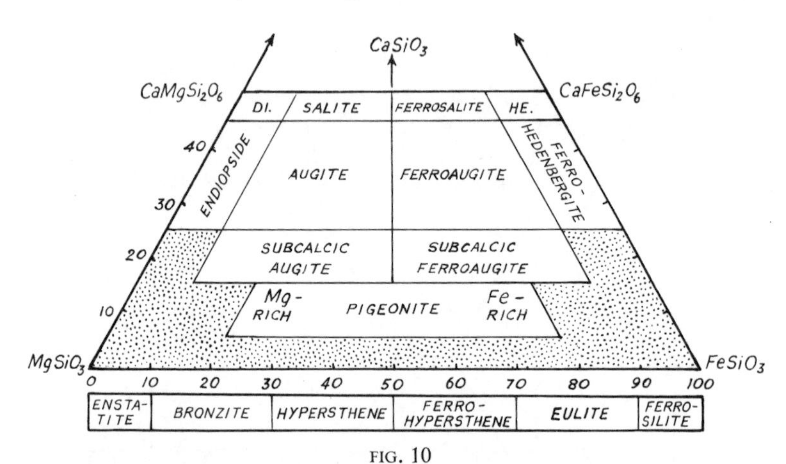

FIG. 10

Classification of the pyroxenes. The scheme for the clinopyroxenes is after A. Poldervaart and H. H. Hess, *J. Geol.* **59** (1951). The stippled area is devoid of clinopyroxenes; but the basal strip between the values of Wo_0 and Wo_5 is occupied by the orthopyroxenes, the names and compositional ranges of which are shown below the base of the triangle.

series contain a small amount, usually between 4 and 5 per cent, of the Wo component.

The mineral **diopside** (Di), with the formula $CaMgSi_2O_6$, is represented by the point half-way between Wo and En; and similarly the analogous silicate **hedenbergite** (He), $CaFeSi_2O_6$, lies half-way between Wo and Fs. It is important to realize that Di and He are points on the compositional triangle corresponding with the theoretical compositions of 'pure' diopside and hedenbergite respectively; but the natural minerals bearing these names cover definite ranges of composition indicated in Fig. 10. A continuously variable series labelled 'salite' in Fig. 10 lies between the diopside and hedenbergite fields. Very few pyroxenes lie above the Di–He line; but between the diopside-salite-hedenbergite series on the one hand and the enstatite-hypersthene-ferrosilite series on the other are two important groups of clinopyroxenes—the augites and pigeonites respectively. The latter are close to orthopyroxenes in composition, but are richer in the Wo component. H. H. Hess suggests that the latter should range between Wo_5 and Wo_{15}.

Thus from the diagram it is seen that there are five series of pyroxenes to be considered:

(*a*) the enstatite-ferrosilite series known collectively as the orthopyroxenes, as they crystallize in the Orthorhombic System;
(*b*) the pigeonite series, restricted to rapidly cooled Basic igneous rocks, especially to some varieties of basalts and chemically related andesites;
(*c*) the subcalcic augite—ferroaugite series;
(*d*) the important and widely distributed augite-ferroaugite series;
(*e*) the diopside-hedenbergite series, rare in igneous rocks, but common in metamorphic rocks.

In the following account the pyroxenes are described in the above order.

(a) Orthopyroxenes: the Enstatite-Ferrosilite Series

Minerals ranging in composition between $MgSiO_3$ and $FeSiO_3$ comprise two series, one crystallizing in the Monoclinic System and distinguished as clino-enstatite, clino-hypersthene, etc. These minerals are stable under high-temperature conditions only: they do not occur in igneous rock but are encountered in meteorites and as products of dry-melt experiments. The second series is by far the more important; these pyroxenes crystallize in the Orthorhombic System and are therefore distinguished as orthopyroxenes. They are widely distributed in a large variety of igneous rock-types. The relationship between these two series has been established experimentally by Bowen and Schairer, who demonstrated that the Monoclinic members invert into their Orthorhombic equivalents at temperatures of the order of 1000° C.[1]

For convenience, the isomorphous series of orthopyroxenes is subdivided into six 'species' defined in terms of the molecular percentages of $MgSiO_3$ and $FeSiO_3$, the divisions between the ranges of composition being drawn at 10, 30, 50, 70 and 90 mol. per cent, as is the common practice for the important groups of the olivines and plagioclase feldspars.

Crystals of orthopyroxenes are not normally available for study; and the student's knowledge of these minerals is based on the examination of cleavage masses or aggregates of hypersthene or bronzite, bounded in part by lustrous bronze-coloured cleavage-surfaces. However, in thin sections particularly of certain lavas, the characteristic shapes of euhedral phenocrysts give the first indication of the presence of orthopyroxene. The most distinctive are basal (transverse) sections, nearly square in shape with the corners truncated, cut across stumpy prismatic crystals. These sections also show well-developed prismatic cleavages, and often in addition an extra single pinacoidal cleavage. The sections show moderate relief: they are either colourless or faintly coloured, and equally faintly

[1] The system $MgO–FeO–SiO_2$, *Amer. J. Sci.*, **29** (1935), 151.

pleochroic, so that this feature may easily be overlooked. In thicker sections the pleochroism is very distinctive from deep pink to bluish green. Enstatite is colourless in thin section, but hypersthene is faintly coloured and feebly pleochroic. Both colour and pleochroism become more intense with increasing Fe-content. The birefringence of orthopyroxenes is weak and the interference colours are limited to shades of grey. Extinction varies, of course, with the orientation of the section. The student should not expect that all sections of these Orthorhombic minerals will show straight extinction: only those sections cut parallel to a crystallographic axis will do so—other sections may extinguish at a wide angle. Similarly the fact that one or two pyroxene sections in a slide show a grey interference colour does not necessarily prove that they are orthopyroxenes: certain sections of clinopyroxene (augite for example) will show this if they are cut perpendicular to an optic axis. Orthopyroxene does not show twinning effects unless it has inverted from pigeonite (see below).

A special feature characteristic of orthopyroxenes occurring in many coarse-grained igneous rocks is the intergrowth with a second (clino-) pyroxene of contrasted composition. The latter is enclosed in the host mineral as regularly orientated plates, in some instances excessively thin, of diopsidic augite, whose identity is easily established by its stronger birefringence as compared with the orthopyroxene host. The theoretical significance of this phenomenon is considered below under exsolution in pyroxenes.

The role of orthopyroxenes as rock-formers is an important one: they occur in lavas, dyke-rocks and plutonites of widely varying composition—from thoroughly Acid to Ultrabasic. This is true of enstatite, hypersthene and bronzite; but iron-rich varieties are less widely distributed and some are very rare. Thus eulite occurs only in rocks of rather problematic origin termed eulysites; ferrosilite is very rare indeed but has been recorded from small steam-cavities in rhyolitic obsidian.[1]

Among occurrences in lavas, andesites provide the most satisfactory sections for studying enstatite and hypersthene (Fig. 111). Among medium-grained igneous rocks certain quartz-dolerites, including the well-known British Whin Sill, contain hypersthene associated with other pyroxenes. In the coarse-grained category orthopyroxenes are most important in Basic and Ultramafic rocks, culminating in the virtually monomineralic bronzitite; the two-mineral rock bronzite-peridotite is another significant type; but the commonest Basic rock in this category is norite in which the dominant pyroxene is hypersthene or bronzite.

The stability relationships between the several pyroxenes can

[1] Bowen, N. L., Ferrosilite as a natural mineral, *Amer. J. Sci.*, **29** (1935), 151.

only be understood when all the pyroxenes have been reviewed: these matters are considered at the end of this account.

(b) The Pigeonite Series

Pigeonites take their name from Pigeon Point, Minnesota. They are stable only at high temperatures and are therefore restricted to lavas and dyke rocks of appropriate composition—dominantly basalts and some basaltic andesites. Under conditions of slower cooling the Monoclinic pyroxene structure inverts to Orthorhombic, giving crystals which are strictly varieties of orthopyroxene but which retain evidence of their Monoclinic parentage. These are known as inverted pigeonites and are characteristic of many gabbros and some relatively slowly cooled dolerites. Details of the phase relations are discussed below.

A certain amount of difficulty arises from the fact that pigeonite hardly ever occurs alone in these rocks: it is usually associated with augite, and sometimes with augite and orthopyroxene, so that accurate identification is essential.

Pigeonite very closely resembles augite in thin section in regard to its general optical characters, but differs in the orientation of the optic axial plane and in the size of the optic axial angle. Many pigeonites are uniaxial but others are biaxial with small 2V (0–25°), so that identification depends upon establishing this fact. The observer should concentrate on grey-polarizing sections—the deeper the grey, the better—and examine these in convergent light arranged for showing interference figures. Once the identity of one grain has been established by this test, very careful comparison will probably enable one to see other pigeonite crystal-grains and to form an estimate of its relative importance in the rock-section.

(c) Subcalcic Augites and Subcalcic Ferroaugites

These pyroxenes are sandwiched between the augites and ferroaugites above and the pigeonites below. Both boundaries are arbitrarily chosen at Wo_{15} and Wo_{25}. The recognition of the subcalcic augite field is a relatively modern refinement of classification: formerly this part of the compositional triangle was regarded as an immiscibility gap between the diopside-augite field above and the pigeonite-orthopyroxene field below. It is sparsely occupied by points representing actual compositions. Subcalcic augites and ferroaugites are apparently restricted to the groundmass of basalts which were chilled rapidly at high temperatures.

(d) The Augite-Ferroaugite Series

In one form or another augite is by far the best known pyroxene. The familiar brown-black crystals similar to that illustrated in

Fig. 11 are among the best known Monoclinic crystals available for study. Common augite is characteristic of Basic igneous rocks and occurs in countless basalts, dolerites and gabbros. The average

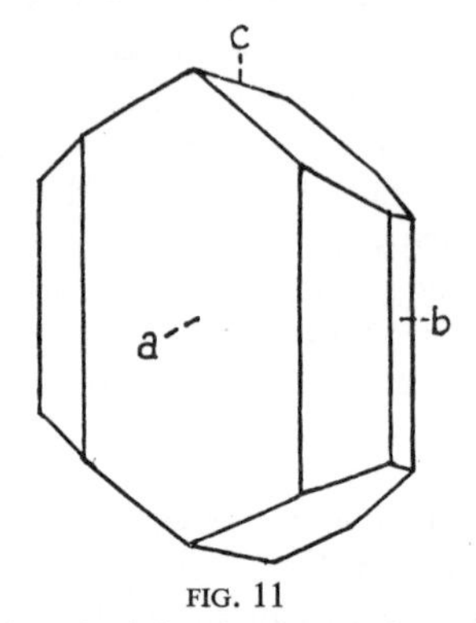

FIG. 11

Augite crystal showing the following forms: front pinacoid (100), vertical prism (110), side pinacoid (010) and hemibiypyramid (111).

Note the orientation of the crystal relative to the crystal axes *a*, *b* and *c*. The *a*-axis slopes downwards *towards* the observer; but the edge between the top hemibipyramid faces (cut by the *c*-axis) slopes downwards *away* from the observer.

composition is represented by the molecular formula $Wo_{40}En_{45}Fs_{15}$, but this ignores a most distinctive feature in the chemistry of augites: in all varieties a proportion of the Si is proxied by Al, so that, instead of the Si_2O_6 of all other pyroxene formulae, that of augite contains $(Si, Al)_2O_6$. Further, cations other than those implied in the above general formula come in, by substitution, and include Fe^{3+}, Na, Ti and additional Al, so that a more realistic formula would be $(Ca,Mg,Fe^{2+},Fe^{3+},Ti,Al)_2(Si,Al)_2O_6$. Augites which crystallize early during the fractionation of a basic magma tend to be Mg-rich; but those which separate later become progressively richer in ferrosilite and are distinguished as **ferroaugite.** Two other common varieties of augite should be noted: titanaugite, obviously so called on account of its appreciable content of titanium; and aegirine-augite, a variety which contains a proportion of the sodic aegirine molecule.

Titanaugite differs in composition from common augite in three respects: most obviously it is rich in titanium (with TiO_2 usually between 3 and 5 per cent, but it may reach 10 per cent); and both Al and Fe^{3+} are high. Most of the points representing the compositions of titanaugites fall within the diopside field on Fig. 10, while a few of them lie above the Wo_{50} line. Titanaugite is readily identified under the microscope: it resembles common augite in all respects but one—it is distinctly pleochroic from lilac to light brown. Titanaugite occurs in alkali-rich Basic igneous rocks, for example in the Scottish essexites and lugarite in which its optical characters are very well displayed.

Aegirine-augite is the most strongly coloured augite: it may be rich green in a slide of normal thickness, and is distinctly pleochroic. In these respects it is much closer to aegirine than to common augite; but is distinguished therefrom by its wide extinction angle.

Aegirine and Acmite. These two Monoclinic pyroxenes, occurring in Na-rich igneous rocks, fill the role of augite in calc-alkaline gabbros and basalts. Aegirine and acmite are the same in chemical composition, $NaFe^{3+}Si_2O_6$; both occur as black lustrous narrow prisms, sharply pointed in the case of acmite. The essential difference in thin section is the tendency for acmite to be brown rather than green. The pleochroism of aegirine is strong, with $\alpha =$ rich green, $\beta =$ yellowish green and $\gamma =$ brownish green. Vertical sections are length-fast and the extinction angle is generally less than 10°.

A common associate of aegirine is nepheline, which is rather elusive in thin sections and may be easily overlooked. The identification of aegirine, which is very distinctive, is a useful hint, therefore, that the presence of nepheline may be suspected. Aegirine is commonly associated, especially in coarse-grained rocks, with even more strongly coloured amphiboles—arfvedsonite or riebeckite—often in reaction relationship. It is instructive to compare the optical behaviour of these two kinds of mineral, particularly as regards pleochroism.

Jadeite can most conveniently be considered at this point as it has the same type of formula as aegirine (acmite) but with Fe^{3+} completely replaced by Al^{3+}. Thus the formula is $NaAlSi_2O_6$. As regards the formula, comparison should be made with nepheline on the one hand and albite on the other (see p. 62). All jadeites so far analysed have compositions close to the ideal formula though there may be a slight substitution of Fe^{3+} for Al.

The pyroxene omphacite shares the chemical characteristics of diopside and jadeite, which fact has led to the statement that 'omphacite contains the jadeite molecule'. This is only a figure of speech, of course; but omphacite results from partial substitution of NaAl for CaMg in diopside, and the introduction of some ferrous and ferric iron, giving (Ca, Na)(Mg, Fe, Al)Si_2O_6. Omphacite is one of the two essential minerals in the interesting rock-type, eclogite. This rock is of particularly deep-seated origin and it is inferred, therefore, that omphacite can crystallize only in environments involving exceptionally high pressures. This is confirmed by the formation of jadeite under high-pressure conditions in the laboratory.

(d) The Diopside-Hedenbergite Series

Ideally the members of this series are represented by points on the left-to-right join half-way between base and apex of the compositional triangle of Fig. 10; actually the compositions of natural specimens lie a little below this line and are gradational into the augites—an arbitrary division between the two series must therefore be drawn, and we follow H. H. Hess in selecting Wo_{45} as the critical composition. This is the least amount of the wollastonite component that diopside may contain.[1]

The Mg-rich members of the series are by far the commonest. Diopside itself is essentially a metamorphic mineral; but diopsidic augite is not uncommon in igneous rocks including some pegmatites. Chrome-diopside and the closely related chrome-augite may be considered together: as the names imply, these clinopyroxenes only differ from diopside and augite in containing a small quantity of Cr_2O_3, which results in a rich green colour both in hand-specimens and thin sections. These minerals are of very limited distribution in igneous rocks, being restricted to certain ultramafites.

Hedenbergite is characteristic of metamorphic, rather than igneous rocks; but ferrohedenbergite, intermediate in composition between hedenbergite and ferrosilite, is quite common in granophyres, for example, forming part of the Skaergaard Complex in Greenland[2] and similar products of extreme differentiation by fractional crystallization. Ferrohedenbergite may be regarded as the lowest temperature member of the crystal fractionation series: augite—ferro-augite—ferrohedenbergite.[3]

The Optical Characters of the Clinopyroxenes

The optic axial plane coincides with the single crystallographic plane of symmetry, parallel to (010) in the clinopyroxenes (Fig. 12), except in some pigeonites in which it is at right-angles to that plane. It follows, therefore, that the maximum extinction angle can be measured in (010) sections. These angles vary with composition and provide an easy means of checking the identity of the various clinopyroxenes as indicated in Fig. 13. The angles indicated in the left top corners of the 'fans' to the left of the middle line are measured between the cleavage traces and the slow vibration, *Z*. The complementary angles, between the fast vibration direction, *X*, and the cleavage traces are obtained by subtraction from 90°. The results would be anomalous if the nature of the vibration under test (whether fast or slow) were ignored; and further, it is only the *maximum* angle that is diagnostic. A section parallel to the front pinacoid (100) of augite extinguishes straight (*i.e.* extinction angle,

[1] Hess, H. H., Pyroxenes of the common mafic magmas. *Amer. Min.*, **26** (1941), 515–35; also *Amer. Min.*, **34** (1949), 621.

[2] Wager, L. R. and Deer, W. A., Petrology of the Skaergaard Intrusion, *Medd. om Grønland*, **105**, No. 4 (1939), 209.

[3] See for example Brown, G. M., Pyroxenes from the early and middle stages of fractionation of the Skaergaard magma. *Min. Mag.*, **31** (1957), 511.

0°). The maximum, measured on (010), is 51°; therefore sections in the zone (100) (010) show angles between these two extremes.

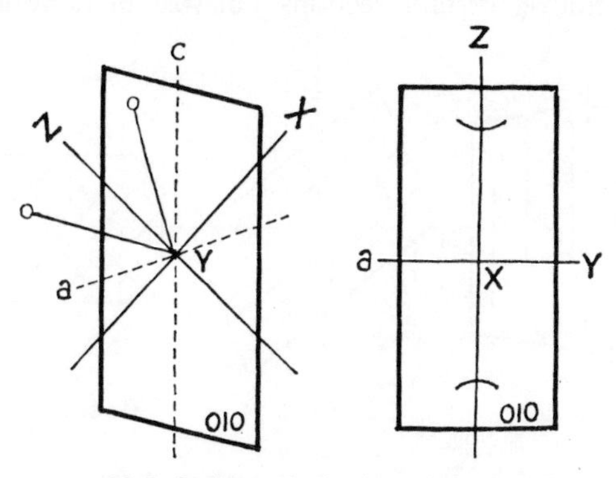

FIG. 12 *Left*, the optical orientation of augite.

The (010) section illustrated is the single vertical plane of symmetry and coincides with the optic axial plane. Note the positions of the optic axes (small circles), and also the extinction angle between Z and c.

Right, the optical orientation of hypersthene.

X (the fast vibration), coincides with the b-axis, Y with the a-axis and Z (the slow vibration) with the c-axis. Positions of the optic axes shown conventionally by small arcs.

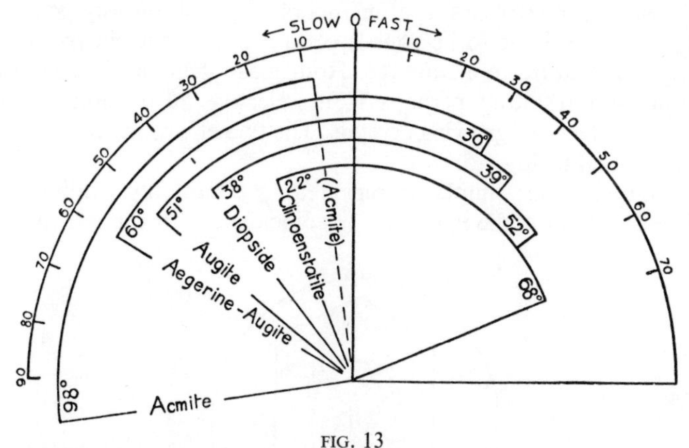

FIG. 13

Diagram of the extinction angles in Monoclinic pyroxenes in sections parallel to the side-pinacoid (010). The vertical line (O) represents the vertical cross-wire; the sides of the fans correspond to the cleavage traces on (010) sections in the extinction positions.

Various sections of augite, drawn in the position of extinction, are shown in Fig. 14. As regards birefringence the range of interference colours appropriate to the top of the first order and bottom of the second order ensures rich colouring between crossed polarizers, strong yellow, blue, green, red and purple being characteristic. As pointed out above, certain sections polarize in dark grey: these

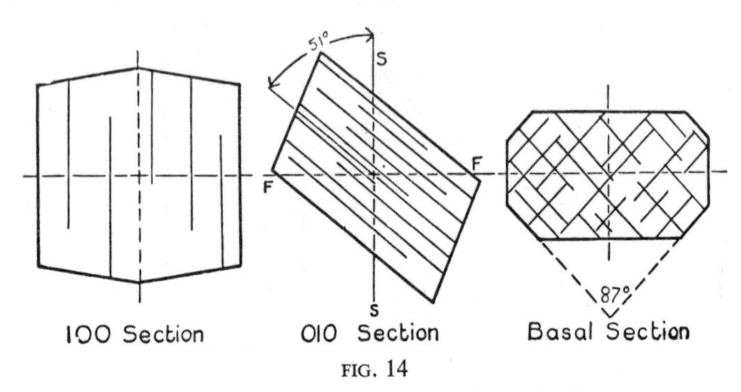

FIG. 14

Sections through an augite crystal drawn in the extinction positions. SS—slow vibration; FF—fast vibration direction. The (100) section extinguishes straight; the basal section, symmetrically; and the (010) section obliquely at the maximum angle.

should be looked at carefully as regards colour, relief and degree of alteration, when it will be realized that such sections are of the same mineral, *i.e.* augite, as the more brilliantly polarizing ones. As most clinopyroxenes are members of continuously variable series, they are liable to be zoned; certain titanaugites in particular display this feature to advantage. Hour-glass structure is a kindred phenomenon resulting perhaps from selective adsorption of ions on different faces of growing crystals. The two effects in combination are illustrated in Fig. 15.

Twinning, either simple or repeated, is commonly exhibited by augite. The twin plane is (100), *i.e.* coincident with the left-to-right

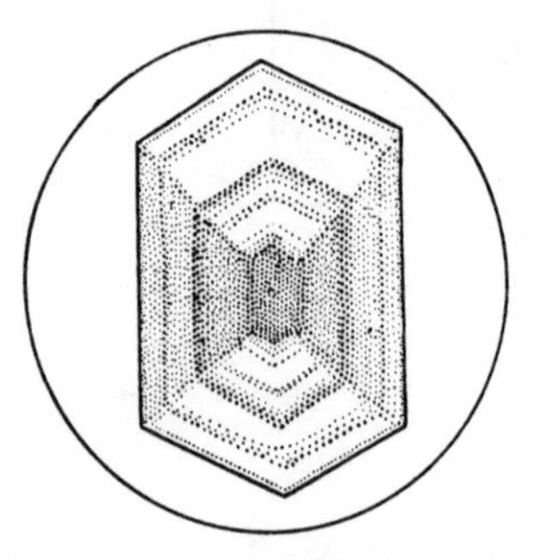

FIG. 15

Zoning and hour-glass structure in augite in lamprophyre, Mühlörzen, Mittelgebirge, Bohemia.

cross-wire in the basal section shown in Fig. 14. More complicated twins, commonly cruciform, are often seen, especially in sections of Basic lavas.

Crystallization of Pyroxenes

In the main series of pyroxenes falling temperature during crystallization causes a shift of composition from left to right on the composition diagram (Fig. 10), *i.e.* from Mg^{2+}-rich to Fe^{2+}-rich types. This progressive iron-enrichment manifests itself in three ways: (1) crystals may be zoned, with Mg^{2+}-rich cores and Fe^{2+}-rich outer layers; (2) early-formed phenocrysts are more magnesian than the crystal grains or microlites occurring in the groundmass of a lava or dyke-rock; (3) in a series of associated rocks representing successive magma-fractions, pyroxenes from early fractions are more magnesian than later ones.

The Monoclinic diopside type of structure is stable over a wide temperature range, and therefore augite is ubiquitous in Basic igneous rocks throughout the whole range of grain-size variation, and is common also in rocks of Intermediate composition, especially in andesitic lavas. Pigeonites and orthopyroxenes are less widely distributed for two reasons: firstly, the former are stable only at relatively high temperatures and invert into the latter on cooling; and secondly, MgFe-rich pyroxenes must be regarded as alternatives to olivines in certain circumstances explained below.

The temperature of inversion from pigeonite to orthopyroxene falls from 1140° C for the Mg^{2+} end member of the series to 980° C for an iron-rich orthopyroxene, say Fs_{80}. The temperature at which pyroxene begins to crystallize from a magma depends upon the composition of the latter. In Mg-rich Basic magmas pyroxene commences to crystallize at temperatures slightly below the clino-orthopyroxene inversion, so that the first precipitated pyroxenes are enstatites and bronzites. On the other hand iron-rich later magma fractions begin to crystallize at temperatures *above* the inversion point, so that pigeonite (usually within the hypersthene range of composition) is first precipitated. In other words the curve representing magmatic temperatures plotted against the composition of the pyroxenes may intersect that of inversion temperatures plotted on the same basis (the Mg : Fe ratio).

It will be apparent from this that careful study of the pyroxenes in a rock may indicate the temperature at which it crystallized. The technique, originally suggested by H. H. Hess,[1] was successfully applied to the Stillwater Complex in Montana. The significant observation is the first appearance of pigeonite in the sequence of

[1] *Amer. Min.*, **26** (1941), 573.

differentiating rock types, since this marks the point in fractional crystallization where the curves intersect. Composition of the pigeonite is determined and the point corresponding with this composition on the inversion curve indicates the magmatic temperature. In the case cited the composition of the first-formed pigeonite was found to be $En_{70}Fs_{30}$, corresponding to a temperature of 1100° C. What happens subsequently depends upon the rate of cooling: if the temperature falls fast, as in the case of a lava, pigeonite may survive in the metastable condition; but with slower cooling, under plutonic conditions, it will invert into orthopyroxene. Such 'inverted pigeonites' may be easily recognized by the inclusion of relatively thick lamellae of exsolved diopsidic clinopyroxene, lying parallel to the (001) plane of the original pigeonite (see below).

Exsolution Phenomena in Pyroxenes

Under conditions involving slow cooling, pyroxenes acquire distinctive structural features not exhibited by their high-temperature equivalents. Firstly, pigeonite cannot survive slow cooling but inverts into orthopyroxene. As indicated on the composition diagram, pigeonite contains a good deal less Mg^{2+} and Fe^{2+} than an orthopyroxene does, but is correspondingly richer in Ca^{2+}: the former contains about 9·5 per cent CaO, as compared with 1·6 in the latter. Therefore, when pigeonite inverts into orthopyroxene there is a large surplus of CaO which cannot be accommodated within the structure of the latter. It is therefore **exsolved.** On account of limited ionic diffusion through solid crystals the unwanted material cannot be completely eliminated, but it is concentrated in parallel, minutely thin lamellae which have the composition of diopsidic clinopyroxene. Two different structures of this kind have been described: the first, named the Bushveld type from its typical occurrence in the complex of that name, involves orthopyroxene as the host mineral, containing exsolved lamellae, excessively thin—from 100 to 250 per mm of thickness—and parallel to the (100) plane (Fig. 71). The second type is named from its typical development in the Stillwater Complex, and differs from the Bushveld type in two respects: the lamellae are thicker (25 to 100 times thicker) and they are parallel to the monoclinic plane (001). Hess has suggested that the coarse (001) lamination resulted from exsolution taking place mainly during the inversion; while the fine (100) lamination was effected with greater difficulty at a temperature *below* the inversion point. When the (001) lamination described above is combined with twinning on (100), the parallel lamellae slope away from the twin plane in a very distinctive manner, giving a 'herring-bone' structure. This proves the orthopyroxene to be inverted pigeonite, for twinning on (100)—the common type of augite twin—

although appropriate to Monoclinic minerals, including pigeonite, is impossible in orthopyroxene.

Augites in Basic gabbroic intrusives show exsolution interlamination exactly comparable with those just described, but with the roles reversed: the host mineral is augite, but the exsolved lamellae are orthopyroxene and are rendered particularly conspicuous by containing large numbers of hair-like rodlets of magnetite or limonite formed from the latter. The rodlets lie in parallel orientation, usually parallel to the *z*-axis of the host pyroxene. They give rise to a characteristic bronzy sheen to the mineral which is said to be schillerized. **Diallage** is the name applied to these schillerized augites. Presumably this is a direct consequence of the change in composition of augites with falling temperature. Augites of early, high temperature formation are richer in $MgFe^{2+}$ and correspondingly poor in Ca^{2+} than those which crystallized later, at lower temperatures. The change in composition involves not only iron-enrichment as noted above, but also enrichment in Ca^{2+}, so that the compositional change moves towards the right and upwards towards the diopside-hedenbergite join in Fig. 10. Early augites contain about Ca_{40} but late augites Ca_{50} per cent of the total metallic cations.

Relationship between Olvines and Pyroxenes

It was pointed out above that a reaction relationship exists between olivines and orthopyroxenes. The former crystallize at higher temperatures than the latter; but early formed olivine may react with the magma by extracting silica from it in sufficient quantity to convert it into the corresponding orthopyroxene. The conversion requires time for its completion, however, and in the case of a lava, cooling may be so rapid that the reaction temperature is passed over too quickly for it to be effected, and some olivine may survive, even when the magma contains sufficient silica to convert it all into orthopyroxene. Normally in the more highly silicated magmas, for example those of andesitic composition, olivine does not occur, though orthopyroxenes, including enstatite and hypersthene, may be well represented and give no indication of having 'evolved' from earlier olivines. The absence of pigeonite from these lavas is a direct consequence of their crystallization from cooler magmas below the pigeonite-orthopyroxene inversion temperature.

Basaltic magmas on the other hand are relatively hot, temperatures of the order of 1100° C having been measured in the basalt 'lava lakes' in Hawaii. They are also more basic and therefore olivine is inevitable unless the magma is unusually siliceous for a basalt. The early crystallization of this mineral withdraws much MgFe from the magma and this naturally affects the composition of the pyroxene which is precipitated in due course: it is a Ca^{2+}-rich clinopyroxene (augite) which occurs in association with abundant olivine in many olivine-basalts. Two essential ions in augite are Ca^{2+} and Al^{3+}; but these are essential too in the plagioclase which accompanies olivine and pyroxenes in these rocks, and which in order to simplify matters we will regard as having the composition of anorthite, $CaAl_2Si_2O_8$. The early separation of anorthite in quantity will deplete the magma in Ca^{2+} Al^{3+}, necessary for the formation of augite, which, therefore, may be suppressed in favour of bronzite or hypersthene. Thus olivine and anorthite work in opposite directions and what exactly happens in any specific case depends upon several factors which it would be unprofitable to examine further at this stage.

AMPHIBOLES

The amphiboles form a large group of complex metasilicates, and are chemically related to the pyroxenes. Any species of pyroxene may contain identically the same elements as the corresponding amphibole, but they are present in different proportions; while a more fundamental difference is the presence in the latter of hydroxyl, represented by (OH), and with a negative charge of one unit. A hydroxyl group is the same size, and functions in the same way, as an oxygen atom.

Chemical Relationships

The X-ray structure consists fundamentally of 'bands' of linked SiO_4-tetrahedra. Each band in effect consists of two pyroxene chains united by shared oxygens as shown in Figs. 16 and 17. The arrangement of the bands parallel to the vertical axis is essentially the same as for the pyroxene chains, illustrated in Fig. 7.

The unit of pattern in Fig. 16 contains $Si_4O_{11}(OH)$; but amphibole formulae are based on twice this unit, *i.e.* $Si_8O_{22}(OH)_2$. The total negative electrostatic charge on this unit is 46, which is only partly counterbalanced by the positive charge of 32, contributed by the silicons. The negative residual charge of -14 units must be balanced by introducing the requisite number of cations into the structure. Seven Mg^{2+} ions would effect a balance and the formula would become $Mg_7Si_8O_{22}(OH)_2$. Alternatively the cations might be Fe^{2+}, giving the formula $Fe_7Si_8O_{22}(OH)_2$. These formulae represent the compositions of kupfferite and grunerite respectively, which are the two end-members of a continuously variable series which includes the Orthorhombic anthophyllite, comparable with orthopyroxene (hypersthene), and Monoclinic cummingtonite similar in composition to anthophyllite.

In Fig. 17, a much-simplified cross-section is presented of the band structure of the amphibole, **tremolite,** in which three different cations occupy particular sites in the structure. At the bottom, two Ca^{2+} ions occupy the so-called *X* position; along the top are

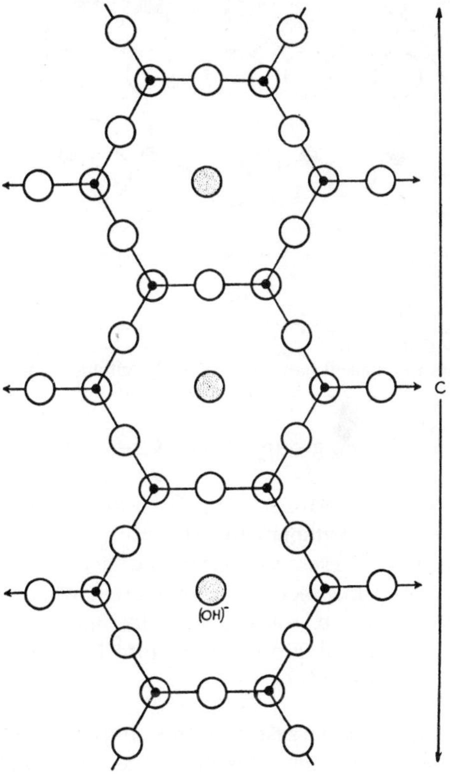

(OH)⁻

c

FIG. 16

Diagram of the atomic structure of the amphiboles: the amphibole 'band'. The arrows indicate the linkage of 'free' oxygen atoms to cations. The oxygens superimposed on the Si atoms are similarly linked to cations, thus isolating each Si-O band from its neighbours.

five Mg^{2+} ions in the *Y* position; while the much smaller silicons occupy the *Z* positions. If the general symbols *X*, *Y* and *Z* are used instead of specific chemical symbols, the unit formula becomes $X_2Y_5Z_8O_{22}(OH)_2$ and all amphibole formulae, no matter how complicated, are built on this plan. Thus the tremolite formula becomes $Ca_2Mg_5Si_8O_{22}(OH)_2$ while the substitution of Fe^{2+} for some of the Mg^{2+}, gives actinolite. Both are essentially metamorphic minerals and only concern us in so far as they occur as late-stage or secondary minerals in gabbroic and doleritic rocks.

One widely used scheme of classification of the amphiboles is wholly chemical. It divides the amphiboles into three groups according to the nature of the cation(s) occupying the *X*-position in the

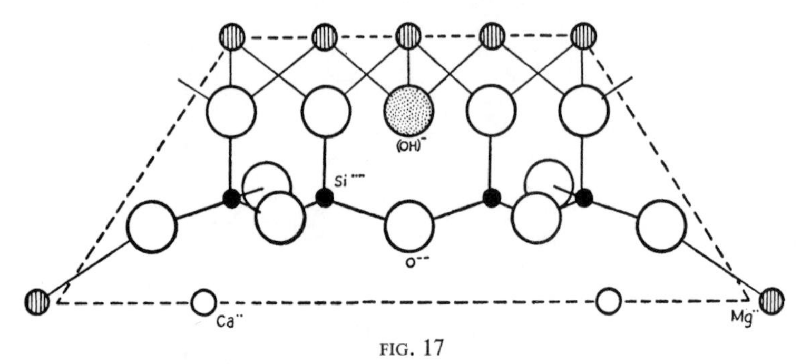

FIG. 17

Diagram of end view of the amphibole band. *b*-axis, left to right; *c*-axis, at right angles to the paper. The two Mg atoms outside the boundary of the band belongs to adjacent bands immediately left and right of the one illustrated. Only some of the bonds are represented.

structure: (1) the MgFe-group, (2) the Ca-group and (3) the alkali (Na) group.

(1) **The MgFe group** comprises grunerite and kupfferite referred to above, together with **anthophyllite** (Orthorhombic) and **cummingtonite** (Monoclinic), these two being intermediate in composition between the two end-members. These minerals, which are analogous with the orthopyroxenes, are essentially metamorphic, though anthophyllite and cummingtonite enter into late-stage intergrowths in certain Basic igneous rocks.

(2) **The Ca-group** is the largest and most important of the three. It includes tremolite and the closely-related actinolite referred to above, both being essentially metamorphic; also common hornblende, the best-known and most widespread of all the amphiboles together with its varieties. The latter include the green (in thin section) pargasite, hastingsite and ferrohastingsite, the first- and last-named being end-members of a continuously variable series; also the brown lamprobolite often referred to as 'basaltic hornblende', and kaersutite which may be thought of as 'titanhornblende' analogous with titanaugite among the pyroxenes. Barkevikite is also included in this group; its colour also is brown, and it is stated by Johanssen to be intermediate in composition between arfvedsonite (a strongly coloured green Na-amphibole) and common hornblende (a Ca-amphibole). Therefore it is on the borderline between the former and the latter groups. This is true also of katophorite which is regarded as an alkali (Na) amphibole; but actually contains both

Ca and Na in the *X*-position, with the former typically in large excess over the latter.

(3) **The Na-group** includes arfvedsonite, the very distinctive glaucophane and riebeckite, again end-members of yet another variable series, and aenigmatite, formerly termed cossyrite, and the only Triclinic amphibole.

Crystallographic and Optical Properties of Monoclinic Amphiboles

The only megascopic crystals of amphiboles which the reader is likely to encounter are brownish black common hornblendes with a rather resinous lustre; prismatic in habit and with terminations like those illustrated in Figs. 18 and 19, which show the ends of a simple crystal and a simple twin for comparison. With rare exceptions the

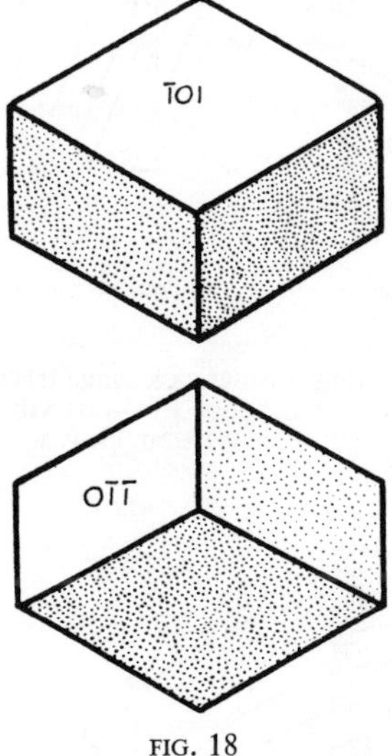

FIG. 18

The two ends of a simple crystal of hornblende, in plan. The forms represented are: clinoprism {011} and hemi-orthodome {101}. The forms in the vertical zone are: prism {110} and side (clino-) pinacoid {010}.

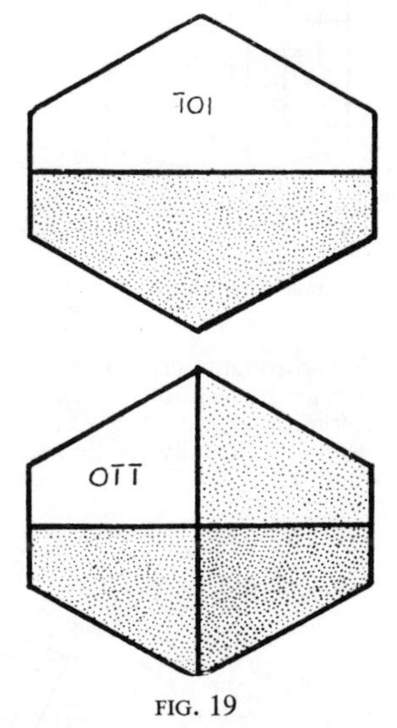

FIG. 19

The ends of a simple hornblende twin, in plan. The upper figure shows the hemi-orthodome faces; the lower figure, the clinoprism faces. The twin-plane is (100).

optical orientation of the amphiboles is the same as that of the pyroxenes: the optic axial plane lies parallel to (010). The details of extinction in variously orientated sections considered above in relation to the pyroxenes, apply equally to the amphiboles. The significant facts are illustrated in Fig. 20. The most obvious means of differentiating between the several members of the group is the size of the extinction angle in (010) sections, as indicated in Fig. 21. The angles read to the left of the middle line are those between the slow vibration direction (Z) and the cleavage traces which mark out the position of the vertical axis. Similarly, those to the right, marked 'fast' on the diagram, are measured between the fast vibration direction (X), and the cleavage traces. One significant point here is that while riebeckite and glaucophane both have much the same extinction angle, in the former the extinction position nearest to the

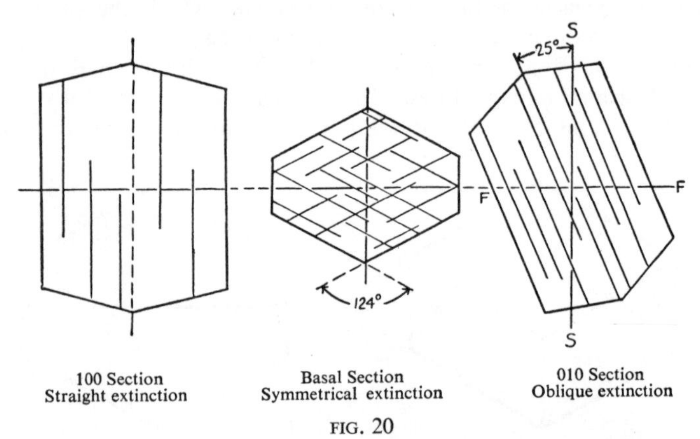

FIG. 20

Sections through a crystal of hornblende, showing prismatic cleavage traces and drawn in positions of extinction. SS—slow vibration, FF—fast vibration. Note that because of obliquity to the plane of the section, cleavage traces are not normally visible in a 100 section.

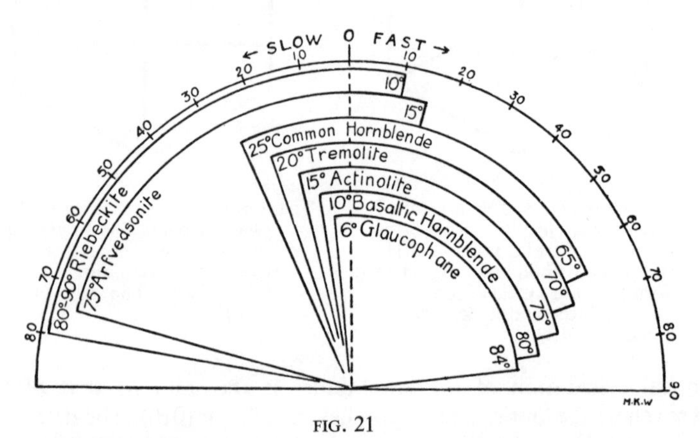

FIG. 21

Diagram showing extinction positions of the amphiboles. The figures in the left-hand top corner of each 90°-fan are the maximum extinction angles shown by (010) sections. The vertical line marked 'O' represents the vertical cross-wire in the eye-piece of the microscope. The sides of the fans mark positions of the cleavage traces in (010) sections in the extinction position.

cleavage traces is the fast vibration, but with the latter it is the slow direction.

Problems of Specific Identification of Amphiboles

It will be appreciated from the foregoing account that several amphiboles are of extremely complex composition. In this connection it is significant that virtually all the chemical components of a Basic igneous rock can, in certain circumstances, become incorporated in a single metamorphic amphibole. With so many variables (due to atomic substitution) it has proved impossible to devise a wholly satisfactory scheme of classification of the amphiboles, and it is impossible to represent the variations in chemical composition on a simple two-dimensional diagram. It follows that it is also impossible to construct a diagram correlating variation in chemical composition with optical properties. The accurate measurement of a single optical parameter, such as refractive index or optic axial angle, can provide a reasonable indication of the composition of, say, an olivine or a plagioclase in which only one variable ratio is involved—Mg/Fe in the former and Na/Ca in the latter; but this technique is impossible in the case of the amphiboles because there are too many variables. Particular optical properties vary widely in different specimens of any one species, with consequent overlap. It would appear from published data that no one chemical characteristic or optical property is diagnostic of a particular amphibole. This must not be taken to mean that optical characters are valueless; but when dealing with strongly coloured amphiboles, particularly brown ones, optical tests prove inadequate and recourse must be made to detailed chemical analysis. Formerly this was impracticable for reasons of time, facilities and cost; but now it is common practice to separate an amphibole from its matrix and submit it to detailed chemical analysis in the course of petrological research.

Characters of Individual Species of Amphiboles

As a matter of convenience we propose to consider firstly the green amphiboles, then the brown ones. This enables us to deal with certain aspects collectively, for purposes of comparison and contrast.

It is appropriate to start with **common hornblende** which we use to illustrate the very extensive atomic substitution which is characteristic of the amphiboles. In the general formula the *X*-position may be occupied by Ca^{2+}, Na^{+} and K^{+} in order of abundance; the *Y*-cations may include, Mg^{2+}, Fe^{2+}, Fe^{3+}, and Al^{3+}; while in the *Z*-position we find Si_6 together with $(Si, Al)_2$ bringing the total number of '*Z*-cations' up to the required total of 8. Note the 'proxy-Al^{3+}' which substitutes for Si to a limited extent, in spite of the difference in valency. Consequently the total charge on '*Z*' is reduced from 32 to 31, and this has to be balanced in another part of the structure, usually by introducing Na^+ in the *X*-position. The complete formula, based on commonly occurring substitutions is:

$$(Ca, Na, K)_{2\text{-}3}(Mg, Fe^{2+}, Fe^{3+}, Al)_5Si_6(Si, Al)_2O_{22}(OH, F)_2.$$

The only further point calling for comment concerns the 2–3 cations in the *X*-position. Although only two are shown in Fig. 17, there are spaces within the structure for three; but in some amphiboles the third is unoccupied: it is a vacant space in the structure.

As regards conditions of formation, because of their complex chemical compositions, the crystallization of amphiboles (including of course, common hornblende) has not been studied experimentally in the same detail as the olivines and pyroxenes; but a good deal may be inferred from relationships visible in thin sections. In slides containing both pyroxene and amphibole, the latter crystallizes after the former—for example, hornblende is seen to be moulded upon augite, and this may be due to the two crystallizing in sequence, or it may be a replacive relationship. There is a third possibility: under plutonic conditions hornblende may be precipitated instead of pyroxene at temperatures up to about 950° C in the presence of adequate water. Amphiboles are products of wet magmas, the essential condition being a high flux concentration. It follows that they are more commonly met with in deep-seated, rather than volcanic rocks. Common hornblende is the commonest coloured mineral in Acid and Intermediate members of the calc-alkali suite, especially granodiorites, tonalites, monzonites and diorites. It is also produced at the expense of original clinopyroxene in Basic rocks (gabbros and dolerites) which have been subjected to regional metamorphism. The resulting rocks are termed amphibolites.

Apart from common hornblende three other amphiboles are green in thin section and call for careful examination: these are **pargasite, hastingsite** (and ferrohastingsite) and **arfvedsonite.** All contain the same elements but in different proportions as shown below:

In pargasite
$X_3 = Ca_2Na$ $\quad Y_5 = Mg_4(Al, Fe^{3+})$ $\quad Z_8 = Si_6Al_2$,
In ferrohastingsite
$X_3 = Ca_2Na$ $\quad Y_5 = Fe_4(Al, Fe^{3+})$ $\quad Z_8 = Si_6Al_2$,
In arfvedsonite
$X_3 = Na_{2\cdot5}Ca_{0\cdot5}$ $\quad Y_5 = (Mg, Fe^{2+}, Fe^{3+}, Al)$ $\quad Z_8 = Si_{7\cdot5}Al_{0\cdot5}$.

One significant point of difference is the reversal of the roles of Na and Ca in ferrohastingsite and arfvedsonite respectively: a 1 : 2 ratio in the former becomes 5 : 1 in the latter. In theory, little difficulty should be experienced in distinguishing between these green amphiboles; but in practice difficulty arises because of the strong colours of ferro-hastingsite and especially arfvedsonite. Contrary to the general rule for amphiboles, in arfvedsonite it is the fast ray $X(=\alpha)$ which is most strongly absorbed giving a rich bluish green. In ferrohastingsite it is the slow ray, $Z(=\gamma)$ which is most strongly absorbed, giving an olive-green absorption colour. In most amphiboles the optic axial plane lies parallel to the plane of crystallographic symmetry (010); but in arfvedsonite it is perpendicular to the latter. Further, the extinction angle is small (less than 15°), while the birefringence is weak (0·005 maximum); but that of ferrohastingsite is relatively strong (0·028).

As regards paragenesis, pargasite is typically metamorphic and is one of the most characteristic products of the thermal metamorphism of dolomitic limestones; but pargasitic hornblende occurs in igneous rocks, notably in granites of rapakivi types in Finland and elsewhere. Hastingsite has been frequently recorded from alkali-syenites; ferrohastingsite, like the analogous pyroxene and olivine (ferrohedenbergite and fayalite respectively) occurs in alkali-rich granitic rocks such as the Tertiary granites of the Marsco area in Skye.[1]

Arfvedsonite is probably the most widely distributed sodic amphibole, being developed in a wide range of sodic syenites. In ditróite, a nepheline-sodalite syenite, it occurs (as a member of a typical sodic assemblage of minerals) in reaction relationship with aegirine-augite.

Until a few years ago it was customary to identify strongly coloured brown amphiboles, with the exception of basaltic hornblende, as barkevikite. The identifications were based on optical techniques, not on detailed chemical analysis. In Britain the suite of rocks forming the differentiated Lugar Sill in Ayrshire, Scotland, came to be regarded as providing rock sections containing typical barkevikite, and these were widely used for teaching purposes. Later, the Ti-rich brown amphibole, kaersutite was identified and described from Kaersut in Greenland; and it has become clear that some former barkevikites fall within the present definition of kaersutite. For example, another of the few British type-rocks—minverite—was for many years thought to contain barkevikite; but re-examination, using modern techniques, has shown the amphibole to be kaersutite.[2] Further, former definitions of certain lamprophyre types included barkevikite as the diagnostic 'dark' mineral: camptonite is a case in point. But now some camptonites containing kaersutite have been recorded from a number of localities,[3] so that the 'barkevikite' in others must be regarded as suspect. To date, kaersutite has been

[1] Thompson, R. N., *Quart. J. Geol. Soc.*, **124** (1969), 349.

[2] Kempe, D. R. S., Kaersutite from the minverite of Cornwall, *Min. Mag.*, **36** (1968), 874–6.

[3] Vincent, E. A., *Quart. J. Geol. Soc.*, **109** (1955), 143.

recorded from a variety of rocks of varying composition including, in addition to those mentioned above, trachybasalts, trachyandesites, trachytes, basalts and associated tuffs. In the last case, the kaersutites are xenocrystic, and would probably have been identified as basaltic hornblende on general grounds.

Lamprobolite is sometimes referred to as 'basaltic hornblende', though the relatively rare hornblendes occurring in basalts are usually xenocrystic and probably do not all belong to the same species. Lamprobolite is widely distributed in lavas belonging to the calc-alkali suite—andesites, latites and dacites. By comparison with other brown amphiboles lamprobolite is characterized by a large content of iron, a high $Fe_2O_3:FeO$ ratio, deficiency in (OH) and by containing oxygen as a substitute for some of the hydroxyl in the $(OH)_2$ bracket, so that the mineral is termed an oxyhornblende. These chemical features result not so much from the kind of magma in which crystallization took place as on the conditions to which the mineral was subsequently subjected. In this connection it is significant that common hornblende can be converted into lamprobolite by heating to about 800° C, and it is therefore inferred that the latter has been 'cooked'. Certainly the crystals show clear evidence of disequilibrium during or subsequent to uprise by the development of a reaction rim of minute granules of magnetite at an early stage; but ultimately the whole crystal may be replaced by iron-ore or by a mixture of the latter with granules of clinopyroxene (Fig. 110).

Katophorite (=catophorite) is the latest brown amphibole to be given a specific name. It has been recorded from Basic, alkali-rich plutonic rocks including theralites and shonkinites together with their volcanic equivalents; and recently has been noted in trachytes, phonolites and notably in pantellerites occurring in the African alkali volcanic province in the Rift Valley region. It is a mineral of high-temperature formation, occurring as phenocrysts in pantellerites as well as being associated with arfvedsonite in the groundmass.

Finally we note **aenigmatite,** probably the most distinctive of the brown amphiboles; it has been with us for a long time, but usually under the invalid name, cossyrite, which was an essential constituent in the original definitions of pantellerite. It is restricted to peralkaline volcanic rocks including phonolites, one variety of which was distinguished as 'apachite' on account of its very abundant aenigmatite. Some varieties of peralkaline syenites also contain aenigmatite, associated with aegirine-augite in some cases, but riebeckite in others.

These brown amphiboles contain approximately the same amount of SiO_2 (40 per cent), but they vary strikingly in certain other details of chemical composition. We may note particularly the high TiO_2 in kaersutite, also the high $FeO:Fe_2O_3$ ratio and high CaO. In katophorite Al_2O_3, MgO and CaO are high; but the FeO : Fe_2O_3 is the reverse of that of kaersutite. Aenigmatite demands for its formation high TiO_2, high Na_2O and *very* high FeO; but it is deficient in Al_2O_3 and MgO.

With regard to the identification of the different brown amphiboles, it is significant that some experts, even after working out optical details and with chemical data available, still think it best to identify a particular specimen as 'a brown amphibole'. Published data show wide variation in the optical characters of brown amphiboles: thus 2V for katophorite ranges from 0 to 50°, thus overlapping with barkevikite (40° to 50°) while lamprobolite and kaersutite are much the same—60° to over 80°. Similarly, birefringence varies from weak to moderate in katophorite (0·007 to 0·021) and from moderate to strong in kaersutite (0·019 to 0·083). Thus, if all specimens of katophorite showed the minimum and all kaersutites the maximum, a simple observation between crossed polarizers would suffice to distinguish between them: the former would show no change in colour on crossing and uncrossing the polarizers; but the latter would show a spangled effect like calcite viewed through a brownish screen. In favourable circumstances this observation would prove satisfactory; but otherwise a check would have to be made, using extinction angles, pleochroism or even paragenesis. Extinction angles are not satisfactory because of strong absorption and, in some cases strong dispersion; but the large angles for kataphorite (36° to 70°) stand in strong contrast to the small angles shown by lamprobolite (average 9°) and the slightly larger angles of barkevikite (11° to 18°). Pleochroism is of very limited value: individual schemes vary as much as do the different species: only aenigmatite appears to be reasonably constant in this respect.

Paragenesis has been covered in the detailed treatment above; but it may be noted that the association of brown amphibole with other Ti-rich minerals such as titanaugite, ilmenite and sphene would suggest kaersutite.

Glaucophane and riebeckite are chemically related sodic amphiboles but differ in paragenesis: the former is strictly limited to low temperature, high pressure regional metamorphism; but the latter occurs in igneous rocks. If, in the formula for tremolite, NaFe is substituted for CaMg, we derive the formula of riebeckite which may be written $Na_2Fe^{3+}{}_2Fe^{2+}{}_3Si_8O_{22}(OH,F)_2$. Similarly, substitution of NaAl for CaMg gives glaucophane; but natural specimens of these minerals lie somewhere between these end-members.

Riebeckite is comparable with aegirine among the pyroxenes and is the most widely distributed of the sodic amphiboles, although, unlike most of the latter, it occurs in over-saturated, quartz-bearing granites, microgranites and rhyolites in certain alkaline complexes, notably in Northern Nigeria.[1] In these complexes, associated granite-pegmatites contain abnormally large crystals of riebeckite. We illustrate riebeckite-microgranite from Mynydd Mawr[2] in North Wales (Fig. 86) and a variant of the same rock-type from Ailsa Craig in the Firth of Clyde, Scotland (Fig. 85). The islet of Rockall west of Scotland in the North Atlantic is built partly of aegirine-riebeckite-granite.

In thin section riebeckite is strongly coloured with distinctive pleochroism. Maximum absorption parallel to $X(\alpha)$ gives deep indigo blue to almost black. Due to strong dispersion it is impracticable to measure extinction angles accurately; but riebeckite is so distinctively coloured, it is identified 'on sight' and confirmatory tests are seldom necessary. Incidentally, although we are not directly concerned with glaucophane, at its best its pleochroism scheme is the most distinctive of all the amphiboles: with X, light yellow, Y, violet and Z, pure Prussian blue.

MICAS

The micas constitute one of the most important groups of mineral silicates. They are chemically distinct from the groups so far considered in two respects: the alkali elements are important in all micas; but unlike certain amphiboles and pyroxenes, calcium is absent. In atomic structure, too, they are distinctive. The atoms are arranged in extended sheets as discussed below.

Atomic Structure and Chemical Relations

In micas and in several other groups of minerals built on the same plan, the SiO_4-tetrahedra are linked to one another at three corners, and thus form indefinitely extended sheets in which the atoms of different kinds are arranged on a hexagonal plan (Fig. 22). The sheets are actually double, as shown in Fig. 23, and are 'staggered' relative to one another, which gives the crystals Monoclinic symmetry, although they are strongly pseudohexagonal. The unit of pattern, outlined in Fig. 22, contains $Si_4O_{10}(OH)_2$; but as a matter of convenience this is doubled and becomes $Si_8O_{20}(OH)_4$. The hydroxyl groups are included in the planes containing the 'free' oxygens and four of these are included in each unit. Using the same convention as for amphiboles we can derive a general formula for micas using X, Y and Z for groups of ions occupying constant positions in the structure. The X position is occupied by the loosely bonded large cations lying between the layers shown in Fig. 22 and

[1] Jacobson, R. E., Macleod, W. N. and Black, R., *Geol. Soc. Special Mem.*, No. 1 (1958).

[2] Pronounced Munith Mower, (rhyming with Tower).

is 'the K layer' in Fig. 23. *Y* refers to the smaller cations in octahedral coordination, stippled in Fig. 23 and labelled Al^{2+}, Mg^{2+}, Fe^{2+}; while *Z* refers to those in tetrahedral co-ordination, the small

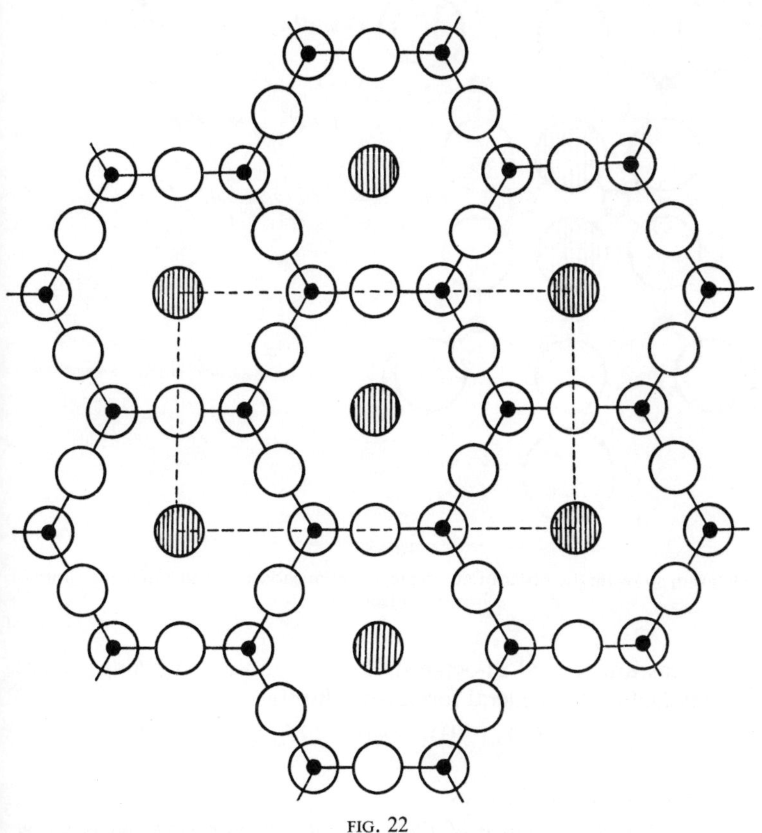

FIG. 22

Diagram showing the extended sheet structure of the micas.

Si atoms, black; O^{2-} atoms, white; hydroxyl (OH), ruled. The characteristic double sheet—the thinnest possible cleavage flake—comprises two of those illustrated, in mirror-image relationship (see Fig. 23).

black cations indicated as the 'Si, Al layer' in Fig. 23. The proportions of the *X*, *Y* and *Z* cations are shown in the general formulae below. There are two alternatives according to whether the *Y* sites are occupied by trivalent cations or divalent cations. As these have to contribute an electrostatic charge of $+12$, there must be four $(Y^{3+})_4$ of the former, but six $(Y^{2+})_6$ of the latter. In all micas *Z* is shared between Si and Al, often in the proportion of 6 : 2. The

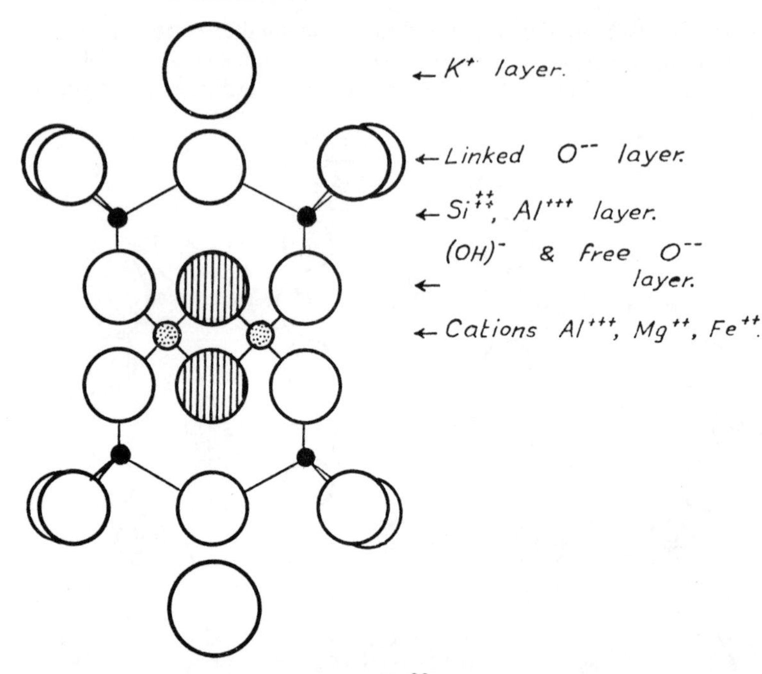

FIG. 23

Diagram showing the order of atomic layers in mica: section through the double sheet.

Y_4 micas are termed di-octahedral, while those containing Y_6 are tri-octahedral, the general formulae being:

$$X_2(Y^{3+})_4Z_8O_{20}(OH)_4 \quad and \quad X_2(Y^{2+})_6Z_8O_{20}(OH)_4.$$

General Properties of Micas

As a direct consequence of their atomic structure the micas show pseudohexagonal symmetry, though actually they are Monoclinic. To the eye they appear to be simple hexagonal crystals of tabular habit; but careful measurement shows that the basal pinacoid is not perfectly at right angles to the prism faces. The angle beta is within a few minutes of 90° however; and similarly the angles between the apparent hexagonal prism faces are nearly, but not exactly, 60°. Occasionally crystals are found which show hemibipyramid and other faces which betray the true symmetry of the mineral: otherwise the percussion figure obtained by smartly tapping a centre-punch placed in contact with the (001) face of the mica, proves that the symmetry is not hexagonal. A six-rayed star is produced in this way, but the rays are not identical: two, which lie in the single

TABLE SHOWING THE COMPOSITION OF THE COMMONER MICAS

	X	Y	Z	
Di-octahedral				
Muscovite	K_2	Al_4	Si_6Al_2	$O_{20}(OH,F)_4$
Paragonite	Na_2	Al_4	Si_6Al_2	$O_{20}(OH,F)_4$
Tri-octahedral				
Phlogopite	K_2	$(Mg,Fe)_6$	Si_6Al_2	$O_{20}(OH,F)_4$
Biotite	K_2	$(Mg,Fe)_{6-4}$ $(Fe,Al,Ti)_{0-2}$	$Si_{6-5}Al_{2-3}$	$O_{20-22}(OH,F)_{4-2}$
Zinnwaldite	K_2	$(Fe,Li,Al)_6$	$Si_{6-7}Al_{2-1}$	$O_{20}(F_{3-2}OH)_{1-2}$
Lepidolite	K_2	$(Li,Al)_{5-6}$	$Si_{6-5}Al_{2-3}$	$O_{20}(OH,F)_4$

plane of symmetry, are more strongly developed than the others. This fulfils another useful purpose, as it serves as a guide when finding the position of the optic axial plane.

The most striking physical property of all micas is the perfect basal cleavage, which takes place between the pairs of sheets of SiO_4-tetrahedra, and in the planes of alkali atoms (Fig. 23). Not only are the atoms fewest in these planes, but the K^+ ions are very loosely bonded to the rest of the structure: they are in twelve-fold co-ordination.

Twinning is not uncommon in micas, on the so-called 'mica-law', in which the twin plane is approximately coincident with (110), while the composition plane may be either (001) or (110). In the latter case the basal cleavages of both parts of the twin are co-incident, but on account of the different orientation of the two parts, there is a striking difference in absorption, best seen when looking obliquely through a cleavage plate towards a good light. Twinning is seldom seen in thin sections.

Optical Orientation

Although the pseudohexagonal symmetry of the mica crystal has been stressed above, examination of its optical properties imme-

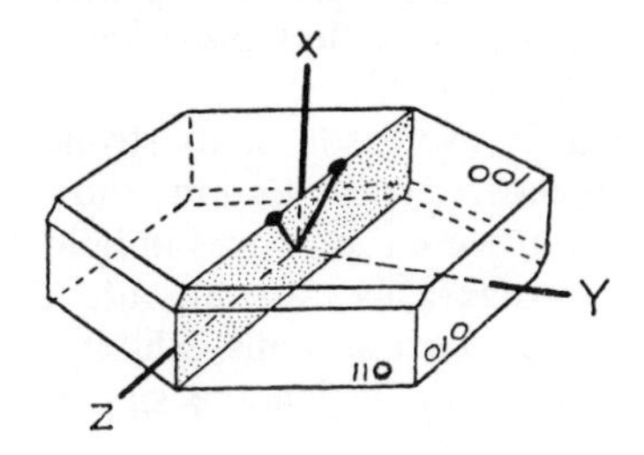

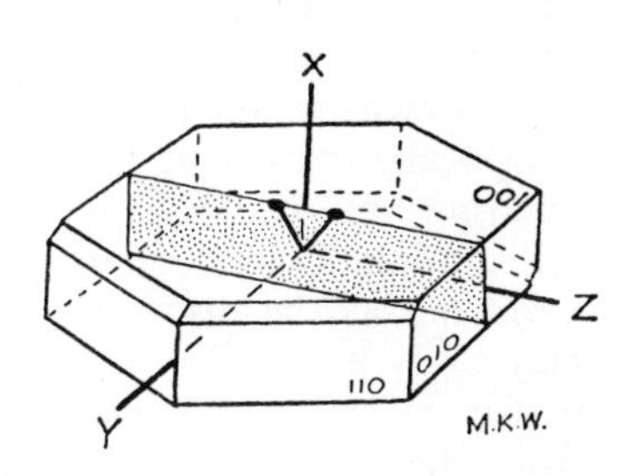

FIG. 24
Diagram showing the optical orientation of the micas. *Left*, the biotite-phlogopite series; *right*, the light micas.

diately demonstrates that it is not hexagonal. As a group the micas are biaxial, though the size of 2V is very variable, and the optic axial plane in some cases is parallel, but in others perpendicular, to (010) (see Fig. 24). This difference in orientation is the basis of the division of the micas into two groups, which essentially comprise the light and the dark micas respectively. In all micas the fast vibration X is perpendicular to the (001) plane, and as X is the acute bisectrix, all micas are optically negative.

THE LIGHT (DI-OCTAHEDRAL) MICAS

(*a*) **Muscovite** is the silvery white mica seen in hand specimens of many granites, particularly the strongly potassic varieties, and in the pegmatites and minor intrusives associated with them. In igneous rocks it is restricted to the most highly silicated types. It is also widely distributed in schists and gneisses.

Muscovite originates in another way, however. The potassic feldspars contain the same elements as muscovite, and are readily converted into the latter as a consequence of hydrolysis. Conversely, when rocks or even sediments containing white mica suffer thermal metamorphism, the mica is converted into orthoclase. A distinction is drawn between primary white mica formed by crystallization from a melt, and the secondary white mica produced by the alteration of alkali-rich silicates. The latter type is **sericite.** Although identical in composition with muscovite, sericite has an entirely different mode of occurrence in rocks, and there appear also to be slight physical differences. Sericite usually occurs in the form of aggregates of minute flakes.

The analogous mica with Na^+ instead of K^+ in the X_2 positions is **paragonite** an important constituent in mica-schists but apparently very rare in other parageneses. By analogy with sericite it might reasonably be inferred that similar micaceous aggregates in sodic plagioclase should be paragonite; but in some cases at least they have been shown to be sericite, formed from the potassic component in albite. Paragonite is uniaxial or nearly so, thus differing markedly from muscovite which is biaxial, with 2V about 40°.

Optically, muscovite is distinctive chiefly by reason of its strong birefringence—0·036. This ensures that most vertical sections show brilliant interference colours; but the basal section polarizes in light grey, and as the distinctive perfect cleavage traces are absent, it may easily be misidentified by the unwary. Such sections exhibit a perfect biaxial interference figure, however, on which the sign is easily checked.

(*b*) **Lepidolite,** a lithium-bearing light mica containing about 5 per cent LiO_2, is attractively coloured lilac in the mass, and although large crystals a foot or more across are obtained from certain complex pegmatites, a more usual mode of occurrence is in the form of aggregates of flakes or scales, as the name implies. Some particularly attractive mineral specimens consist of lepidolite acting as matrix to brightly coloured lithium tourmalines. Although the general optical orientation is the same as for muscovite, the angle 2V is very small, and in some varieties the mineral is sensibly uniaxial. Lepidolite is a characteristic constituent in some complex pegmatites, in close association with Li-bearing tourmalines.

It will be seen from the statement of the compositions of micas that lepidolite is intermediate between di -and tri-octahedral, because of the substitution of tetravalent Ti for some Al in the *Y* positions. Among other substitutions are Sr, Rb and Cs for part of K in the *X* positions. The presence of strontium and rubidium in lepidolite is important in radiometric age determinations involving the measurement of the $^{87}Rb : ^{87}Sr$ ratio.

(*c*) **Zinnwaldite** is also a lithium-bearing mica (about 3 per cent LiO_2) but is iron-rich, whereas lepidolite is iron-free. It also occurs in pegmatites and is important in greisens, and associated cassiterite ('tinstone') veins, hence the name zinnwaldite from German *zinn* = tin.

THE DARK (TRI-OCTAHEDRAL) MICAS

These stand in strong contrast to the group considered above, not only in appearance and optical orientation, but also in mode of occurrence and distribution in rocks.

Biotite and **phlogopite** are end-members of a gradational series between a hypothetical biotite in which all the *Y* sites are occupied by (Mg^{2+}, Fe^{2+}) and hypothetical phlogopite which is wholly magnesian in this sense. The division between them is arbitrarily drawn where the ratio of Mg to Fe is 2:1; but in practice the two are distinguished on the basis of colour. Biotite is black, but phlogopite is reddish brown (bronze). Biotites are of extremely complex composition as suggested in the Table, but this gives only a hint of the substitutions which may occur—for K, small amounts of Ca and Na may be substituted; for *Y*, Fe^{3+}, Al, Mn and Ti commonly accompany Fe^{2+} and Mg^{2+}. **Lepidomelane** is a variety of biotite particularly rich in total iron and occurring in quartz-rich granitic rocks and alkali-rich syenites.

The optical orientation of biotite is as shown in Fig. 24. The angle 2V is very small, so that on rotation of the stage the isogyres must be closely watched to see that they do actually separate. Further, very strong illumination must be employed, as the absorption is particularly strong in basal sections, so much so that even in thin sections no light is allowed to pass in some cases. Vertical sections are almost violently pleochroic, the vibration parallel to *X* yielding a light (straw-) yellow absorption tint, while the *Y* and *Z*

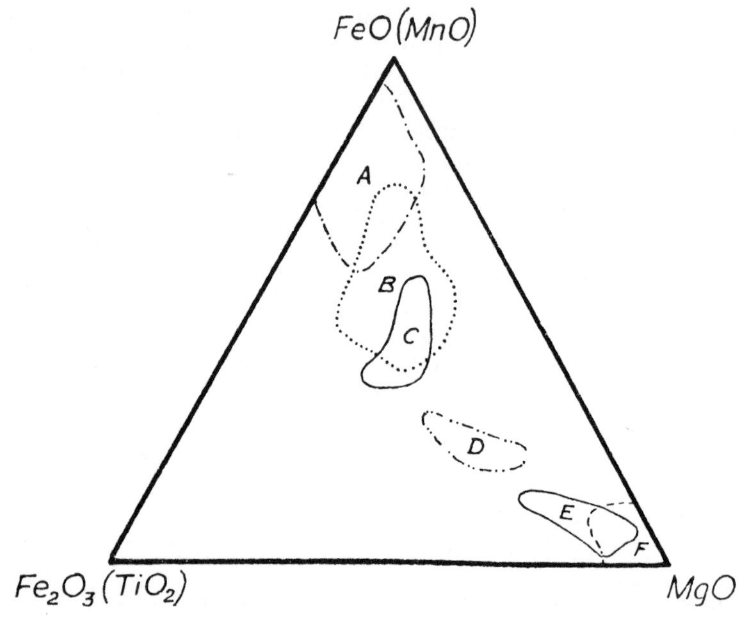

FIG. 25

Diagram showing the variation in composition of micas of the biotite-phlogopite series in rocks of different kinds.

A—Dark micas from granite-pegmatites; B—Dark micas from granites, granodiorites, etc.; C—Dark micas from tonalites and diorites; D—Dark micas from gabbros; E—Dark micas from ultrabasic rocks; F—Phlogopites from metamorphosed magnesian limestones. (*After E. W. Heinrich, 1946*).

vibrations are usually indistinguishable, giving equally dark brown, sometimes nearly black. It follows that a basal section is non-pleochroic: there is no perceptible change on rotating such a section over the polarizer. It should be noted that unaltered biotite is sometimes green in thin section, and pleochroic in shades of green and yellow instead of the more familiar deep brown to yellow. Biotite frequently shows better than any other mineral the effects of internal bombardment by alpha particles emanating from radioactive elements in zircon inclusions. This bombardment causes concentric zones of discoloration round the inclusions, and is seen in thin sections as pleochroic haloes. Biotite alters rather readily into a light green chlorite, first along the cleavages, but ultimately the whole crystal may be replaced.

With regard to paragenesis biotite is familiar in mica-schists and occurs also in igneous rocks covering the whole range of compositions from Acid to Ultrabasic. We find a parallel with the olivine, pyroxene and amphibole groups in that the most magnesian micas occur in the most Basic rocks—in peridotites and pyroxenites, among which kimberlite is an outstanding example (see p. 430). These micas are phlogopitic; so too are those occurring in some lamprophyres in which they may be perfectly idiomorphic and strongly zoned. Ordinary biotite occurs in a wide range of Acid and Intermediate igneous rocks, notably granites in which it may be closely associated with muscovite—in the so-called two-mica granites. Micas are not commonly found in Acid lavas: they are much more characteristic of plutonic and pegmatitic phases. In some Indian pegmatites crystals up to a yard across occur.

CHAPTER II

THE FELSIC MINERALS

THE term felsic is mnemonic in the sense that it covers the minerals in the feldspar and feldspathoid groups, together with the silica group. For ease of reference the anhydrous alumino-silicates occurring in rocks may be listed in a manner which displays their chemical relationships, as under:

Potassic		*Sodic*	
$KAlSiO_4$	kalsilite	$NaAlSiO_4$	nepheline
$KAlSi_2O_6$	leucite	$NaAlSi_2O_6$	jadeite[1]
$KAlSi_3O_8$	orthoclase, etc.	$NaAlSi_3O_8$	albite, etc.

It will be noted that the addition of silica (SiO_2) to kalsilite and to nepheline gives the second mineral in each column; and similarly the addition of silica to leucite and jadeite gives orthoclase and albite respectively. The SiO_4 and Si_2O_6 minerals in the list are unstable in the presence of free silica and are therefore said to be unsaturated. By contrast orthoclase and albite are stable under these conditions and are described as silica-saturated.

On account of their outstanding importance as rock-formers the feldspars are considered first, followed by the feldspathoids.

FELDSPARS

The feldspars are quantitatively the most important of the rock-forming silicates. They are all aluminosilicates—of potassium, sodium, calcium and rarely barium, and crystallize in the Triclinic and Monoclinic systems. Although there must be differences in the details of the atomic structure of the Triclinic and Monoclinic feldspars, fundamentally they are much alike. The three-dimensional framework of linked SiO_4 tetrahedra may be resolved into 'chains' of a special type, which are joined to one another by sharing oxygens[2] (Figs. 26, 27). The chains are aligned parallel to the *a*-axis

[1] Jadeite is included to show its chemical equivalence to leucite among the potassic minerals: it is not a feldspathoid and is actually related to the pyroxenes.

[2] Although we use the word 'chain' for ease of interpretation, this is not comparable with the chains and bands of pyroxenes and amphiboles respectively, which were self-contained and joined to one another through the medium of the cations. In the feldspar structure, however, the 'chains' are linked together, left and right, in front and behind, by sharing oxygens.

(clino-axis) of the crystal, and two of these, in mirror-image relationship across a plane parallel to (010), occupy the unit cell.

Each single link of the chain consists of four SiO_4-tetrahedra; but

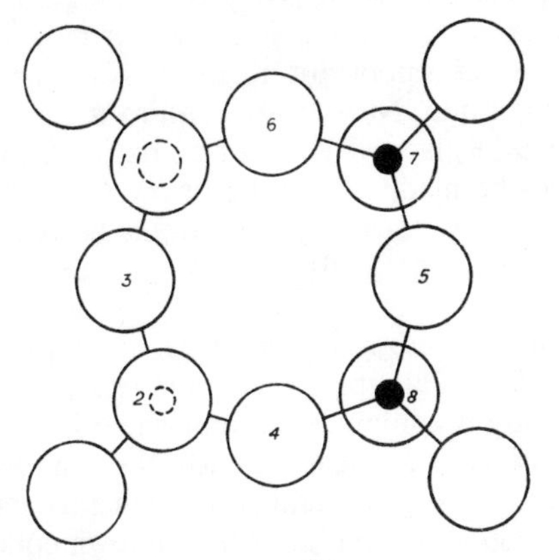

FIG. 26

A single link in the feldspar 'chain': with eight oxygen atoms lying in the plane of the paper, two (nos. 1 and 2) above, and two (nos. 7 and 8) below the plane. Three silicons occur in the link (one hidden by oxygen 2), and one proxy-aluminium (behind oxygen 1).

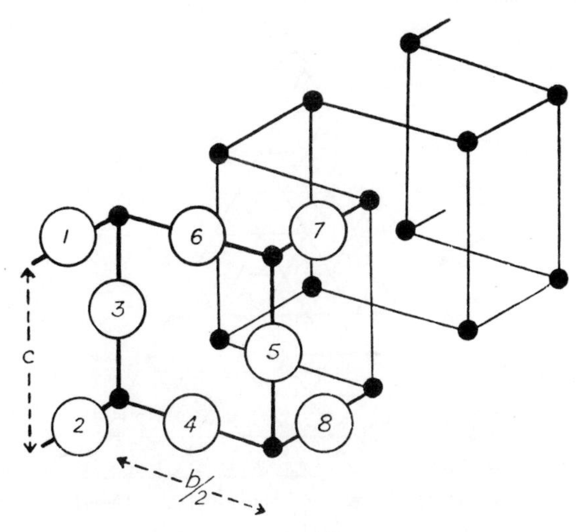

FIG. 27

A formal diagram of the 'chain' structure of feldspar. Silicon and proxy-aluminium shown in black. To avoid overcrowding the diagram, only the 'shared' oxygens in the single link, represented in Fig. 26, are included. The chain is half the width of the unit cell, and runs from back to front, parallel to the *a*-axis.

as all the oxygens are shared, it may be said to contain Si_4O_8. Further, in each of these units, one Si is displaced by a proxy-Al, so the unit formula becomes $(AlSi_3)O_8$. This leaves a surplus negative charge, which in one important group of feldspars is balanced out by the addition of a K^+ ion giving the formula of orthoclase: $KAlSi_3O_8$. In another important member Na ions occur in the place of K, giving albite, $NaAlSi_3O_8$. A third member is anorthite, which contains Ca in place of K and Na; but as Ca is divalent, an adjustment has to be made. A second proxy-Al^{3+} is introduced in place of another Si^{4+}, so the complete formula becomes $CaAl_2Si_2O_8$. Finally, in rare cases Ba^{2+} fills the role of Ca^{2+}, giving celsian, $BaAl_2Si_2O_8$.

Now as the K and Ba atoms are of approximately the same radius, they are capable of mutual replacement, and feldspars such as hyalophane occur, having a composition somewhere between orthoclase and celsian. Similarly, Ca^{2+} can replace Na^+ in the structure to almost any extent, giving a number of minerals intermediate in composition between albite and anorthite, known collectively as the plagioclase series.

By contrast, Na cannot replace K in a manner which will retain the original structure, and therefore homogeneous Na-K feldspar comparable with plagioclase does not occur, except at high temperatures (see below).

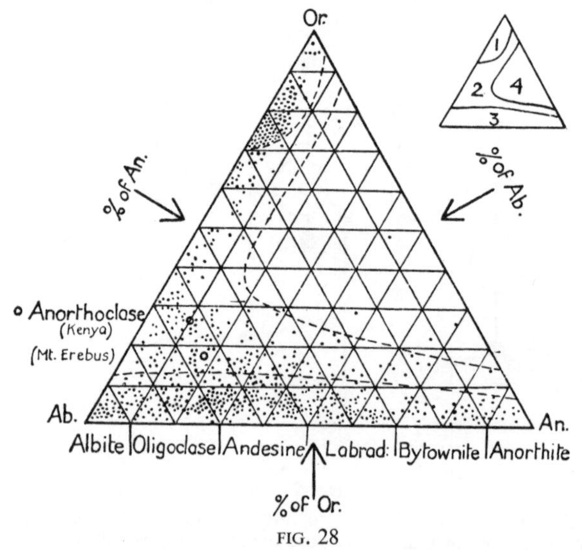

FIG. 28

Diagram illustrating composition of natural feldspars in terms of orthoclase, albite and anorthite. Note maximum concentration in fields 1 (orthoclase and microcline) and 3 (plagioclases).

The formulae so far considered are those of 'pure' feldspars, of ideal composition, and in the account which follows are represented by Or, Ab and An. Natural feldspars are actually ternary systems, consisting of various proportions of Or, Ab and An, and may therefore be represented by points on a triangle which has these three components at its apices. The three components are measured in the directions of the arrows in Fig. 28. The dots on the triangle represent the compositions of several hundred feldspars. The maximum concentration near the Or apex indicates the range of composition of the natural K-feldspars. The average K-feldspar can be seen to have the composition approximately of $Or_{76}Ab_{20}An_4$. The marked concentration of points just above the base-line indicates the range of composition of the plagioclases. It will be noted that on average the plagioclases, although fundamentally Na-Ca feldspars, contain about 5 per cent (or more) of Or. The absence of points in field 4 (Fig. 28) reflects the fact that Ca^{2+} and K^+, on account of dissimilarity in atomic radius, are not mutually replaceable in the atomic structure.

General Characters of the Feldspars

In the Monoclinic feldspars the simplest combination of faces consists of the basal pinacoid (001), side-pinacoid (010) and prism (110). A slightly more complicated crystal is illustrated in Fig. 29.

In the Triclinic feldspars although crystal faces are developed in analogous positions, and although they superficially resemble the Monoclinic crystals, the lower symmetry causes the Monoclinic prism to be represented by two complementary hemi-prisms, 'm' and 'M'. Similarly the hemibipyramid of the Monoclinic feldspars becomes the quarter-bipyramid of the Triclinic types. The better cleavage in all feldspars is parallel to (001); but that parallel to (010) is little inferior. Naturally these two cleavages are at right angles in Monoclinic feldspars, but intersect at angles between approximately 93° and 94° in the plagioclases.

THE ALKALI-FELDSPARS

In this category are included the potassic (K^+) feldspars, the sodic (Na^+) feldspars and the several intergrowths between them, termed, generically, perthites.

(*1*) The Potassic Feldspars

Several distinctive species or polymorphs of potassic feldspar exist, all having the same essential composition, $KAlSi_3O_8$, but with slight differences in the positions of the ions in the atomic structures, which are determined mainly by the temperature of crystallization.

Sanidine, orthoclase and adularia are Monoclinic, while microcline is just Triclinic, as the name implies.

(*a*) **Orthoclase.**—This form of $KAlS_3O_8$ occurs in a variety of crystal habits. The two commonest are the so-called Carlsbad habit (Fig. 29) in which the crystals are tabular, flattened parallel to the

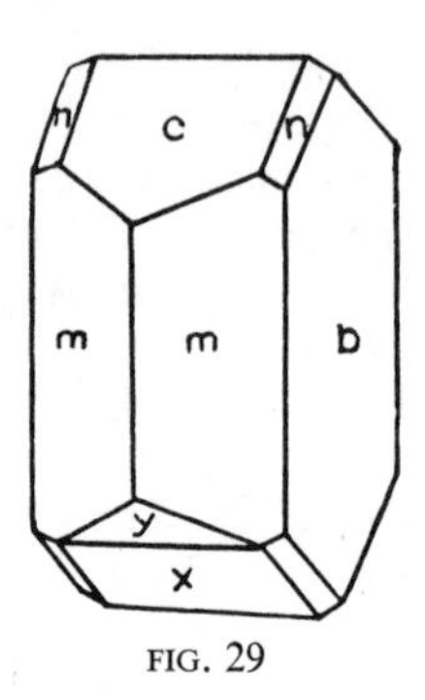

FIG. 29

Crystal of orthoclase (Carlsbad habit).

c Basal pinacoid (001); *b* Clinopinacoid (010); *m* Prism (110); *x* and *y* Hemi-orthodomes ($20\bar{1}$) and ($10\bar{1}$); *o* Hemi-bipyramid ($11\bar{1}$), in zone with *x* and *b*; *n* a clinodome (or clinoprism) (011).

side-pinacoid faces (010); and the Baveno habit in which the crystals are elongated parallel to the inclined *a*-axis. A third habit is characteristic of the feldspar adularia, and crystals of this type are prismatic, while the side-pinacoid faces are virtually suppressed. These crystals have a roof-like termination formed by the basal pinacoid and the hemi-orthodome, dipping in opposite directions: they can be easily identified provided the cleavage traces are visible (see Fig. 30).

Orthoclase crystals are frequently twinned in distinctive fashion (Fig. 31), on one or other of the following laws:

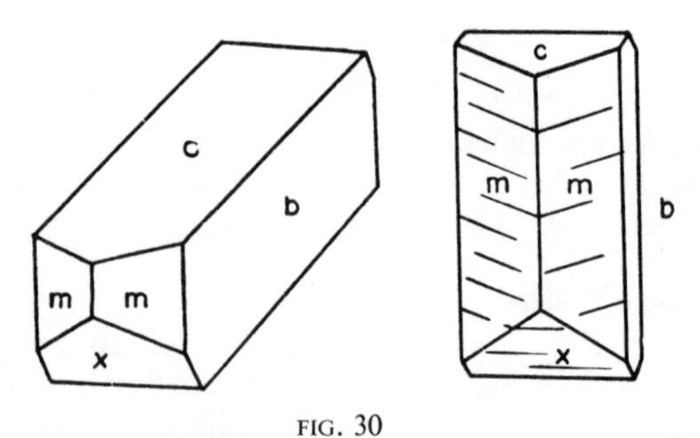

FIG. 30

Left, Orthoclase of Baveno habit. *Right*, Adularia.

Carlsbad, usually interpenetrant and resulting from rotation through 180° about the *c*-axis;

Baveno, in which twin-plane and composition-plane are parallel to a clinodome (011);

Manebach, in which twin-plane and composition-plane are parallel to (001).

Of these the first is much the commonest; Manebach and Baveno twinning are uncommon in orthoclase, but less rare in sanidine.

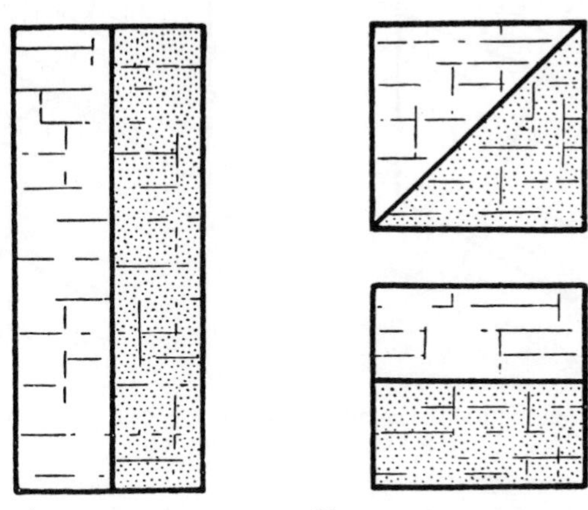

FIG. 31

Vertical sections through twins of orthoclase.

Left, Carlsbad; *Right top*, Baveno; *Right bottom*, Manebach.

Optical Characters

The refractive indices of orthoclase are low: γ, 1·525; α, 1·519, both well below Canada balsam. Birefringence also is weak (0·007)—a little lower than the quartz by which it is normally accompanied in igneous rocks. The optic axial plane is perpendicular to the plane of crystallographic symmetry (010), and is inclined at 5° to 8° to (001). Therefore the extinction is oblique to this extent in (010) sections. The angle 2V is large; but, on heating, it progressively diminishes, and becomes 0° at a certain temperature, above which the axes open out in the symmetry plane (010) (see Fig. 32).

(*b*) **Sanidine.**—On heating, orthoclase inverts into sanidine at the inversion temperature, 900° C. As might be expected, therefore, sanidine is the form of K-feldspar[1] occurring in quickly-cooled lavas and some dyke-rocks, including rhyolites, trachytes, pitchstones,

[1] The high temperature structure of sanidine allows Na and K ions to occupy equivalent positions, so that, although most sanidines are dominantly potassic, they may contain appreciable Na.

etc. It may persist indefinitely in these rocks, in the metastable state; but with the slow cooling characteristic of deep-seated rocks sanidine inverts into orthoclase. The only significant difference

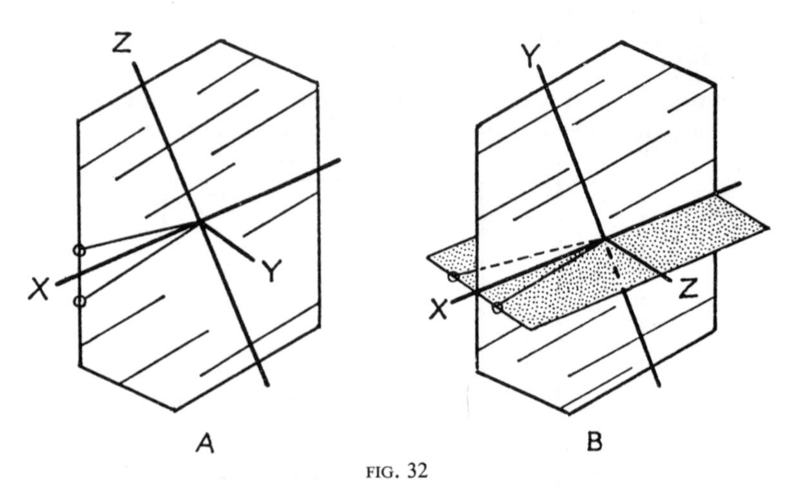

FIG. 32

Diagram showing optical orientation of sanidine (A), and orthoclase (B). Optic axial plane stippled in B.

between the two minerals in thin section is the difference in orientation of the optic axial plane, noted above, and the small, to very small, 2V in sanidine. When unweathered, sanidine is clear and glassy.

(*c*) **Adularia.**—This is the name of a crystal habit rather than a mineral species, though it has been found that adularia is a variety of microcline (rather than orthoclase) containing about 10 per cent of albite. It is of prismatic habit (Fig. 30) and commonly occurs in mineral veins and as incrustations in vugs. It may be glassy like sanidine or opaque like orthoclase.

(*d*) **Microcline** is the Triclinic form of K^+-feldspar, but the departure from Monocline symmetry is very slight, the angle between the basal and side pinacoids being 89° 50′, so that there is no visible difference between orthoclase and microcline in the hand specimen, even if a contact goniometer is available for measuring.

Under the microscope, however, microcline is very distinctive by reason of a complex system of spindle-shaped twin lamellae in two sets approximately at right angles in (001) sections. The twinning may affect only part of a crystal, or it may be absent altogether, but when found, it is completely diagnostic. Refractive indices (γ, 1·529, α, 1·522) and birefringence (0·007) are very similar to those of orthoclase, but on account of the Triclinic symmetry, the extinction in a (001) section is oblique (extinction angle 15° to the (010) cleavage traces).

Although most types of microcline are coloured red or brown, specimens from pegmatites from Colorado and elsewhere are bright green in the variety known as **amazonstone.** This contains the rare elements caesium and rubidium, and the depth of colour is said to be proportional to the amount of the latter.

Microcline is practically restricted to the highly potassic granites and pegmatites among igneous rocks; but it occurs in many metamorphic rocks, often in large porphyroblastic crystals formed by potassium metasomatism. In granites it is rarely pure microcline but contains the substance of albite intergrown as perthite, while even in those specimens which appear homogeneous, analysis often proves some substitution of Na^+ for K^+.

The extent to which the barium feldspar, **celsian** (Cn), $BaAl_2Si_2O_8$, enters into the composition of potassic feldspars is not known with certainty because in many analyses barium is not separately determined. It is known, however, that certain types of feldspathoidal lavas are relatively barium-rich, and the barium is presumably in the alkali-feldspar. For example 'sanidine' occurring in a rock of this type has the following composition $Or_{73\cdot0}Ab_{22\cdot9}An_{0\cdot3}Cn_{3\cdot8}$.[1]

(*2*) The Sodic Feldspars

As in the case of the potassic feldspars, there are several structural states in which $NaAlSi_3O_8$ can occur, dependent upon the conditions of crystallization. The differences between the varieties are very slight, however, and not nearly so apparent as those, for example, between orthoclase and microcline. Two varieties of different habit may be recognized in albite and clevelandite: the former is of outstanding importance.

(*a*) **Albite** —The status of albite is somewhat anomalous: it is the end-member of the important plagioclase series and is therefore indubitably a plagioclase; but in composition it is analogous with orthoclase, with which it is commonly intergrown in varying proportions. Natural albite invariably contains some potassium substituted for sodium. Therefore it is important to regard albite as an alkali feldspar rather than as a plagioclase.

Albite is stable under widely varying conditions. It occurs as a primary constituent in many rock-types, particularly in alkali-granites and syenites: indeed one variety of the latter consists of nothing but albite and is therefore termed albitite. Secondary albite is also widely distributed in rocks of a wider range of compositions which have suffered the type of change referred to as albitization.

[1] Anal. J. H. Scoon, quoted from Tilley, C. E., *Amer. Min.*, **43** (1958), 758–61.

Albite is also an important constituent of some varieties of crystalline schists. In its crystallographic and other physical properties albite is like any other plagioclase, as described below.

(*b*) **Clevelandite.**—This mineral bears the same relationship to albite as adularia bears to orthoclase: it occurs chiefly in pegmatites and sometimes in mineral veins, in aggregates of tablet-like crystals flattened parallel to (010). Again it is the name of a crystal habit rather than of a distinct mineral species.

(*3*) Sodipotassic Feldspars: the Perthites

As noted above both orthoclase and microcline are liable to form intergrowths with albite. The latter occurs in the form of films and, in the coarser types, less regular plates or veins. The films are usually of constant orientation dependent upon the structure of the host mineral (Figs. 33 and 34). The following types of perthite have

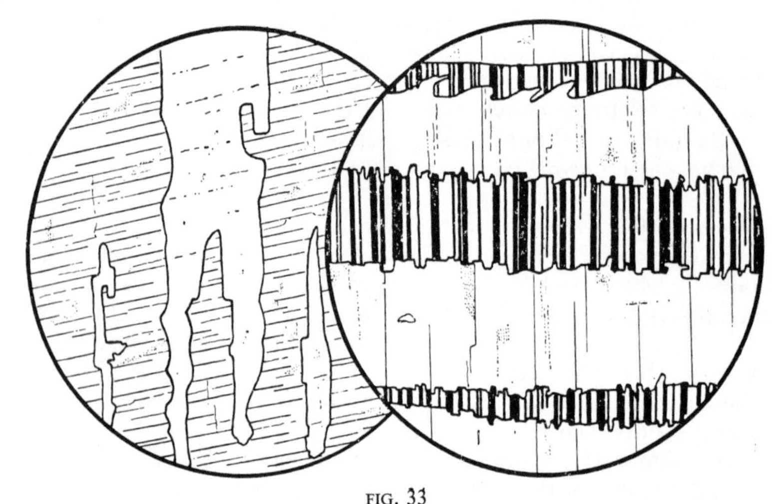

FIG. 33

Formalized sections through vein perthite (albite veins in microcline). *Left*, a (010) section in plane polarized light; (001) cleavage shown. *Right*, a (001) section showing the nature of the margins of the albite veins against the host-microcline, and the (010) cleavage-traces.

been described and figured: vein-perthite (probably the commonest), film-perthite, and braid-perthite in which the albite sheets are arranged parallel to the prism faces ($1\bar{1}0$) and (110), and thus give a braided appearance in a basal section. Less regular intercalations are known as patch- and string-perthite which, on account of the lack of orientation, may have originated differently. The intergrowths may be megascopic, microscopic, or on so fine a scale as to be beyond the resolving power of the microscope, though 'visible' to X-rays. In the case of microcline being the host feldspar, the mineral name is added as a qualifier: thus, 'microcline-microperthite' indicates an

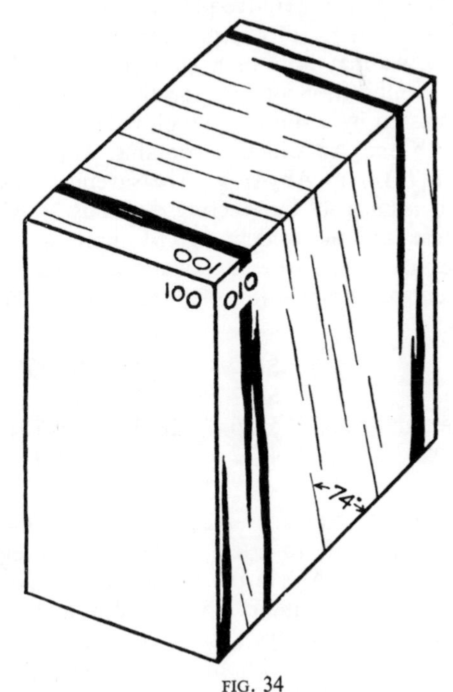

FIG. 34
Block-diagram showing relation of film- to vein-perthite in K-Na feldspar.

intergrowth of microcline and albite on a microscopic scale, the albite lamellae varying from 5 to 100 microns in thickness. If albite is the host mineral to inclusions of potassic feldspar, the term anti-perthite is used. The significance and mode of origin of perthitic feldspars are discussed below under the heading 'Stability Relationships'.

The terms soda-orthoclase and soda-microcline are sometimes given to cryptoperthites of the appropriate composition: they are chemically alike.

A striking optical effect is produced when the intergrowth is on a very small scale (cryptoperthitic). Thus **moonstone,**[1] a semi-precious gemstone, is prized for its bluish sheen which results from the submicroscopic interlamination of two feldspars with slightly different refractive indices: the effect is termed chatoyancy.

[1] Spencer, E. E., *Min. Mag.*, **22** (1930), 291.

THE PLAGIOCLASE SERIES

It has long been known that the members of this series show a gradational variation in composition between the two end-members, albite and anorthite. For descriptive purposes it is convenient to use names to fix certain ranges of composition. Using the convention already described, we may note that, ignoring the small amount of Or present, the composition of any specific plagioclase may conveniently be expressed by a simple symbol, indicating the molecular percentages of Ab and An which it contains. Thus the full range extends from $Ab_{100}An_0$ to Ab_0An_{100}. The several kinds of plagioclase are defined arbitrarily by erecting divisions at Ab_{10}, 30, 50, 70 and 90. These are the widely agreed limits between albite, oligoclase, andesine, labradorite, bytownite and anorthite respectively. The percentage composition of three selected plagioclases is tabulated below:

	SiO_2	Al_2O_3	Na_2O	CaO
Albite (Ab)	68·7	19·5	11·8	0·0
Labradorite ($Ab_{50}An_{50}$)	55·6	28·3	5·7	10·4
Anorthite (An)	43·2	36·7	0·0	20·1

Since the ratio of the molecular weight of Ab to An is as 1:1·061, it is approximately accurate to calculate the relative weights of the two components (Ab and An) in the proportion of the number of molecules. Thus $Ab_{60}An_{40}$ contains Ab three-fifths and An two-fifths by weight. These partial analyses stress the fact that Na^+-rich plagioclases are Si-rich, but Al-poor, while Ca^{2+}-rich types are correspondingly poorer in Si but richer in Al.

Formerly it was believed that the plagioclases provided the perfect example of an isomorphous series, involving progressive replacement of Na^+Si^{4+} (100 per cent in albite) by $Ca^{2+}Al^{3+}$ (100 per cent in anorthite). This has proved to be an over-generalization: only parts of the full compositional range are isomorphous. Two breaks occur, at approximately An_{30} and An_{70}. Plagioclases less calcic than An_{30} are isomorphous except for a limited range of composition, between An_5 and An_{17}, which is a zone of unmixing occupied by the so-called **peristerites**—beautifully schillerized albite-oligoclase feldspars. Similarly plagioclases more calcic than An_{70} (that is, those covered by the terms bytownite and anorthite) are isomorphous. Between the two breaks, and therefore covering the range andesine–labradorite, the plagioclases are heterogeneous: they consist of *two* feldspars having the composition and structure of albite and anorthite respectively. The interlamination of the two is submicroscopic, but can be inferred from X-ray analysis.

But in spite of these breaks the overall effect is to produce a series of feldspars which, for all practical purposes, are isomorphous; so that the optical properties with which the student is mainly concerned vary progressively and systematically from one end of the series to the other (see p. 76).

Crystallographically the plagioclases differ from orthoclase by having the basal pinacoid (and the cleavage parallel to it) inclined to the side-pinacoid (010) at an angle of 86° to 87° instead of 90°; otherwise they are closely similar in appearance and cleavage. Plagioclase may be twinned in any of the ways in which orthoclase twins, but in addition exhibits two special types which are distinctive. Plagioclase of any composition may be twinned once or repeatedly on the Albite Law, in which twin plane and composition plane are (010). Secondly, in twinning according to the Pericline Law, twin plane and composition plane vary systematically in the manner shown in Fig. 35. The pericline, like the albite twinning, is

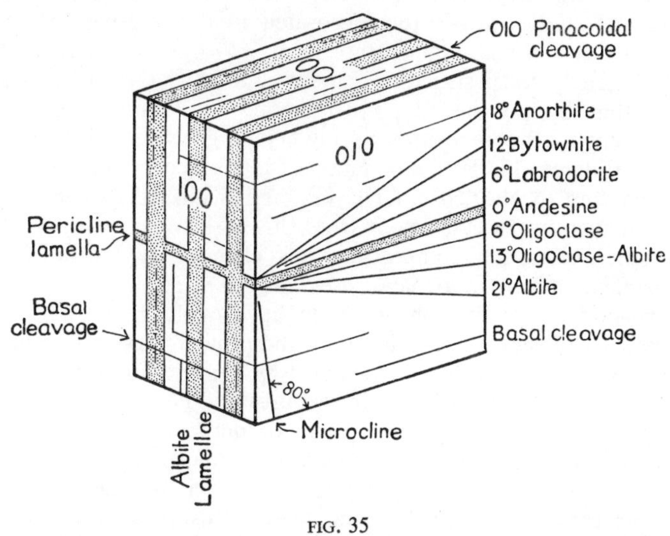

FIG. 35

Block-diagram showing relation between albite and pericline twinning and the cleavages in plagioclase feldspars.

commonly repeated, and in both cases the twin-bands are thin, and give a characteristic lamination between crossed polarizers. Both types may be associated with simple twinning on the Carlsbad Law. Sometimes these features can be seen with the naked eye, but more readily with a lens, while they make the plagioclases unmistakably distinctive in thin sections.

Statistical studies of the frequency of occurrence of the several types of twins shows that normally twinning is more complex in plagioclase of calcic, rather than sodic composition. Complex twinning involving combinations of simple (Carlsbad and/or Manebach) and lamellar (albite and/or pericline) twinning is commonly encountered in sections of igneous rocks particularly those of Basic composition.

Plagioclase crystals often contain minute inclusions which may be arranged irregularly, or in definite planes parallel to the periphery of the crystal. The latter is commonly the case with phenocrystic plagioclase in basalts, when the inclusions may be glass, devitrified glass or small patches of the groundmass. In deep-seated rocks, such as certain gabbros and norites, the plagioclase may be strongly schillerized and exhibits a striking play of colour in the hand-specimen. In thin slice this is seen to be due to myriads of orientated rod-like inclusions, apparently of iron-ore. It seems probable that the iron was in solution in the feldspar at high temperatures (proxying for Al^{3+}). On heating, the schillerization due to these inclusions disappears and the iron is reincorporated in the feldspar, which is converted into a high temperature form.

A definitely zonal structure is more commonly shown by plagioclases than by any other type of mineral.[1] This results from variation in chemical composition during crystal growth, and may be of several different kinds: the commonest involves a gradation from a Ca-rich core to Na-rich outer layers, and is termed *normal zoning*. Less frequently the converse is true: this is *reverse zoning*; while in yet other cases the composition has changed rhythmically and repeatedly, causing *rhythmic* or *oscillatory zoning*.

Explanations of this phenomenon have ranged from a kind of pulsatory diffusion of ions reaching the growing crystals, to movements of crystals for example, carried by currents to different parts of a magma chamber where temperatures and hence the stability of the precipitating phase would vary. In our view the most likely explanation is fluctuation of water-vapour pressure which, as described on p. 169, has a marked effect on the stability of the plagioclase which is being precipitated. This process is effective without any change of temperature or the need to move the growing crystal rapidly from one environment to another.

Plagioclase is liable to alteration, and occasionally suffers complete replacement by secondary white mica, scapolite, zeolites, or minerals of the epidote group, often associated with calcite. The last type of alteration results from dynamothermal metamorphism, and results in the separation, as it were, of the Ab from the An. The former is stable; but the latter changes readily under these conditions into zoisite, clinozoisite or epidote, embedded in a 'background' of

[1] Phemister, J., *Min. Mag.*, **23** (1934), 541; Fries, C., *Amer. Min.*, **24** (1939), 782; and Hills, E. S., *Geol. Mag.*, **73** (1936), 49.

secondary albite. The name **saussurite** is sometimes applied to plagioclase showing such alteration.

Status and Distribution of Plagioclase in Rocks

Plagioclase of one kind or another occurs in representative members of all the main rock groups. Rarely it may make up nearly the whole of the rock: albitite, oligoclasite, andesinite and labradoritite (under the name anorthosite) have all been described, though only the last is other than very rare.

Apart from these monomineralic types, however, plagioclase is an important—often a dominant—component in many Intermediate and Basic rock-types. In general, in passing from more Acid to more Basic types, the plagioclase becomes progressively richer in An: thus in syenitic rocks oligoclase is commonly found: in dioritic (including andesitic) types, it is andesine typically; while in gabbroic (including noritic and basaltic) rocks, labradorite or bytownite occur. Anorthite is less common: indeed, though it does occur in some gabbroic rocks (such as allivalite, for example), it is more typical of metamorphic rocks. An argillaceous limestone might be expected under thermal metamorphism to give rise to anorthite, among other Ca-rich minerals.

Identification of Feldspars in Thin Section

The identification of feldspars in rock sections involves techniques which must be mastered by students at an early stage of their training as such identification lies at the root of rock classification and naming.

It should be realized at the outset that some grains may be so orientated as to make summary identification impossible. Twinning is often sufficiently diagnostic, particularly the distinctive 'cross-hatching' of microcline and the lamellar twinning of plagioclase; but not all microcline, and not all plagioclase is twinned in this way. Further, any kind of feldspar may be untwinned; or may be lying with the twin-plane parallel to the plane of the slide, in which case the crystal is apparently, though deceptively, simple. Such 'simple' crystals must be viewed with suspicion: they may be orthoclase, sanidine, microcline or plagioclase, and before they are identified as, say, untwinned orthoclase, a confirmatory test must be made. The Becke Test is most useful in this connection, and is quickly carried out, preferably on grains lying on the edge of the slide, so that direct comparison may be made between the refractive indices of the grain and of the embedding medium.[1] The indices of

[1] Remember that the bright line moves into the substance of *lower* refractive index on racking *down*, with the light well diaphragmed down to ensure maximum definition.

orthoclase and microcline are well below that of the embedding medium: the table in Fig. 36 shows the range of refractive indices of the plagioclases.

Actual measurement of the refractive indices of feldspars in a rock

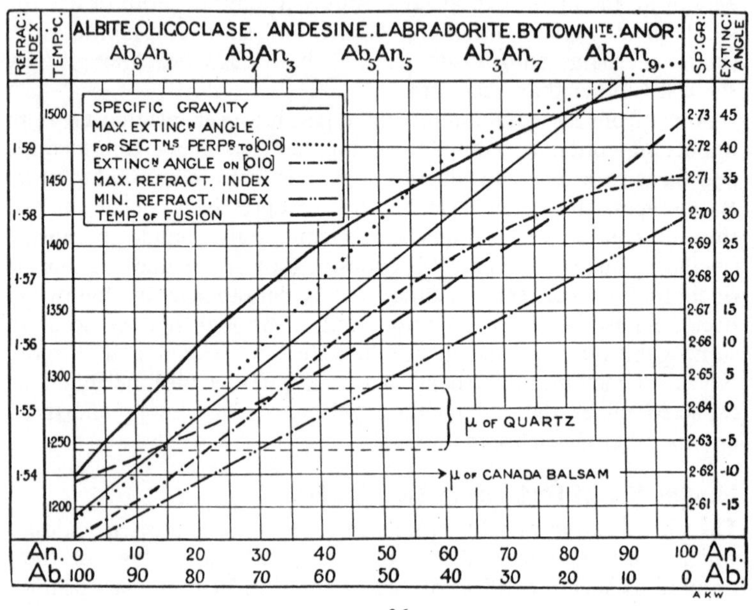

FIG. 36

The optical characters of the plagioclase feldspars.

must be carried out on grains separated from a crushed specimen. Careful measurement gives results comparable with those obtained by chemical analysis.

If, on inspection of a slide, it is believed that two feldspars are present, it is useful to realize that it is virtually impossible for both to have the same 'habit' and to be altered to the same extent and in the same way. Figs. 82 and 113 are convincing in this respect.

Given that plagioclase has been recognized in a rock, it remains to discover what kind of plagioclase before the rock can be accurately named. Restricting ourselves to optical methods using the ordinary petrological microscope, the composition of the plagioclase is indicated by the maximum extinction angles obtained from suitably orientated sections. A useful method makes use of sections at right angles to the Albite-twin lamellae. Such sections show equal illumination, and therefore the same shade of grey, when the twin planes lie parallel to the vertical cross-wire. They also extinguish symmetrically: the angles obtained on either side of the cross-wire should be approximately the same. Several measurements should be made on different grains, as only the *maximum* angle is diagnostic. Reference to the line on Fig. 36 marked 'maximum extinction angles for sections perpendicular to (010)' will then give the composition of the plagioclase. In the special case of oligoclase (Ab_{80} An_{20}) the extinction of both sets of twin lamellae is straight. As the composition changes towards albite, the extinction angle increases to a maximum of 20°, while andesine (Ab_{62} An_{38}) also gives this angle. For angles less than 20°, therefore, a confirmatory test must be applied. This is easily done by means of the Becke Test; the refractive indices of andesine are above, while those of albite are below, Canada balsam.

Extinction angles measured on complex Carlsbad-Albite twins can also be used to determine composition: the method is fully described—together with several others—in all determinative mineralogy textbooks, to which the reader is referred.

Stability Relationships

The equivalence of sanidine (high-temperature K^+-feldspar characteristic of quenched rocks) and orthoclase (low-temperature K^+-feldspar of the same composition but occurring in coarse-grained and more slowly cooled igneous rocks) has long been known. Similarly it has been realized that the alternative crystallization of either orthoclase or microcline must be controlled by the physical conditions of temperature and pressure (doubtless by little understood kinetic factors of crystallization), as indicated by the fact that microcline is the typical K-feldspar of low-temperature granites, granite-pegmatites and a wide range of metamorphic schists and gneisses. Because temperature-controlled differences of atomic structure are much more subtle in their effects in the plagioclase series, it was only comparatively recently that they have been recognized. Differences were first suspected because of the lack of agreement in detail between the optical orientation of plagioclase phenocrysts from quenched lavas and those from coarser-grained plutonic rocks. These differences are greatest for sodic plagioclases; but even so, they are so slight that they can be detected only by accurate Universal Stage measurements. X-ray investigation is needed to establish the structural state of the feldspars; and the complexity of the problem is indicated both by the enormous amount of research that has been devoted to it and by the fact that interpretation of much of the data is still in doubt.

Details of the changes affecting the feldspar structures subjected to variations of temperature are imperfectly known and largely of interest to specialists[1]; but one of the main factors involved can be easily understood by reference to the simplified diagram of atomic structure, Fig. 26. This shows the cation positions occupied by Si^{4+} and Al^{3+}. In the diagram the latter is shown in the top left position; but it may occupy any of the four cation positions. In high-temperature feldspars the arrangement of the Al^{3+} ions is random, and the structure is described as 'disordered'. By contrast, in low-temperature feldspars the structure is 'ordered'—the Al^{3+} ions occupy the same relative positions in all the ($AlSi_3$) units.

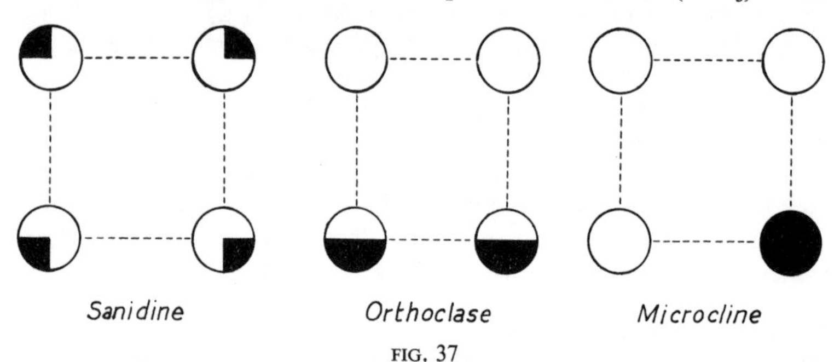

FIG. 37

Conventionalized diagram to illustrate the statistical distribution of Al ions among equivalent sites in the atomic structure analogous to the positions numbered 1, 2, 7 and 8 in Fig. 26. Note the loss of symmetry resulting from the ordered state of Al in the microcline structure.

In the case of the end-members of the feldspar group, pure K-, pure Na- and pure Ca-feldspar, the possible structural changes are limited largely to the degree of ordering of the Al^{3+} ions. It is only in the case of the K-feldspars that the effects of this can be detected easily from the optical differences between K-sanidine and orthoclase, and by the optical and crystallographic differences between these two minerals and microcline. In this series orthoclase has a structure which is transitional between the disordered one of sanidine and the ordered one of microcline. Orthoclase actually grades into microcline with an increase in the so-called triclinicity of the crystals as the ordering of the structure becomes more perfect. Generally the maximum triclinicity of the lowest temperature microcline is accompanied by the most perfect development of cross-hatched twinning: crystals which are transitional between orthoclase and microcline may have the twinning only patchily developed. At the highest temperatures, above 800° C, the stable form of K-feldspar is 'high sanidine', which is different in some details of optical orientation from 'low sanidine', stable between 800° C and 650° C, which is approximately the temperature at which

[1] Of the many papers on feldspar structure, see *e.g.* Ferguson, R. B., Trail, R. J. and Taylor, W. H. *Acta Cryst.*, **11** (1958), 331–48.

sanidine inverts into orthoclase. The corresponding inversion temperature between high and low albite is 720° C. These figures are approximate only and are affected by pressure, flux concentration and the amount of K-feldspar in solid solution. They are included only to give the reader an idea of the temperatures involved.

Discovery of transitional optical and crystallographic features is of great petrological significance: it indicates that the crystal concerned was originally precipitated at a high temperature and that, due to the sluggishness of the transformation, relics of the high-temperature state have survived. A completely stable, low-temperature state is much less informative. This may arise either (*a*) because very slow cooling has allowed an originally high-temperature crystal to invert completely, or (*b*) because the original crystallization took place below the inversion temperature. There is no way of distinguishing between these alternatives. Exactly the same considerations apply to high- and low-albite, although, as noted above, the differences between the structural states in this case are so slight that they can be distinguished only by specialized tests.

When we come to consider the feldspars of mixed composition—the normal condition—the opportunities for differences to develop in the degree of ordering of the structures are greatly increased. Not only can Al^{3+} occupy various positions as outlined above: but this also applies to the other cations. The effects are most marked in alkali-feldspars. At high temperatures the Na^{+} and K^{+} occupy similar and interchangeable positions in the structure, which is therefore disordered. This gives a single type of homogeneous structure for the whole of the alkali-feldspar (sanidine-anorthoclase) series. As the temperature falls, interchangeability of the alkali ions becomes restricted so that the Na^{+} and K^{+} are gradually forced to occupy different parts of the structure. Eventually this gives rise to unmixing (see below) and the growth of separate structural units of sodic and potassic feldspars, each in a more perfectly ordered state than previously.

Unmixed alkali-feldspars are **perthites** which have been described above. Perthites in which either orthoclase or microcline form the host mineral with inclusions of albite, are the slowly cooled and low-temperature equivalents of sanidine; while antiperthites with sodic plagioclase as host mineral are equivalent to anorthoclase.

The process of unmixing may start when the temperature falls below a critical value represented by the highest point on the solvus curve[1] (such as shown in Fig. 73 to which the reader may refer at

[1] The position of this curve, representing the phase boundary limiting the exsolved or unmixed phases (*i.e.* the limits of solid solution), is by no means certain. Heating natural perthites until they become homogeneous gives a very different maximum for the solvus temperature than that obtained with synthetic crystals.

this point); the first structures to form are sub-microscopic and detectable only by X-ray study. Whether or not further unmixing occurs depends upon the rate of cooling: if this is rapid it may enable sanidine to survive as a metastable mineral. The cooling rate is one of the most important factors in controlling the effectiveness of unmixing, and in general the coarsest intergrowths occur in very slowly cooled rocks.

Unmixed intergrowths are all perthites; but varying degrees of coarseness may be distinguished by prefixing a suitable qualifier: thus **X-ray perthites** are apparently homogeneous even under high-power magnification, though their true nature is revealed by X-ray analysis. **Cryptoperthites** (the root meaning 'hidden') are also sub-microscopic, but produce distinctive optical effects through diffraction as in moonstone, a beautiful variety of alkali-feldspar which shows an attractive bluish chatoyance, and peristerite, a sodic plagioclase which displays a beautiful iridescence, especially on (010) faces. The term **'microperthite'** is self-explanatory. The coarsest intergrowths, visible to the naked eye should, by analogy, be termed **macroperthites** but generally they are referred to merely as perthites, without qualification.

The finest intergrowths (X-ray and cryptoperthites) can be produced experimentally by slowly cooling a homogeneous sanidine to below the unmixing temperatures, and conversely, such perthites can be homogenized on reheating. Unmixing of the coarser perthites is not reversible under experimental conditions. This is due primarily to the time factor: because of the very slow rates of ionic diffusion involved in the unmixing process, structures developed in natural perthites which have crystallized and cooled during periods of tens, hundreds or even thousands of years cannot be put into reverse by reheating for the duration of a laboratory experiment.

Tuttle[1] has put forward convincing arguments for believing that the unmixing process does not stop at the formation of perthites, and he has suggested the addition of two further stages to the series listed above. In the first the Na ions migrate to the margins of the host crystals of K-feldspar, to form marginal zones or rims of albite. In the final stage, the completely unmixed albite and K-feldspars may recrystallize as adjacent discrete crystals. Thus it is possible that the combination of separate crystals of Na- and K-feldspars found in some granites may have developed by a long process of unmixing and recrystallization in the solid state from an original single phase of sodi-potassic feldspar.

This interesting hypothesis cannot be proved experimentally for

[1] Tuttle, O. F., *J. Geol.*, **60** (1952), 107–52; also Tuttle, O. F. and Bowen, N. L., Origin of granite in the light of experimental Studies in the system $NaAlSi_3O_8$-$KalSi_3O_8$-SiO_2-H_2O. *Geol. Soc. Amer. Mem.*, **74** (1958), 17.

the feldspars; but its truth is suggested by analogy with some sulphide systems in which the rates of unmixing are much faster. Mixtures of bornite and chalcopyrite heated to the melting point and then quenched at 600° C form homogeneous solid solutions. Heating again to 600° C and cooling under controlled conditions, ranging from five minutes to twenty-four hours, results in the development of a series of structural arrangements exactly analogous to those described above for the alkali-feldspars, including intergrowths of varying degrees of coarseness; and with more protracted cooling, rims of the minor component around cores of the more abundant mineral; and finally aggregates of discrete grains of bornite and chalcopyrite in close association, but yielding no trace of having evolved from high-temperature solid solutions. Perhaps the final observation is the most significant for petrologists.[1]

From this very brief survey it will be appreciated that the interpretation of the history of crystallization of a feldspar from its present state is beset with difficulties. This is emphasized by the fact that Tuttle and Bowen list fourteen possible phase combinations that may be found in alkali-feldspars. Even a single crystal may contain four recognizable phases, as for example the combination: high-albite, low-albite, orthoclase and microcline.

THE FELDSPATHOID MINERALS

In this group are included several minerals which are closely related, as their name implies, to the feldspars. They contain the same elements as the latter, though in different proportions, and are notably poorer in silica.

Leucite, the potassic feldspathoid, resembles orthoclase in composition. The SiO_4-tetrahedra are linked corner to corner, and if we consider three of these, the unit contains Si_3O_6. But in each unit of structure one proxy-Al takes the place of one Si; so that the formula of the unit becomes $(AlSi_2)O_6$. Balance of the total positive and negative charges is brought about by adding one K^+-ion, giving the ideal formula, $K(AlSi_2)O_6$. In natural leucites some Na ions invariably replace some of the potassium, so the formula is more accurately represented as $(K, Na)AlSi_2O_6$.

It is difficult to write about the crystallography of leucite in unequivocable terms. In many mineral collections large crystals occur in the form, apparently, of the simple unmodified icositetrahedron; but the optical properties of the crystals prove that they cannot belong to the Cubic System: the form is a *pseudo-icositetrahedron*,

[1] Schwartz, G. M. Intergrowths of bornite and chalcopyrite, *Econ. Geol.*, **26** (1931), 186–201.

resulting from complex twinning of Tetragonal crystals which, in the untwinned condition, are unknown in Nature.

In thin slice leucite normally shows characteristic eight-sided sections, when cut through centrally. The refractive index, 1·508, is one of the lowest among common rock-forming minerals. Cleavage is absent; but small inclusions may be arranged in zones or tangentially, especially in the minute leucites occurring in the groundmass of some lavas. The most significant feature, however, is the twinning (Fig. 38). In a perfect, centrally-cut section twin lamellae

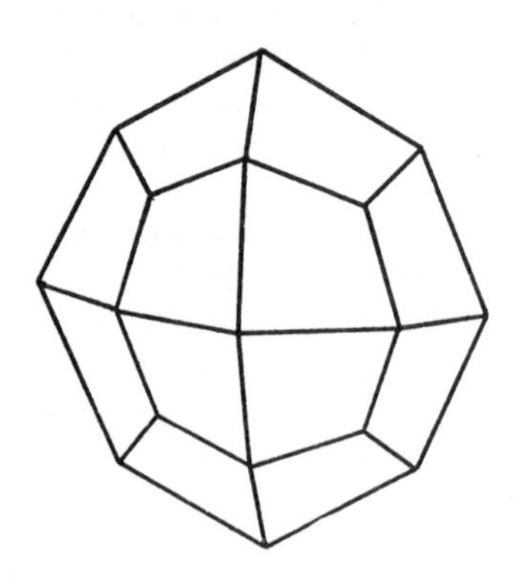

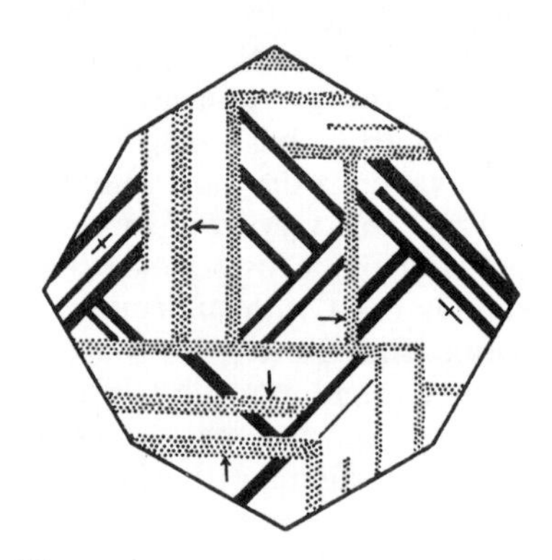

FIG. 38

Left, crystal of leucite showing pseudo-icositetrahedral form. *Right*, central section through a leucite crystal between crossed polarizers to show twinning in six directions. Dip arrows and signs for verticality indicate the attitude of the twin lamellae in relation to the section.

in six directions (parallel to the faces of the pseudo-rhombdodecahedron) give a particularly striking appearance, though, on account of the abnormally weak double refraction (0·001), very strong illumination is necessary to make this feature really convincing.[1] The mere fact that the twin planes are parallel to directions which, in a Cubic crystal, would be planes of symmetry, proves that such leucites are not Cubic.

As a fact, the small leucites in the groundmass of a lava do appear to be isotropic between crossed Nicols, and therefore are presumably Cubic. This is notably the case with lavas in which the groundmass crystals are embedded in a base of glass. These facts establish the existence of two forms of leucite: a high-temperature Cubic form, only seen in Nature's quenching experiments, and which inverts at about 625° C into the low-temperature pseudocubic form occurring as phenocrysts, and exhibiting the twinning described above.

[1] The twinning effect is brought into prominence by using a gypsum plate when viewing the section between crossed polarizers.

Remarkable pseudomorphs after leucite have been described from strongly potassic rocks in the medium and coarse grained categories from several different localities, one (doubtfully) British but notably from a leucite-phonolite dyke in the Bearpaw Mts. in Montana. In some instances the original icositetrahedral shapes have been perfectly preserved; but at the other extreme only a vague suggestion of the original form remains and poorly defined, rounded, light-coloured masses are *inferred* to be **pseudoleucites** from their mineral composition. These pseudomorphs consist of orthoclase or sanidine, nepheline and a small amount of adventitious material representing original impurities. The composition of the Bearpaw Mts. pseudoleucites was found to be: sanidine 66 per cent, nepheline 30 per cent and aegirine 3 per cent[1] .The composition is, of course, consistent with that of the original mineral, and is evidently an equal volume change resulting from complete reconstruction during slow cooling. We would emphasize that this change has nothing whatever to do with weathering: it represents an adjustment to changed environmental conditions and is, in effect, unmixing carried to the ultimate limit. Pseudoleucite represents the ultimate mineral association stable under the existing low-temperature, atmospheric pressure conditions.

These pseudoleucites demonstrate one of the most important facts regarding the stability of leucite: it cannot survive under conditions involving the retention of volatiles. As the concentration of the latter increases, the stability field of leucite diminishes, as shown in Fig. 73 and is ultimately completely eliminated, irrespective of the degree of undersaturation of the magma. If the latter is sufficiently undersaturated to yield *potential* leucite, the place of the latter must be taken by some other mineral or combination of minerals: biotite, by virtue of its composition, is one of those best suited to fill this role.

Leucite is thus virtually confined to rocks that have crystallized under low-pressure conditions, *i.e.* lavas of the appropriate composition which have been quenched. It is one of the first minerals to be precipitated from a magma of the right composition (*i.e.* potash-rich) as indicated by its invariable occurrence as euhedral phenocrysts. If cooling is sufficiently slow, reaction with silica in the melt converts some or all of it into K-feldspar. This is the reverse of the process known as incongruent melting of orthoclase which involves the reaction:

$$KAlSi_2O_6 + SiO_2 \leftrightharpoons KAlSi_3O_8.$$

On heating orthoclase to approximaely 1170° C it has been shown

[1] Zies, E. G. and Chayes, F., Pseudoleucites in tinguaite from the Bearpaw Mts., Arkansas. *J. Petrol.*, **1** (1960), 86–98.

experimentally[1] to melt incongruently to leucite crystals in liquid which contains the excess SiO_2 liberated by the reaction. It is important to realize that at high temperatures leucite can exist in the presence of free silica. If at this stage quenching ensues, these leucites will survive as phenocrysts, while the liquid congeals as glass. With slow cooling, however, at the appropriate temperature the leucites will be made over into sanidine—provided the magma contains sufficient silica. If it does not, some leucite will survivive in association with orthoclase. The relationship between leucite and orthoclase is thus analogous in all respect to that between olivine and pyroxene.

Summarily, therefore, there are three factors which control the formation and survival of leucite: (1) the degree of saturation of the potassic magma; (2) the rate of cooling; (3) water-vapour pressure. It may be noted that the occurrence of nepheline is dependent upon only the first of these factors: therefore it is much more widespread, in a greater variety of rocks in all grain-size groups, than leucite. Leucite is best developed in certain lava-types occurring in volcanic regions of the Roman province in Italy, the Leucite Hills in Wyoming, the Kimberley District in Australia and lava fields in Central and Eastern Africa. The mineral lends its name to two rock-types, leucitite and leucitophyre. Certain Vesuvian lavas are well known by reason of the perfectly formed phenocrysts of leucite which they contain embedded in a dark basaltic-looking matrix: leucite-basanite is one such type. Italite is a unique volcanic rock consisting almost entirely of closely packed leucite crystals which are easily separated from the meagre matrix.

The least siliceous of the alkali aluminosilicates include two end-members $KAlSiO_4$ (= Ks) and $NaAlSiO_4$ (= Ne) connected by a series of solid solutions of intermediate composition. The potassic end-member is the mineral **kalsilite** and the sodic analogue is **nepheline.** Although pure Ne does not apparently occur in natural rocks, the name nepheline is applied to solid solutions involving Ne and Ks in widely varying proportions, and in this sense is reminiscent of the alkali-feldspars. Most nephelines from plutonic environments have a composition consistently close to $Na_3KAl_4Si_4O_{16}$, *i.e.* containing Ne and Ks in the ratio of 3:1. In volcanic environments the composition varies considerably: nephelines which crystallized out of a strongly sodic magma and occurring for example in phonolite may contain Ne up to over 80 per cent. Conversely, crystals in the Ks–Ne series formed from a highly potassic magma are correspondingly rich in Ks, and in this case there is no limitation: natural kalsilites occurring in certain rare volcanic rocks in Uganda and other

[1] Morey, G. W. and Bowen, N. L. The melting of orthoclase. *Amer. J. Sci.*, (1922), 1–22.

parts of Africa contain 98 per cent Ks. As a point of detail it may be noted that in nepheline and kalsilite (except the theoretical end-members of the series) analyses show a slight excess of SiO_2—up to 6 per cent in nephelines and about 2 per cent in kalsilite.

Kalsilite is one of the very few rock-forming minerals with a really helpful name: the first five letters are taken from the chemical formula. It was originally misidentified and described as nepheline, quite understandably, as it shows identical crystallographic characters and is optically indistinguishable: X-ray analysis was necessary before its identity was established.

It is known to occur in only four localities, three of them in Central Africa and one in Italy, and in all cases the mode of occurrence is the same: it forms complex phenocrysts closely associated with nepheline in 'perthite' together with leucite and sometimes sodalite in rare ultrabasic lavas. The same relationship occurs between nepheline and kalsilite as between albite and orthoclase: both pairs form mix-crystals at high temperatures, and with falling temperature unmixing occurs resulting in the formation of a complete series of intergrowths, from X-ray perthites, cryptoperthites, to microperthites. A further degree of unmixing gives rise to euhedral rimmed crystals, the core consisting of kalsilite and the rim of nepheline. Neither mineral is 'pure': the core contains Ks_{80}, while both the rim and groundmass nepheline contain between a quarter and a half of kalsilite. Using the same convention regarding composition as for the feldspars, the groundmass nephelines range from $Ks_{25}Ne_{75}$ to $Ks_{45}Ne_{55}$.[1]

Diagnostic Characters of Nepheline

Nepheline crystals which the student is likely to encounter during his training are restricted to thin rock-sections. They are known to belong to one of the less symmetrical classes of the Hexagonal System; but in thin section they look like simple stumpy hexagonal prisms, about as broad as they are tall. Consequently square (vertical) and six-sided (basal) sections are typically well displayed. The former display first order (generally grey) interference colours, and of course extinguish straight; the latter are isotropic and give a negative uniaxial interference figure. The latter is one of the most useful characters which distinguish nepheline from untwinned feldspar for which it may quite easily be mistaken on account of the similar relief ($\alpha = 1{\cdot}534$, $\gamma = 1{\cdot}537$) and birefringence. Cleavage is variable. In some cases the square sections show cleavage traces, the hexagonal ones do not: therefore the cleavage is pinacoidal, parallel to the basal plane (0001). In other cases comparison of the vertical and

[1] Sahama, T. G., Kalsilite in the lavas of Mt. Nyiragongo, Belgian Congo. *J. Petrol.*, **1** (1960), 146.

basal sections demonstrates clearly that the cleavage is prismatic. Usually the cleavage traces are very feebly developed, but they may be made clearer by alteration. This may occur in a variety of ways. Nepheline may be replaced by an aggregate of white mica flakes or by zeolites or cancrinite. The latter is very distinctive and when it rims nepheline is a very useful aid in confirming the diagnosis of nepheline in doubtful cases.

As might be expected nepheline crystals are frequently zoned, though it needs expert handling of the microscope to make the zoning visible. It can be rendered much more striking if the nepheline is stained with fuchsine after treatment of the rock-section with weak acid.[1]

The Role of Nepheline in Igneous Rocks

Nepheline occurs in rocks characterized by an abundance of alkalies and alumina, but low in silica. Thus it is never found in Acid rocks, but may be abundant in those of Intermediate to Basic composition. In an Acid magma containing potential free silica combination takes place between the components of nepheline and the silica, thus:

$$Na(AlSi)O_4 + 2SiO_2 = Na(AlSi_3)O_8 \quad \text{(albite)}.$$

Consequently nepheline tends to take the place of albite in magmas deficient in silica, and in certain types of nepheline-syenites, for example, nepheline may be the dominant felsic mineral. In hand-specimens of nepheline-syenites and nepheline-gabbros the mineral may be more obvious even than in thin slice, particularly on weathered surfaces. In fresh specimens the nephelines are grey, with a rather silky texture and greasy lustre; but weathering develops a strong red or reddish-brown colour.

Among corresponding lavas, a well-characterized type is phonolite (Fig. 104); while among the Basic lavas nepheline occurs in tephrites and basanites. The term nephelinite is applied to similar lavas in which the nepheline is unaccompanied by feldspar of any kind.

Cancrinite is related to nepheline in composition, and in a sense may be regarded as 'nepheline-carbonate'. Fundamentally it consists of $NaAlSiO_4$ but with a proportion of the Na^+ ions substituted by Ca^{2+} and with additional anions (CO_3), (SO_4) and Cl. Cancrinite is hexagonal, but rarely forms definite crystals. Under the microscope it occurs usually in shapeless masses, either in, or marginal to, nepheline. It is colourless, with both refractive indices below balsam: $\alpha = 1{\cdot}496$, $\gamma = 1{\cdot}519$. It has a perfect prismatic cleavage, and as its birefringence is high $(0{\cdot}023)$ it may resemble muscovite, but is easily distinguished therefrom by its negative relief, by its associates,

[1] Shand, S. J. Staining of feldspathoids and on zonal structure in nepheline. *Amer. Min.*, **24** (1939), 508.

and by showing a uniaxial negative interference figure. Occasionally it is a primary constituent of nepheline-syenites, and in so-called cancrinite-syenites is an essential component. More often it fills a minor role, as an alteration product of nepheline, or arises by reaction between the latter and included grains of calcite (Fig. 97). In Britain cancrinite has been recorded from a pegmatitic facies of borolanite, at Loch Borolan, Assynt.[1] It is a characteristic product of fenitization, for example in the Alnö and Fen Complexes; and occurs also in carbonatites. It is a primary mineral in these occurrences.

Three other members of the Feldspathoid group of minerals are often associated with nepheline and are closely related in chemical composition. They are sodalite, nosean and hauyne.

Sodalite.—Sodalite is a Cubic mineral which is normally seen as grains or interstitial patches in certain types of nepheline-syenites. Occasionally it ranks as an essential mineral and then may form bright blue patches in the hand-specimen.

In thin section sodalite is invariably colourless, and has a particularly low refractive index (1·48). This is considerably lower than the indices of the feldspars and nepheline with which the sodalite is normally associated. An imperfect dodecahedral cleavage may be visible in some sections. Between crossed polarizers the mineral is, of course, isotropic.

On account of its rather negative characters, sodalite is never easy to identify with certainty, and in some cases a micro-chemical test is necessary. Small quantities of the mineral have been discovered in certain rocks by using ultraviolet light, when sodalite exhibits a brilliant yellow fluorescence.

The composition is represented by the formula $Na_8(AlSiO_4)_6Cl_2$ which, to aid the memory, may be rewritten as six nephelines combined with the components of two of NaCl.

Nosean or **Noselite** is also Cubic, and crystallizes in the same form as sodalite, *i.e.* the rhombdodecahedron. The crystals are commonly very light yellowish-grey in thin section, and although they may show the characteristic six-sided sections to be expected on account of the crystal form, they are often strongly corroded, with a heavy dark margin (Fig. 39). As the refractive index is low (1·495) negative surface relief is exhibited. Curious canal-like channels cross the surface, and traces of a dodecahedral cleavage can frequently be discerned.

In composition nosean is basically like sodalite, but with the anion $(SO_4)^{2-}$ substituted for Cl^-. Thus the formula becomes $Na_8Al_6Si_6O_{24}.SO_4$ which may be rewritten six nephelines combined with Na_2SO_4.

[1] Stewart, F., *Min. Mag.*, **26** (1941), 1.

Nosean appears to be restricted to volcanic undersaturated Intermediate rocks and is characteristic of some phonolites and leucitophyres (Fig. 39). In Britain a well-known nosean-phonolite builds the Wolf Rock off the Cornish coast.

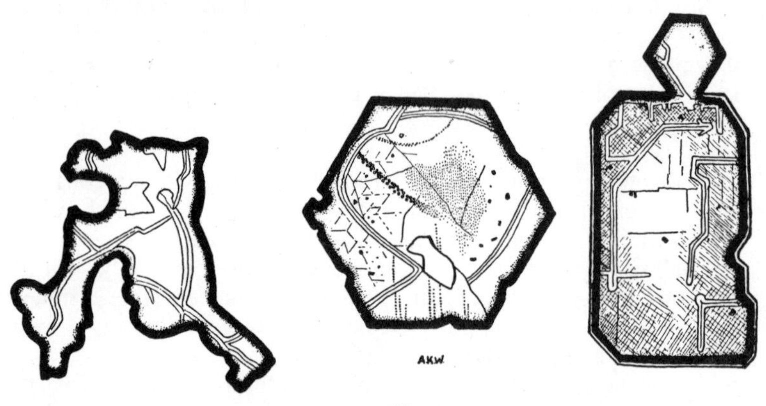

FIG. 39

Sections of nosean from nosean-leucitophyre, Rieden, Eifel, showing effects of magmatic corrosion, the characteristic canal-like markings, trains of gas bubbles, dodecahedral cleavage and (*right*) schiller structure.

Hauyne or **Hauynite.**—This mineral also crystallizes in the same form as sodalite and nosean—the rhombododecahedron. Hauyne too, like sodalite and nosean, contains the constituents of 6 nephelines in the unit cell, together with (SO_4) anions. In hauyne a large proportion of the Na^+ ions in nosean are replaced by Ca^{2+}, and to effect a balance either one or two SO_4 anions may be included as suggested by the formula: $(Na, Ca)_{4-8}(Al_6Si_6O_{24})(SO_4)_{1-2}$.

Again, like nosean, hauyne is restricted to quickly cooled volcanic rocks undersaturated with silica and may occur in phonolites, for example in the Roman volcanic province. In addition it is associated with melilite in particularly Basic lavas and dyke-rocks.

With virtually identical atomic structure and closely similar compositions, it is natural that hauyne crystals should resemble those of nosean, very closely, even to points of detail such as the 'shaded' margins. Both may be colourless, in which case positive identification can be made only after a microchemical test; but typically hauyne crystals are clear, light sky-blue.

ANALCITE

Analcite (syn. analcime) is a mineral with a dual personality: it fills two completely different roles. From its chemistry it finds a natural place among the family of hydrated alumino-silicates of the alkalies—the zeolites; and in association with other zeolites is widely distributed in geodes and vesicles in lavas, especially basalts and andesites. In addition, however, analcite occurs in a variety of igneous rocks, rarely as phenocrysts, but more widely in the groundmass of certain lavas, also as an essential constituent of some plutonic Basic igneous rocks, notably teschenite. We have to decide, therefore, whether to deal with analcite at this point as if it were related to the feldspathoids, or to postpone considering it until the next chapter, among the other zeolites. On balance we think it is more appropriate to deal with it now, largely because at least in part it is more than a mere accessory, and because it has already been brought into the picture.

Analcite may be represented by the formula $NaAlSi_2O_6 \cdot H_2O$ and therefore, apart from the water, is like leucite, with the K^+ ions completely substituted by Na^+. Actually in analcite some substitution of K^+ for Na^+ occurs. Analcite is Cubic, apparently holosymmetrical, the only form commonly encountered being the icositetrahedron. This is the high temperature form which, like leucite, inverts into a pseudocubic, either Tetragonal or Rhombohedral form without change of shape, though with leucite-like optical anomalies—weak birefringence and lamellar twinning. This aping of leucite is not unexpected as the two minerals have the same essential atomic structure; and it is not surprising that analcite can pseudomorph leucite. Potassium analcite has been synthesized. It must therefore be regarded as a distinct possibility that phenocrysts of analcite in lavas were originally leucites; but this would appear to be incapable of proof, one way or the other.

Italite, referred to under leucite, is known from only one occurrence—in the Roman volcanic province—where it occurs in the form of blocks ejected explosively from a volcanic vent. By what would appear to be an extraordinary coincidence, an exactly comparable rock, but containing up to 90 per cent of phenocrystic *analcite* was explosively ejected by a Cretaceous volcano in Alberta, Canada. Is this rock—'blairmorite'—a 'substituted italite'?

As already noted analcite and leucite are the same shape, they have the same lustre, both may be translucent and colourless, though commonly off-white; but analcite may be pink, or, as in the blairmorite locality, quite strongly coloured red. However, while leucite phenocrysts are available in quantity, isolated euhedral analcites are rare: much more commonly the mineral occurs in aggregates associated with other zeolites.

In thin section analcite may be difficult to identify with certainty, on account of its rather negative character. Typically it is colourless and transparent, of low relief (R.I., 1·487). Careful manipulation of the iris diaphragm may reveal traces of cubic cleavage, and make it easy to apply the Becke test, which will confirm its low refractive index. The optical anomalies visible between crossed polarizers have already been mentioned. In altered rocks analcite may be replaced by a finely crystalline aggregate of usually unidentifiable material.

The associates of analcite in teschenites and related rocks are ilmenite, labradorite, lilac-coloured titanaugite, and red-brown barkevikite. In all these rocks the analcite is a primary mineral, though of late formation. Therefore it normally occurs in interstitial patches between the earlier formed constituents; but with increasing amounts it spreads into the adjacent plagioclases along cleavages and veins, and progressively replaces them. This *analcitization* is a late-stage replacement comparable in its effects with albitization.

MELILITE

This name is applied to a series of uncommon, somewhat complex silicates of calcium, aluminium and magnesium. Any one specimen may be regarded as having a composition that can be expressed in terms of the two end-members of the series, akermanite ($Ca_2MgSi_2O_7$) and gehlenite ($Ca_2Al_2SiO_7$). The former is reminiscent of diopside, and the latter, of anorthite, though there are significant differences. The former may be broken down into: $CaSiO_3.MgSiO_3.CaO$. In the presence of free silica the CaO would combine to form $CaSiO_3$, thus increasing the wollastonite component in the clinopyroxene. Similarly the gehlenite formula may be written thus: $CaSiO_3.CaO.Al_2O_3$; and in the presence of free silica the $CaO.Al_2O_3$ would combine to form anorthite. Therefore akermannite may be regarded as the unsaturated equivalent of Ca-rich clinopyroxene, and gehlenite is, in part, undersaturated anorthite (in plagioclase).

Whether melilite should be grouped with the feldspathoids or with the mafic minerals is a moot point. In the sense that melilite of any composition is heavily biassed towards pyroxene it would be reasonable to regard it as a mafic mineral. On the other hand, part of the melilite composition is the unsaturated equivalent of anorthite and is thus analogous to the true feldspathoids in relation to the alkali-feldspars. It is still useful to consider melilite in the context of feldspathoids for petrochemical reasons. One can say that melilite represents the ultimate in silica-undersaturation, and occurs in very Basic to Ultrabasic rocks, adequately rich in CaO, in which all other kinds of under-saturation—for example in relation to alkalies and the FeMg components—have already been achieved.

Melilite crystallizes in one of the less symmetrical classes of the Tetragonal System, but crystals are rare, and the student is likely to encounter the mineral only in thin sections. It is virtually restricted to certain Ultrabasic lavas and dyke rocks. Many of the known occurrences of melilite-bearing rocks are found in the great lava fields in central and eastern Africa, where the mineral is associated with leucite, kalsilite, nepheline and the accessory, perovskite—all unsaturated minerals.

In thin section melilite is very distinctive. The sections are roughly rectangular with rather irregular margins; the relief is moderately high, but the chief diagnostic feature is the anomalous birefringence. The typical interference colour is a deep inky blue, irregularly distributed: the marginal parts tend to be lighter than the central zone which may be nearly isotropic. There is no other mineral with which melilite may be confused. Its constant companion is **perovskite** ($CaTiO_3$), occurring in small octahedra, with exceptionally dark borders due to its very high refractive index.

The genesis of melilite has been studied experimentally and is discussed in the petrographical section of the book.

THE SILICA GROUP

Several distinct mineral species consist of pure silica, SiO_2; some of these occur as euhedral crystals, others as microcrystalline or

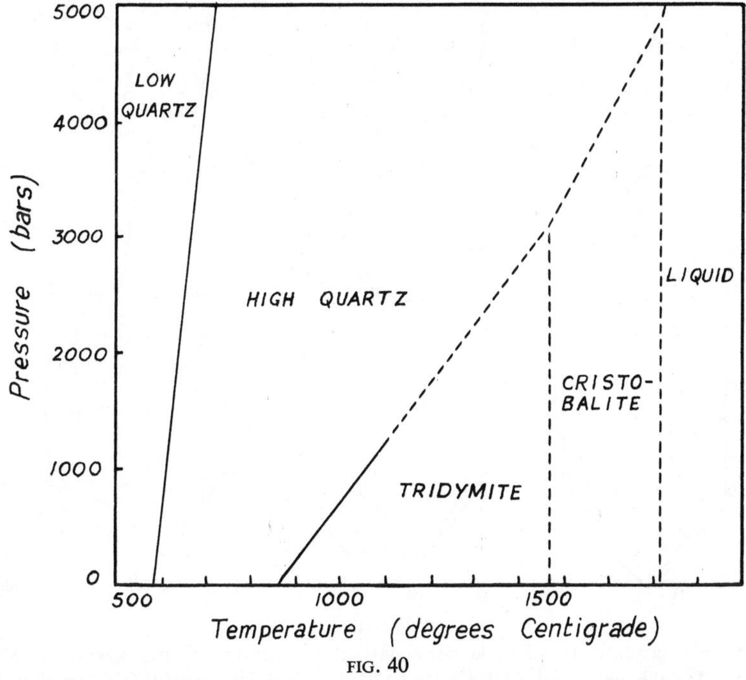

FIG. 40

Stability relationships of polymorphs of SiO_2. (*After Tuttle and Bowen*, 1958.)

cryptocrystalline aggregates, while amorphous silica also occurs naturally.

Each of the three mineral species, quartz, tridymite and cristobalite, occurs in both high- and low-temperature modifications, distinguished as β- (or high-) and α- (or low-) quartz, etc. The complete range of six minerals forms a series stable under varying physical conditions.

In all these silica minerals SiO_4-tetrahedra are linked to one another by all their corners. The actual arrangement of the atoms is complicated and impossible to illustrate simply. The important fact is that tetrahedra are linked spirally in both forms of quartz, the arrangement being somewhat more symmetrical in the high- than in the low-temperature form: thus while α-quartz crystallizes in the holo-axial (trapezohedral) class of the Trigonal System, β-quartz belongs to the corresponding class of the Hexagonal System.

α-quartz is the quartz of mineral veins and vugs: it is usually a product of inversion in igneous rocks. Its crystal characters are indicated in Fig. 41. Twinning is apparently ubiquitous in α-quartz,

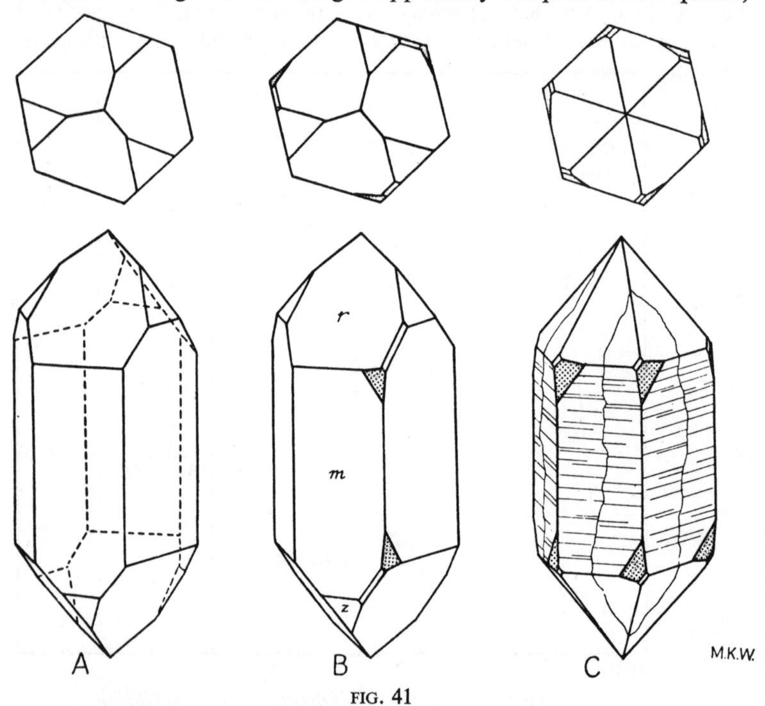

FIG. 41

Quartz crystals: (A) Simple combination of positive (*r*) and negative (*z*) rhombohedra and vertical prism (*m*); (B) A right-handed crystal showing the same forms, with trapezohedron (51$\bar{6}$1) and trigonal bipyramid: (C) A double left-handed Dauphiné twin. *Trigonal trapezohedron faces stippled.*

according to a variety of laws; but it can only be demonstrated by means of external characters in some cases—notably by the occurrence of the form ($51\bar{6}1$) on contiguous faces, instead of on alternate ones (Fig. 41C). Otherwise the twinning can be demonstrated by special optical tests[1]—not by the examination of sections of normal thickness.[2]

β-quartz[3] shows a characteristic Hexagonal bypyramidal form with sometimes a poorly developed Hexagonal prism in addition: these are forms characteristic of the holosymmetric class of the Hexagonal System; but actually it is known from the internal structure of the mineral that it belongs to the trapezohedral class of that system.

The study of twinning in β-quartz has been somewhat neglected as the phenomenon produces no visible effects in thin rock-sections: but it has been shown to be quite normal. Collections of milky crystals of β-quartz from Cornish quartz-porphyries contain many twins, the twin-plane being either a face of a rhombohedron or a trapezohedron. The angle between the *c*-axes of the two crystals involved varies in different types between 43° and 84°. The so-called Japanese twins are of these types.

The inversion of quartz from the 'high' (β) to the 'low' (α) form is achieved by only a slight distortion of the atomic structure, and so occurs readily at a fixed temperature which may be used as a point on a geological thermometer. At atmospheric pressure this temperature is 573° C; it rises slightly with increasing pressure. All quartz is now in the low (α) condition: it may have crystallized originally in this form in quartz veins and in some pegmatites; or it may have inverted from the high form subsequently to its primary crystallization. It has been discovered that there is a measurable and very significant difference between primary α-quartz and inverted β-quartz. On reheating the former, inversion occurs approximately one degree above the temperature at which the latter inverts. The material used in these experiments included crystals from cavities in limestones which are primary α-quartz (known to be so by their crystallographic character), and hexagonal bipyramidal β-quartz phenocrysts from rhyolites. When subjected to the same heat-treatment quartz from granites behaved in some cases like the former, but in others, like the latter. The significance of this is discussed in the chapter on granites.

[1] By immersing a thick basal section of quartz in oil of the same refractive index and viewing it along the direction of the optic axis between crossed plates of polaroid.

[2] For much interesting information on the twinning, etc., of quartz, see Symposium on Quartz Oscillator Plates, *Amer. Min.*, **30** (1945), p. 205.

[3] Drugman, J., 1927, On β-quartz twins from some Cornish localities, *Min. Mag.*, **21**, 336.

Despite the many interesting and indeed unique crystallographic and electrical properties of quartz, the optical characters lack distinctiveness. It is uniaxial and optically positive. Its refractive indices are low (1·553 and 1·544),[1] with consequent absence of surface relief; it has no cleavage, and the birefringence (0·009) is weak. It even lacks distinctive alteration products, for it is completely stable. These details apply to both the α- and β-phases, and there are, in fact, no definite tests available for differentiating between the two in thin section. Both phases of quartz are liable to contain inclusions of several different kinds, sometimes in such quantity as to render the crystals opaque or at least 'milky'. Frequently the inclusions are minute gas- or fluid-filled cavities (Fig. 42) often

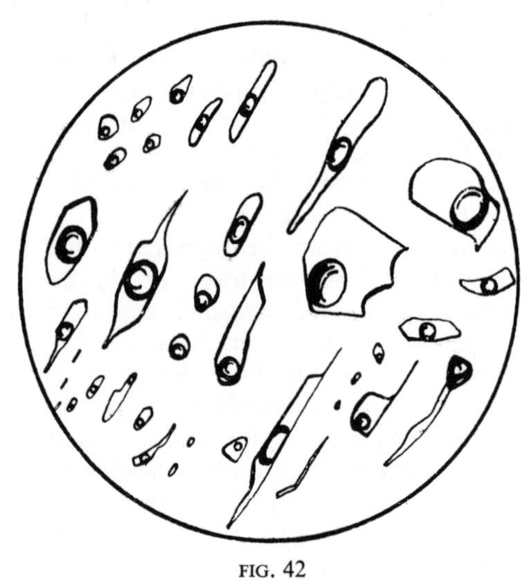

FIG. 42
Fluid cavities containing gas bubbles in α-quartz—highly magnified.

lying in curved planes that evidently represent resealed fractures. Each fluid cavity may contain a gas bubble, mobile in some cases, or even a minute apparently cubic crystal. In some instances the cavity is bounded by plane surfaces giving a *negative crystal* of minute size but perfect form. In other cases acicular needles of rutile are embedded in the quartz. In the granitic rocks termed charnockites, they are exceedingly minute and occur literally in myriads. They produce an optical blue clour in the crystals in which they occur. In smaller numbers they can often be seen, in some Scottish

[1] Because quartz is the only common rock-forming mineral which has a constant and unvarying composition, its refractive indices and birefringence provide valuable standards for comparison.

granites, for example; but they are so minute that it is only by analogy with the much stouter prisms that sometimes occur, that they can be identified as rutile.

Tridymite is the form of SiO_2 stable above 870° C. At its best it forms Orthorhombic tabular plates of almost perfect hexagonal shape; but more commonly occurs as twinned groups of minute crystals, in fine-grained rhyolitic rocks, the best known British examples occurring in the Tardree rhyolite, where the tridymite lines small cavities. Optically tridymite differs from quartz in its lower refractive indices (1·473 and 1·469) and weaker birefringence (0·004); but its most distinctive characters are the narrow, lath-like crystals and the twinning.

Tridymite is not an obvious mineral in thin sections, and it is probably more widely distributed in Acid lavas than is commonly supposed. It is a product of vapour-phase crystallization and is normally restricted to originally gas-filled cavities in rhyolitic lavas including those of ignimbritic type (p. 253).

Cristobalite has long been known from its development in silica-bricks used for lining furnaces. It is the highest temperature polymorph of silica and is therefore a typical phase in high grade metamorphism of siliceous rocks. In relatively rare occurrences in igneous rocks—usually of Acid composition—it is normally cryptocrystalline and would escape detection even under high magnification with a petrological microscope; but its presence is revealed by X-ray analysis. It is a constituent of spherulites in rhyolitic lavas, for example in the Yellowstone Park; it occurs as half-millimetre cubes and rare cube-octahedrons on fracture surfaces of olivine-basalt from Plumas Co., California; and has been reported in the groundmass of certain basaltic lavas. A significant occurrence of current topical interest concerns one of the moon rocks—an ophitic fine-grained dolerite or basalt which is normal and typical so far as the main constituents are concerned; but the rock contains obvious cristobalite filling angular interspaces between the plagioclase and pyroxene crystals. It may be significant that both tridymite and cristobalite have lower densities and less closely spaced structures than quartz, so that their formation is favoured by conditions of low pressure.

Cryptocrystalline silica occurs in several differently coloured named varieties used for ornamental purposes. Some of them occur as vesicle- and geode-infillings, chiefly in lavas.

Chalcedony bears a very ancient name; but in spite of long acquaintance mineralogists have been baffled until very recently in their efforts to discover the nature of the mineral(s) of which chalcedony is composed. Anomalies in certain physical properties and careful measurement of the optical constants suggested that the mineral could not be quartz; but X-ray analysis proved there to be no difference between chalcedony and quartz. The electron microscope has solved the problem: the mineral *is* quartz in the form of minute interlocking crystals containing micropores only 0·1 μm in diameter. Chalcedony, if pure, is soft yellow in colour due to the scattering of light by the micropores; but on account of its porosity the mineral readily absorbs colouring materials. The familiar agate is rhythmically colour-banded chalcedony and is usually stated to contain some opaline silica.

CHAPTER III

ACCESSORY AND SECONDARY MINERALS

IT is impracticable to draw a hard-and-fast line between these two categories of minerals as some play a dual role. Thus the important chlorites are seen to have replaced such mafic silicates as augite, hornblende and biotite in some rocks and are therefore indubitably secondary and replacive; but in other rocks chlorite is evidently a primary constituent, though of late formation.

Indeed in a few instances we have done less than justice to the status of some minerals included in this chapter as accessories, for, although this is their normal role, occasionally they attain to the status of essential minerals. Thus garnets are encountered as accessories in a number of igneous rocks, but eclogite is a two-mineral rock consisting of omphacite and red garnet only, so that the latter is very much an essential. Similarly apatite and sphene are widespread as accessories; but a unique rock-type consisting solely of sphene set in a matrix of apatite occurs in an alkali-complex in the Kola Peninsula, U.S.S.R.

THE SPINEL GROUP

One of the most important groups of accessories comprises the spinels, including magnetite and chromite. The former especially is ubiquitous in igneous rocks of widely varying compositions. Chromite is restricted to coarse-grained Ultrabasic rocks and locally occurs in exploitable quantities. In addition to these quantitatively important members of the group there are other spinels which are occasionally encountered in Basic and Ultrabasic rocks.

The spinels crystallize in the holosymmetric class of the Cubic System, but it is rare to find forms other than the octahedron represented.

The spinels of simplest composition are: the semi-precious 'ruby' spinel, $MgAl_2O_4$ magnetite, $Fe^{2+}Fe^{3+}{}_2O_4$, a ubiquitous iron ore, described in the next section; and chromite, $FeCr_2O_4$. Picotite (= chrome spinel) is $(Mg, Fe^{2+})(Fe^{3+}, Al, Cr)_2O_4$ It should be noted that 'chrome spinel' is not chromite.

Chromium occurs as a trace element in most Basic igneous rocks, sometimes in such minerals as chrome-augite or chrome-diopside, more often in accessory chromite. In a cooling Basic magma the necessary concentration of Cr is attained at a high temperature and a swarm of euhedral chromite crystals is precipitated, and on account of high specific gravity these tend to settle in the magma body forming a basal layer of the kind named chromitite, described in Chapter 14. These are of great economic importance as providing the only commercially exploitable sources of chromium.

With regard to the identification of spinels in thin sections, it is useful to remember that it is highly unlikely that they will be found in rocks other than those which are olivine-rich; therefore a slide of peridotite or similar rock-type should be searched for spinels which are easily identified by (1) very strong surface relief due to high refractive index; (2) strong and distinctive absorption in deep green, coffee brown or plum purple; and (3) perfect isotropism.

IRON ORES

The opaque iron-ores such as magnetite as well as the other oxide- and sulphide-minerals occurring in igneous rocks can be adequately studied only with the aid of a metallurgical, as distinct from a petrological, microscope. Identification depends upon colour, reflectivity and optical reactions under *reflected* light. These properties are just as distinctive for this class of mineral as are the familiar optical properties of transparent minerals viewed by *transmitted* light. This is not the place to describe the techniques involved; but if the apparatus is available the student is strongly advised to master the techniques, since these are not inherently difficult and are well worth while, if for no other reason than that the opaque minerals display textural relationships which are both interesting and significant in the full understanding of the crystallization history of a given rock. In the absence of the necessary apparatus the common ore minerals may be identified by illuminating the upper surface of the slide, at the same time diverting the transmitted light.

The iron ores that occur as constituents of igneous rocks are magnetite, titanomagnetite, ilmenite, pyrite and rarely pyrrhotite. The first three are normal accessories in a wide range of rock-types, and in some cases occur in such amounts that they must be regarded as important essential components.

Magnetite is the most widely distributed accessory in igneous rocks: it is the 'iron ore' of countless petrographic descriptions. Magnetite is invariably opaque, even in the thinnest sections, and in the absence of crystal form, it may not be easy to distinguish it from other opaque ores; but by oblique reflected light it has a characteristic steely metallic sheen. It is the most strongly magnetic of the iron ores, and may be separated from the others by means of a magnet.

In many igneous rocks, especially the more Basic, magnetite may

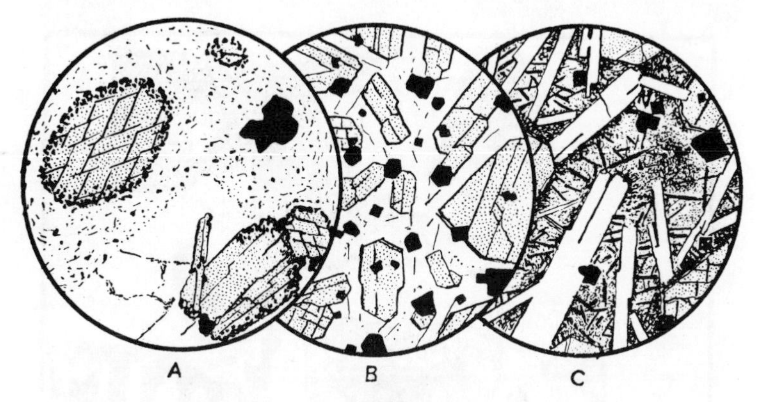

FIG. 43

Iron ores in thin section. (A) Magnetite grains produced round lamprobolite phenocrysts by magmatic corrosion, in hornblende-andesite, Siebengebirge (×32); (B) Early crystallized, euhedral magnetite grains in dolerite (×32); (C) Early formed euhedral crystals, and very late iron ore embedded in glassy groundmass, in olivine-basalt, Hawaii (×60).

occur as phenocrysts or microphenocrysts of early formation—the characteristic, often perfectly formed octahedrons occurring plentifully in thin sections of basalts and other lavas, are of this type; but in some basalts it is one of the latest constituents to crystallize, and forms intricate dendritic growths in the interstitial glass (Fig. 43C). Further, magnetite is produced at different stages in the history of a rock by alteration of iron-bearing silicates. A familiar instance is the serpentinization of olivine; while the conversion of amphibole into pyroxene also liberates magnetite (Fig. 43A).

The occurrence of magnetite in layered basic complexes is described later, under 'Monomineralic rocks'.

Ilmenite.—Ideally ilmenite is titanite of iron, $FeTiO_3$, but in naturally occurring specimens some substitution of Ti by Fe has taken place, so that up to 30 per cent of Fe_2O_3 may be shown on analysis. Ilmenite crystallizes in the Trigonal System, but crystals are rare, and normally the mineral is massive. It is quite opaque, and is therefore sometimes difficult to distinguish from magnetite, especially when quite fresh. Alteration renders the task easier, however, for ilmenite is progressively converted into **leucoxene**. In an early stage, the change is superficial only, and the grains appear white by reflected light. In an advanced stage of alteration the mineral becomes translucent and light brownish-grey, while the Trigonal symmetry is emphasized by thin parallel black bars crossing in three directions (Fig. 44). At one time leucoxene was thought to be secondary sphene in the process of formation, but it has been shown[1]

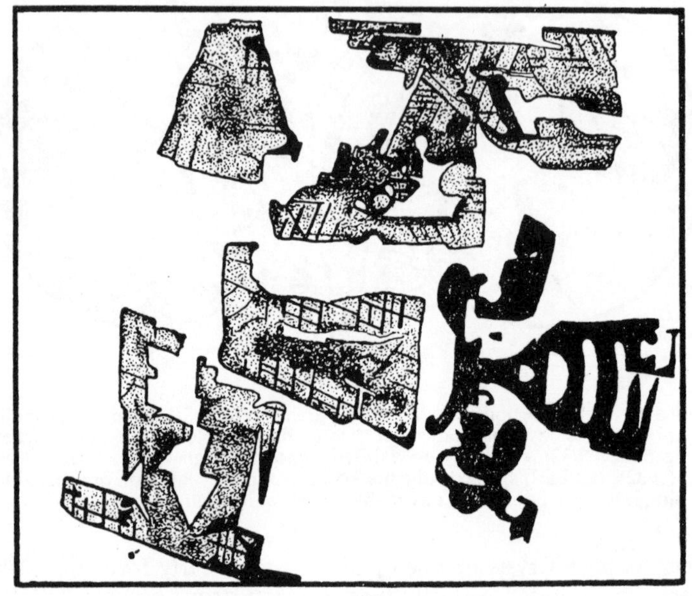

FIG. 44

Ilmenite and its alteration products; from lugarite, Lugar, Ayrshire. The one skeletal crystal is fresh; the others have been altered into grey-brown translucent leucoxene traversed by bars of iron-oxide.

to be hydrated oxide of titanium, $TiO_2 . nH_2O$.

Compared with magnetite, ilmenite is less variable in mode of occurrence. It normally occurs as an accessory mineral, particularly in the more basic, coarse-grained igneous rocks of the gabbro type. In such rocks, indeed, ilmenite is an essential constituent, and occasionally increases in amount, particularly towards the base of layered intrusions. In this country the Carrock Fell gabbro is noted for its richness in ilmenite.[2]

Having described the properties of magnetite and ilmenite it is necessary to consider the close association of these two minerals, particularly in coarse-grained Basic igneous rocks. The nature of the association can be fully appreciated only after examining polished specimens with the metallurgical microscope, when it is commonly seen that apparently single crystal grains are in fact aggregates con-

[1] Edwards, A. B., *Min. Mag.*, **26** (1942), 273.

[2] Harker, A., *Natural History of Igneous Rocks*, (1909), p. 133.

sisting of magnetite and ilmenite in widely varying proportions. The grain boundaries are generally angular or smoothly curved. In addition to such aggregates of discrete crystal grains of the two ores, either may contain inclusions of the other. They take the form of very thin plates or lamellae, the directions of which are determined by the crystal structure of the host mineral. Thus ilmenite lamellae in magnetite are parallel to the octahedron faces of the latter, and on a polished surface form a graticule of either two or three sets of lamellae according to the orientation of the crystal. Magnetite lamellae in ilmenite, however, are confined to the single direction, parallel to the basal pinacoid (0001) of the ilmenite. These intergrowths are exactly analogous with the perthitic intergrowths in feldspars and pyroxenes, already described, and like them are the result of unmixing during slow cooling.[1] This conclusion is suggested by the textural relationships between the two ore minerals; it is confirmed by the fact that in many basalts, which are the high-temperature equivalent of gabbros, only a single, homogeneous iron-ore phase occurs—and is a titaniferous magnetite or **titanomagnetite.**

Ilmenomagnetite is the unmixed equivalent of titanomagnetite as described above. Other oxide minerals including haematite and spinel enter into the ilmenomagnetite but only in very small quantities. They are visible only under high magnification, and their identification is a matter for the specialist.

Of the two sulphide ores with which we are concerned, pyrite (= pyrites) is by far the commoner. It is unlikely that it should occur as a primary constituent in igneous rocks; but as a secondary product of alteration of iron-bearing silicates it is widely distributed, associated with calcite, chlorite, secondary quartz, etc., notably in areas affected by solfataric activity. For example, in some of the areas of Ordovician volcanic activity in N. Wales not only do the slates contain large numbers of euhedral pyrites but the latter occurs as perfect 'cubes' in the dyke rocks also.

Pyrite is used in mineralogy courses to demonstrate the crystallography of one of the less symmetrical classes of the Cubic System; but it may also be 'massive', in which case it can be identified by its light brassy appearance on examination in reflected light.

Pyrrhotite has the same chemical composition (FeS_2) as pyrite, but is a Hexagonal mineral, though actual crystals are seldom encountered. When fresh, pyrrhotite resembles pyrite, though somewhat more strongly coloured. Exposure causes it to tarnish, and ultimately it assumes a characteristic bronzy appearance. A further point of difference between these two sulphides is that pyrrhotite is magnetic,

[1] Buddington, A. F., Fahey, J. and Vlisidis, A., Thermometric and petrogenetic significance of titaniferous magnetite, *Amer. J. Sci.*, **253** (1955), 497.

though variably so. Pyrrhotite is an important accessory in certain Basic igneous rocks, it has been recorded from some nepheline-bearing rocks and, rarely, from granitic rocks. The best-known occurrence is the Sudbury 'norite' in Ontario, in which the pyrrhotite appears to be a mineral of late crystallization, although it is believed that the sulphidic material separated at an early stage from the silicate-melt as a consequence of immiscibility. This interpretation is based largely on the mode of occurrence of the pyrrhotite, much of which has the form of composite globular aggregates which are interpreted as immiscible droplets.

A British example at the other end of the compositional range is the Shap Granite, in which pyrrhotite is an accessory: the opaque grains seen in thin sections of this well-known rock are pyrrhotite—not magnetite.

In favourable circumstances pyrrhotite may be used in geological thermometry: it is the high-temperature form of FeS_2, and it is therefore important to differentiate between it and pyrite.

THE ZIRCONIUM-BEARING ACCESSORIES

Zircon, a silicate of zirconium, $ZrSiO_4$, is one of the best-known minerals which crystallize in the Tetragonal System. Typical crystals are illustrated in Fig. 45.

Zircons are common in granitic, syenitic and dioritic rocks, but

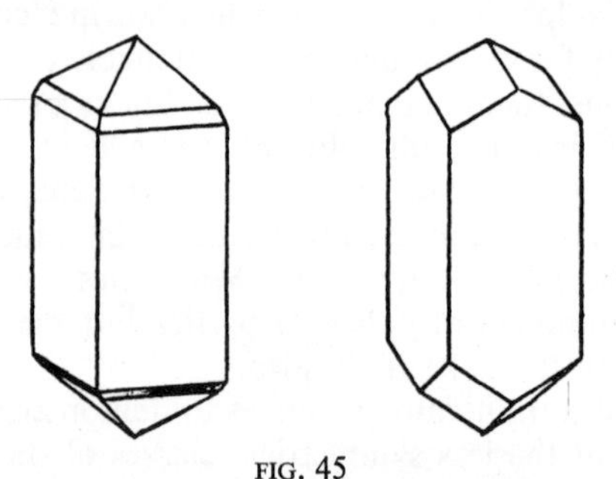

FIG. 45
Crystals of zircon.
Left, vertical prism (110) combined with Tetragonal bipyramids (111) and (331); *Right*, prism (100) with bipyramid (111).

appear to be rarer in those of more Basic composition. As a general rule they are of small (microscopic) size, colourless with very high refractive indices ($\alpha = 1 \cdot 927$, $\gamma = 1 \cdot 982$). The birefringence is likewise strong ($0 \cdot 055$). As they crystallize at a high temperature, they are liable to be enclosed in minerals of later formation.

In certain types of nepheline-syenites the zircons are of much larger size and have the status of an essential constituent, particularly in the coarse-grained 'zircon-syenite-pegmatites'. Zircons are interesting from several points of view: they are important accessories in many igneous rocks, and frequently in the systematic separation of the heavy minerals during the study of an igneous rock a 'flood' of perfectly formed small zircons is apparent. These crystals are in the main colourless and transparent; but purple, yellowish or brownish translucent crystals may also occur.[1] By reason of its radioactivity zircon was the mineral used in pioneer attempts to estimate the age of igneous rocks, using for the purpose biotite-granites which contain zircons surrounded by the well-known **pleochroic haloes** when embedded in dark mica. The possibility that careful study of shape-variation of zircons might throw light on petrogenetic problems, particularly when the possibility of replacement is involved, has been realized by some petrologists; but the results are difficult to interpret. In some cases rocks of igneous aspect have been found to include rounded zircons, presumably of sedimentary origin, which might be taken to imply that the rock-body has been produced by *in situ* replacement of the country rock; but it might equally well mean that some zircon-containing sediment had been incorporated in magma and completely assimilated except for the zircons which are chemically stable. On the other hand the several members of a great batholythic complex have been shown to contain euhedral zircons which evidently crystallized at high temperatures as normal accessories in an igneous environment. The crystal habit was found to be constant within each member of the complex, but they were distinctive individually, reflecting slight differences in the cooling conditions.[2]

Eudialyte[3] is one of a number of complex zircono-silicates which occur as accessory minerals in nepheline-syenites, particularly of pegmatitic facies. It increases in amount in more basic nepheline-bearing rocks and may attain to the status of an important essential constituent, as in the so-called eudialyte-syenites in southern Greenland. One rock near Julianehaab, in this area, contains eudialyte to the extent of a third of the whole rock. It is also important in similar rocks in the Kola Peninsula, U.S.S.R. In Britain it has been recorded from quartz-syenite veins in limestone at Barnavave, Ireland.

Eudialyte is a Trigonal mineral. It is commonly bright red with vitreous lustre in the hand specimen, and therefore very distinctive in appearance. In thin section it is often colourless, but may be

[1] The highly prized blue zircons of gemstone quality do not occur naturally in this colour: they are heat-treated.

[2] Larsen, L. H. and Poldevaart A., *Min. Mag.*, **31** (1957), 544.

[3] The composition of eudialyte is indicated by the following formula, based on a recent analysis: $(NaCa)_5ZrSi_6O_{17}(O,OH,Cl)$, with 0·6 per cent of $(Nb,Ta)_2O_5$.

pink, and then exhibits a pink to yellow pleochroism. Refringence is moderate, with $\alpha = 1 \cdot 609$ and $\gamma = 1 \cdot 611$. The birefringence is notably weak—about 0·002.

TITANIUM-BEARING ACCESSORIES

In igneous rocks any available TiO_2 may be incorporated in a number of mafic minerals as we have seen—there are TiO_2-bearing pyroxenes, amphiboles and micas; but in addition it may be concentrated in accessory minerals, in oxide form in rutile, anatase and brookite; in silicate form as sphene ($CaTiSiO_5$) if sufficient CaO and silica are available. If not, perovskite ($CaTiO_3$) will take the place of sphene in undersaturated rocks. In addition TiO_2 may occur in combination with iron in titanomagnetite and ilmenite.

(i) **Rutile, Anatase** and **Brookite.**—**Rutile** is the most stable of the three crystalline forms of TiO_2—rutile, anatase and brookite—and is the one which most commonly occurs as an accessory mineral especially in the more Acid rocks. It forms slender Tetragonal prisms and 'needles' which are frequently twinned to give geniculate and heart-shaped forms. Both in hand-specimens and under the microscope, rutile is coloured yellow or red. It is remarkable for its exceptionally high refractive indices (2·61 and 2·90), the highest for any of the normal rock-forming minerals.

In pneumatolytized granites, rutile is often present as a product of the breakdown of complex titanium-bearing minerals, such as biotite and sphene. Under these circumstances it may form an interlocking structure of regularly orientated needles, known as sagenite. Quartz crystals in certain granites contain myriads of exceedingly minute needles believed to be rutile, by analogy with the megascopic, reddish-brown acicular crystals of more robust habit occurring in the quartz of certain pegmatites. Relatively large crystals also occur in eclogites.

Anatase and brookite are normal accessories in granites, though a few crystals may be dispersed throughout a large volume of granite, so that the chances of encountering them in a thin slice of granite are very remote. Both are familiar 'heavy minerals' in concentrates derived from stream sands in granitic terranes. In Britain the Dartmeet sands on the outcrop of the Dartmoor granite provide typical examples of both minerals, which display features of interest out of all proportion to their small size.

Anatase varies considerably in crystal habit as shown in Fig. 46. The anatase crystals vary from colourless to rather dull shades of blue, green or brown. The refractive indices are very high (2·49 and 2·56), and the birefringence is also strong (0·07). In view of its distinctive appearance a confirmatory test is seldom necessary, but the tablets show a uniaxial (positive) interference figure in convergent light.

Brookite occurs as thin platy light brown or yellow Orthorhombic

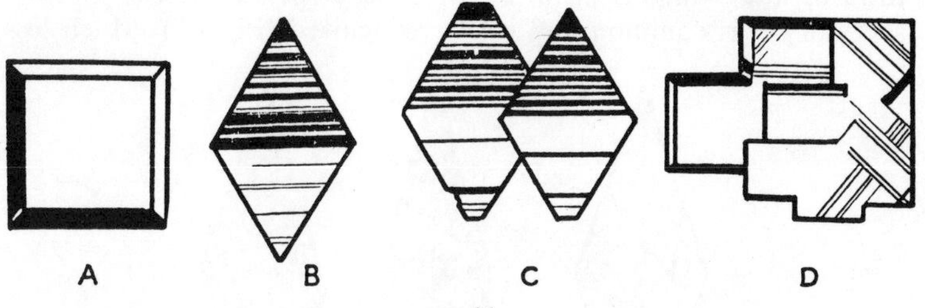

FIG. 46

Crystals of anatase: (A) and (D) of tabular habit, with characteristic surface markings in the latter case; (B) and (C) of bipyramidal habit, and, in the latter case, crystals in parallel growth.

crystals dominated by the front pinacoid, (100), the faces of which are closely and regularly striated due to oscillation between the pinacoid and the vertical prism sometimes developed on the edges of the crystals. The really distinctive feature, however, is the failure to extinguish, on account of crossed axial-plane dispersion. The effect of this property is to cause the section to display a succession of interference colours, in the order of Newton's scale, when the stage is rotated between crossed polarizers.

(ii) **Sphene** or **Titanite.**—Sphene is a silicate of titanium and calcium, $CaTiSiO_5$, occasionally seen in mineral collections as small wedge-shaped crystals, but much better known as a distinctive and widespread accessory mineral in rocks of many different kinds. Sphene can sometimes be seen even with the naked eyes, and easily with a lens, in hand-specimens of granodiorites, syenites and diorites, for example. The crystals, which belong to the Monoclinic System, are brilliantly lustrous and range in colour from light yellow to green or brown.

In thin section sphene may be colourless, but is commonly a shade of greyish-brown, and is slightly pleochroic. The refractive indices, 2·01 and 1·90, are notably high, so that surface relief is strong. The birefringence also is particularly strong, so that the interference colours are very light—the so-called 'high order whites'—and these, superimposed on the normal absorption tint of the mineral, make no appreciable difference. Thus, provided the section is not in the position of extinction, crossing and uncrossing the polarizers makes no difference to the colour of the section. This observation provides a unique test for the identification of sphene.

As an accessory in igneous rocks sphene normally occurs as small, scattered crystals; but a rock-type discovered in the Kola

Peninsula, U.S.S.R., consists of crowded crystals of sphene embedded in a matrix of apatite—a truly extraordinary rock-type.

In different rocks sphene may show very different relationships towards the minerals with which it is associated. As a normal accessory it is common and easily recognized (Fig. 47). Much less

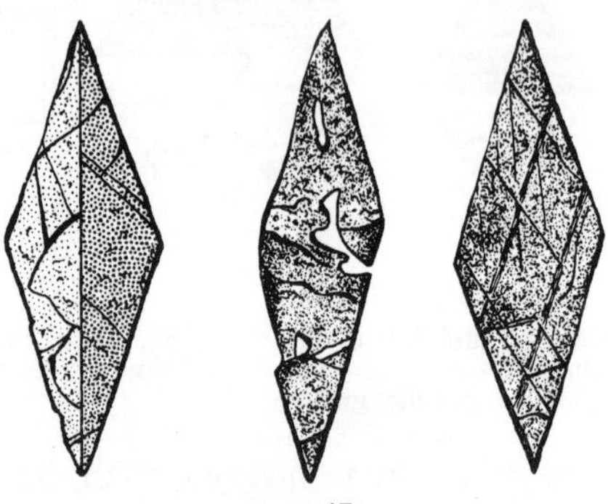

FIG. 47

Sections of sphene: two on right from ditroite, Ditro, Transylvania, one showing cleavages, the other corrosion inlets; on the left a simple twin from leucite phonolite, Perlerkopf, Brohltal.

commonly it is moulded upon the minerals in which it is normally embedded: instead of being euhedral it is xenomorphic, and in a dioritic assemblage is bounded by the plane faces of plagioclase and amphibole.[1] This sphene is primary, but belongs to a late stage in the crystallization sequence. Finally, sphene may occur in association with other secondary minerals pseudomorphing one of the primary constituents. For example, certain dioritic rocks from Jersey, Channel Islands, contain pseudomorphs after euhedral Ti-rich amphiboles consisting of penninitic chlorite containing much sphene, the latter representing the titanium and calcium which could not be incorporated in the chlorite. In the rare rock, borolanite, garnets occur which contain appreciable amounts of sphene of almost the same colour as the melanite in which it is embedded. Evidently in this case after the initial crystallization of a Ti-rich garnet, presumably schorlomite, owing to a change in environmental conditions it became unstable and inverted to a Ca- and Ti-poor garnet intimately associated with sphene.

(iii) **Perovskite.**—Among the rarer accessory minerals perovskite fills an important role in some uncommon rock-types. It is calcium titanate, $CaTiO_3$, and crystallizes in the Cubic System, the octahedron being the only form normally developed. The small crystals

[1] Wells, A. K. and Bishop, A. C., *Quart. J. Geol. Soc.*, **111** (1955), 143–66, especially Figs. 5, 10 and 12.

are isotropic, and Cubic beyond question; but larger crystals exhibit optical anomalies and, it is recorded, quite considerable birefringence. The outstanding optical property is a very high refractive index (2·38), which makes the small crystals stand out strongly, despite their size. In fact very small crystals have such a heavy margin that they appear almost opaque. Cleavage is stated to be perfect, parallel to cube faces.

Perovskite is characteristic of, and restricted to, under-saturated lime-rich rocks, and is a constant associate of melilite with which it occurs as small grey octahedrons and twinned groups. Probably the most notable occurrences so far described are in ultrabasic lavas from Uganda. The perovskite in these lavas occurs in unusually large quantities, up to 6·2 per cent in one type. In these rocks the perovskite is golden-brown, yellow or green instead of the commoner grey.

APATITE

Phosphorus, estimated as P_2O_5, is apparently present in all igneous rocks, and all of this element occurs in apatite, a ubiquitous accessory. **Apatites** are essentially phosphates of calcium, with small amounts of fluorine, chlorine, and/or hydroxyl. According to the dominance of either fluorine or chlorine, two varieties, fluor-apatite and chlor-apatite are distinguished: the former is much the commoner in rocks. Apatite crystallizes in the Hexagonal System, in the class in which there are no vertical planes of symmetry; but usually the crystal development is so simple—a combination of prism in zone with a bipyramid and basal pinacoid—that this passes unnoticed. Apatite may be colourless, but is often bluish, brown or green in hand specimens. Commonly it occurs as minute euhedral crystals of prismatic habit which may be seen in thin sections of most igneous rocks. Relatively large crystals occur in coarse-grained syenites, especially syenite-pegmatites; at the other extreme are hair-like 'needles' scarcely visible under ordinary magnification. Although normally colourless in thin section they may appear bluish, and quite commonly have cloudy coloured cores (Figs. 95 and 96).

Apatites may be recognized by their form, moderate relief (refractive indices about 1·63), very low order weak-grey interference colours and isotropic basal sections. Crystals are often fractured across by a basal parting. Apatite impresses viewers by the uniformly perfect shapes of the crystals seen in thin rock sections. From the fact that it freely penetrates into all of the other components, it might be inferred that it was the first mineral to crystallize. However, apatite is seen to have crystallized from late-stage residua like those occurring in quartz-dolerites, for example. The euhedrism of apatite reflects an inherent tendency to succeed in building perfect crystals in competition with other, less well endowed minerals, by reason of its 'pressure of crystallization'.

Beryl is a silicate of beryllium and aluminium which crystallizes in the holohedral class of the Hexagonal System, and is the mineral generally used to demonstrate the crystallography of that class, since beautiful crystals, obtained from granite pegmatites, are available. Some beryls of phenomenal size have been mined from certain pegmatites; very much smaller crystals are found occasionally in cavities in granites, including those of the Mourne Mts in north-eastern Ireland and Lundy Island. It is worth noting that even the smaller crystals represent a remarkable feat in terms of natural concentration of beryllium. The latter is one of many so-called 'trace-elements', which may exist in rocks in concentrations of only a few parts per million. Normally these minute amounts are accommodated in the structure of common minerals; but in beryl and other, even rarer, pegmatite minerals, the trace-elements are sufficiently concentrated to form independent compounds.

GARNET

Garnets are silicates of aluminium, iron, manganese, chromium, calcium and magnesium. The crystal structure consists of separated SiO_4 groups, and there are three of such groups in the unit of the structure, giving Si_3O_{12} which appears in all the formulae. In this structure there are twelve free links to which other ions are attached according to the scheme:

$M^{2+}{}_3R^{3+}{}_2(Si_3O_{12})$, where M is a divalent, and R a trivalent element. The varieties shown in the Table may be distinguished. These are largely hypothetical: natural garnets are more complex in composition than any of these 'pure' garnets and in any specific case may be regarded as containing *either* the first three, *or* the second three components in varying proportions. Thus a common red

	Formula	Colour	Refr. index
Pyrope	$Mg_3Al_2Si_3O_{12}$	blood red	1·705
Almandine	$Fe_3Al_2Si_3O_{12}$	dark red	1·830
Spessartine	$Mn_3Al_2Si_3O_{12}$	red	1·800
Uvarovite	$Ca_3Cr_2Si_3O_{12}$	green	1·870
Grossularite	$Ca_3Al_2Si_3O_{12}$	green or orange-red	1·735
Andradite (Melanite)	$Ca_3Fe_2Si_3O_{12}$	black	1·895

garnet on analysis would be found to contain Mg^{2+}, Fe^{2+} and Mn^{2+}, and according to the amounts present might be expressed in terms of the molecular percentages of Py, Al and Sp respectively. Both groups of three form continuously variable series, but grossularite and pyrope, for example, mix only with difficulty under high pressure.

In thin section all garnets are characterized by high relief, on account of the high refractive indices; but the figures quoted above must be regarded as approximations only, on account of the complex composition of natural garnets. As might be expected, the sections often exhibit zoning, reflecting changes of composition during growth (Fig. 48). Garnets alter in a variety of ways. Thus in rocks

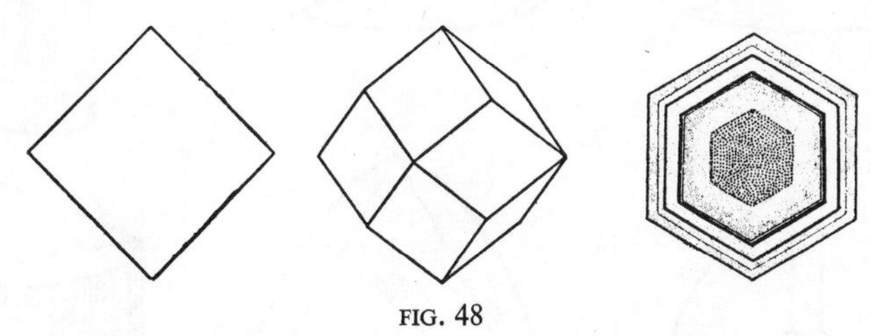

FIG. 48

Garnet: *Centre*, the simplest form, the rhombdodecahedron (110); *left*, a section through centre of crystal, parallel to one of the axial planes of symmetry (100), (010) or (001); *right*, section through centre of crystal parallel to one of the twelve dodecahedral faces. Zoning is formally shown.

which have suffered contact metamorphism, garnet may be pseudomorphed by cordierite. The commonest alteration product is probably chlorite; but garnet carrying titanium (schorlomite) may alter into sphene embedded in chlorite.

Garnet may occur as a normal accessory in a wide range of igneous rocks. In granitic dyke-rocks a garnet of the first group, often rich in manganese, occurs in this way. Syenitic rocks containing feldspathoids such as nepheline and leucite, not infrequently contain garnets near to melanite (andradite) in composition. As might be anticipated, accessory garnets are found in Ultrabasic igneous rocks such as peridotites, and are rich in magnesium. They tend to survive, as distinctive red crystals, when such rocks are converted into serpentinites (see below).

THE PNEUMATOLYTIC MINERALS

Tourmaline.—Tourmalines are complex hydrated borosilicates of aluminium, magnesium and sodium, with iron, manganese, calcium and small quantities of potassium, lithium and fluorine. They all contain about 10 per cent of boric acid, and from 3½ to 4 per cent of water. Tourmaline crystals are Trigonal, and are illustrated in Fig. 49.

From many points of view, tourmaline is a mineral of outstanding

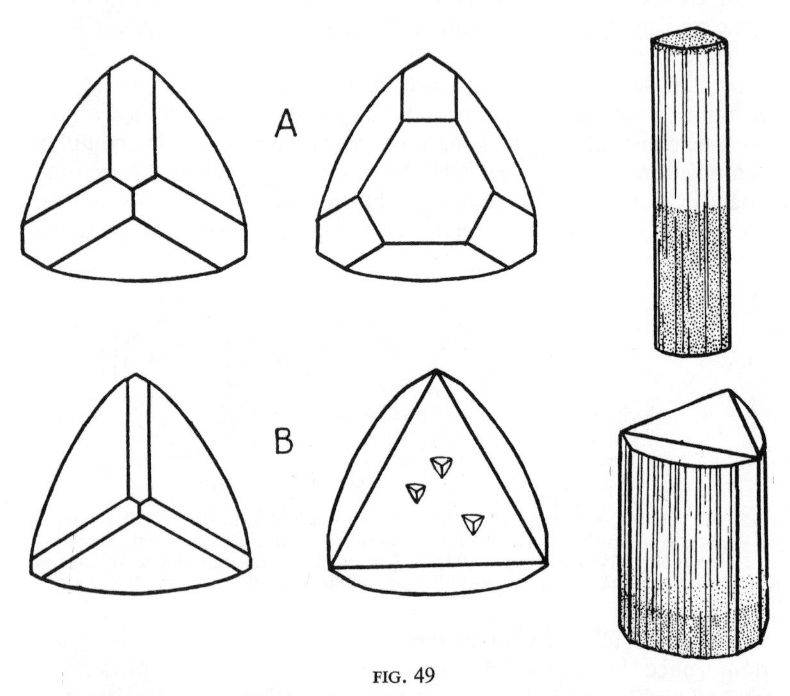

FIG. 49

Tourmaline crystals: (A) The two differing terminations of one crystal, showing combinations of positive and negative trigonal pyramids at one end, and similar forms with the addition of a pedion (0001) at the other. (B) Similar combinations shown by a crystal of slightly different habit, with the characteristic shape of etched figures added diagrammatically. The crystals (*right*) show polar colour variation and vertical striations. Tourmalines of gem quality from Minas Geraes, Brazil.

interest. Crystallographically it provides striking examples of polar symmetry, the terminations of the same crystal being obviously different; while if the crystals are so simple as to show a single horizontal plane at both ends, the complete independence of these pedions is proved by physical differences between them, one being as lustrous as glass, while the other may be lustreless or strongly etched. This polar symmetry is in many instances shown by an unsymmetrical distribution of colour along the length of the crystal.

In thin section tourmaline may occasionally be colourless; but normally it is blue, brown or green, and strongly pleochroic. The absorption of the ordinary ray is much stronger than the extraordinary, so that prismatic sections show their deepest colour when the principal axis lies at right angles to the vibration direction of the polarizer in the microscope.[1]

Basal sections, often of modified triangular cross section, are frequently zoned, even if the crystal is minutely acicular. They yield a negative uniaxial interference figure in convergent light.

Tourmaline is essentially a component of the so-called pneumatolytic rocks and of granite-pegmatites. Thus it is widely distributed in and around the West of England granites. In part it is a primary mineral, but in large measure it has been produced at the expense of pre-existing mica and feldspar, in which case it commonly assumes a distinctive habit, much-elongated acicular crystals with a radial disposition, or being so closely packed as to form a felt-like aggregate (see under 'Pneumatolysis', and Fig. 81).

Topaz is fluosilicate of aluminium, $Al_2(F,OH)_2SiO_4$, and crystallizes in the holosymmetric class of the Orthorhombic System. Typical crystals are illustrated in Fig. 50.

Topaz of gemstone quality from certain well-known localities in Brazil has a distinctive honey-brown colour; but other varieties

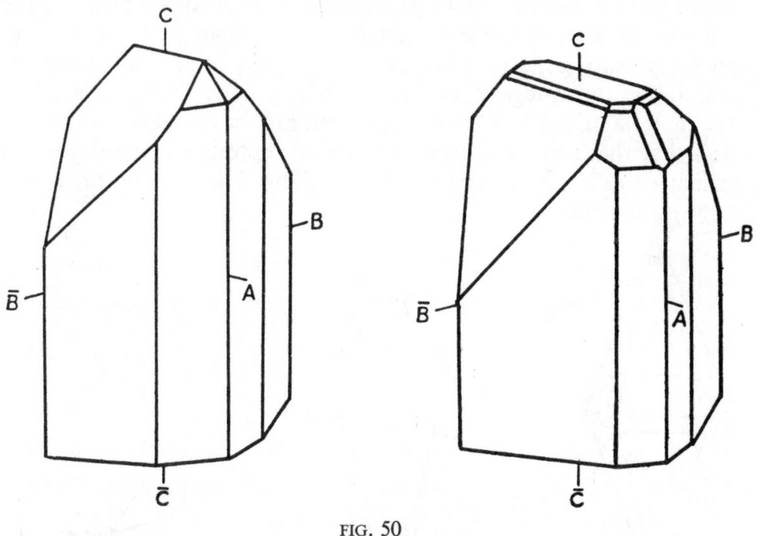

FIG. 50

Crystal habits of topaz from Klein Spitzkop, South-West Africa.

Combinations of vertical prisms (110) and (210); the dominant 'dome' (021), the 'dome' (101) and bipyramids. The crystals are single-ended, being bounded below by basal cleavage planes. The crystallographic axes marked A, B, C, should be *x*, *y*, *z* according to current practice.

[1] The vibration of the polarizer lies parallel to the cleavage traces in a vertical section of biotite, when the stage has been turned to the position in which the mineral shows its deepest absorption tint.

may be faintly tinted blue, though many are quite colourless. The single cleavage, parallel to the base of the crystal, is perfect, though difficult to produce artifically.

Optically topaz is characterized by moderate refractive indices, 1·615 for α and 1·625 for γ. Thus the birefringence, 0·009, is almost identical with that of quartz, though the relief is considerably stronger.

Topaz occurs in irregular grains and spongy masses in pneumatolytic rocks, notably in greisens. In such rocks the amount of topaz may be very large: in a 'topazfels' from Schnechenstein in Saxony, 80 per cent of the rock is topaz. The mineral also occurs rarely as a primary mineral in granites, though in Britain it appears to be restricted to irregular druses in the Mourne Mountains and Lundy Island granites.

CONTAMINATION ACCESSORIES

Under this heading we include certain minerals which normally occur in metamorphic rocks. However, they also occur in igneous rocks which have assimilated, and therefore have been contaminated by, xenolithic sedimentary material. Such contamination is shown by the occurrence of corundum in certain Intermediate to Basic igneous rocks, or andalusite in others. Cordierite and some kinds of spinel also fall in this category. In view of their composition it is clear that the material assimilated must have been highly aluminous sediment, and that the 'contamination accessories' represent surplus alumina which could not be incorporated to form feldspars or other Al-bearing silicates.

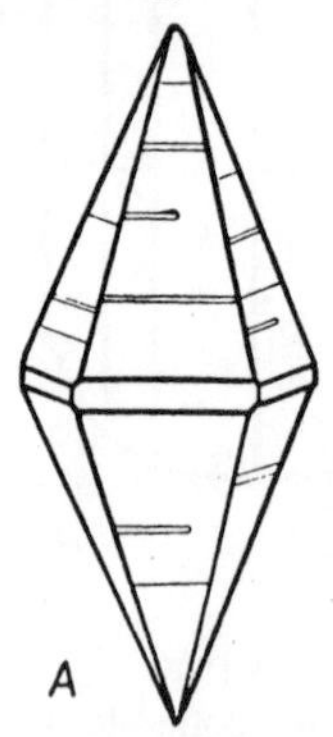

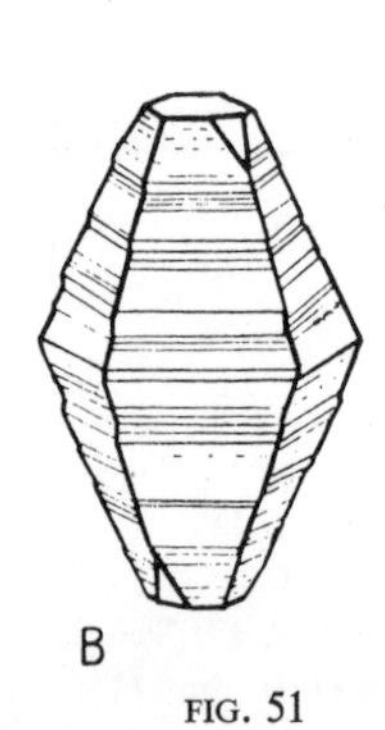

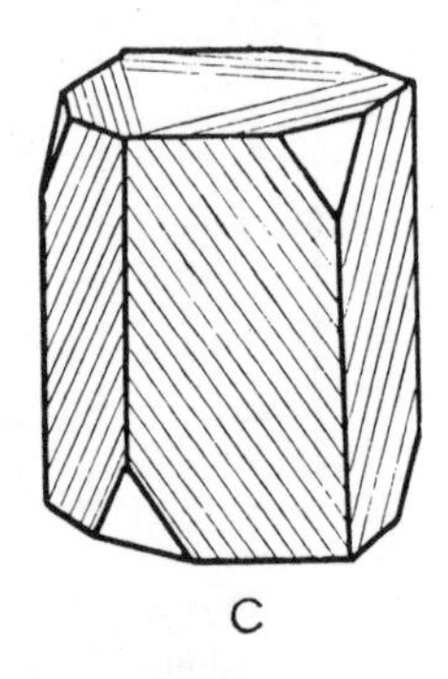

FIG. 51

Three crystal habits of corundum.

(A) Sapphire: combination of hexagonal prism and bipyramid. (B) Opaque red corundum, Ceylon: combination of heavily striated bipyramid, rhombohedron and basal pinacoid. (C) Common corundum, Pietersburg, Transvaal, South Africa: combination of hexagonal prism, rhombohedron and basal pinacoid, with characteristic surface markings.

Corundum.—There are three very distinct varieties of this mineral—common corundum, and the clear gem-stones, ruby and sapphire. These all crystallize in the rhombohedral class of the Trigonal System, but the crystal habit is very variable. Common corundums may be tabular, bipyramidal with very irregular faces (Fig. 51B) or prismatic (Fig. 51C).

In composition all varieties are essentially alike, only differing in the minute amounts of trace elements that act as pigments to the coloured varieties. Apart from these, corundum is just crystallized alumina, Al_2O_3.

Corundum is notable for its extreme hardness (9 in Moh's scale). It has a parting parallel to the (0001) face. Optically corundum is distinguished by its high refractive indices ($\alpha = 1 \cdot 760$, $\gamma = 1 \cdot 768$) and weak birefringence (0·008), almost the same as quartz. Normally corundum is colourless in thin section, but a strongly coloured sapphire is light blue, and of course slightly pleochroic. Corundum is uniaxial negative.

Although corundum is typically a product of the thermal metamorphism of argillaceous sediments, it is a rare constituent of some types of igneous rocks. In the form of small blue sapphires, corundum has been described from argillaceous xenoliths in Basic in-

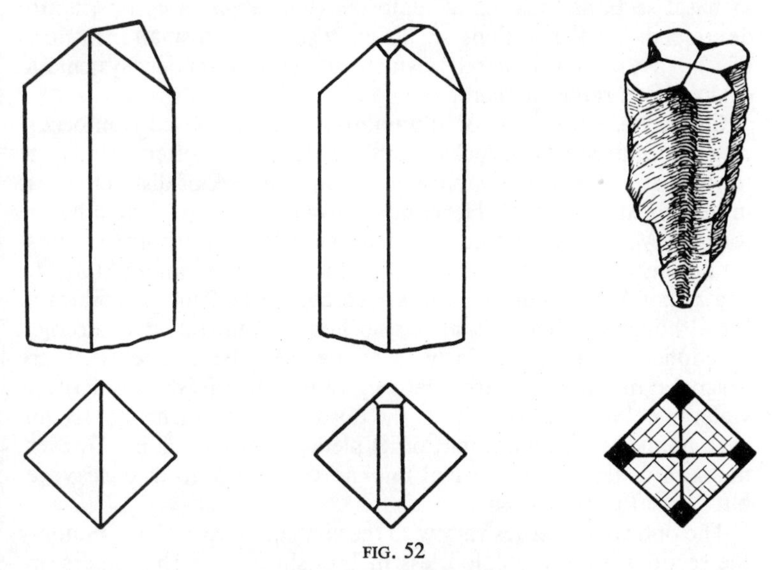

FIG. 52

Andalusite and chiastolite crystals.

Left: Combination of prism (110) and 'dome' (011). *Centre:* A more complicated crystal showing the forms (001) and (101) in addition. *Right:* A weathered crystal of chiastolite. *Below:* The ideal cross-section showing prismatic cleavages and characteristic carbonaceous inclusions.

trusions of Tertiary age in Mull and Ardnamurchan. Corundum occurs only in quartz-free igneous rocks, and it appears to be restricted to those rocks which have suffered desilication: that is, they have contributed silica by transfer to the adjacent wall-rock. Thus certain syenitic and dioritic rocks from the Bancroft area in Ontario and from some South African localities, are corundum-bearing.

Andalusite is one of a number of silicates of aluminium but is the only one occurring in igneous rocks. It may be represented by the formula Al_2SiO_5, being thus identical in composition with sillimanite (which may be regarded as the high-temperature equivalent of andalusite), and kyanite, its stress equivalent. Andalusite is Orthorhombic, but exhibits strong pseudo-tetragonal symmetry.

In thin section andalusite may be distinctive, largely by reason of its pleochroism. At its best the colour corresponding to the slow vibration, γ, is peach pink, while α and β may both be colourless, or there may be a trace of bluish-green. This pleochroism scheme is, in fact, closely similar to that of hypersthene. In many rocks, however, the andalusite is non-pleochroic. The refractive indices are moderate, with $\alpha = 1{\cdot}635$ and $\gamma = 1{\cdot}643$ (average values); while the birefringence is weak, varying in different specimens from a little below to a little above quartz. Traces of prismatic cleavage, in basal sections crossing at almost a right angle, are consistently developed, but the sections are strongly sieved with small inclusions of the associated minerals, which sometimes render systematic examination rather difficult.

Cordierite.—A silicate of aluminium, magnesium and iron occurring in a wide variety of rocks of different kinds; it is often a challenge to the skill of the petrographer on account of lack of distinctiveness in its optical characters. The composition may be represented by the formula $Mg_2Al_4Si_5O_{18}$; but this may be written in a form reminiscent of beryl, except that the Si_6O_{18} of the latter becomes $(AlSi_5)O_{18}$ —a proxy-Al replacing a Si atom—in cordierite. This is significant, for although cordierite is an Orthorhombic mineral, it is strongly pseudohexagonal, particularly when twinned. Natural crystals are prismatic in habit, but are relatively rare. When fresh, cordierite is violet in colour and pleochroic, the obsolete name 'dichroite' having been applied to it on account of its pleochrosim, which is, however, not seen in sections of normal thickness. There is no true cleavage, but a basal parting is characteristic.

The optical properties vary with the content of iron; but commonly the sections are quite colourless in thin slice, while the indices are low—$\alpha = 1{\cdot}535$, $\gamma = 1{\cdot}544$. Thus one index is the same as quartz and the surface relief is the same as that of the latter mineral, so is the birefringence ($0{\cdot}009$); and it is therefore often very difficult to distinguish between the two minerals. Under the most favourable conditions the twinning is diagnostic. Ideally the twins comprise three or six sectors; but actually the number of sectors varies (Fig. 53) and it may be complicated by lamellar twinning superimposed on the sectorial. As the twin planes are prism faces, vertical sections are similar to simple twins of feldspar, while the lamellar twinning may cause an uncomfortably close resemblance to plagioclase. However, cordierite is characteristically closely associated with spinel, often in swarms of small dark green octahedra; and yellow pleochroic haloes occur round minute inclusions of zircon and rarer accessories. Finally, the crystals may show any degree of alteration to an aggregate of white mica and chlorite, termed 'pinite'.

As regards mode of occurrence, cordierite is typically a product of thermal metamorphism of argillaceous sediments. It occurs, also, though much less commonly, in Basic and in Acid igneous rocks, in which it undoubtedly represents imperfectly digested xenoliths of argillaceous material. The cordierite-garnet-norites of Aberdeenshire provide good examples.[1] The Land's End granite contains

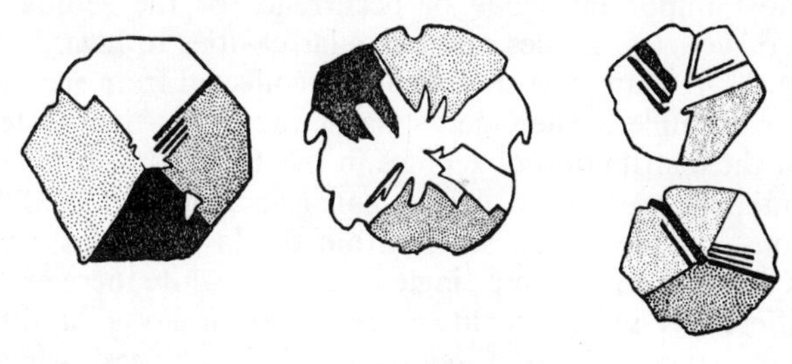

FIG. 53

Sections of cordierite between crossed polarizers, showing sectorial twinning. From cordierite-norite, Arnage, Banffshire.

euhedral greenish-grey pinitic pseudomorphs after cordierite, up to an inch or so in length. More rarely the mineral is found—again usually in xenoliths—in lavas such as basalt and andesite. Cordierite crystals or aggregates, often of large size, occur also in pegmatites in several localities. As the mineral is not usually present in the wall-rock, which in different instances is igneous (either Acid or Basic) or metamorphic, it is inferred that the cordierite has been formed from the pegmatitic liquid, *not* from caught-up (xenolithic) material.

THE ZEOLITES

The zeolites constitute a group of hydrated silicates of aluminium with Na^+, K^+, Ca^{2+} and more rarely Ba^{2+}, so that they are closely

[1] Read, H. H., The geology of Banff, Huntly and Turriff, *Mem. Geol. Surv.*, (1923), 128–37.

comparable in composition with the feldspars and feldspathoids. They are commonly formed from the latter by hydrothermal alteration; and conversely, on heating under conditions of thermal metamorphism zeolites alter very readily into feldspars of the appropriate composition. There are more named zeolites than there are species in any other mineral group so far dealt with; but in spite of their numerical superiority, with one exception they are of minor significance among the rock-forming minerals.

Their chief interest is centred in their crystallographic features; but unfortunately space does not allow these to be enlarged upon. Many zeolites are pseudosymmetrical: they appear, on superficial examination, to belong to one crystal system; but careful measurement shows them to belong to another, of lower symmetry. Several species form fibrous aggregates, each 'fibre' being a much-elongated prism. In hand specimens these look much alike; but they may be identified optically. Other zeolites are easily identified by their crystallographic features, as detailed in mineralogy textbooks.

The most important mode of occurrence of the zeolites is as infillings of vesicles, geodes and irregular cavities in lavas, notably basalts (p. 145). Many fine specimens are collected from such occurrences, for example at the Giant's Causeway in Antrim. Systematic studies of the distribution of zeolites in the Tertiary volcanic region of Antrim[1] has brought to light a zonal arrangement of different associations, on the largest scale within the lava field as a whole, and on a small scale within a single lava flow; while there are significant differences in the zeolite-assemblages in lavas of different petrographic types. These assemblages define 'zones' which are mappable and which were evidently dependent upon the temperature involved. Zeolitization is believed to have been caused by the circulation of water, largely of meteoric origin, and heated during a period of vulcanicity. Lower zeolite zones reached a higher temperature than higher zones, but pressure also was involved, for it is believed that the zeolites were deposited under a cover of 500 to 1000 feet of overlying lavas.

To emphasize the zonal relationship the zeolites may be described as they occur in these natural associations.

1. The **chabazite–thomsonite** association. The former zeolite occurs as aggregates of Trigonal crystals of rhombohedral habit, looking almost like cubes, while thomsonite is one of the fibrous zeolites and one of the most widely distributed. It is a 'soda-lime' zeolite and one which might be expected to result from hydrothermal alteration of plagioclase.

2. The **stilbite–heulandite** association. Although these two minerals may occur separately, they are usually seen in close association

[1] Walker, G. P. L., *Min. Mag.*, **32** (1960), 503.

in beautifully crystalline aggregates. Their most striking physical character is a pearly lustre resulting from a single cleavage direction, parallel to the side pinacoid. Stilbite characteristically forms composite crystals, pinched in at the middle giving a general resemblance to a wheatsheaf. Heulandite, on the other hand, usually forms single perfect Monoclinic crystals of a coffin shape. The stilbite–heulandite association occurs typically in the reddened top, highly vesicular part of olivine-basalt flows, the central more compact parts containing thomsonite instead.

3. The **natrolite–analcite** association. These two minerals of similar composition define a zone, though they are more or less mutually exclusive. Natrolite is a fibrous zeolite, essentially sodic and therefore closely similar to albite and nepheline. When natrolite occurs in sufficiently robust crystals, these look to be simple Tetragonal prisms capped by the unit bipyramid, but the crystals are actually Orthorhombic. The mineralogy of analcite, including its chemical composition and optical properties have been dealt with (p. 88). The important feature of analcite in its zeolite role is its crystallographic distinctiveness. Even in a finely crystalline aggregate lining a cavity in, say, basalt, the recognition of a single trapezoidal face of an icositetrahedron is sufficient to establish the identity of analcite.

THE EPIDOTE GROUP

This group of minerals comprises the Orthorhombic zoisite, and the Monoclinic clinozoisite, epidote (pistacite), piedmontite and orthite (allanite).

The composition of zoisite may be represented by the formula $(OH)Ca_2Al_3Si_3O_{12}$. If any iron replaces aluminium, the substance tends to crystallize in the Monoclinic System as clinozoisite. The latter is one end member of a continuous series, the other being the hypothetical 'iron-epidote' $(OH)Ca_2Fe_3Si_3O_{12}$. If the latter does not amount to more than 10 per cent of the whole, the mineral is clinozoisite; if between 10 and 40 per cent, the term epidote is applied. No natural epidote contains more than 40 per cent of the iron-bearing component. Two members of the group are notably rich in manganese: thulite is a manganiferous zoisite; while piedmontite is a manganiferous epidote with the formula $(OH)Ca_2(Al,Mn)_3Si_3O_{12}$. The variety withamite is a poorly manganiferous piedmontite. Finally allanite, sometimes called orthite, is essentially a variety of epidote containing the rare elements cerium and yttrium.

Clinozoisite–Epidote.—All Monoclinic members of the group occur in crystals elongated along the *b*-axis, and difficulty is experienced in sorting out the faces in the principal zone parallel to this direction.

The optical and other physical properties vary systematically from clinozoisite to ferriferous epidote. The minimum and maximum refractive indices and the birefringence are tabulated below:

	clinozoisite	ferriferous epidote
α	1·725	1·732
γ	1·740	1·781
$(\gamma-\alpha)$	0·015	0·049.

Thus the highest interference colours in a slide of normal thickness (30 μm) should be a pale third order yellow for iron-rich epidote. The value of the birefringence rises rapidly with increasing iron content. When the latter is low, the sections are colourless; but increasing iron causes a light yellow colour and slight pleochroism. A noteworthy feature of epidote and other minerals in the group is the noticeable variability in birefringence even within the limits of a small crystal. The birefringence of clinozoisite is notably anomalous, on account of strong dispersion: the interference colours are rich dark-blue or brown, like those of penninitic chlorite.

The plane of optical symmetry is perpendicular to the length of the crystal, therefore all sections lying with the principal crystallographic *b*-axis in, or parallel to, the plane of the slide, will exhibit straight extinction. Sections cut perpendicular to the length, however, show oblique extinction up to 30 degrees measured between the slow vibration direction (*Z*) and the traces of the (001) cleavage.

The optical orientation of epidote is such that one optic axis emerges at right angles (approximately) to the (100) face, on which a detrital grain or small crystal will normally lie; while the other optic axis is almost perpendicular to the basal cleavage. Therefore the distinctive optic axis figure appropriate to such a section is seen in convergent light both in crystals of normal habit and in cleavage flakes.

Allanite or **orthite** is also distinctively coloured in thin section, and strongly pleochroic, from buff to deep red-brown.[1] Twinning, usually simple, is common on the usual epidote plan. Identification of allanite is aided by its common association with another member of the epidote group—sometimes even in zonal association.

Mode of Occurrence and Origin

Excluding zoisite, all the chief members of the group are produced by, and characteristic of, dynamothermal metamorphism, in the course of which coloured silicates and plagioclase break down into

[1] Owing to the strong absorption, it is often difficult or impossible to check the character of the vibration, whether fast or slow, in the usual manner using a quartz wedge.

new stable associations, prominent among which is a member of the epidote group. Plagioclase may be represented by perfect pseudomorphs consisting of packed aggregates of zoisite prisms in slightly divergent groups; or the central parts of a large crystal may be composed of zoisite crystals or granular aggregates, embedded in, and surrounded by, albite.

Clinozoisite may be regarded as a high-density representative of anorthite. Obviously, unless there has been an actual influx of Fe ions, anorthite can be replaced only by zoisite or clinozoisite (typically the latter). Similarly, both common augite and hornblende contain the components of epidote, and the latter is produced, usually in association with chlorite, under the conditions that result in the replacement of anorthite. This type of replacement is often termed **epidotization.**

In addition to this secondary development, *primary* epidote occurs in some pegmatites, and less frequently even in granites and other normal igneous rocks.

In many other rocks occurring in areas which have *not* been subjected to regional metamorphism, epidote or its varieties may be common, having been formed by late-stage (deuteric) readjustments. The substance of this epidote has been derived from the mafic minerals, including hornblende and augite, both of which are often pseudomorphed by penninitic chlorite. The unwanted Ca^{2+} ions (together with the other necessary components) go to form epidote embedded in the chlorite.

THE CARBONATES

The rhombohedral carbonates with their chemical formulae and certain of their optical properties are tabulated below:

		γ	α	$\gamma - \alpha$
Calcite	$CaCO_3$	1·658	1·486	0·172
Magnesite	$MgCO_3$	1·700	1·509	0·191
Siderite	$FeCO_3$	1·875	1·633	0·242
Dolomite	$CaMg(CO_3)_2$	1·680	1·501	0·179
Ankerite	$Ca(Mg, Fe)(CO_3)_2$	—	—	—

Analyses show that there is a considerable degree of miscibility between the various end-members, and also with $MnCO_3$ (rhodochrosite).

In thin sections it is impracticable to differentiate between the several members of the group by ordinary methods, but it is relatively easy to do so using microchemical tests which are described in the appropriate textbooks on petrographic methods. As shown in the above table, the double refraction is outstandingly strong. Therefore in sections of normal thickness they polarize in high order colours and often appear dappled with very 'watery' pink and green. In calcite, one of the indices is below, the other well above that of Canada balsam, so that any section of the mineral (other than the basal one which is, of course, singly refracting), changes in appearance as it is rotated over the polarizer. In certain positions the refractive index of the light passing through it will equal that of the balsam, so that the calcite practically disappears. When turned through a right angle, however, the outline becomes bold and the surface relief strong. Thus the rapid appearance and disappearance of strong surface relief on rotation of the stage constitutes a unique test for calcite—the 'twinkling' test.

The carbonates are characteristic alteration products of igneous rocks, particularly the more Basic ones. In extreme cases little but the original texture remains: all the component minerals have been replaced by calcite, with small amounts of other secondary minerals. Normally calcite arises by weathering or hydrothermal alteration of calcium-rich silicates, notably the more basic plagioclases. In addition, calcite is a very common associate of zeolites, chlorite, chalcedony, etc., in vesicles and amygdules in lavas.

In two groups of rocks the presence of carbonates is of special significance. In some nepheline-syenites calcite occurs and has all the appearance of a primary mineral: it occurs in anhedral grains surrounded by and embedded in other primary minerals, which appear to be of later formation. This fact is significant in connection with the problem of the origin of this type of igneous rock, and is more fully considered in due course. Intrusive veins, dykes and more particularly plugs of crystalline carbonate rocks occurring in close association with nephelinic rocks forming ring-complexes are of special interest, and their origin is a major petrological problem. The intrusive carbonate rocks are termed *carbonatites*: they are described and discussed in a later chapter.

The only other igneous rocks which contain carbonates in significant amounts are certain lamprophyres in which calcite may be sufficiently abundant to cause effervescence on treatment with dilute acid. This calcite is of late-stage, deuteric origin, and frequently pseudomorphs the primary silicates.

THE SERPENTINES AND CHLORITES

Strictly speaking, from the point of view of their atomic structure, serpentines and chlorites belong to separate mineral groups. However, in view of their close chemical affinity and the fact that chlorites and serpentines frequently occur in intimate association in imperfectly and finely intergrown aggregates it is convenient to treat them collectively. Both groups are sheet silicates with atomic structures based fundamentally on layers of linked SiO_4 tetrahedra comparable with those of the micas. However, serpentines are devoid of aluminium which is an essential component of the structure of micas, and both serpentines and chlorites are devoid of potassium. Serpentines are essentially hydroxyl-bearing Mg-silicates with compositions approximately $Mg_6(Si_4O_{10})(OH)_8$. The structure[1] can be envisaged as made up of alternations of the layers of SiO_4 tetrahedra and layers of Mg and (OH) ions, which, because of their composition, are commonly called 'brucite layers'. The unit cell dimensions of the alternate layers, measured parallel to the layers, are not quite equal, and the structure is consequently slightly curved rather like the bending of a bi-metallic strip when two different coefficients of expansion are involved. Unless this distortion in the minute units of the structure were compensated in some way it would be impossible to build up larger crystals because of the over-all degree of strain involved. Nature, in fact, has adopted three ingenious ways of compensating the strain, leading to the formation of three polymorphs of serpentine composition. In **antigorite** the curvature, which is concave towards the brucite layers, is reversed in alternate parallel strips of the structure to give an effect like corrugated iron, if one could see it enormously magnified, but averaging out into flat sheets at normal magnification. Antigorite thus has a flaky or scaly crystal habit. It frequently occurs in intimate association with fibrous sepentine, **chrysotile,** in which the curvature of adjacent strips is not reversed so that the strips close round on themselves to form tube-like structures with axes parallel to the megascopic fibres. Finally, in **lizardite** the strain that would be involved in crystal growth is limited by the size of the 'crystals' which are of submicroscopic dimensions and give an amorphous effect in aggregate. The serpentine minerals are generally pale green in hand-specimens and thin sections, though they discolour very easily due to oxidation, and may be stained yellowish-green, red or brown, lending much to the ornamental value and attractiveness of serpentine-rock (serpentinite). Because of their generally minutely crystalline or even amorphous character, identification of serpentine minerals is really a matter for the expert. However, considered collectively they can often be recognized most easily from the evidence of olivines and pyroxenes (dominantly) which they are replacing. Details of this replacement and discussion of the stability of the serpentine minerals are given under the heading 'Serpentinites', p. 439. **Bastite** is the name given to sepentine pseudomorphs after orthopyroxene which are sometimes highly distinctive in shape and lustre.

[1] Deer, W. A., Howie, R. A. and Zussman, J., *Rock-forming Minerals*, vol. 3. (1962), p. 170.

Chlorites have double sheets of SiO_4 tetrahedra linked by Mg ions between the apical oxygens and hydroxyls as in the mica structure, shown in Fig. 23, but with brucite layers taking the place of the K ions. If Mg were the only metallic cations present, the composition of chlorite would be identical with that of antigorite. In natural chlorites, however, there is always some Al and Fe, the former substituting partly for Si and partly occupying normal Mg positions; and the latter substituting for Mg. Chlorites can be regarded therefore as formed of three end-members, 'antigorite' $Mg_6(Si_4O_{10})(OH)_8$; amesite, $(Mg_4Al_2)(Al_2Si_2O_{10})(OH)_8$; and daphnite, $(Fe_4Al_2)(Al_2Si_2O_{10})(OH)_8$. As might be expected, crystal form and cleavage of chlorite are comparable with those of mica, crystals being Monoclinic but pseudohexagonal in the (001) planes of cleavage.

Penninite is one of the commoner chlorites, being a widely distributed alteration product of micas, amphiboles and pyroxenes. Penninite is an iron-poor chlorite, relatively deficient in aluminium, and may be expressed as dominantly Ant_{60-80} Am_{40-20}. It is the chlorite nearest in composition to the serpentine antigorite. In thin section the mineral appears light green and slightly pleochroic. Frequently it is fibrous, and shows a radial to spherulitic structure. With indices $\alpha = 1{\cdot}575$ and $\gamma = 1{\cdot}576$, the birefringence is very weak—0·001. Coupled with strong dispersion, this gives an anomalous interference colour described as 'ultra blue'. It is a deep inky blue usually, but deep brown, or rich violet are also sometimes exhibited. As the acute bisectrix, *Z*, is perpendicular to (001), and therefore to the cleavage traces, all sections showing the cleavage traces will prove to be fast along the cleavages.

Clinochlore or **Clinochlorite.**—This chlorite has the same limited iron-content as penninite, but is richer in Al. The composition may be expressed as Ant_{60-40} Am_{40-60}, that is, about equal amounts of antigorite and amesite molecules. The general optical orientation is the same as penninite but clinochlorite may show slightly oblique extinction (2 to 9°) measured to the cleavage traces, hence the name of the mineral. The surface relief is much the same as for penninite, but the birefringence is slightly greater (0·004 to 0·011).

Iddingsite and **Bowlingite.**—In certain types of basaltic rocks the olivine phenocrysts are partially or completely converted into a strongly coloured yellowish or reddish-brown lamellar substance, known under one of the above names. Their exact composition seems uncertain, but they are easily altered into limonitic pseudomorphs.

Iddingsite has been proved by X-ray analysis to be a complex alteration product, not a single mineral. The only crystalline phase in iddingsite is goethite, a hydrated oxide of iron, the rest is amorphous. Bowlingite may well be similar: it is more strongly coloured than iddingsite, in rich red shades.

Chlorophaeite.—Under this name a strongly coloured chloritic mineral has been described from certain of the basaltic and doleritic rocks of Carboniferous age from the Midland Valley of Scotland. Its most distinctive feature is a rich green colour when fresh; but it is prone to very rapid oxidation, when it turns brown.

PART II

GEOLOGICAL SETTING AND MODES OF OCCURRENCE OF IGNEOUS ROCKS AND CONSOLIDATION OF MAGMA

CHAPTER I

THE GEOLOGICAL SETTING OF IGNEOUS ACTIVITY

Introduction

It is hard to realize that the concepts of sea-floor spreading and plate tectonics which have revolutionized geological thinking and provided for the first time a coherent picture of the structural evolution of the earth have evolved almost entirely since the last edition of this book was published. It would be inappropriate here to review the geological and geophysical evidence on which these concepts are based: in any event this is scarcely necessary because they have received wide publicity in books, review articles and on television.[1]

Out of the exciting picture of the structure of the earth which is now emerging we shall pick out two themes which are of major significance in relation to igneous petrology. The first concerns the formation of magma by partial melting of parts of the mantle (see Fig. 54); and the second concerns the factors which govern the distribution of igneous activity, confining it very largely to the vast and mostly submerged chains of volcanic mountains in the Atlantic, Indian and Pacific Oceans, or to the orogenic belts of the Americas and of the island arcs encircling the Pacific (Fig. 55).

Global Distribution of Igneous Rocks

Common volcanic rocks can be divided broadly into four major groups: basalts, andesites, trachytes and rhyolites. Of these, basalts contain the lowest percentages of silica (averaging a little below 50 per cent) and relatively high percentages of Fe, Mg and Ca; while rhyolites are the most siliceous volcanics, containing about 70 per cent of SiO_2, and are relatively rich in Na and K, but poor

[1] See for example: Bott, M. H. P,. *The Interior of the Earth*, (1971), essentially a geophysical account; Hart, P. J. (Ed.), *The Earth's Crust and Upper Mantle*, Amer. Geophys. Union Monogr. **13** (1969); Isacks, B., Oliver, J., and Sykes, L. R., Seismology and the new global tectonics. *J. Geophys*, *Res.*, **73** (1968), 5855–99; Bullard, E. C., Continental drift. *Quart. J. Geol. Soc.*, **120** (1964), 1–33; and review articles in *Scientific American*, including Wilson, G. T., Continental drift, No. 868 (1963); and Heirtzler, J. R., Sea-floor spreading, No. 875, **219**, (1968), 60–70.

in Fe, Mg and Ca. Andesites and trachytes are both intermediate between basalts and rhyolites; but while andesites are calcic and

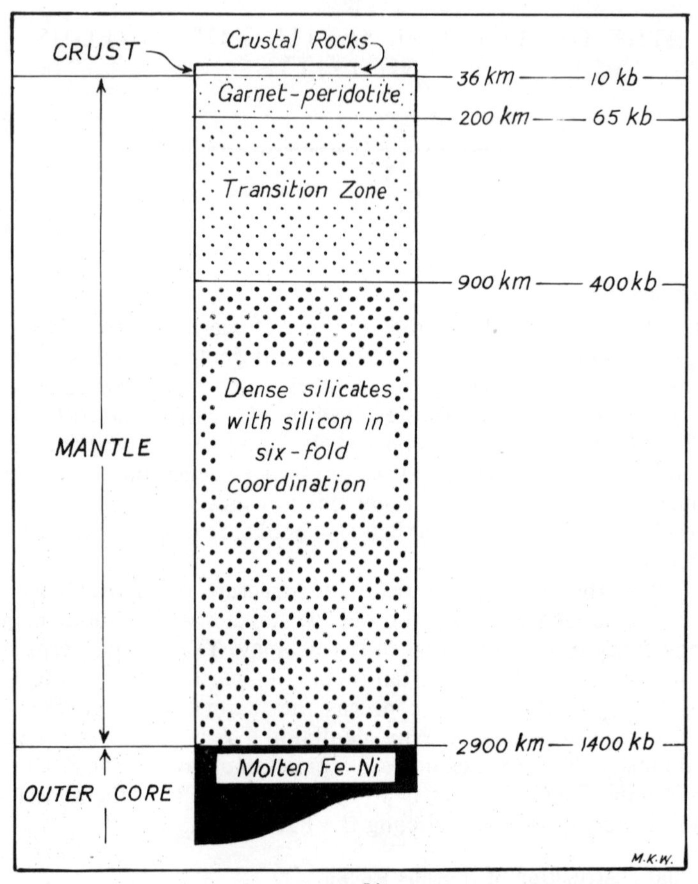

FIG. 54

Diagrammatic section of the Earth showing the inferred relationship between crust, mantle and outer core. (*After F. R. Boyd, Science*, **145** (1964).

contain an abundance of plagioclase, trachytes are alkali-rich and contain predominantly alkali-feldspars.

An excellent comprehensive review of the abundance and distribution of the various kinds of igneous rocks was provided by R. A. Daly in 1933 in his book *Igneous Rocks and the Depths of the Earth.*[1] He pointed out that, considering the earth as a whole,

[1] This is recommended reading for anyone who wishes to see modern developments in regional petrology and petrogenesis in historical perspective.

basaltic lavas predominate over all other types. They are overwhelmingly the most abundant volcanic rocks in the ocean basins in which, if one excludes island arcs, it is true to say that there are no andesites (in the sense in which these rocks are defined by most authorities), no rhyolites and only minor amounts of trachytes and related rocks. Basalts also occur abundantly on the continents, particularly in non-orogenic environments. Basaltic outpourings on the grandest scale occupy vast areas in Idaho, Oregon and Washington, in the Parana Basin of Brazil and in the Deccan in India where the lavas known as the Deccan Traps cover an area of over 500,000 km^2 and reach an aggregate thickness of more than 1000 m. Such occurrences of flood or plateau basalts have been built up by large numbers of flows, individually of no great thickness (2 to 50 m) but of great lateral extent, which have been erupted through fissures, and therefore fed by dykes. This form of extrusion, known as fissure eruption, has taken place within historical times in Iceland. The basalts erupted in this way are generally of very uniform composition, predominently tholeiites (see p. 362) of a relatively siliceous character.[1] Flows of a different composition—say rhyolite—occur only very occasionally. It is obvious from the dimensions of the flows and the marked absence of pyroclastic rocks of explosive volcanic origin that the basaltic magmas were very fluid at the time of eruption.

One can infer from the position of Iceland, athwart the Mid-Atlantic Ridge (Fig. 55) that the style of eruption of most submarine basalts is analogous to that of flood basalts on land. Recent volcanic activity is concentrated in a central rift-zone in Iceland[2] (witness the creation of the volcanic island of Surtsey in 1963[3]) and evidence from many sources shows that an analogous volcanically active rift-zone extends along the crest of the mid-oceanic ridges.[4]

The earth's crust beneath the oceans is of the order of 6 to 10 km thick. So far, knowledge of the crustal rocks occurring above the Mohorovicic discontinuity is limited to data concerning their densities (averaging about $2 \cdot 8$ g/cm^3) and earthquake *P*-wave velocities (increasing with depth from a little less than 5 km/sec to about $6 \cdot 8$ km/sec), augmented by sampling of rocks lying on the sea-bed or obtained by deep-sea drilling. The latter at present

[1] Kuno, H., *Plateau Basalts*. Amer. Geophys. Union Monogr., **13** (1969), 495–501.

[2] See for example Walker, G. P. L., Evidence of crustal drift from Icelandic geology. In *A Symposium on Continental Drift, Phil. Trans. Roy. Soc.*, **258** (1965), 199–204.

[3] This is impressively illustrated in *Surtsey, The New Island in the North Atlantic*, by S. Thorarinsson. (1969).

[4] See for example, Aumento, F., Longcarevic, B. D., and Ross, D. I., Hudson geotraverse geology of the Mid-Atlantic Ridge at 45°N. *Phil. Trans. Roy. Soc., Lond. A.*, **268** (1971), 623–50.

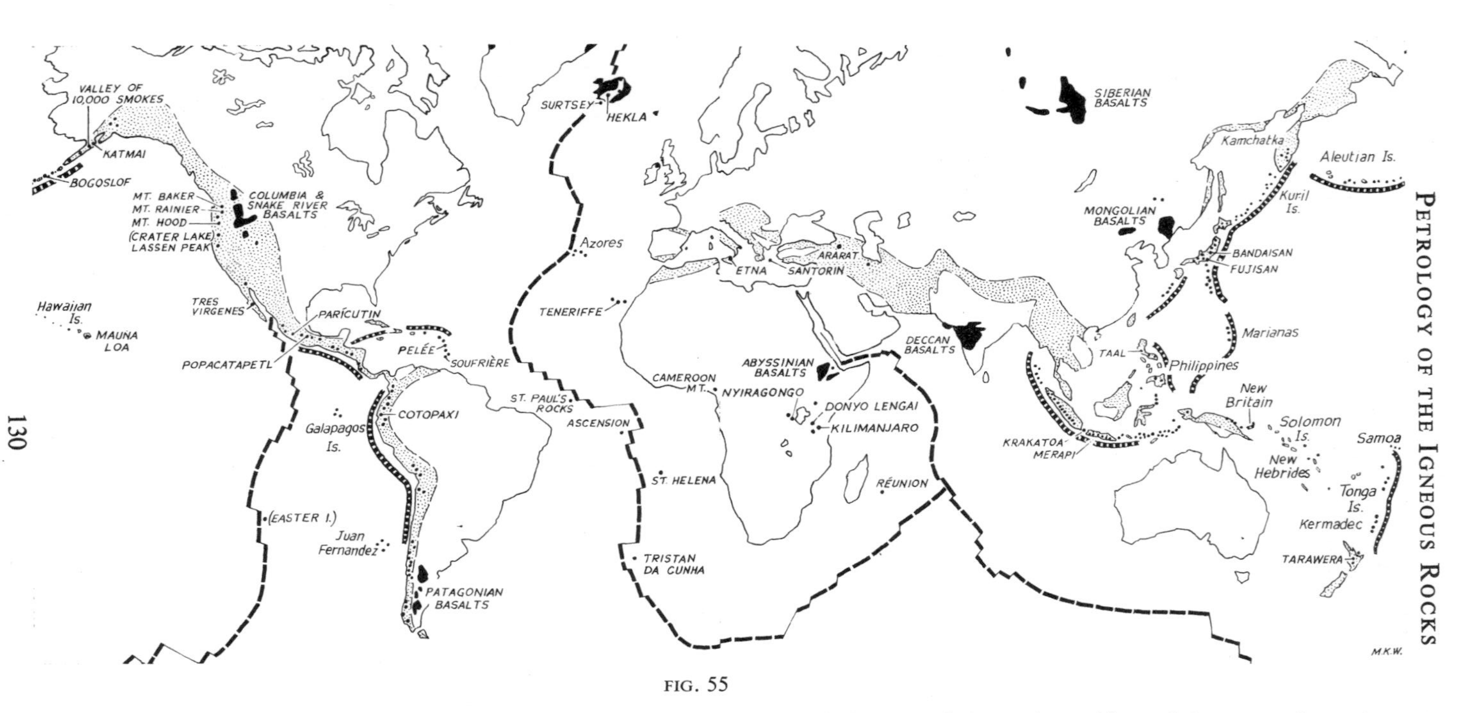

FIG. 55

Map showing the distribution of volcanic activity in relation to the major tectonic features of the earth: positions of the crests of oceanic ridges are indicated by a heavy line, broken to show displacements due to transverse faults; the relationship between oceanic trenches (*black-and-white ornament*) and the belts of Tertiary to Recent folding of the continents and island arcs (*stippled*) are indicated. Dots indicate the positions of only a selection of Recent of currently active volcanoes. The main areas of Tertiary of young flood basalts are shown in black.

is able to penetrate only a short distance; but fortunately Nature occasionally provides samples from deeper levels in the form of rock fragments (xenoliths) incorporated in the lavas erupted by oceanic volcanoes. All this evidence combines to show that the oceanic crust is composed broadly of rocks of basaltic composition: in the upper layers basaltic lavas predominate, giving place to intrusive rocks of similar composition at greater depths. Towards the base of the crust the rocks will be hot—though just how hot is a matter of debate—and will be in various states of metamorphism depending upon their geological history.

Concepts of sea-floor spreading (evolved from the work of a great many geophysicists and geologists, but associated especially with the names of Hess, Vine and Matthews) are in such excellent accord with many different lines of evidence that they may be accepted as providing a factual basis for the consideration of igneous activity in the ocean basins. In essence one can say that the oceanic crust is being formed at the present time by the addition of igneous rocks injected as dykes or other forms of intrusion, or erupted as submarine lavas, in or near the median zone of the oceanic ridges. This zone of relatively intense volcanic and seismic activity is under tension, as the oceanic crust on either side of the median line moves away at a rate of 1–6 cm per year. The crust on each side of the rift zone forms the uppermost part of a so-called 'plate' of crystallized rock, the lithosphere, which is of the order of 100 km in thickness.[1]

Although the rate of growth of the oceanic crust is imperceptibly slow, extended over a few hundred millions of years its cumulative effects are enormous. This is reflected in the fact that volcanoes which were active in the rift-zone in the distant past are now found scattered widely over the floors of the oceans as sea-mounts, and in general those which are situated farthest away from the present rift-zones are older than those nearer at hand. The most precise data concerning the extent and rate of sea-floor spreading are provided by a pattern of magnetic anomalies developed symmetrically on either side of the median line of an oceanic ridge (e.g. south of Iceland), and resulting from the influence of periodic reversals of the earth's magnetic field on the igneous rocks consolidating in the rift-zone.[2]

[1] The lithosphere comprises a limited thickness of crustal rocks (with seismic velocities less than 7 km/sec) above the M-discontinuity, and a much greater thickness of equally crystalline mantle rocks (seismic velocities greater than 7 km/sec) below the discontinuity.

[2] Details of the phenomenon are omitted for reasons of space; but the reader will experience no difficulty in finding descriptions in a wealth of articles and books written since 1963. See for example, Vine, F. G., Sea-floor spreading. In *Understanding the Earth*, ed. I. G. Gass, P. J. Smith and R. C. L. Wilson, (1971), 233–49.

Igneous activity in the orogenic belts and island arcs is governed by a tectonic regime which is in strong contrast with that of the ocean basins. Instead of new crust being formed in regions of tension as plates gradually separate from one another, in an orogenic belt the crust is being shortened and thickened as plates of the lithosphere converge on one another. Movement of the plates is not stopped by head-on collision; but instead, an oceanic plate on reaching a position marked by one of the deep trenches in the ocean floor (Fig. 56) is bent downwards and passes beneath the opposing plate.

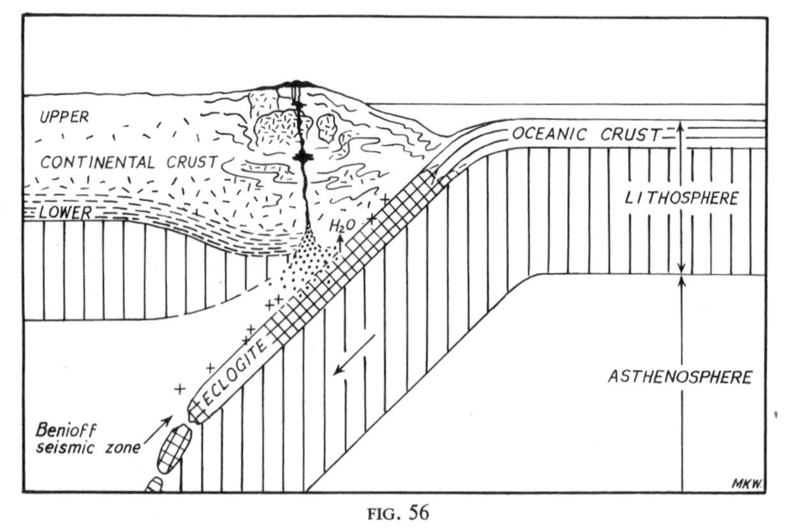

FIG. 56

Schematic section across an orogenic belt showing the bending of an oceanic plate passing beneath a continental plate. Probable sites for the generation of andesitic magma are stippled and earthquake foci suggested by crosses.

The two plates grind past one another on steeply inclined planes and in the process give rise to numerous earthquakes in the so-called **Benioff zone** of seismic activity.

This tectonic environment favours the formation of a higher proportion of relatively siliceous rocks than occur in the ocean basins so that, although basalts may be erupted under favourable conditions, they are generally very subordinate to andesites and rhyolites. As a result of the isostatic uplift and subsequent erosion, intrusive equivalents of the volcanic rocks are extensively displayed in many orogenic regions. Commonly the intrusions take the form of massive bodies of coarsely crystalline igneous rocks which constitute the so-called plutonic class of major intrusions described in the next chapter. Most of the intrusive rocks which occur in this manner contain considerable amounts of quartz and are often collectively referred to as 'granites', though in fact they cover wide ranges of composition similar to those of orogenic volcanic rocks.

The history of the evolution of a sector of continental crust in an orogenic belt may extend over a period of several tens of millions of years in the course of which it may develop an extremely complex structure and heterogeneous composition. On average the density of the igneous, metamorphic and sedimentary rocks which constitute the continental crust amounts to about 2·85 g/cm^3. Densities and velocities of earthquake waves increase downwards in the crust reaching values in the lower levels which compare with those of the 'basaltic' crust beneath the oceans, so that formerly it was customary to recognize a distinctive layer of more Basic composition, termed the **sima** by Daly, and lying beneath what was often referred to as the granitic layer, or **sial.** Actually very little is known, or indeed can be known, concerning the nature of the deep crust beneath the continents; but present knowledge suggests that the densities and elastic properties are dependent upon the high pressures and temperatures acting on rocks of various compositions occurring throughout the crust, rather than to the existence of distinctive layers of granitic and basaltic composition respectively. The continental crust has an average thickness of about 30 km, but may exceed 50 km in an orogenic belt. Deep erosion of the roots of an ancient mountain chain shows how varied the geological conditions may be in both space and time in such a thickened crustal segment. Some idea of this complexity is indicated in the discussion on the origin of granites, in which it is pointed out that the temperature in the deeper parts of the crust may rise to the point where some of the component minerals melt to form granitic magmas in an environment of very high-grade metamorphism. With the passage of time the geological conditions may change and the crust may become relatively cool and brittle so that magma generated below the crust may penetrate to the surface via dyke fissures similar to those which must be the general rule for the oceanic environment.

The study of granitic batholiths and other kinds of major plutonic intrusions shows that conditions favouring movement and uprise of the igneous rock-forming material, that is, magma in the broadest sense, may vary widely in the continental crust. Great masses of partially crystalline magma, amounting in some instances to volumes measurable in hundreds of cubic kilometres can rise in the crust only exceedingly slowly before consolidating as major intrusions. Space must be made for the intrusive material by a combination of different processes, which include *mechanical displacement* of the invaded rocks—commonly termed 'country rocks'; and *replacement* involving chemical reaction and interchange between the magma and its surroundings.

As might be expected, the very wide range of geological conditions which influence igneous activity in an orogenic region causes considerable diversity in the composition and character of the rocks produced.

Causes of Diversity of Composition among Igneous Rocks

If the compositions of all the igneous rocks in a given region of igneous activity are plotted on some form of variation diagram (Fig. 75) it is generally apparent that the rocks are chemically related to one another: apart from odd exceptions the compositions lie on or close to smooth variation curves and are not randomly scattered over the diagram. The chemical affinity shown by the rocks indicates that they have all been derived from a common source, *i.e.* from the same parent magma, or that they have originated by the progressive operation of some common process. The results of analysing large numbers of rocks from all over the world and comparing the results by the construction of variation diagrams reveals that the rocks of particular provinces belong in the main to one of two suites:

(1) the **Calc-Alkali** (or **Sub-Alkali**) **Suite** characteristic of orogenic regions and comprising a series of volcanic rocks: basalt, andesite, dacite and rhyolite, together with their intrusive equivalents; and

(2) the **Alkali Suite,** comprising the volcanic rocks: alkali olivine-basalt, hawaiite, mugearite and trachyte together with their intrusive equivalents. These rocks are particularly characteristic of the ocean basins and have hence been referred to as the 'oceanic suite'. In fact they occur also in continental, non-orogenic regions and are particularly well prepresented in areas of rift-valley tectonics, as in East Africa.

Evidence from many sources (the relationships shown by associated rocks in the field, petrographic details of individual rocks and experiments on crystallization of silicate melts) show very clearly that much of the diversity shown by igneous rocks is due to two processes which affect the compositions of magmas as they rise through the crust, namely, magmatic differentiation and syntexis.

Magmatic differentation is the term covering the production of a number of different magma fractions appearing high in the crust or erupted at the surface and derived from a single parent magma in depth. Our understanding of the processes involved owes much to the great pioneer experimental petrologist, N. L. Bowen,[1] who, as explained in Chapter 3, demonstrated the importance of fractional crystallization among the various groups of rock-forming silicates during cooling and crystallization of a magma. As the temperature falls chemical interchange (reaction) occurs between the minerals

[1] See *Evolution of the Igneous Rocks*, Princeton, (1928).

precipitated at high temperatures and the surrounding liquid. If early-formed crystals can be separated by any means from the magma in which they crystallized, the remaining liquid, depleted of the crystalline components, has a composition different from that of the original, or parent, magma. Commonly, the early-formed crystals are denser than the liquids in which they grew and, given favourable conditions, may sink towards the bottom of a magma reservoir. This is probably the most important means whereby crystals can be separated from a magma, and its effects can frequently be seen within the confines of an individual intrusion. One of the best-known illustrations of this process is the Palisades Sill, near the floor of which is a concentrated layer of olivine crystals which settled out of the magma so that the upper part of the sill became enriched in silica[1]. This is a simple illustration of differentiation involving the subtraction of only one mineral, olivine, and is limited in effectiveness by the size of the intrusion. It is reasonable to suppose, however, that in their passage through the crust magmas might undergo progressive differentiation involving fractionation of a wide range of minerals. According to Bowen's original thesis the whole range of chemical variation shown by the rocks in an igneous province may be due to differentiation of the parent magma of basaltic composition.

The second major factor affecting the composition of magma is **syntexis**—the general name for any kind of reaction between a rising body of magma and crustal rocks with which it comes into contact. This is in direct contrast with differentiation in that two parent sources are involved (magma and wall- or roof-rock) which combine to give a single magma of intermediate composition. Evidence of such interaction is particularly abundant for andesites and their intrusive equivalents, diorites.

Theories of igneous petrogenesis prior to the 1950s were based almost entirely on these two processes. The starting point of petrological thinking was a ready-made magma situated at an unknown depth in a 'magma-chamber' or 'magma reservoir'. The latter was an almost entirely mythical concept invoked by petrologists as a last resort when arguments based on all the observable evidence had been exhausted. Data necessary for controlled speculation on the origins of magmas and the real nature of magma reservoirs were just not available.

The Formation of Magmas by Partial Melting of Pre-existing Rocks

Progress towards the development of new ideas concerning the origin of magmas by partial or complete melting of pre-existing

[1] Walker, F., Differentiation of the Palisades diabase, New Jersey, *Geol. Soc. Amer. Bull.*, **51** (1940), 1059–106.

rocks was stimulated by research on the origin of granites. This fascinating and controversial subject is fully discussed in a later chapter: it must suffice for the moment to say that the field relations of granitic rocks exposed in deeply dissected terrains (*e.g.* Precambrian shield areas) combined with experimental work and geophysical data suggest that, in the presence of adequate water, granitic liquids may be formed at relatively low temperatures by melting rocks in the deeper levels of the continental crust. A considerable time elapsed before the full implications of the process, seen to operate in the case of granites, were explored for their applicability to more Basic magmas originating at greater depths. This may seem strange, for once geophysical data had shown that the crust and mantle were crystalline and contained no permanent layers of non-crystalline material that could act as a ready-made source of magma, it followed that the latter must originate by rock-melting. Logically, partial melting must be regarded as of fundamental significance in petrogenesis, and for this reason we have included a brief outline of some of the factors involved, particularly in relation to the origin of basaltic magmas (Chapter 8).

It is important to realize that igneous rocks which crystallize on or near the surface carry virtually no evidence relating to the origin of their parent magmas. The evidence which the rocks do provide relates almost entirely to the later events in their magmatic histories and these, of course, are our principal concern. The difficulty arises partly from the actual nature of the fractional melting process. When a rock is heated and its components begin to melt, the successive fractions of liquid produced by the melting may be indistinguishable in their chemical character from liquids derived by differentiation of an initial magma: only the sequence in which the liquids are produced is reversed. This is due to the fact that the relationships between crystals and liquids are governed by identical laws of physical chemistry, and the phase equilibria in operation are theoretically the same whether a rock is being melted or a magma is being cooled.[1]

It is impossible to judge, simply from the range of compositions of a suite of volcanic rocks, how much of the chemical variation is due to differentiation and how much to differing conditions of partial melting controlling the compositions of the parent magmas. Fortunately geological and geophysical experience is available to help in assessing the possibilities.

[1] Studies in the phase relationships in silicate melts are based almost entirely on the crystallization of melts: the reverse process presents serious practical difficulties. The reader may find it helpful to trace the course of fractional melting in one of the silicate systems described in Chapter 3, for example, the plagioclase system, by 'reading' the phase diagram in the reverse of the usual order, *i.e.* with rising temperature, starting with a crystalline solid.

The Mantle as a Source of Magmas

Much of the evidence concerning the composition of the mantle is necessarily indirect, combining the results of high pressure and high temperature experiments extrapolated from the known petrology of crustal rocks and geophysical data on densities, earthquake wave velocities, etc. Direct evidence is confined to occasional fragments of rock which, from their mineralogy, are seen to have originated at great depths before being carried to the surface as xenolithic inclusions (sometimes called 'nodules' on account of their rounded shape) and to intrusive bodies of peridotites of so-called 'Alpine type'. The problem with both the xenoliths and the peridotite intrusions is to sort out the rocks which are truly representative of the mantle from those produced in other ways. The xenoliths in a basalt, for example may have been derived from levels in the mantle where the magma originated, they may be of cognate origin, *i.e.* natural concentrates of early-formed, high temperature constituents, or in a continental environment they may be metamorphic rocks of crustal origin.[1] It is generally agreed that the mantle is composed largely of olivine-rich rocks which may contain varying proportions of pyroxenes and garnets. Petrographically such rocks

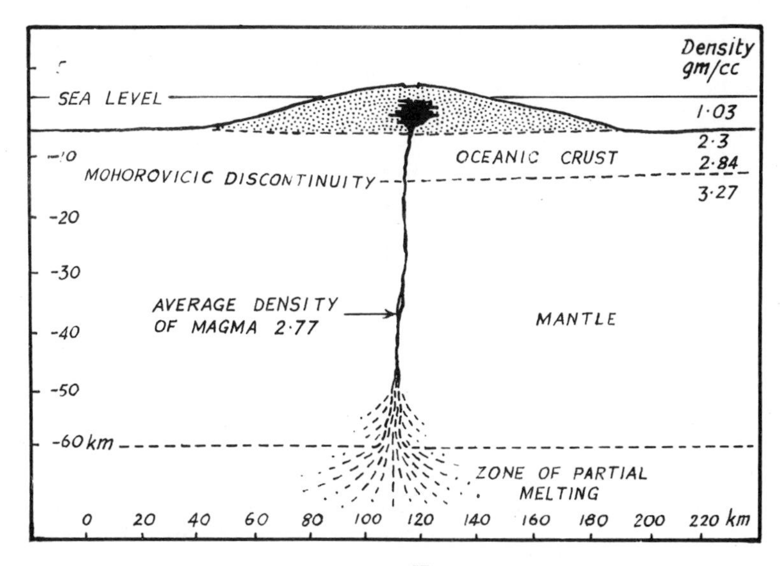

FIG. 57

Schematic section of an idealized Hawaiian volcano showing the depth at which magma segregates from the mantle, and the position of a high-level magma reservoir. (*After Eaton and Murata* (1960).)

[1] Davidson, C. F., The so-called cognate xenoliths of kimberlite, in *Ultramafic and Related Rocks* (1967), pp. 342–6. Ed. P. J. Wyllie.

are varieties of peridotites, described in Chapter 14.[1] Evidence from xenoliths suggests that the uppermost parts of the mantle beneath ancient continental crust may be appreciably different from that beneath the oceans, with eclogite extensively developed. Evidence relating to the depths at which partial melting occurs to form magmas in the mantle is largely seismological. Seismic wave velocities in the crust and uppermost parts of the mantle tend to increase with depth. However, at an ill-defined depth of the order of 50 km the

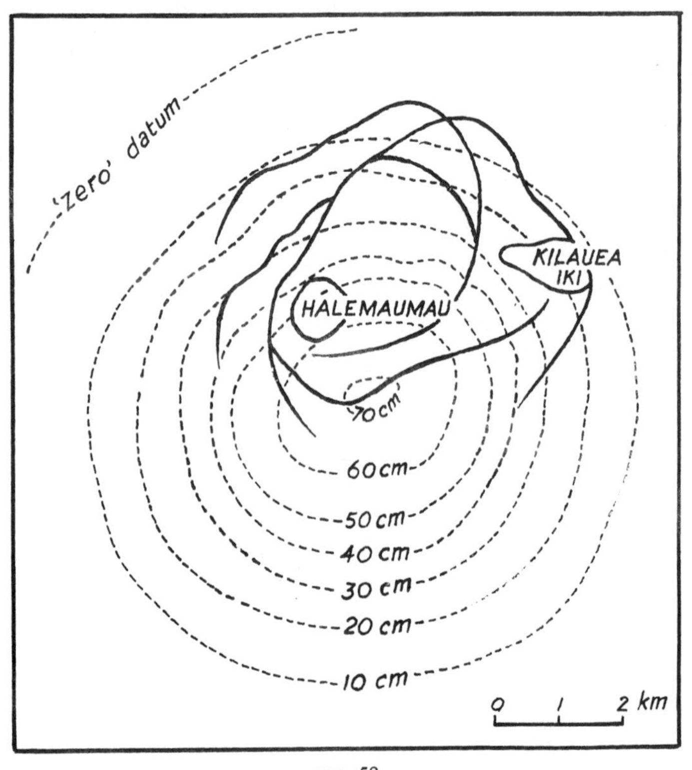

FIG. 58

Diagram showing the amount of uprise of the summit area of Kilauea for the period January 1966 to October 1967. Measurements are made against an assumed zero datum located on the adjoining Mauna Loa volcano. Contour interval, 10 cm. The heavy lines show the positions of the marginal structures (faults) of the complex Kilauea caldera, evolved during earlier episodes of volcanic activity. Current episodes of eruption are accompanied by the formation of a lava lake in the Halemaumau 'pit'. (*After Fiske and Kinoshita* (1969).)

[1] See Kuno, H., Mafic and ultramafic inclusions in basaltic rocks and the nature of the upper mantle, in *The Earth's Crust and Upper Mantle*, ed. P. J. Hart, Amer. Geophys. Union Monogr., **13** (1969), 507–13.

velocities are reduced somewhat. This reduction marks the top of a layer—the *asthenosphere*—some tens of kilometres thick, in which the rocks are capable of deformation by plastic flow. The reduction in seismic velocities and strength can both be explained if the rocks are nearer to their melting temperatures in this layer than is the case for rocks above or below it. The asthenosphere therefore provides conditions under which partial melting can most easily take place. The surest indication of actual melting and the formation of a body of magma capable of injection into the overlying crust is provided by seismic evidence of another kind. By detailed and continuous observation of seismic activity associated with Hawaii, it has been possible to detect the first stirrings of magma entering the roots of the volcano at a depth of about 60 km, and to trace its ascent over a period of several months until it finally erupts at the surface.[1] As it approaches to within a few kilometres of the surface the magma tends to spread out into a complex of minor intrusions, building up what amounts to a high-level magma-chamber, situated only 2 to 3 km below the summit of the volcano (Fig. 57). The expansion of this chamber causes actual inflation of the summit of the volcano as shown in Fig. 58. When eruption finally occurs the summit rapidly deflates, returning almost, but not quite, to its former shape.[2] High-level magma reservoirs of the kind indicated by the Hawaiian evidence are of great interest to geologists because they represent the formative stages of what may eventually be exposed as a complex of sub-volcanic intrusions comparable with some of the central complexes described in the last section of this book.

Having brought magma up from the depths of the mantle, as it were, we turn in the next chapter to consider its eruption at the surface or its consolidation in various intrusive forms.

[1] Eaton, J. P., and Murata, K. J., How volcanoes grow, *Science*, **132** (1960), 925–38.

[2] Fiske, R. S., and Kinoshita, W. I., Inflation of Kilauea Volcano prior to its 1967–8 eruption, *Science*, **165** (1969), 341–9.

CHAPTER II

MODES OF OCCURRENCE OF IGNEOUS ROCKS

Introduction

DESPITE the advances which have been made in the study of rocks in the laboratory, the proper solution to any problem in petrology involves a careful assessment of the rocks in the field. It is important to know the conditions under which a volcanic rock was erupted and how it is related in time and space to other products of volcanic activity in the area. In the case of an intrusive rock it is desirable to know the shape, size and mode of emplacement of the intrusion, and whether the latter is homogeneous in composition or varies because of internal differentiation or reaction between magma and country rock. In an extreme case it may require very detailed field mapping to establish whether a given rock crystallized from a magma or formed by the complete metasomatic transformation of a pre-existing rock.

A logical starting point for the study of the modes of occurrence of igneous rocks should be the active volcano, because it is only here that igneous rocks may be observed in the making. However, space considerations allow us to touch upon only one or two general aspects of volcanicity[1] which have a special bearing on the conditions under which volcanic rocks consolidate, and hence on their ultimate petrography. It seems more appropriate to concentrate rather on features associated with ancient volcanoes, since it is these that the reader is most likely to encounter in the first instance. Britain provides unique opportunities for the study of the exposed roots of ancient volcanoes provided by the ring-complexes of Tertiary age in West Scotland and Northern Ireland.

The Nature of Volcanic Activity

The character of volcanic eruptions is governed primarily by the composition of the magma involved, particularly by the amounts of SiO_2 and H_2O it contains. Viscosity of magmas tends to increase

[1] The subject lends itself particularly well to photographic illustration: see for example, *The Earth and its Satellite* ed. J. E. Guest (1971). An excellent introduction to the subject is provided by Chapter 12 in Arthur Holmes' *Principles of Physical Geology*, 2nd. edn. (1965).

with increasing silica-content. Basaltic magmas are generally erupted as fluid lava which spreads relatively easily over considerable distances. The extreme illustration of this is provided by flood-basalts, already referred to, in which the lava quietly wells out from a fissure to form what is virtually a lake of basalt with a flat upper surface.

Successive flows of basalt in eruptions of central type may build up a **shield volcano** like that of Mauna Loa (Fig. 57).

Andesitic magma, being more siliceous and hence more viscous, is often erupted with explosive violence, so that, in addition to lava, fragmental volcanic debris, loosely known as volcanic ash, may be ejected. Volcanic ash is one kind of the various pyroclastic rocks described in a later chapter. It collects round the source of eruption and, lying at its angle of rest, helps to build up the conical shape one normally associates with a volcano. Commonly eruptions of lava alternate with pyroclastic rocks in building such volcanoes.

The most siliceous magmas, of rhyolitic composition, are so viscous that they are incapable of forming lava-flows: they tend to consolidate in, or in the near neighbourhood of the vent; but the greater part of rhyolitic ejectamenta are pyroclastic and include ignimbrites or ash-flows described fully in a later chapter.

Volcanoes vary greatly in size and complexity. The smallest and simplest are normally produced during a single phase of eruption, fed through a pipe or volcanic neck of roughly circular cross-section. With the cessation of activity the cross-section may survive long after the erupted materials had been removed by denudation. The material filling the vent may be either a plug of lava or pyroclastic rock varying between an accumulation of large blocks—an agglomerate—or volcanic ash, or both.

Basaltic plugs are seen in front of the Campsie Fells faultscarp in the Midland Valley of Scotland and must originally have filled the vents of ash-cones. Trachytic plugs of the same nature form familiar landmarks in the same part of Scotland, and are exemplified by the Bass Rock and N. Berwick Law. The vent-intrusions are usually lithologically and texturally indistinguishable from surface lava-flows.

In this country parts of the Midland Valley of Scotland are noteworthy by reason of the large numbers of small volcanic vents which have been located, especially in Fifeshire and Ayrshire. In the former locality many of the vents are seen in cross-section on the coast. An outstanding example, occurring on the outskirts of Edinburgh, forms Arthur's Seat, a very prominent landmark over a wide stretch of country (Fig. 59).

The structure of giant volcanoes like Mauna Loa and Etna are vastly more complex than those we have just been considering. They are products of numerous eruptive episodes spread over hundreds of thousands of years. A characteristic feature of most volcanoes of this calibre is collapse of the summit region perhaps repeatedly,

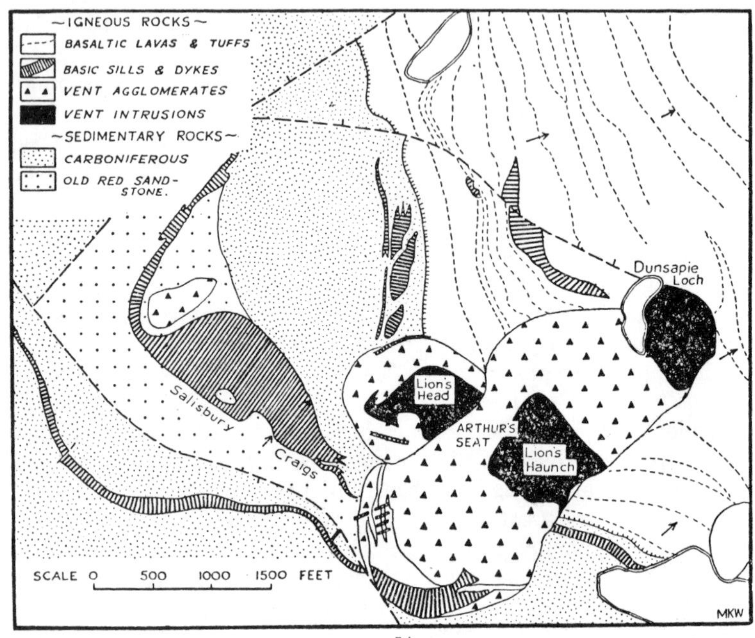

FIG. 59

Simplified map of the Carboniferous lava flows, intrusions and volcanic vents in the vicinity of Edinburgh. (*After H.M. Geol. Survey.*)

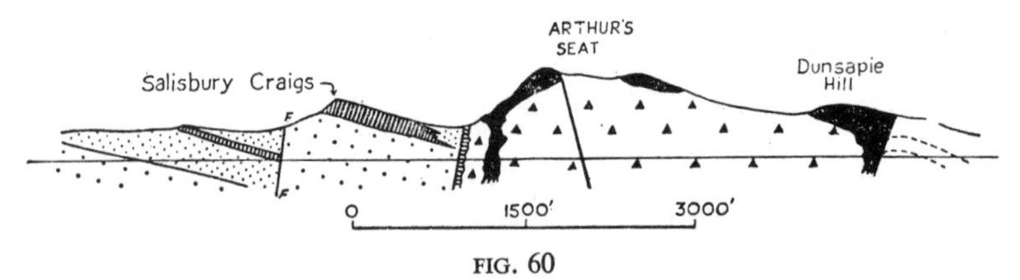

FIG. 60

Cross-section of the Arthur's Seat vents. The key to the ornament is given in Fig. 59.

to form a caldera, with a sunken block, often of the order of 5 to 10 km in diameter, bounded by an arcuate system of ring-faults similar to those depicted for Kilauea (Fig. 58). It is not difficult to correlate these in the mind's eye with ring complexes of the kind represented in Mull. It is fascinating to speculate on what kind of intrusive complex would be exposed by a few thousand metres of erosion on Kilauea: how would today's high-level magma chambers compare with tomorrow's intrusions?

We pass now to consider the kinds of structures observable in individual lava-flows, particularly those of basaltic composition.

The Structure of Lavas

From direct observation it is possible to distinguish a number of types of lava-flow. Under subaerial conditions the flow may resemble a tumbled mass of clinker or slag. This is distinguished as **block-lava**, and results when the volatiles in the magma are boiled off in the vent, before eruption. With more rapid uprise and less loss of volatiles the solidified lava shows on the surface contorted, snaky folds, suggestive of irregular viscous flow. This is **ropy lava.** A feature highly characteristic of submarine lavas of all ages is **pillow structure.** Lavas exhibiting this structure consist of isolated pillow-shaped masses piled one upon another, the intervening spaces being filled with sedimentary material, sometimes chert, sometimes limestone, sometimes hardened shale (Fig. 61). Internally the pillows are characterized by concentrically arranged vesicles, and occasionally there is a central ovoid cavity. Pillow structure is shown by basaltic and andesitic lavas, though flows of spilite are particularly

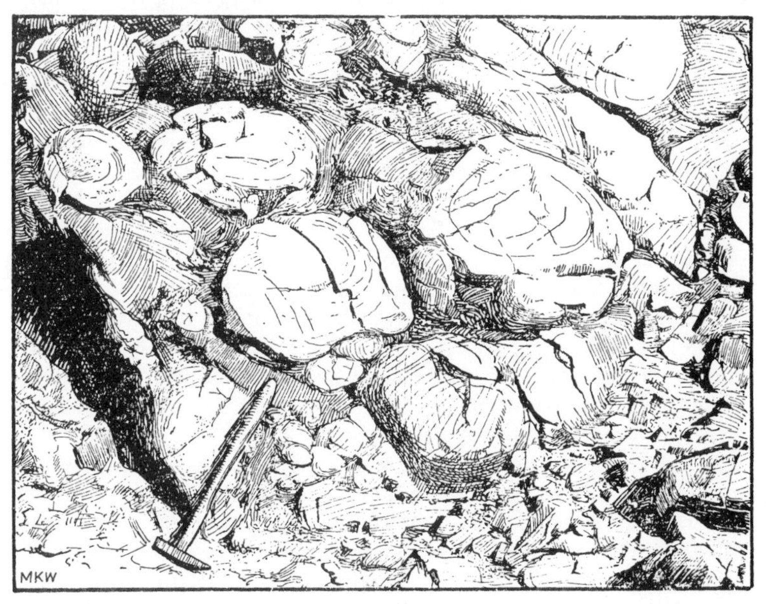

FIG. 61

Pillow-lavas (spilites) with jasper between some of the pillows, Anglesey. (*M. K. W. after photo by H.M. Geol. Survey.*)

noteworthy in this respect. Instead of parting at once with the steam incorporated in the lava, when erupted on the sea-floor, the pillows swelled up like a sponge and retained sufficient steam in the vesicles to drastically reduce the density. Moreover, such lavas on eruption seem to have been in the spheroidal state, and since each spheroid or pillow as it became detached was jacketed in steam, the lava flowing over the sea-floor formed a mobile sheet of rolling spheres, seldom touching one another until they had cooled down. Subaerial lavas cannot show typical pillow structure and the latter may be regarded as a safe criterion of submarine eruption, although a Tertiary basaltic pillow-lava in Mull which resulted from eruption into a crater lake suggests wider possibilities. Beautifully fresh tholeiitic pillow-lavas have been erupted, presumably from fissures associated with the rift-valley in the Mid-Atlantic Ridge. The margins of some of the pillows were chilled to glass (tachylyte) though others are crystalline showing the quench texture illustrated in Fig. 119. Under-water photography is providing striking illustrations of these phenomena. In Britain the best examples of pillow-lavas are of Pre-cambrian and Ordovician age.

As a direct consequence of the conditions of outpouring, certain lavas are distinguished by marked flow- (or fluxion-) structure, due to the rolling over and pulling out of the semi-solid mass. This may result in very distinctive internal structures described in a later chapter. Certain lavas are characterized by a cellular or scoriaceous structure due to the elimination of water vapour and other gases during eruption.

Basaltic magma normally has a relatively low viscosity so that upon eruption, when pressure on the magma is reduced, the dissolved gases are largely able to escape into the air. They are in effect boiled out of the lava. However, the loss of volatile substances (which act as fluxes in their dissolved state) causes a rapid increase in viscosity. Many of the gas bubbles thus become trapped by the congealing lava, particularly towards the top of a flow. The gas cavities or **vesicles** are spherical or ellipsoidal, but may be flattened and elongated by flow movements of the lava. Diameters commonly range from a few millimetres to one or more centimetres, though larger, less regular cavities also occur.

In all recently erupted lavas the vesicles are gas-filled cavities; but in older basalts they are commonly filled-in by low-temperature minerals such as calcite, chalcedony and zeolites. Aggregates formed by these light-coloured minerals are frequently of the approximate shapes and sizes of almonds so that they are called amygdales.[1]

Amygdaloidal basalts are frequently conspicuous among the early

[1] Latin *amygdalus*—almond.

members of a lava succession, and it is probable that the amygdale minerals are precipitated from circulating ground-water which becomes heated volcanically under an insulating blanket formed by overlying flows. Statistical studies of the distribution of the vesicle-infilling minerals have demonstrated a zonal distribution within a particular volcanic area, presumably reflecting temperature control. The zones are characterized by different mineral assemblages; while, as might be expected, different types of lava give rise to distinctive assemblages of vesicle minerals.[1] In Britain fine mineral specimens, particularly of zeolitic assemblages, are obtainable from the lavas of Devonian age in Scotland (Old Dumbarton), and of Tertiary age at the Giant's Causeway, County Antrim.

FIG. 62

Amygdaloidal basalt, Butaure, Tyrol.

The amygdales are less regularly shaped than usual and were infilled with various minerals deposited in zonal sequence. In the large, central amygdale calcite forms the outer zone, followed inwards by chlorite and chalcedony showing agate structure.

Pipe amygdales have been described from lavas in many parts of the world. They commonly occur at the base of the flow, and have the form of long narrow cones, tapering upwards, up to a foot in length, and perhaps half an inch in diameter. They are filled with the same kinds of minerals as occur in the more normal ovoid amygdales. They appear to have resulted from the uprise of steam from the moist surface over which the lava flowed, for they commonly occur where lava was erupted over wet mud.[2]

[1] Walker, G. P. L., The amygdale minerals in the Tertiary lavas of Ireland, *Min. Mag.*, **32** (1960), 503.

[2] Du Toit, A. L., Pipe amygdaloids, *Geol. Mag.*, **4** (1907), 13; Bailey, E. B., Geology of Knapdale etc., *Mem. Geol. Surv.*, (1911), 69.

As a consequence of contraction due to cooling many igneous rocks come to be traversed by regularly arranged systems of intersecting joints. In its most perfect development—in certain Basic lavas and minor intrusions—the resulting structure is aptly termed **columnar structure.** The columns, which are very often long and regular, are bounded by three, four, five or six planes, producing triangular, quadrangular, pentagonal, and hexagonal prisms. Where the rock-texture is homogeneous the six-sided prisms are most prevalent, for of all the cases in which the centres of contraction are equidistant, and the angles of the prisms fit together without any intervening space, the hexagonal arrangement gives the highest ratio of area to periphery. The long axes of the columns are perpendicular to the retreating isotherms during cooling. In accordance with this law, the columns are vertical in horizontal sills and flows; while in dykes they are horizontal if the walls are vertical. In many lavas, particularly those of Basic composition, three roughly parallel layers may be distinguished: an upper slaggy and vesicular portion; a central zone with somewhat irregular columns of small cross-section, and a basal layer, with more massive, regular, hexagonal columns. These latter have grown upwards from the slowly cooling base while the thin 'wavy' columns grew downwards from the more rapidly cooling upper surface. The columns are divided into segments by cross-jointing, and usually such segmentation is accompanied by a spheroidal tendency, producing *ball-and-socket joints.* Spectacular examples are seen at the Giant's Causeway in Antrim, by far the finest columnar basalt in Britain[1] and at Fingal's Cave in the Hebrides.

Intrusions: General Considerations

Intrusions vary widely in their relationship to the country-rock, in size and in shape, depending on the composition of the intrusive magma and on the geological and structural setting. They have been classified very broadly according to size, into 'major' and 'minor' intrusions, and by reference to the depth at which they occur into so-called plutonic (abyssal) and hypabyssal. None of these distinctions is entirely valid: indeed, the fact of categorizing them in this way has hindered progress in thinking. Perhaps the most valid distinction that can be made is between the kinds of intrusive rock-bodies typically found in an orogenic environment, emplaced under conditions generally of crustal compression; and those intruded under anorogenic conditions of crustal tension. In the former case, as we have already seen, massive bodies of crystallizing magma intrude into higher levels of the crust, sometimes forcefully pushing

[1] Tomkeieff, S. I., The basaltic lavas of the Giant's Causeway of Northern Ireland, *Bull. Volc., Naples,* series 2, **6** (1940), 89.

aside the rocks lying in their path. The characteristic form of intrusion in this environment may be regarded as the batholith. By contrast, in the latter environment crustal conditions allow relatively easy access to fluid magma giving rise to regional dyke swarms on the largest scale. Typically intrusions in this environment have parallel-sided, sheet-like forms. In some cases it appears as if localized tensions and faulting of the crustal rocks had prepared a space for the magma to flow into, giving rise to what has been referred to as a 'permitted' intrusion (H. H. Read), in contrast to a 'forceful' one. It must be emphasized, however, that these are broad generalizations: the manner of intrusion in any particular case is governed by the *local* structure and stress conditions in the country rock, and by the pressure and other physical conditions obtaining in the intruding magma. So we often find dykes and other forms of minor intrusions in an orogenic environment, and conversely 'plutonic' characteristics may be adopted by intrusions in a non-orogenic, volcanic environment.

In the following pages the different kinds of intrusions are grouped together for purposes of description primarily on the basis of their shapes and mechanism of emplacement.

SHEET INTRUSIONS

(*a*) **Dykes.**—In this category we include all intrusions of sheet-like form which are vertical or nearly so, at the time of intrusion. Dykes vary in thickness from a fraction of an inch to hundreds of feet, but the average width is probably between one and three feet. As a consequence of their attitude the outcrops of dykes are not affected by the topography of the country in which they occur: they therefore appear as nearly straight lines on geological maps, maintaining a uniform direction, sometimes for long distances. They can often be picked out in the field because they weather differently from the rocks they penetrate: in easily eroded sediments dykes may project like walls; or if occurring in more resistant rocks they may form ditch-like depressions. In the latter case their preferential erosion may well be due to cross jointing which most dykes display, the joints being perpendicular to the chilled marginal surfaces.

Although in any igneous environment where magma is able to penetrate a local crack or fissure in the country rock a small dyke will result, it is hardly appropriate to refer to the very large members of a regional dyke swarm as 'minor intrusions'. The Cleveland dyke, for instance, a member of the Mull dyke swarm extends right across the North of England for a distance of 130 miles. The swarms of Tertiary dykes shown in Fig. 151 have been studied in greater detail than those of any other part of the World and in some coastal sections the dykes are so well exposed that quantitative data have been obtained. In one such section 115 dykes belonging to the Mull swarm occur in just over a mile; and this amounts to a 9 per cent extension of the crust.[1]

One can infer from the geometry of the swarm that the dykes were fed from a ridge-like magma body located where the general level of melting in the mantle rose closest to the surface, immediately beneath the axis of the swarm. Radiometric dating of the dykes shows that their emplacement extended over a long period: some of the earliest acted as feeders for the plateau lavas, while the latest post-date the central intrusion complexes of the region, as explained in our final chapter. Around intrusive/volcanic centres the pattern of dyke intrusion may be much more varied; and the pattern may indicate that the rising magma body exerted an outwardly directed pressure on the surrounding rocks, leading to the injection of radially disposed dykes as clearly demonstrated in the Island of Rhum (Fig. 156). Various kinds of intrusions with a concentric pattern of distribution are described later but it may be noted at this point that a cone-sheet is a kind of dyke—albeit inclined rather than vertical, the attitude of which is determined by the geometry of the magma body.

Successive injections of the same type of magma into the same dyke-fissure form **multiple dykes.** On occasion the same fissure has been followed by two or more injections of contrasted types, giving **composite dykes.** The most general case is that in which an initial injection of basaltic magma appears to have been followed by a later injection of granitic material: the former is typically black and finely crystalline in the hand specimen, while the latter is often much lighter coloured, often red, and may carry relatively large crystals of quartz and feldspar. Examples of both multiple and composite dykes of Tertiary age occur in Arran[2] and Skye[3] and are well displayed also in the north-easterly dyke-swarm of southern Jersey, Channel Islands.

Most petrologists agree that dykes are injected during periods of stress or tension in the earth's crust. The first stage in the emplacement of a dyke swarm must obviously be the development, in the area affected, of a series of parallel fractures, which admittedly might well be produced by shearing; but whether or not, the fractures must open to allow ingress of the basaltic material, and this is most readily visualized under tensional conditions operating at right

[1] Sloane, T., Speight, G. M., and Skelhorn, R., The structure of the Tertiary dyke swarms of Skye, Mull and Ardnamurchan, *Proc. Geol. Soc., Lond.* **1658** (1969), 199–202.

[2] Gregory, J. W., and Tyrrell, G. W., *Proc. Geol. Assoc.*, **35** (1924), 413 and Pl. 26.

[3] Harker, A., *Mem. Geol. Surv.* (1904), 201–7.

angles to the given fissure. In the aggregate the 'stretching' of the original belt of country is equal to the total thickness of all the dykes involved, and in some cases amounts to a most impressive figure.

(*b*) **Sills.**—The three-dimensional form of a sili is the same as that of a dyke; but the attitude is different. Sills are formed when the structure and stresses in the crust make it easier for the magma to spread laterally rather than vertically. This situation arises most frequently when ascending magma enters a succession of stratified rocks near the earth's surface. The magmatic pressure at a certain level may exceed the load-pressure exerted by the overlying rocks, so that it becomes easier for the magma to spread out laterally

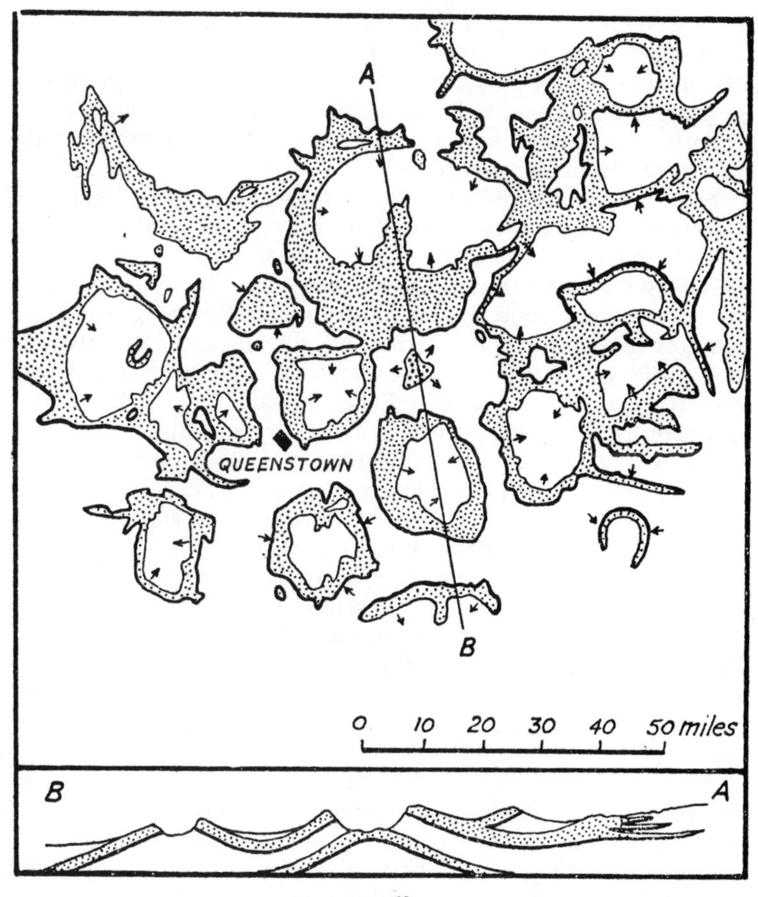

FIG. 63

Sketch map and section (vertical scale exaggerated) of thick Karroo dolerite sills intruded into sediments of the Beaufort Series, South Africa.

rather than force its way upwards towards the surface. Sills are essentially *concordant* with the stratification of the country-rock, and therefore occur as sensibly horizontal sheets in untilted strata, though subsequent tilting or folding may throw them into any attitude—including the vertical. It thus behaves at outcrop as if it were a part of the stratigraphical succession, and difficulty may be experienced in deciding whether a given sheet is sill or lava-flow. In some cases the mass-characters of the igneous rock may be sufficiently distinctive (see p. 143); otherwise the following observations may be helpful. Although sills may keep to one horizon for considerable distances, they do sometimes transgress suddenly to a higher or a lower horizon, in a manner which would be impossible for a lava-flow. Similarly vein-like offshoots, particularly into the rock *above* the igneous sheet, would indicate that the latter is a sill rather than a flow, though sometimes sills have been injected into mud on the sea-floor, and may simulate lava-flows very closely as regards such features as vesicularity and contact alteration of the adjoining sedimentary materials.

Some sills are massive enough to display considerable lithological variation in two respects: (1) in grain-size of the component rocks; and (2) in composition due to differentiation either *in situ* or prior to intrusion, in which case the sill would be composite. It would be very easy to quote specific illustrations; but the general case requires little explanation. A massive sill intruded into relatively cold rocks will be completely enveloped in a chilled facies of fine-grained rock. In the case of a Basic sill, this would be basalt, indistinguishable lithologically from a surface flow. Inwards the grain would progressively coarsen and the basalt would grade into dolerite, possibly even into gabbro. Diversity would be increased however, by two processes: firstly, the upward streaming of volatiles, chiefly water, and these would be trapped some distance below the upper surface, and here the water-enriched fraction would crystallize to form a pegmatitic phase of particularly coarse grain. At the same time there would be movement in the opposite direction, of early-formed olivine crystals gravitating downwards to accumulate as an olivine-enriched layer in the lower part (though not at the actual base) of the sill. The Palisades Sill, as we have already seen, is a case in point.

As with dykes, sills may be multiple and, in other instances, composite. Again they consist dominantly of consolidated basaltic magma, and sills occur extensively underlying the areas of vast basaltic outpourings, noted above. They are also noteworthy in the Karroo in South Africa, not only on account of their dominance of the characteristic scenery, but also of the extraordinary extent of some of the individual sills: some extend over areas of from 3000 to 5000 square miles. The dissected outcrop of one of these giant sheets is shown in Fig. 63. It should be noted that the remarkable undulations, causing basin- and dome-like structures are not characteristic of sills in general.

In England, the most extensive sheet is the Great Whin Sill, which underlies much of north-eastern England, and is well exposed in Teesdale where its outcrop forms the falls known as High Force. The most impressive sill known to the writers forms a thick, quasi-horizontal sheet of almost black dolerite intruded into bright red granite which forms spectacular cliffs in the north of Jersey, Channel Islands. Other better known sills include those of Tertiary age, consisting of 'quartz-porphyry' (a granitic rock) and outcropping on the coast of southern Arran (the Dippin Sills) in western Scotland.

LENS-SHAPED INTRUSIONS

(a) **Laccoliths.**—These result when the intrusion of viscous magma along a plane of weakness has caused an uparching of the strata above the intrusive mass. Ideally the floor remains flat; and in the classic description of intrusions of this type, from the Henry Mountains, Utah,[1] it was suggested that the laccolith was fed by means of a central pipe. This is, however, purely hypothetical, and it is just as likely that the intrusion was fed in the same manner as a sill. The hydrostatic pressure of the magma is believed to have caused the uparching of the roof. Fundamentally a laccolith has the shape of a plano-convex lens, lying flat-side down.

In sills the lateral dimensions are very great compared with the thickness; but in the laccolithic type of intrusion, the latter varies between one-third and one-seventh of the diameter. Laccoliths may be multiple, as in the case of the so-called cedar-tree laccoliths.

(*b*) **Phacoliths.**—This name is applied to lens-shaped intrusions of concavo-complex shape situated in the core of an anticline. The phacolith owes its shape to intrusion into pre-existing folded strata, while the laccolith, in theory, *causes* the arching of the roof. The type phacolith forms the Corndon, a gabbroic or doleritic intrusion occurring in the Shelve district of Shropshire.[2]

(*c*) **Lopoliths.**—Intrusions of this type are the converse of phacoliths, as they have the concavo-convex shape, but are convex downwards, that is, the central part of the intrusion has been warped downwards. Some of the greatest Basic intrusions in the world are of this type. The Sudbury (Ontario) lopolith and the Duluth (Minnesota) gabbroic complex are outstanding examples.[3] The term

[1] Gilbert, G. K., *U.S. Geol. Surv., Washington* (1877).

[2] Harker, A., *Natural History of Igneous Rocks* (1909), 77; and Blyth, F. H. G., *Quart. J. Geol. Soc.*, **99** (1943), 169.

[3] Grout, F. F., *Amer. J. Sci.*, **46** (1918), 516.

lopolith was first applied to the Duluth intrusion which forms a crescent-shaped outcrop at Duluth, bounding the western tip of Lake Superior. The crescentic outcrop resulted from tilting of the lopolith; but no such tilting has affected the Bushveld Complex, which is the largest Basic-Ultrabasic intrusion in the world.[1] It extends east-to-west for 300 miles and occupies a surface area of 20,000 square miles within which an extraordinary variety of Basic and Ultrabasic rock-types are magnificently displayed (Fig. 64). The

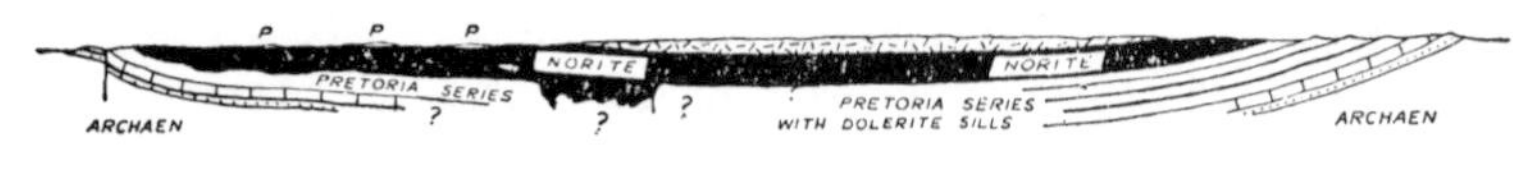

FIG. 64

Diagrammatic E–W section across the Bushveld complex, Transvaal. A red 'granite' overlies the Basic rocks (chiefly norite). Various members of the Transvaal System, into which the lopolith is intruded, are indicated as follows: *stippled*, Black Reef Quartzite; *cross-ruled*, Great Dolomite; outcrops of Pretoria Series above intrusion, P P P. (*After Daly, simplified*)

Complex is strikingly layered, and differential weathering has etched out the structure so that the harder rocks form strike-ridges approximately parallel to the boundary of the lopolith and also to the strike features formed by massive quartzites which dip beneath the floor of the intrusion. The thickness of the lopolith may be as much as 6 miles. The total volume of magma which must have risen from the mantle up to a high level in the crust is therefore enormous; and it is not surprising that the isostatic balance of the crust was disturbed, allowing the central part of the lopolith to sag downwards. How the lopolith was fed is unknown.

Some major intrusions of layered gabbros do, however, possess a funnel-like shape, both as regards the attitude of layering and the form of the intrusion. This applies, for example, to the Freetown Complex in Sierra Leone.[2] The dip of the layered structures becomes increasingly steep towards the centre of the intrusion, and it may be inferred from geophysical evidence and borehole data that the floor is similarly inclined. It may be noted that towards the outer margin of the Complex where the layers are more gently inclined, they form scarp features comparable with those of the Bushveld Complex. Although the Freetown Complex shares certain features

[1] Hall, A. L., *Mem. Geol. Surv., South Africa,* **28** (1932). Petrological accounts of the Bushveld, Stillwater (Montana) and other lopolithic intrusions are contained in *Layered Igneous Rocks* by L. R. Wager and G. M. Brown (1968).

[2] Wells, M. K., and Baker, C. O., The anorthosites in the Colony Complex near Freetown, Sierra Leone, *Col. Geol. Min. Res. Bull.*, **6**, no. 2 (1956), 137–58; Wells, M. K., Structure and petrology of the Freetown layered Basic Complex in Sierra Leone, *Overseas Geol. Min. Res.* (*London*), *Bull. Suppl.*, **4** (1962), 1–115.

with the great Basic lopoliths, it seems best to regard it as a funnel intrusion.

Funnel intrusions[1] are generally much smaller than the Freetown example, having circular outcrops a few miles in diameter. The type example is provided by the Cortlandt or Peekskill Complex of New York, in which there are three foci towards which the layered structures converge. Other intrusions of the same general type have been noted in New Hampshire, and for these it has been suggested that the magma was forcefully emplaced, making room for itself by pushing aside and doming the country rock. Some of the Tertiary gabbroic intrusions in Scotland may have been emplaced in a similar manner.

The most spectacular and the most completely studied intrusion known to have a funnel shape is that of Skaergaard, Eastern Greenland.[2] The greater part of the complex is occupied by a layered series of gabbros, varying considerably in composition from below upwards. The significance of the variation in composition and of the layering displayed is discussed in due course. It is probable that space was made for the intrusion initially by volcanic activity of the explosive type. The diameter of the complex is comparable with that of many volcanic calderas.

So-called funnel intrusions thus vary widely in character: at one extreme they may be scarcely distinguishable from lopoliths of sheet-like form; while at the other, steep and discordant contacts suggest affinity with ring complexes described below.

RING COMPLEXES

Just as lavas can be referred to two main types of extrusion, the one of a widespread character (plateau lavas), and the other of a central and localized type (cone or central volcanoes), so with intrusive complexes. The dyke swarms and widespread sills are of the first kind and stand in strong contrast to the central type complexes, in which the full force of igneous activity was brought to a focus over a small area. The classical area for studying this type of complex is in the Inner Hebrides of Scotland, the chief individual centres occurring in Mull, Skye and Ardnamurchan. From the example in Mull, it is inferred that the intrusive complex represents the basal wreck of a large volcano of central type.

The crustal forces operating in central complexes result in the formation of characteristic crescentic intrusions. Only their essential characters are considered here, since examples are described in greater detail in the last section of this book.

[1] Wager, L. R., and Brown, G. M., Funnel-shaped intrusions, *Bull. Geol. Soc., Amer.*, **68** (1957), 1071–5.
[2] Wager, L. R., and Deer, W. A., *Med. om Grønland*, **105,** no. 4 (1939).

(*a*) **Ring-dykes.**[1]—This term covers intrusions which have arcuate outcrops, the radius of which seldom exceeds two or three miles. The inner and outer walls of the intrusion may be parallel at ground

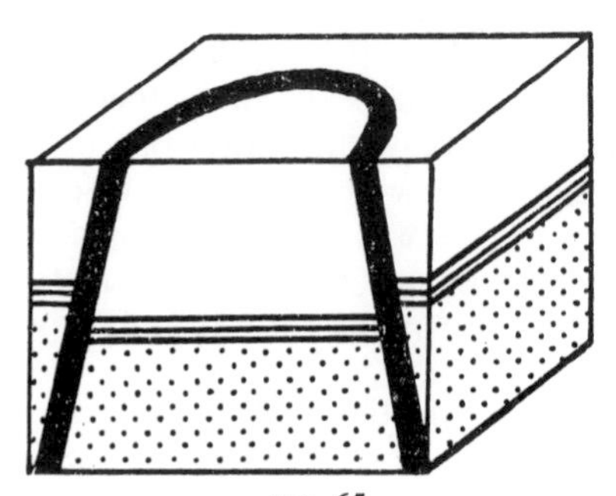

FIG. 65
Block-diagram of a single ring-dyke injected through horizontal strata.

level, or they may have different degrees of curvature so that when traced along the strike they converge and eventually intersect, forming a crescentic outcrop. Very occasionally the dykes form complete rings as in the case of the Ossippee Mountains ring-dyke in New Hampshire.[2]

Ring-dyke walls are usually steeply inclined or may be vertical, though evidence is necessarily scanty. The Hebridean ring-dykes are believed to have steeply inclined, outward-dipping walls, and to have been emplaced by down-faulting of a central mass of rock enclosed within a ring-fault. Such down-faulting would provide a potential cavity into which magma could be drawn. This mechanism, first proposed by Clough, Maufe and Bailey for the Devonian ring-intrusion of Glencoe, Scotland, and elaborated particularly by Richey as an explanation of the manner of intrusion of the many Tertiary ring-dykes in Scotland, neatly provides a solution to the space problem, particularly if it can be demonstrated that the walls dip outwards. However, there are instances for which alternative explanations must be sought. Some ring intrusions in the Hebridean Province have widths up to a mile, and furthermore possess internal flow-structures inconsistent with a simple hypothesis of subsidence; in some significant cases the walls locally at least dip inwards. It is possible that ring-faulting may be accompanied by the formation of a zone of fracturing and brecciation into which magma may be able to stope its way.

[1] Richey, J. E., Tertiary ring structures in Britain, *Trans. Geol. Soc., Glasgow*, **19** (1931–2), 45.

[2] Billings, M. P., Mechanics of igneous intrusion in New Hampshire, Daly Volume, *Amer. J. Sci.*, (1945), 40, with references to literature.

Ring-dykes may occur singly, but frequently they form **ring-complexes** consisting of several individual intrusions arranged concentrically about a common centre. The order of intrusion of the several members of the Complex may be inferred on the evidence of included fragments of an earlier member in a later one. Chilling of one member against another may occur but is apparently rare, presumably because the several members of the Complex were emplaced within a relatively short interval of time. Thin partitions or 'screens' of country rock may separate neighbouring ring-dykes.

Although the rocks forming ring-dykes are often coarse-grained and therefore by definition fall in the 'plutonic' category, there is no doubt that many, if not all ring-dyke complexes, were directly connected with surface vulcanicity, particularly with calderas. The near-surface character of some of the rocks involved is proved by the occurrence of such features as fine-grained texture, flow banding and explosive brecciation.[1]

Theoretically any kind of igneous rock capable of intrusion may, on occasion, form a ring-dyke; but actually granitic rocks are far more commonly represented among such rock-bodies than those of other types. This may possibly be a consequence of the relatively low specific gravity of granitic magma. Among noteworthy examples of ring-complexes are those discovered and (in part) described within recent years in several different parts of Africa. Many of these include varieties of riebeckite-granites,[2] carbonatites and nepheline-bearing rocks. In Britain the triple ring Complex of the Ardnamurchan Peninsula in western Scotland is of special interest as among the rocks involved are Basic gabbros and 'eucrites', Intermediate monzonites, and more granitic types including tonalite.

(*b*) **Cauldron subsidence intrusions; bysmaliths.**—One further type of intrusion must be mentioned here, for it also is connected with a type of ring-faulting. As defined by Iddings[3] a bysmalith is an injected body, having the shape of a cone or cylinder, which has either penetrated to the surface or terminates in a dome of strata like that over a laccolith. The 'plutonic plug' of Russell[4] is a similar conception. Vertical displacement with faulting is the characteristic of this method of intrusion. Mount Holmes, in Yellowstone Park, is cited by Iddings as a type.

It is clear that if the roughly cylindrical fracture of Fig. 66 had failed to reach the surface, or alternatively, if the sinking block had

[1] Richey, J. E., Association of explosion-brecciation and plutonic intrusion in the British Tertiary igneous province, *Bull. Volc.*, series 11 (1940), 157.

[2] Jacobson, R. R. E., Macleod, W. M., and Black R., Ring-complexes in the younger granite province of Northern Nigeria, *Mem. Geol. Soc., Lond.*, **1** (1958).

[3] *J. Geol.*, **6** (1898), 707.

[4] *Ibid.*, **4** (1896), 23–43.

been terminated upwards by a horizontal plane of weakness (X in Fig. 66), a space would be formed *above* the sinking block and into it magma would be drawn. On consolidation this intrusion would

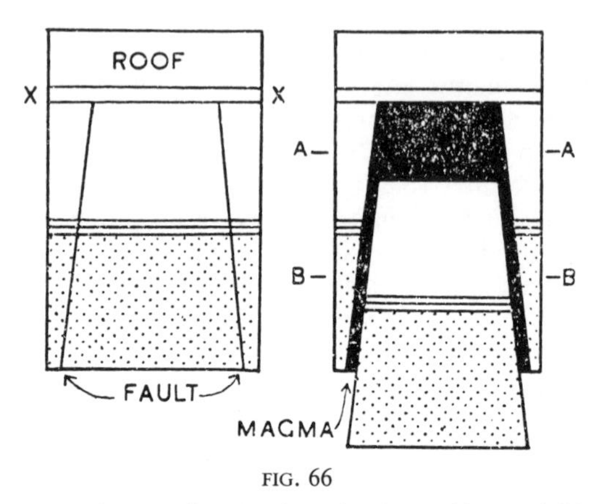

FIG. 66

Diagram sections to illustrate intrusion by cauldron subsidence.

be of cylindrical form, of circular cross section (as at level A in Fig. 66); it would have a flat floor, and at a certain level (as at B) it would give place to a ring-dyke. Such movements are referred to as 'cauldron subsidences' and they are due to 'piston faulting'. The classical British example occurs at Glencoe in South-West Scotland, described by Bailey, Clough and Maufe. An intrusion may be inferred to have been emplaced in this manner if the following conditions are satisfied: (1) the environment is one in which there is a regime of ring-faulting; (2) the outcrop of the intrusion is approximately circular; (3) the diameter is of the order of 5 to 15 miles; (4) the contacts are sharp and steeply inclined outwards, though they may be vertical; (5) the surrounding rocks are undisturbed, or show a terminal curvature in the sense consistent with the sinking of the central cylindrical block. If, in addition, part of the floor is exposed, or if there is a sharp turn-over from steep walls to flat roof, the evidence is much strengthened.

(*c*) **Cone-sheets.**—The Tertiary intrusions of the Hebridean Province provide many examples of basaltic 'dykes' aptly termed cone-sheets by E. B. Bailey. This three-dimensional term replaces A. Harker's two-dimensional term, 'inclined sheets'. Cone-sheets are normally only a few feet thick. No individual cone-sheet forms a completely circular outcrop as suggested in Fig. 67; but the diagram does embody the essential features of this type of intrusion: a curved outcrop at ground level, this being a segment of the circle shown in the diagram, and an inward dip towards a common focal point. In any area mapped in detail the extraordinary parallelism of the outcrops, their concentric arrangement about a centre and their constant dip are outstanding features. By extrapolation the

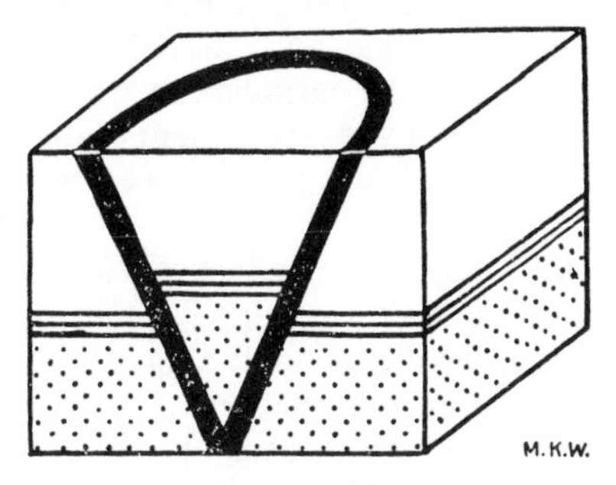

FIG. 67

Diagrammatic representation of a single cone-sheet, idealized to show the essential features of its shape, outcrop, and relationship to adjacent rocks. When comparing Figs. 66 and 67, remember the difference in relative dimensions of ring-dyke and cone-sheet. The converging walls of the latter doubtless reach the 'magma reservoir' at some point *above* the apex of the cone.

dip observed at the surface may be extended downwards, and indicates the depth of the magma reservoir from which the material of the cone-sheet swarm was derived. Estimates relating to Hebridean centres range from 3 to 5 miles below the present surface.

The mechanism of intrusion of a cone-sheet is implicit in the diagram. The upper surface ('hanging wall') is separated from the lower surface ('foot wall') by the thickness of the dyke-rock, the intrusion of which must have involved an uplift which is measurable in favourable circumstances. If the cone-sheet were of the ideal form, injection would involve an upward displacement of the cone of country rock lying within its outcrop. This applies to all members of such a cone-sheet complex, and in the aggregate the total amount of uplift may be of the order of some thousands of feet. To achieve this the magma must have been capable of exerting a powerful upward thrust. The conditions are thus exactly the opposite to those operating in the case of ring-dyke injection, for which a deficit of magma-pressure is essential. An interesting feature of some of the Hebridean volcanic centres is an alternation of ring-dyke and cone-sheet injection.

Cone-sheet complexes do not appear to be common outside the Hebridean volcanic province; but relatively recently cone-sheets of unexpected material have been discovered, forming parts of complexes involving nepheline-bearing rocks and carbonatites. The latter in some case form a central plug of crystalline carbonate which occurs also in associated cone-sheets. These are identical in form and attitude with the better-known basaltic examples, though the material of which they are composed is so strikingly different. The Alnø Complex in Sweden and certain African complexes display these features. In these occurrences it is probable that the concentric conical fractures into which the carbonatites were injected resulted from explosive volcanic activity, consistent with the high concentration of volatiles necessary to produce this association of rock-types.

The large ring-dykes and granitic masses which have been intruded as a result of displacement following ring-faulting, form a transition between the intrusions described so far, in which the form and mechanism of intrusion are reasonably well understood, and the great subjacent intrusions, with steeply plunging contacts which continue downwards to unknown depths, and whose three-dimensional shapes and mode of origin have occasioned so much controversy. There is certainly scope for wide differences of opinion on this major problem of petrology, and for markedly different interpretations of the field evidence.

SUBJACENT PLUTONS

Subjacent plutons are major intrusions which have no visible floor. The 'walls' are generally steeply inclined and within the limits of observation these rock-bodies tend to increase in size with depth. The rocks occurring in these plutons are mostly quartz-bearing types and can be classified as 'granites' in the widest sense. The following description may therefore be regarded as relating to the various modes of occurrence of granitic rocks, in contrast to Basic intrusives which more commonly occur as dykes, sills, funnels and lopoliths described above. The distinction is not, of course, exclusive.

(*a*) **Batholiths.**—This term was introduced by Suess to connote the major, deep-seated intrusive masses of very large size occurring typically in the great mountain ranges, and generally elongated parallel to their tectonic trend. The most spectacular batholiths occur in the western fold-mountain ranges of North America. The Coast Range batholith of Alaska and British Columbia extends along the strike for approximately 1100 miles and varies in width between 80 and 120 miles; it thus occupies a surface area of some 110,000 square miles. The Sierra Nevada batholith measures 400 miles by 40 to 70 miles in width. A large part of the problem is concerned with the actual origin of the 'granitic rocks' within the intrusions, and this is discussed in the chapter devoted to this subject. At this point we have to consider evidence relating to the physical characteristics of the whole rock unit. The complete batholith is built up of a number of distinctive members, each with its own petrographic characteristics. The sequence of emplacement of these separate intrusive bodies can quite often be worked out from field relationships—the inclusion of xenoliths of an earlier in a later member; apophyses from a younger member penetrating another; the occurrence of chilled margins of one against another, etc. Valuable information concerning the actual shape, mechanism of emplacement and cooling history of each unit may be gained from a study of the internal flow- and joint-structures described below.

Geological investigation is always limited, of course, by the degree of exposure. In the extreme case the main intrusion may not be visible at all, though its existence some distance below the surface may be suspected from the evidence of metamorphism and perhaps mineralization of the overlying rocks.

The stripping off of the 'roof' by denudation may, in the early stages, disclose isolated, irregular outcrops of igneous rock, which, apart from their petrographic similarity, convey no hint of the immensity of the pluton lying beneath. At a later stage, the isolated outcrops connect up, while the intervening stretches of highly metamorphosed roof-rock are reduced to small islands (roof-pendants). In addition 'cupolas' may rise from the hidden batholith, as satellitic intrusions. There is convincing evidence that the separate granite outcrops in S.W. England (Fig. 149) are such cupolas, connected underground to a single batholithic intrusion.

In regions of great relief such as the Rocky Mountains and the Alps granite batholiths may be seen to maintain their characteristics to depths of the order of 10,000 to 15,000 feet. Further evidence of the downward extension of batholiths and other plutons has been obtained by geophysical means. The evidence available to geologists gives the impression that batholiths extend downwards almost indefinitely into the deepest parts of the crust, and that deeper erosion would expose ever greater proportions of granitic rocks. Geophysical investigations are gradually correcting this impression. Granitic rock has a low density compared with most other rocks, and this produces a deficiency of mass which can be easily detected by gravity surveying. Using this evidence[1] in conjunction with seismic data[2] it has been shown that the S.W. England granite batholith extends down to a depth of only 10 to 12 km, and is underlain by rocks of much greater density. We may claim fairly confidently in this case, therefore, that the batholith takes the form

[1] Bott, M. H. P., Day, A. A., and Masson-Smith, D., The geological interpretation of gravity and magnetic surveys in Devon and Cornwall, *Phil. Trans. Roy. Soc., Lon.*, **251** (1958), 161–91.

[2] Bott, M. H. P., Holder, A. P., Long, R. E., and Lucas, A. L., Crustal structure beneath the granites of South West England, in *Mechanism of Igneous Intrusion*, ed. G. Newall and N. Rast, (1970), 93–101.

of a mass of granite, elongated parallel to the Armorican structural trend and restricted to the upper part of the continental crust. The nature of the rocks underlying the floor or roots of the batholith remains obscure; but it is reasonable to infer that they would include schists and gneisses similar to those occurring in deeply eroded shield areas. The authors responsible for the geophysical surveying suggest that these rocks in part consist of unmelted residue remaining in the lower crust after mobilization of the granitic components. A similar picture is presented by Hamilton and Myers in a broad survey of batholiths in Western America,[1] which are envisaged as being suspended, as it were, in the upper part of the crust, having risen to their present position because of the low density of granitic magma.

(*b*) **Stocks** differ from batholiths only in size, and may be defined as subjacent bodies less than 40 square miles in area. Like batholiths they have steeply plunging contacts and no visible floor.

(*c*) **Boss** is the term applied to stocks of circular cross-section. Some bosses and stocks have undoubtedly been emplaced by the cauldron-subsidence mechanism.

Internal Structures in Subjacent Intrusions

Two structural elements are involved: firstly, the lineation of individual crystals and inclusions due to flow movements; and secondly, jointing due to fracturing after consolidation. Study of these phenomena was pioneered by H. Cloos[2] and the techniques have been summarized by R. Balk.[3]

Flow structures develop as a consequence of the alignment of crystals and xenoliths during the act of intrusion. With regard to the first, the phenomenon is most clearly demonstrated by crystals of flattened, tabular habit, notably by feldspar phenocrysts. In this country the West of England granites, for example the 'giant granite' of Dartmoor and Land's End, are typical in this respect. As regards xenoliths, these are often softened sufficiently to be drawn out into **'schlieren'** which appear as dark streaks in the normal granite. As a consequence of these flow structures, layers are formed which are normally parallel to the walls of the intrusion. In an intrusion with a domed roof these flow-layers or **platy flow-structures** dip outwards from the centre of the mass, and plunge steeply downwards in conformity with the dip of the actual surface

[1] Hamilton, W., and Myers, W. B., The nature of batholiths, *U.S. Prof. Paper*, **554** (1967), C1–C30. See also Fyfe, W. S., Some thoughts on granite magmas, in *Mechanism of Igneous Intrusion*, ed. G. Newall and N. Rast (1970), *Geol. J. Special Issue*, **2**, 201–16.

[2] Cloos, H., *Abh. Preuss. Geol. Landesnast.*, **89** (1925), 1.

[3] Balk, R., *Geol. Soc., Amer., Mem.*, **5** (1937).

of contact with the wall-rocks. They may thus be almost vertical in steep-walled stocks and bosses.

It is possible for the streaking out of these flow structures in the marginal zone of the intrusion to cause heterogeneous mineral banding similar to that commonly regarded as due to regional metamorphism. In the case of such primary gneissic banding, however; other metamorphic features, such as evidence of stress and cataclasis, should be absent.

It may be possible to detect the actual direction of flow of the magma by means of the parallel orientation of crystals of prismatic habit, such as hornblende, for example, or by the softening and elongation of xenoliths. These **linear flow structures** may be combined with layering, or they may occur independently; but in any case three-dimensional exposures are necessary to distinguish between the two types. It must be realized that these flow-structures are marginal phenomena; they tend to die out towards the centre of the mass, where the rock becomes massive, and even the feldspar phenocrysts, if present, become disorientated.

Joint structures.—Once the outer crust of the mobile mass has solidified, no further flow is possible; but beneath the crust the magma may still exert a pressure on the crust and, by stretching it, produce tensional or 'cross joints'. Since the stretching follows the same direction as the linear flow of the previous plastic stage, the cross-joints are developed perpendicular to the linear flow direction of the latter.

There are, however, several sets of joints in a granite mass, and enough has already been said to show that their proper interpretation is dependent upon the recognition of the flow structures. Some of the more important primary jointing systems are:

(*a*) Cross joints, already described;

(*b*) Longitudinal joints, striking parallel to the trend of flow lines, and probably developed as a result of weaknesses parallel to the aligned minerals;

(*c*) Diagonal joints at angles of approximately 45° to the trend of the flow lines, formed as a result of the compression which operates at right angles to the flow.

(*d*) Flat-lying joints, of which the origin is still obscure.

These joint directions may be followed by veins of aplite and pegmatite, or the joint planes may be coated with veneers of minerals of hydrothermal origin, thus indicating their primary origin. In the absence of such mineralization it may be impossible to distinguish between primary joints connected with the act of intrusion, and those of later origin. Prominent among the latter are planes which develop parallel to the surface of the ground and named, rather misleadingly, exfoliation or bedding joints. In fact they may be highly perfect joint planes spaced widely apart, and occurring to a depth of many feet in the rock. Naturally only primary joints are of value for

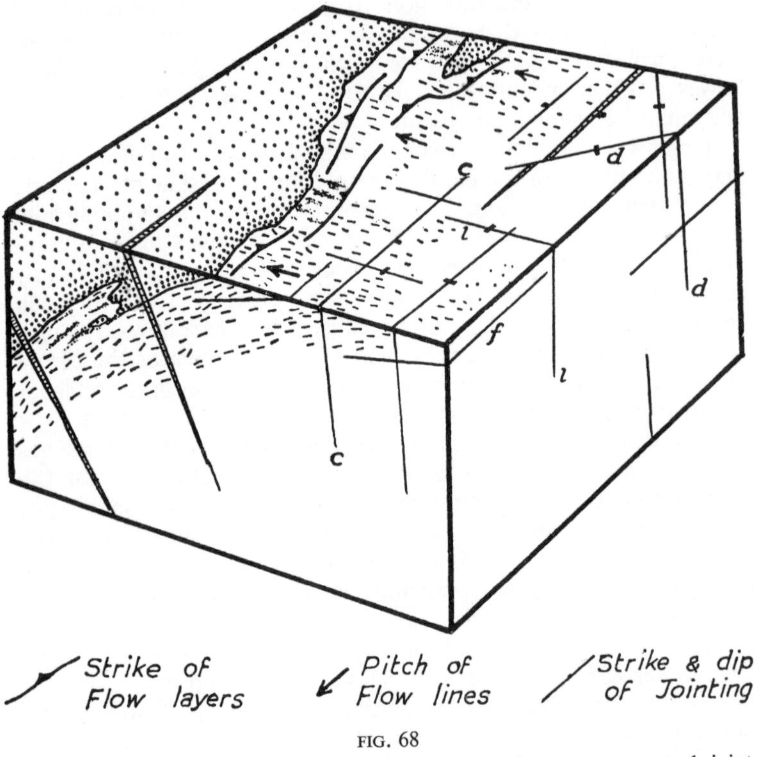

FIG. 68

Block diagram showing the relationship between flow structures and joint systems in an intrusive mass. (*c*) cross joints; (*d*) diagonal systems; (*f*) flat-lying joints; (*l*) longitudinal joints.

determining the possible shape of the rock-mass, and in practice most of such information is obtained from the study of flow structures combined with cross jointing alone.

Weathering takes place particularly easily along these various joint planes, so that exposed surfaces appear broken into angular blocks. On a freshly exposed surface it may be almost impossible to determine any directional properties in the rock, which appears massive and structureless. Nevertheless, planes of weakness, generally three almost rectangular sets, are normally present and can be detected by the experienced quarryman.

CHAPTER III

THE CONSOLIDATION OF MAGMA

THE original raw material of igneous rocks, or magma, may have a wide diversity of chemical and physical properties. Some of these have already been described when the nature of igneous rocks was discussed. For the purposes of the present chapter, however, it is assumed that a magma is originally a hot fluid which is essentially a melt of complex silicates and oxides, or more correctly, one capable of crystallizing out to yield these components. In addition, small amounts of ordinarily volatile components such as H_2O, Cl, CO_2, etc. are present. The proportions of the various elements vary greatly, mainly according to whether the magma is highly siliceous (Acid), or poor in silica (Basic); but always the chief components include the following: Si, Al, Fe^{2+}, Fe^{3+}, Mg, Ca, Na, K, and oxygen. It so happens that in the silicates, the ratio of the other elements to oxygen is always such that their formulae may be written down in terms of oxides; for example, orthoclase $KAlSi_3O_8$ may be written as $K_2O.Al_2O_3.6SiO_2$. For this reason the convention has been adopted of stating the analyses of rocks in terms of the percentages of various oxides. It is well to bear in mind, however, that these 'oxides' are not really present as such.

The cooling and crystallization of a melt of such complex composition is inevitably a very complicated process; but by careful study of the textural interrelationship between the minerals in rocks, it is possible to establish an order of crystallization. It is evident that the first minerals to crystallize are those which can be precipitated from a completely, or almost anhydrous melt, at high temperatures. These are the so-called **pyrogenetic minerals**, and include the majority of the silicates found as primary constituents in the Basic rocks—olivines, most pyroxenes, the calcic plagioclases, etc. The separation of these pyrogenetic minerals leaves the liquid relatively enriched in H_2O and various other components of low atomic and molecular weights, which are known as the volatile, hyperfusible, or fugitive constituents. Several rock-forming minerals require for their formation a high concentration of these volatiles. These **hydatogenetic minerals** depend more on concentration of volatiles than on high temperature for their formation. Most of the alkali-rich minerals and those containing hydroxyl fall in this category.

The history of the cooling and crystallization of a magma can be divided into a number of stages, based largely on the dominance of the roles of temperatures and concentration of volatiles.

The **orthomagmatic stage** covers the separation of the pyrogenetic minerals, and in the case of a Basic rock, accounts for the crystallization of the greater part of the component minerals.

This is followed by a stage during which the portion still in the fluid condition has extremely low viscosity on account of the increasing concentration of volatiles, while the temperature is still fairly high, perhaps between 400° and 600° C. This leads to the development of crystals of exceptional size, distinctive of the **pegmatitic stage** of crystallization. It is possible that the more volatile fractions may sometimes separate out as independent gaseous phases under these conditions, especially if there is a local reduction of the external pressure within the system. Certain minerals, notably tourmaline, topaz and fluorite, are especially characteristic of such gaseous, or **pneumatolytic** conditions. The products of the pegmatitic stage are commonly segregated into veins and dykes, in which case they form highly distinctive rocks.

Finally the residual fluid may gradually pass towards the condition of low temperature aqueous solutions, and any deposition or replacement occurring at this stage is said to be **hydrothermal**.

The cooling history of a magma has been divided into more elaborate stages by some authors, who base their divisions on temperature limits. There is, of course, no direct way of determining the temperature at which any particular constituent has crystallized, but there are several mineral changes, for example the inversion of β- to α-quartz, which are known to occur at fairly constant temperatures. The presence of these critical minerals, therefore, allows one to estimate approximate limits for the several stages.

These points on the 'geological thermometer' are, however, too few and too much subject to variation under natural, as distinct from experimental conditions, to provide a wholly satisfactory basis on which to establish a complicated system of stages of crystallization.[1] In any case, the processes are actually continuous, overlapping of the stages is inevitable, and exact limits cannot in practice be assigned to them. In particular, it is often impossible to distinguish between phenomena produced during the pegmatitic-pneumatolytic stage and the hydrothermal stage. In some cases, indeed, it is possible only to differentiate between the products of

[1] A valuable summary and criticism of the many terms which have been so used is given by Shand, S. J., The terminology of late-magmatic and post-magmatic processes, *J. Geol.*, **52** (1944), 342.

the orthomagmatic, primary crystallization on the one hand, and the late-stage modifications on the other. Some of the more important aspects of these two main, broad divisions are considered below.

PRIMARY, ORTHOMAGMATIC CRYSTALLIZATION

From an early date in the history of petrology much attention has been paid to the apparent order of crystallization of the minerals in rocks. It was observed that in a large number of cases there was a definite sequence. Applying the principle that if mineral 'A' encloses or is moulded upon mineral 'B' the latter must be of earlier formation than the former, the commonly observed sequence was found to be: accessory minerals, ferromagnesian minerals, feldspars, quartz.

However, many exceptions have been observed, and the principle is liable to misinterpretation since the observed order is that of completion, not commencement of crystallization. The so-called law depends on the premise that, since the intitial composition of magmas varies only within quite narrow limits, there can be little variation in the order of crystallization.

In fact, however, the components in a complex silicate melt such as occurs in Nature mutually lower one another's freezing temperatures to a considerable degree. Further, the order of crystallization is dependent upon the concentration of a given component in the melt: no mineral will crystallize from a melt until the latter is effectively saturated with it, under the prevailing conditions.

An understanding of the laws governing the crystallization of minerals from a silicate melt has been made possible only by the experimental researches carried out at such specially equipped institutions as the Geophysical Laboratory, Washington. These experiments involve a high degree of specialist skill. Although the early experiments of this kind dealt only with the simplest silicate components at atmospheric pressures and under dry-melt conditions, the range and complexity of the experiments is being extended all the time, particularly in the field of high pressures involving volatile constituents, so that now even the most complex minerals are being synthesized, as the conditions of the experiments are brought ever closer to those of Nature.

The crystallization of a rock consisting of two components is instructive, and two possible cases will be considered: first, that in which the components are incapable of forming solid solutions (*i.e.* they are immiscible in the solid state); and secondly, that in which the components are capable of forming a continuous series of solid solutions. In addition, the crystallization of simple three-component systems will be examined.

The crystallization of a pair of minerals which do not form mixed crystals can be exemplified by means of a temperature-concentration diagram, in which the relative proportions of the two components are shown in percentages by abscissae, and the temperatures by

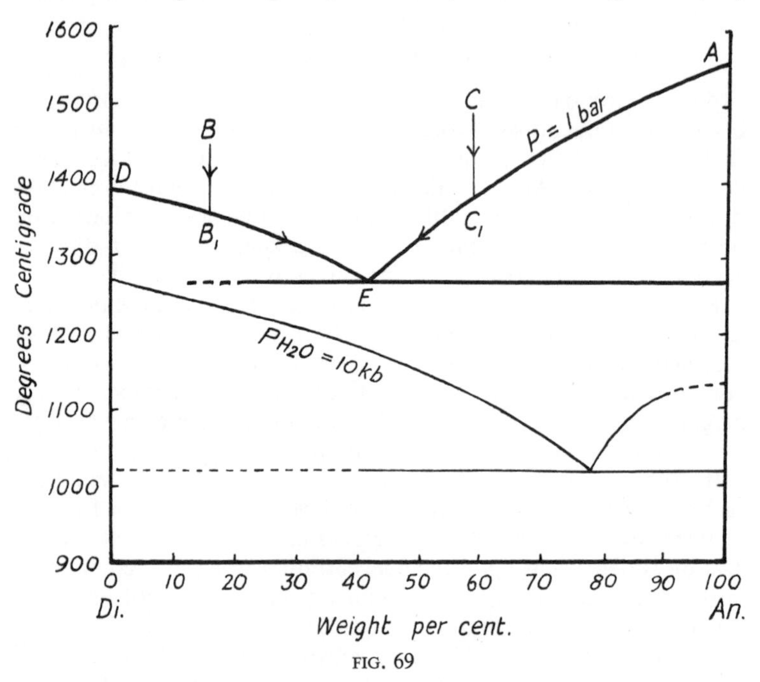

FIG. 69

Phase relationships in the system Diopside-Anorthite. Note the depression of eutectic temperature and change of eutectic composition resulting from high water-vapour pressure. (*After Yoder*, (1969).)

ordinates. The case of an anorthite-diopside melt is illustrated in Fig. 69.

Starting with pure anorthite at the point A, the addition of an increasing proportion of diopside lowers the anorthite freezing-point as shown by the curve AE, and similarly the addition of anorthite to diopside produces the curve DE. The point at which the two curves intersect is the **eutectic point,** E.

Given a 'melt' represented in composition and temperature by the point B, and subject to a falling temperature, no crystallization will take place until the temperature reaches B_1. At B_1 diopside (which is in excess of the eutectic proportion) begins to separate in the pure form. Its removal from the liquid causes the composition of the latter to change with falling temperatures towards the point E. On reaching this point diopside and anorthite crystallize together until all the liquid is exhausted. If, on the other hand, we start at a point C, representing a liquid in which anorthite is in excess, a similar result is obtained, but pure anorthite first separates. It is to be noted that the solid formed by freezing at the eutectic point is a mechanical mixture of the two minerals, and not a solid solution. Thus it is seen that when a pair of minerals incapable of forming solid solutions (*mix-crystals*) are cooling together from a molten condition, there is a certain definite ratio of the one to the other in which they will crystallize out simultaneously in intimate admixture. Whichever mineral is in excess of this (the eutectic) proportion, will crystallize out first. It is important to realize (1) that the order of separation of the two components is independent of their freezing-points, and (2) that over a definite interval of time the two minerals are forming simultaneously.

The eutectic ratios for certain pairs of minerals have been established. Thus from an exhaustive study of gabbroic and noritic rocks Vogt has demonstrated that for hypersthene (or diallage) and labradorite the ratio is 35 : 65.[1] Investigations of the equilibrium conditions governing the crystallization of simple silicate melts have shown that for diopside and anorthite the ratio is 58 : 42, and for diopside and forsterite 88 : 12.[2]

An excellent illustration of the relation explained above is afforded by a case recorded by Harker from the Isle of Rhum. In a series of rocks, consisting essentially of anorthite and olivine, the former is found to have crystallized out first when the rock is rich in that mineral, while in varieties rich in olivine the reverse holds good. Only when the two minerals occur in the eutectic ratio have they crystallized simultaneously.[3]

If in addition to the two minerals a third be present, say pyroxene in the case quoted, Nernst's law of the reduction of solubility between the substances having a common *ion*, appears to govern the order of crystallization. Thus olivine and pyroxene have the ion (Mg,Fe) in common, consequently the solubility of the olivine is much reduced, and it invariably crystallizes before the feldspar, even when not present in very large amount.

The crystallization of two minerals that form a continuous series of solid solutions is well illustrated by the plagioclase feldspars.[4] In such a system, the solid and liquid in equilibrium at any temperature are of different composition, the solid being always richer in the component with the higher freezing-point. In Fig. 70, the curve

[1] Vogt, J. H. L., Physical chemistry of the crystallization and magmatic differentiation of igneous rocks, *J. Geol.*, **29** (1921), 441.

[2] Bowen, N. L., The System diopside, forsterite, silica, *Amer. J. Sci.*, **38** (1914), 209.

[3] Harker, A., The geology of the Small Isles of Inverness, *Mem. Geol. Surv.* (1908), 85.

[4] Bowen, N. L., *Amer. J. Sci.*, **35** (1913), 583.

marked 'solidus' traces the change of composition of the solid with falling temperature, while that marked 'liquidus' shows the corresponding change in composition of the liquid. Temperatures

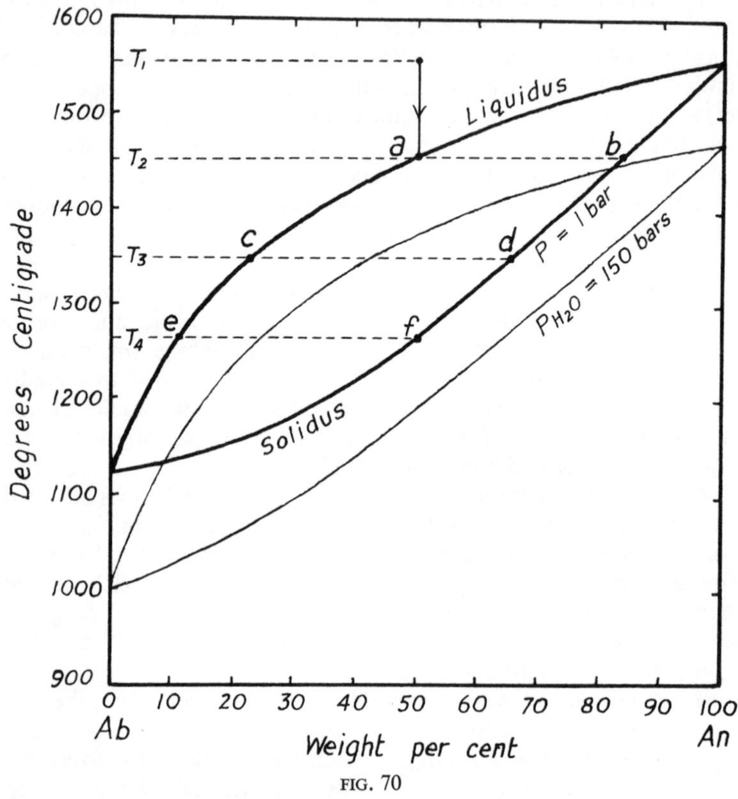

FIG. 70

Diagram to illustrate the crystallization of the plagioclase feldspars, in the absence of water (*thick lines*); and under a water-vapour pressure of 150 bars (*thin lines*). *After Bowen and Yoder.*

are plotted vertically, while the compositions in percentages of anorthite (An) and albite (Ab) are measured on the horizontal 'composition line'.[1] The course of crystallization is as follows: Starting with a melt of composition $Ab_{50}An_{50}$ at a temperature T_1, no solid will be deposited until the point (*a*) is reached. At this temperature (T_2) a small amount of solid of composition (*b*) will be deposited. This solid (*b*) is in equilibrium with liquid (*a*) at temperature T_2 (and similarly for every other pair of points on the two

[1] The composition of the solids and liquids represented by the points (*a*)–(*d*) are found by dropping perpendiculars from the point under consideration to the composition line.

curves cut by horizontal lines representing successive temperature changes). The separation of anorthite-rich solid has, however, altered the composition of the liquid phase which is richer in albite than originally. Let us suppose that the fall of temperature is temporarily arrested at T_3, when solid (*d*) is in equilibrium with liquid (*c*). The earlier formed solid (*b*) will react with this liquid, the excess of anorthite will be leached out, and equilibrium will be restored when all the solid has the composition (*d*). But the fall of temperature is continuous, the change in composition of the liquid is continuous, so the solid also gradually changes in composition, approaching closer and closer to $Ab_{50}An_{50}$. The whole mass will have this composition at the temperature of final consolidation, T_4.

During the cooling of such a system, only a single liquid phase can exist; while if equilibrium is fully established, then there is also only one homogeneous solid phase. The difference of temperature between the commencement and completion of the act of crystallization is called the 'crystallization interval' (T_2–T_4). An interesting corollary is that the freezing point and melting-point of such a system are not the same: freezing begins at T_2 while melting begins at T_4.

Hitherto we have assumed that cooling was taking place very slowly and that equilibrium was fully established at every stage. Such, however, is not invariably the case. It frequently happens that cooling is so rapid that, on account of the high viscosity of the melt, equilibrium between solid and liquid is not fully established, and hence the growing crystals vary in composition, and in optical and other physical characters from the centre outwards. Each successive layer or zone deposited on the nucleus retains its original character: there is no time for reaction with the liquid portion of the system to take place. In this way **zoned crystals** are formed, and are specially common in the plagioclase and pyroxene groups. In the former case the nucleus of a crystal having the bulk composition of labradorite may be almost pure anorthite, $Ab_{20}An_{80}$ (*b*), while the last drop of liquid, *i.e.* the last zone added to the growing crystals, is nearly pure abite, $Ab_{88}An_{12}$ (Fig. 70).

Even a small concentration of water in the melt causes a considerable lowering of the solidus and liquidus temperatures. If therefore, the water-vapour pressure fluctuates, due for example to the periodic escape of volcanic gases, the composition of the crystals separating from the liquid at a given temperature will also fluctuate, becoming more calcic under high water-vapour pressures. Within the limits shown in Fig. 70, anorthite content would vary by more than 10 per cent. This is the most likely cause of oscillatory zoning in plagioclase.

Although we have considered the plagioclases only, the importance of the principles illustrated by this one example may be gauged from the fact that with the exception of quartz, all the important rock-forming silicates are members of similar solid solution series and their crystallization must follow a similar course: the comparable diagram for the olivine series is illustrated in Fig. 4.

The course of crystallization is naturally more complicated when dealing with two components, both of variable composition, and we choose the pyroxenes to illustrate the principles involved. It was pointed out in the section on pyroxenes (p. 41) that during cooling of a melt the compositions change progressively as the temperature falls, the changes in general involving iron-enrichment. Further, in normal Basic magmas over the higher temperature range, two series of pyroxenes of contrasted compositions crystallize simultaneously. When the compositions achieve a certain degree of iron-enrichment, however, the two pyroxenes give place to a single pyroxene-phase, corresponding in composition with ferroaugite.

To understand more fully the course of crystallization at any point along the fractionation series it is necessary to consider a cross section at right angles to the latter (as shown in Fig. 10), through the mid-point of the En–Fs join. It lies well within the two-pyroxene field mentioned above. The composition of the melt may lie anywhere within the system, of course, but arbitrarily it is assumed to be represented by the point *o* in Fig. 71. Cooling of the melt to a point on the liquidus phase boundary causes an augite to crystallize. Its composition is obtained by tracing along an isotherm (the horizontal broken line *p*–*q*), then dropping a perpendicular from *q* on the solidus curve to the base line, where the composition corresponds with a_1. This is the first augite to crystallize from a melt of composition *o*, and the crystal *q* is in equilibrium with the liquid *p* at temperature t_1. Removal of this early augite enriches the melt in MgFe; and as cooling proceeds the composition changes until the point *r* is reached. This corresponds with the eutectic point of the previous diagram (Fig. 69); but we are now dealing with two isomorphous components and the curved lines meeting in *r* represent a section across a valley, the 'long profile' of which extends at right angles to the diagram. The line of the valley may be referred to as the **cotectic** line.

At the temperature t_2 two pyroxenes crystallize simultaneously, augite of composition a_2 being joined by pigeonite. No further fall of temperature occurs until all the liquid is used up. Such further changes as do occur involve the crystals already formed and may be referred to as sub-solidus changes. Inversions and exsolution phenomena come into this category. The space below the augite and pigeonite solidus curves and between the augite and hypersthene field boundaries constitutes an immiscibility gap. The field boundaries referred to are called solvus curves, and it will be noted that they diverge downwards. At the temperature t_3 augite (v), more calcic than a_2, is in equilibrium with pigeonite (u) richer in MgFe than that which first formed. At temperature t_4 still more calcic augite (x) is in

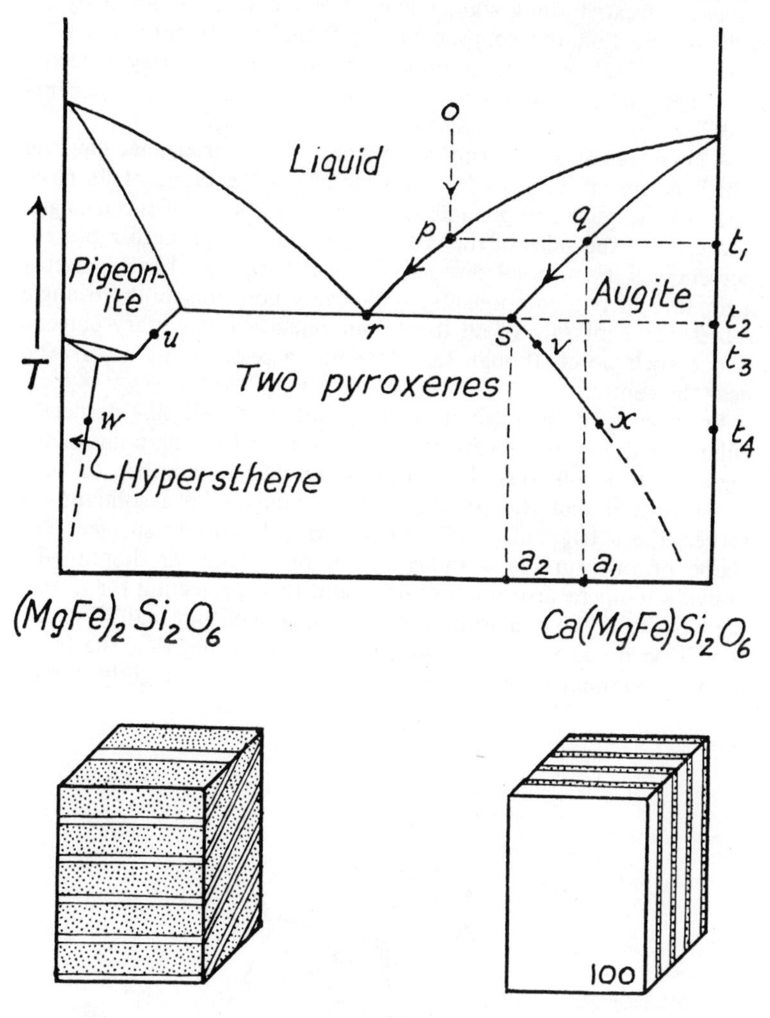

FIG. 71

Diagram illustrating the course of crystallization of pyroxenes; and (below) rectangular block-diagrams showing the orientation of exsolution lamellae. On the left are shown lamellae of augite parallel to the Monoclinic (001) planes, with orthopyroxene as host mineral. On the right, lamellae of orthopyroxene parallel to (100) are enclosed in augite.

equilibrium with hypersthene. These changes in composition as cooling proceeds are, of course, consistent with the tendency of both minerals to exsolve as explained under 'pyroxenes' in the mineral section of the book.

The order of crystallization in a system of three components can be easily treated along similar lines. For the sake of simplicity we will assume that the components A, B and C do not form solid solutions. Each pair of components can form a binary eutectic, and there is in addition a ternary eutectic of three components ABC.

A system of this kind can be represented by a **triangular diagram** which is the projection of a solid model on the plane of its base. Each of the pure components is placed at a corner of the triangle, and each of the sides of the latter is divided into parts to represent percentages. Each point on the sides of the triangle then represents a mixture of two components; while every point inside the triangle represents a mixture of all three components. The ternary eutectic is one such point, though this does not necessarily lie at or even near the centre.

From each of these points representing compositions, perpendiculars are drawn proportional in length to the freezing-point of the mixture. It is obvious that there is an infinite number of such mixtures and that the perpendiculars would make a solid model rather like a trigonal prism with an irregular upper surface. The shape of this surface is indicated on the triangular diagram by drawing temperature-contours on it and then projecting them onto the base, as in the preparation of an ordinary contour map (Fig. 72). It will be noted by those familiar with contouring that the topography represented by this 'map' comprises three slightly curved

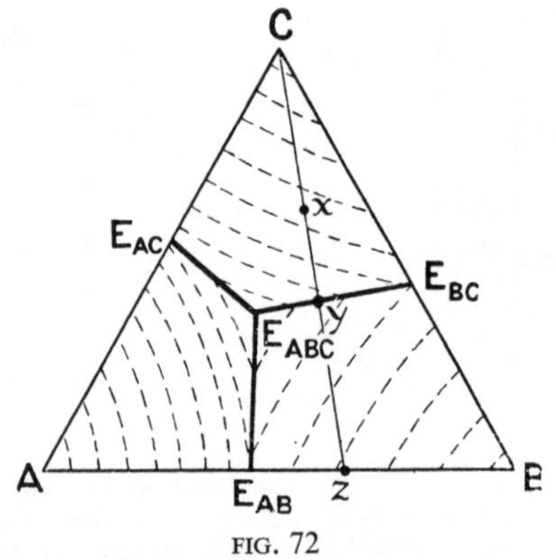

FIG. 72

Triangular diagram to illustrate the course of crystallization in a three-component system.

surfaces meeting in three valleys (shown by heavy black lines) converging on the lowest point E. The valley floors are phase boundaries which divide the whole triangle into three fields of existence for A, B and C respectively, called the A-field, etc. Any point in the triangle represents a particular composition and a particular temperature. The course of crystallization of a mixture represented by the point x will be as follows: since the mixture clearly contains the component C in excess of the binary eutectic ratio BC, this excess begins to separate out, and the composition of the mixture changes along the line joining C to x and continues until it strikes the cotectic-line E_{BC} E_{ABC}, at y. The solution is then also saturated for the component B, and B and C separate together as the binary eutectic. The composition then changes along the line y-E_{ABC}, till finally the ternary eutectic is reached at E_{ABC}. For any other point the course of crystallization is similar, except that, if the original mixture lies within the B-field, B crystallizes first; if it lies within the A-field then A begins to crystallize first. A line such as $Cxyz$ in Fig. 72 illustrates an important principle which is useful in the consideration of such diagrams, namely that every point on it represents a constant ratio of A to B, the proportion of C only varying (from 0 per cent at z to 100 per cent at C).

When a solid model is made on the principle outlined above, it is readily seen that each of the vertical planes standing on the lines AB, BC and CA is a binary eutectic diagram like Fig. 69.

The outstanding points of interest are (1) that the order of separation of the minerals is determined chiefly by their relative concentration in the mixture, and (2) that the periods of crystallization of the several components overlap; during a definite period two components, and at a slightly later stage three components, form simultaneously.

The theoretical example of a three-component system just described can be regarded as an aid to understanding the more complex system described below: this involves the silica minerals, feldspars and feldspathoids (Fig. 73). The system is of vital interest in petrology since it covers the field of composition of alkali granites and also has an important bearing on the origin of feldspathoidal rocks. The continuous lines in all the diagrams correspond with phase-boundaries. In A the boundary surface between melts and the several minerals is contoured with broken lines, the figures indicating hundreds of degrees Centigrade. Under dry-melt, experimental conditions the minerals formed include high-temperature forms such as carnegeite (car.), cristobalite (cr.), and tridymite (tr.). These are metastable under conditions of slow cooling and are converted into their more familiar low-temperature equivalents, nepheline and quartz.

The relatively large field of stability of leucite at high temperatures should be noted. A contoured diagram of this kind indicates the nature of the crystals first precipitated from a melt of any composition covered by the triangle. Separation of the crystals causes

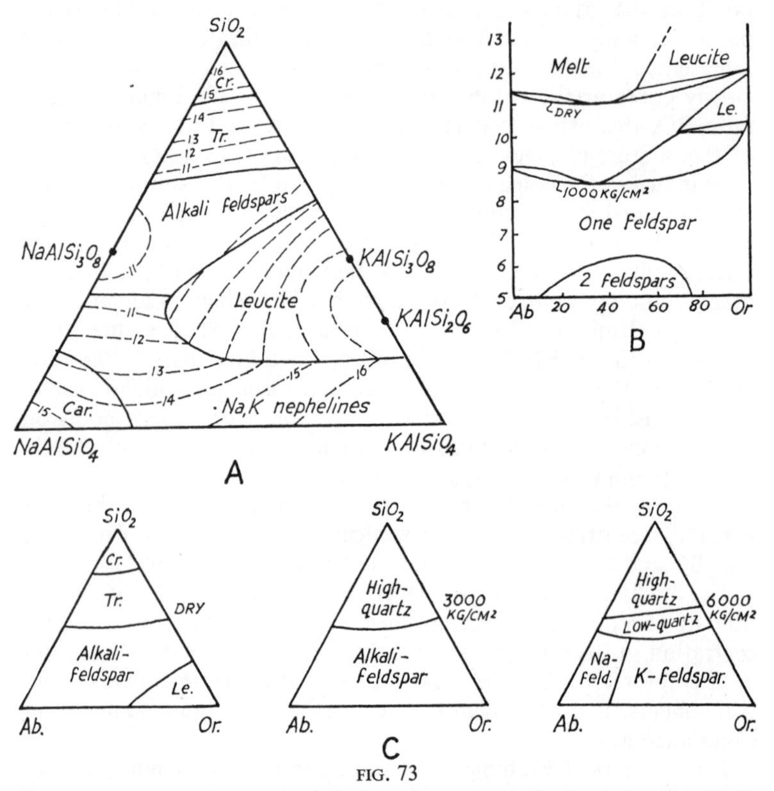

FIG. 73

Phase equilibrium diagrams relating to the occurrence of quartz, the alkali-feldspars and feldspathoids. (*A after Schairer and Bowen* (1935); *B and C after Tuttle and Bowen*, (1958).) For explanation see text.

the composition of the melt to change, the change always following a 'downhill' route on the contoured surface. The ultimate destination is one of the thermal 'valleys' following the cotectic boundaries between either quartz and the alkali-feldspars, or the latter and nepheline.

Fig. 73B may be regarded as a vertical section along the join between albite and orthoclase. The diagram also shows the solidus curves, and the effects of reaction can be assessed between the crystals first precipitated and the liquid. For example, leucite precipitated from a potassic melt of appropriate composition reacts with the liquid at a lower temperature, being converted into orthoclase, if cooling is sufficiently slow.

The relations appertaining to dry-melt conditions should be compared with those relating to an aqueous melt. Even if the water-vapour pressure is only moderate (equivalent to 1000 kg/cms^2) the melting temperatures are lowered by hundreds of degrees. Another noteworthy effect is the reduction of the leucite field. Ultimately leucite disappears altogether if the water content is sufficiently high. This is the main reason why leucite is absent from plutonic rocks, regardless of whether the latter are saturated with silica or not.

Fig. 73B also shows the sub-solidus (or so-called solvus) curves of unmixing of the synthetic alkali-feldspars.[1] A high-temperature homogeneous feldspar, for example sanidine, will unmix under conditions of sufficiently slow cooling to form various kinds of perthitic or antiperthitic intergrowths, the compositions of the sodic and potassic phases at any given temperature below 660° C being indicated by the solvus curve. Only the top of the curve is shown in the diagram; but the divergence continues for temperatures below those indicated, so that efficient unmixing, seen in the most slowly cooled rocks, results in a relatively 'pure' K-feldspar host containing lamellae of Na-feldspar or vice versa.

Fig. 73C shows the phases that occur in the system SiO_2-albite-orthoclase under various conditions of water-vapour pressure. The first triangle, showing the dry-melt conditions is merely a duplication of the upper part of A, inserted for comparison. The second shows the effect of lowering the liquidus temperatures with sufficient water-content to eliminate leucite and to produce β-quartz. With very high water-vapour pressure the liquidus surface is so far depressed that it intersects the solvus and therefore two alkali-feldspars are simultaneously precipitated from the melt.

Since the pioneer researches of N. L. Bowen, much emphasis has been placed upon the **reaction principle** in rock-building, that is on the mutual interaction which takes place in a cooling magma between the solid and liquid phases. This reaction may be continuous, giving homogeneous solid solutions, as in plagioclases, or it may be discontinuous, taking place only at definite temperature intervals. The latter type is illustrated by the replacement, completely or in part, of olivine by pyroxene, and of pyroxene by primary hornblende. Such a series of minerals, when arranged in the order in which the transformations occur, constitutes a *discontinuous reaction series.* Bowen has drawn attention to the following series: olivine → orthorhombic pyroxene → common augite → hornblende

[1] The solvus curve for the alkali feldspars may be compared with that of the pyroxenes shown in Fig. 71: the main difference lies in the fact that the pyroxene solvus is 'beheaded' by intersection with the solidus curves.

→ biotite; leucite → orthoclase; anorthite → albite; common augite → sodic augite; common hornblende → sodic hornblende → lepidomelane.

It should be realized that almost all the rock-forming silicates are members of isomorphous groups, each one of which is a continuous reaction series; that each of these in turn participates in the changes hinted at above. Further, taking the simplest possible view of the course of crystallization in a natural magma, it has been established that there are at least two series of changes taking place concurrently, and to some extent overlapping and interdependent. The one involves the coloured silicates (the mafic constituents) and the other the felsic components. N. L. Bowen[1] represents them as in the table.

FORSTERITE (1890° C)
OLIVINE ↘
FAYALITE (1205° C)
↘ PYROXENE
ENSTATITE
ORTHO-
↘ HYPERSTHENE
MG-CA-CLINOPYROXENE
↘ AMPHIBOLES
↘ BIOTITE

ANORTHITE (1540° C)
↙ BYTOWNITE
↙ LABRADORITE
↙ ANDESINE
↙ OLIGOCLASE
↙ ALBITE

↘ ORTHOCLASE
↓ QUARTZ

Table showing discontinuous reaction series (left-hand side) and continuous reaction series of the plagioclase feldspars (right-hand side).

In the discontinuous series represented on the left-hand side of the table above, the first silicate minerals to appear are the olivines, and of these, magnesium-rich precede iron-rich types[2]. At a certain lower temperature these react with the magma and tend to be 'made over' into magnesium-rich pyroxenes. Actually the sequence of precipitation of minerals is dependent on the total composition of the magma fraction involved. Although 'Mg-Ca-rich clinopyroxene' follows orthopyroxene in Bowen's list, in many Basic rocks both pyroxenes may crystallize together simultaneously (p. 170).

Meanwhile at high temperatures calcic plagioclase will crystallize and with falling temperature will react with the magma, changing its composition steadily towards the sodic pole—the albite end of the continuous series. The interrelation of the two series of changes

[1] Bowen, N. L., *The Evolution of the Igneous Rocks* (1928).
[2] Bowen, N. L., and Schairer, J. F., The system MgO-FeO-SiO_2, *Amer. J. Sci.*, **29** (1935), 151.

is suggested by the fact that the conversion anorthite → albite releases CaAl, while the change in the pyroxenes involves the addition of CaAl.

It may be noted that there is a continual increase in complexity of atomic structure in passing from olivine to biotite. In olivine the structure is as simple as possible, the crystals being constructed of separated SiO_4-tetrahedra. In the pyroxenes they are linked into chains; in the amphiboles they form bands; and in the micas they form extended sheets. The successive conversions involve abrupt changes in crystal atomic structure: each change is a passage from simple to complex.[1]

The Later Stages of Crystallization: Deuteric Phenomena

Once we get down to the amphiboles and micas in Bowen's reaction series, the complexity of composition had until the 1950's made experimental work almost impossible. This is true of all the hydroxyl-bearing silicates, and to those which have crystallized from a fluid enriched in volatiles, or fugitive constituents. The very term 'fugitive' implies, as Shand meant it to do, that these constituents do not, in many cases, remain as part of the ultimate rock body. They may take part in several reactions and make others possible, and yet leave no trace of their passage. Thus it is necessary to face the task of interpreting an important part of the crystallization of an eruptive rock without knowledge of either the quality, the quantity or the ultimate whereabouts of the agents involved.

The important role of the fugitive constituents can be judged from the following facts.

1. When volatile-rich magma is erupted at the surface as lava, the fugitives are enabled to escape rapidly, and in so doing they immediately increase the viscosity of the lava, which solidifies rapidly, either as glass or as a cryptocrystalline aggregate. This rapid crystallization, coupled with the violent oxidation of escaping gases, accounts for the rise in temperature, amounting in some cases to a hundred or two degrees Centigrade, which is almost certainly responsible for the corrosion of previously formed phenocrysts. The significance of the vesicular and amygdaloidal structure of certain lavas, and even minor intrusives that consolidated near the surface, is too obvious, in this connection, to need stressing.

2. In some active volcanic regions there exist powerful reminders of the activities of fugitive constituents in the form of **fumaroles,** which give off intermittent or continuous streams of gases, some of which form mineral deposits by sublimation. These volatiles also cause considerable alteration of the rocks surrounding the vent.

[1] Brammall, A., Mineral transformations and their equations, *Sci. Prog.*, no. 120 (1936).

From this visible evidence provided by volcanic activity, we turn to consider the comparable effects produced in a deep-seated environment. Here direct observation is impossible, but it is reasonable to infer the probable course of events. One of the chief difficulties is to decide whether the active solutions arise entirely as a residuum left over after the completion of orthomagmatic crystallization, or whether they have been derived from some extraneous source. When the active solutions are directly of magmatic origin, all changes in mineral composition or in texture produced by them are termed **deuteric.** This term was introduced by J. J. Sederholm,[1] and may cover crystallization or alteration phenomena appropriate to both the pegmatitic and the hydrothermal stages. In the second case the active solutions are derived usually from a later intrusion, and any modification they effect is distinguished as **metasomatic.** Not infrequently it is impossible to distinguish between deuteric and metasomatic phenomena. We may thus be guilty of extending Sederholm's most useful term beyond its original meaning. The very acid and the highly alkaline rocks are usually products of magmas rich in H_2O and other fugitive constituents, and therefore tend to exhibit deuteric phenomena on an extensive scale, and in a variety of ways.

Since deuteric phenomena arise when the rock is very nearly solidified, it is natural that they should consist to a large extent of veining and replacement of earlier formed minerals. Several of these late-stage replacement phenomena are sufficiently distinctive and important to have names of their own: albitization, analcitization and chloritization are among the more important examples, while tourmalinization and silicification might also be considered in this category on occasion.

The term **albitization** covers a wide range of phenomena, though in essence, of course, it is simply the partial or complete replacement of earlier formed plagioclase or potassic feldspar by albite, which is stable in the presence of volatile-rich, lower temperature residual solutions. At this point we consider only those aspects of albitization which may legitimately be regarded as deuteric: we omit for the time being the widespread development of albite-rich rocks on a regional scale, exemplified by the occurrence of the spilitic suite; and similarly the conversion of basic plagioclase into albite charged with inclusions of lime-rich minerals, due to low-grade metamorphism of the so-called regional type. Deuteric albitization, on the other hand, involves, at an early stage, the development of the less regular types of perthitic structure. The 'patch-' and 'injection-perthites' are presumably formed in this way, by the action of

[1] On synantetic minerals and related phenomena, *Bull. Com. Geol. Finlande.*, no. 48 (1916), 134.

residual solutions on orthoclase or microcline. In slightly different circumstances the albite forms water-clear crystals interstitial to the minerals of earlier formation, or it may be deposited as a mantle around earlier feldspars. If the latter is orthoclase or microcline, there will be a striking contrast in appearance between the core and the external zone of the crystals, and this may be obvious in hand-specimens. If the core consists of plagioclase, however, there may be an almost imperceptible gradation towards the outer albite rim. The result may be a zoned crystal, indistinguishable from one produced by rapid cooling of an anhydrous melt. Thus there is no hard-and-fast dividing line between orthomagmatic and late-stage crystallizations: it is often impossible to say where one ends and the other begins. In fact, the distinction between the two stages only becomes apparent if the deuteric minerals are of a different nature from those of earlier formation, and if the former demonstrably replace the latter.

Analcite behaves in much the same manner as albite in many cases, especially as regards veining and progressive replacement of feldspars. But, like all late-stage minerals, it need not necessarily be replacive, but may crystallize in the interstices between crystals of earlier formation. When this is the case, although the crystal boundaries may be plane, with no sign of corrosion or embayment, there is clear evidence of the chemical activity of the residual solutions from which the analcite crystallized, particularly when the surrounding crystals are pyroxenes. For example in teschenites the pyroxene is a typical lilac-coloured titanaugite except where it is in contact with such 'pockets' of analcite: here it is rimmed with bright green aegirine-augite. Thus analcite plays a dual role. That which occupies these interstitial areas must be regarded as the last of the primary minerals to crystallize out; but that which so clearly veins and replaces earlier feldspar crystals is just as definitely secondary by definition, although the division is arbitrary and artificial.

This double role is also played by **chlorite.** The several members of the chlorite group may replace and pseudomorph pyroxenes, amphiboles and micas, generally with the development of a fibrous, and occasionally a spherulitic, habit. In certain rocks, however, pellucid grains of chlorite occur, bounded by crystal faces of the very minerals which elsewhere are replaced by it. Obviously such interstitial chlorite is primary—it has replaced nothing. Once again we are seeing the result of overlap between the final primary crystallization and deuteric replacement.

Because of their late-stage origin, there is a widespread tendency to call such minerals as chlorite and analcite secondary, whatever their mode of origin. The only logical use of this term, however, is to restrict it to minerals which demonstrably replace others of earlier formation. If one of these late-stage minerals has an interstitial mode of occurrence, there is no alternative to calling it primary unless, of course, there are actual relics of an earlier mineral visible.

Two particularly interesting and much debated phenomena remain for consideration: they are (1) interstitial micrographic intergrowth of alkali-feldspar and quartz occurring in certain Basic rocks; and (2) the peculiar quartz-plagioclase intergrowth, known as myrmekite, which is relatively common in Acid and Intermediate rocks.

In regard to the first, various hypotheses have been suggested to explain the intergrowth, which is commonly called **micropegmatite.** Vogt[1] claimed that micropegmatite results from the simultaneous crystallization of orthoclase and quartz in eutectic proportions in residual solutions. Other workers have found evidence that similar structures arise from the introduction of alkalies and lime into highly siliceous rocks; that is, micropegmatite may result from the feldspathization of quartz.[2] The converse process—silicification of feldspar—has also been suggested as a further possibility.

Both silicification of feldspar and feldspathization of quartz undoubtedly form types of quartz-feldspar intergrowth, though it is doubtful if the resulting structures are as geometrically perfect as those resulting from simultaneous crystallization. In the case of interstitial micropegmatite, however, any replacement hypotheses appear to be untenable, since the intergrowths are moulded upon, and radiate from, unaltered and euhedral feldspar crystals. In other words the micropegmatite crystallizes, without any replacement, from an interstitial residuum.

Myrmekite is quite different, both in composition and in mode of occurrence. It consists of lobate patches, often described as cauliflower-like in form, of plagioclase riddled with small 'vermicules' of quartz. The lobes normally have grown on plagioclase into microcline (Fig. 74). F. Becke[3] has estimated that the amount of quartz in the intergrowth increases as the host plagioclase becomes more basic. This fits in well with the hypothesis that the quartz is released as a result of the replacement of potassic feldspar (the formula of which may in this connection be represented by $K_2O.Al_2O_3.6SiO_2$), by plagioclase which contains the anorthite component, represented by $CaO.Al_2O_3.2SiO_2$.

Besides releasing silica, this reaction would also release potassium which is displaced by the incoming sodium and calcium ions. The

[1] Physical chemistry of the crystallization and magmatic differentiation of igneous rocks, *J. Geol.* **31** (1923), 245.

[2] Reynolds, D. L., Demonstration in petrogenesis from Kiloran Bay, Colonsay, *Min. Mag.*, **24** (1936), 367.

[3] Uber Myrmekit, *Tsch. Min. Petr. Mitt.*, **27** (1908), 377.

potassium probably contributes to the formation of mica as a by-product of myrmekite formation.

The intergrowths are most abundant along plagioclase-orthoclase,

FIG. 74

Myrmekite intergrowths of plagioclase (with twinning indicated) and vermicular quartz (*black*), penetrating microcline (*stippled*). The latter contains two rounded blebs of clear quartz. Dancing Cairn, Aberdeen.

or plagioclase-microcline boundaries, so that J. J. Sederholm[1] has suggested that the juxtaposition of these two contrasted feldspars is essential for the formation of myrmekite. He calls such reaction products formed between two neighbouring solid phases **synantetic.** Almost certainly the necessary reagents migrate along intercrystal boundaries, and it is possible that they may not all be entirely deuteric, for myrmekite is often found in granite adjacent to later Basic intrusions, and thus it may be partly of metasomatic origin.

Myrmekite is just one of several phenomena which arise by interaction between adjacent mineral grains. Corona structures, which have a similar origin and are most commonly developed in the gabbroic rocks, are described in due course.

[1] On synantetic minerals and related phenomena, *Bull. Com. Géol. Finlande*, no. 48 (1916), 134.

PART III

PETROLOGY OF THE IGNEOUS ROCKS, THEIR SIGNIFICANCE, DISTRIBUTION, AND ORIGINS

CHAPTER I

PRINCIPLES OF ROCK-CLASSIFICATION AND NOMENCLATURE

(1) General Considerations

ONE major purpose in attempting to devise a scheme of rock-classification is to ensure uniformity in nomenclature: it is essential that all petrologists should call the same rock by the same name. Petrological knowledge is growing faster than at any previous period, chiefly because of a large increase in the number of personnel involved; but the introduction of new techniques and the availability of more sophisticated apparatus also play their part. Formerly the study of rocks was largely qualitative, but recently emphasis has shifted, and quantitative data gained from studies in the field as well as the laboratory are becoming increasingly available. Statistical studies of the geographical distribution of different kinds of igneous rocks are increasingly important; but those whose research lies in this field are seriously handicapped by the lack of an internationally agreed system of classification and of uniform terminology. Even today a given rock name may have very different meanings for petrologists of different nationalities; and even for students in the same country but trained in different 'schools' following different schemes of classification. Thus, for example, 'dolerite' of British petrologists is 'diabase' to the Americans. It seems extraordinary that up to the present time it has proved impossible to reach general agreement on the exact definitions of even some of the most widespread igneous rock-types such as granite and basalt.

No one scheme of classification can be regarded as ideal for all purposes. Schemes designed for use in petrological research are generally too sophisticated and rely on characteristics that are unsuitable for everyday practical use. In this category we would include the more refined details of chemical composition which may be of critical importance in establishing magmatic relationship between igneous rocks.

The scheme which we advocate is essentially that which we have used formerly. It is designed to meet the needs of field geologists and students of geology; but it can readily be made more elaborate if and when greater precision in rock-naming is necessary for petrological research. The classification is based on mineral, as distinct from chemical, composition because the former is tangible —the mineral composition can be directly observed and most coarse-, and many medium-grained igneous rocks can be at least broadly identified by their megascopic characters, seen by the naked eye or, in more difficult cases with the aid of a petrological microscope.

A classification based solely on the chemistry of rocks is bound to include in the same category rocks differing widely not only in mineral composition but also in appearance, notably as regards degree of crystallization. Thus a granite and an obsidian may have identical compositions; but it would be ridiculous to give the same name to these two natural objects which are as different in appearance as any two rocks can be.

Several hundred rock-types have been described and named, many of them are widely distributed, and have names that are well known and acceptable to the majority of petrologists. An agreed scheme has to incorporate some of them, re-defined, if necessary, to fit the classification framework. Except in so far as general principles are concerned, no one scheme is applicable to all igneous rocks: a scheme based on the proportions of feldspars is obviously inapplicable to feldspar-free rocks; and the details appropriate to rocks consisting largely of felsic minerals cannot be expected to be entirely suitable for rocks which are largely mafic.

Consideration of Chemical Composition

By long-standing tradition rock analyses are generally stated in weight percentages of the nine major oxides: SiO_2, Al_2O_3, Fe_2O_3, FeO, MgO, CaO, Na_2O, K_2O and H_2O. Most analyses also include some of the commonly occurring minor components such as TiO_2, MnO and P_2O_5. In addition analyses may include some 'trace' elements such as B, F and Cr, taken from a long list of possibilities. These elements are present normally in such minute concentrations that they do not give rise to separate mineral phases, but substitute for appropriate major elements in the main rock-forming minerals.

Complete rock analyses are cumbersome and difficult to comprehend, so that for many purposes, both in petrological research and in the literature, it is convenient to select certain components or groups of components for comparison. The data so obtained may be presented diagrammatically. One of the most widely used **variation diagrams** of this kind was devised by Harker[1] and comprises the percentages of individual oxides plotted against SiO_2-percentage for each member of a suite of magmatically related rocks (Fig. 75).

[1] *The Natural History of Igneous Rocks* (1909).

Figure 112 illustrates another common kind of diagram showing variations of three groups of components, magnesium, total iron and total alkalis.

It will be appreciated that, although these ways of manipulating

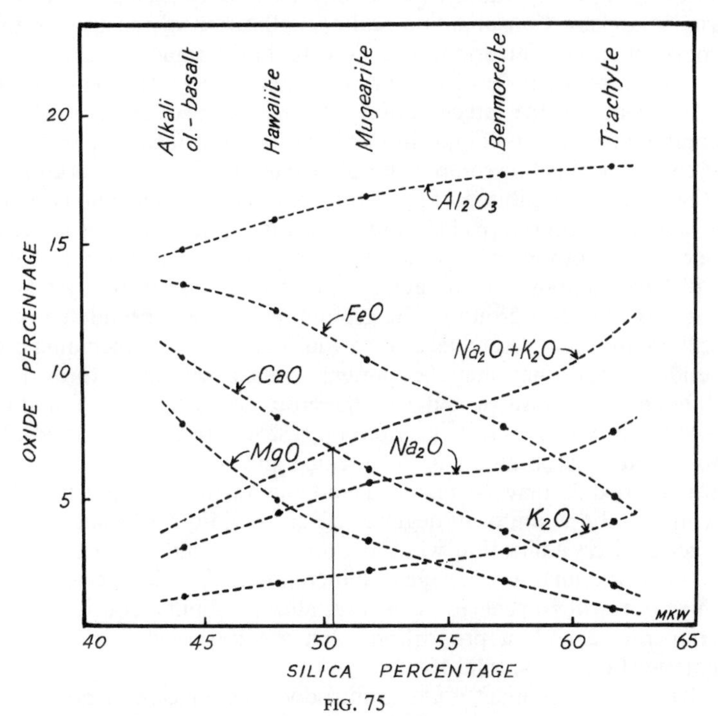

FIG. 75

Variation diagram for Hawaiian lavas of the Alkali Suite (constructed from average compositions quoted by G. A. Macdonald, *Geol. Soc. Amer. Mem.* 116 (1968), table 8, p. 502). Vertical line drawn from the intersection of the CaO and ($Na_2O + K_2O$) curves gives the so-called alkali/lime index, for these rocks about 51.

chemical data have little bearing on the problems of rock classification, it is ultimately the chemical composition that determines the mineralogical composition and most of the properties of a rock on which a classification must be based. Bulk chemical compositions or data abstracted from chemical analyses are therefore valuable in dealing with petrogenetic problems relating to magmas; but for the identification and classification of an igneous rock one needs either the actual mineral content or some form of equivalent 'mineralogical' composition that is calculated from the rock analysis. The former is known as the **mode** of a rock and the latter is the **norm.**

The Relationship between Chemical and Mineralogical Composition

The chemical compositions of most igneous rocks provide the closest one can get to the compositions of the fractions of magma from which they consolidated. As we have seen in the previous chapter, many factors affect the consolidation of a given magma. With loss of volatiles and very rapid cooling this may be quenched to form a volcanic glass. Consolidation and crystallization of most lavas and rocks in minor intrusions occurs with fairly rapid cooling over temperature intervals of one or two hundred degrees Centrigrade. This leads to the precipitation of early-formed crystals which, compared with the composition of the original melt, are enriched in the higher-melting components of their appropriate mineral groups. Reactions are inhibited both in the continuous isomorphous series described by Bowen (p. 176) and in the discontinuous series between one mineral group and another. Phenocrysts are generally zoned and will have compositions different from second-generation crystals of the same mineral group in the groundmass. The consolidated rock will comprise an assemblage of various components—including any residual glass that may be present—which will have formed at different temperatures and will therefore be in a state of mutual disequilibrium. This state may be contrasted with the close approach to equilibrium obtaining in deep-seated and slowly-cooled rocks, in which crystals may be unzoned and the various components will tend to show uniform degrees of crystallinity. Although some fractional crystallization will have occurred in the earlier stages of cooling, the final assemblage of minerals will show little evidence of this, but will appear to have crystallized simultaneously at one temperature and in proportions that are governed solely by the composition of the total rock.

This is a convenient point to introduce the concept of normative compositions, or norms, noted briefly above. A norm comprises the proportions of standard or normative 'mineral' components calculated from the chemical composition of the rock. It represents an assemblage that *could* crystallize under ideal equilibrium conditions from a magma of the same composition as the rock. It is important to realise that it is a calculated and therefore hypothetical assemblage. The value of this kind of calculation in petrology lies in the fact that a norm depends solely on chemical composition and is independent of the factors described above which affect a magma during its consolidation. Provided two rocks have identical compositions, even though one may be a glass and the other coarsely crystalline, their norms will be identical. A norm is thus essentially a statement of *magma*-composition expressed in terms of potential minerals.

The norm concept was introduced by the American petrologists Cross, Iddings, Pirrson and Washington[1] who devised an entirely

[1] *The Quantitative Classification of Igneous Rocks* (1903).

novel and elaborate scheme of classification based on proportions of the normative components. Although the attempt to cut adrift from the ambiguities of all previous schemes was entirely laudable, the classification has become obsolete because of the artificial nature of the scheme, which was too elaborate to be accepted by most petrologists.

However, although the classification has been abandoned, the calculation of normative compositions is still widely practised because of its value in dealing with many problems of petrology. A good example is provided by the division of basalts into two main types, those whose compositions yield normative hypersthene and those with normative nepheline (see p. 364). These normative minerals represent incompatible phases which cannot coexist in the one rock. It has been found that basaltic magmas with normative hypersthene give rise to suites of differentiates very different from those containing normative nepheline. This quite fundamental distinction may not be apparent simply by comparing analyses because the compositions of the two types of basalt can be closely similar. The mode of a basalt may be no more helpful, since accidents of the cooling histories of different basalts frequently prevent either hypersthene or nepheline from appearing among the minerals which actually crystallize. In a great many cases, therefore, the distinction between basalt types can only be made on the basis of their norms.

Only specialists need to be able to follow all the details of the calculation of a norm[1] but it is worth while for any student to master the underlying principles because they are helpful in understanding the chemical factors governing crystallization of minerals in igneous rocks.

The list of normative minerals is largely confined to the simplest end-members of the main groups of anhydrous rock-forming silicates, together with quartz and accessory minerals such as apatite, magnetite and ilmenite. Hydroxyl-bearing components such as hornblende and biotite are excluded because their compositions are too complex and variable for routine calculation. For ease of calculation, the compositions of the normative minerals are expressed in terms of the combining proportions of oxides, as in the following examples:

Orthoclase	$K_2O.Al_2O_3.6SiO_2$	(1:1:6)	Or
Albite	$Na_2O.Al_2O_3.6SiO_2$	(1:1:6)	Ab
Anorthite	$CaO.Al_2O_3.2SiO_2$	(1:1:2)	An
Wollastonite	$CaO.SiO_2$	(1:1)	Wo
Enstatite	$MgO.SiO_2$	(1:1)	En
Ferrosilite	$FeO.SiO_2$	(1:1)	Fs
Forsterite	$2MgO.SiO_2$	(2:1)	Fo
Fayalite	$2FeO.SiO_2$	(2:1)	Fa

[1] Full details and the necessary tables are given in *Petrographic Methods and Calculations* (1930), by A. Holmes, and in vol. I of *Descriptive Petrography of Igneous Rocks* (1931), by A. Johannsen.

As a first step in a norm calculation it is necessary to convert the weight percentages of the oxides (excepting H_2O) into molecular proportions by dividing each value by the appropriate molecular weight.[1]

The second step involves allocation of appropriate amounts from the available oxides to build up the normative minerals. Most of the oxides are shared between a number of minerals; but for the purposes of calculation, certain oxides can be regarded as entering into the composition of only one mineral. Thus all the TiO_2 is allocated to ilmenite (combined with an equal allocation of FeO); all the Fe_2O_3 to magnetite and all the P_2O_5 to apatite. These are therefore the first 'minerals' to be calculated.

The sequence of calculations can be illustrated by the relatively simple case of a rock of andesitic composition such as the examples quoted on p. 321. In addition to the accessory minerals listed above, the norm in this case includes the felspars, a calcic clinopyroxene of diopside type, a Ca-free orthopyroxene and quartz. The minerals are calculated in the order listed. In the case of an andesite magma one can assume for the purpose of calculation that all the alumina enters into the composition of the felspars, though in the actual rock some Al would occur in the augite and in any hornblende or biotite that might be present.

If x equals the total proportion of Na_2O available, the allocation for albite requires the addition of an amount x of Al_2O_3 and $6x$ of SiO_2. Similar allocations are made for orthoclase on the basis of the total K_2O, and anorthite is determined by the amount of Al_2O_3 that remains. This leaves some CaO, all of which is then combined in a 1:1 ratio with (FeO + MgO) to make normative diopside. The remaining FeO and MgO are allocated to an orthopyroxene molecule. The ratios of FeO to MgO found in naturally co-existing clino- and orthopyroxenes are closely comparable, so it is convenient to use identical ratios in calculating both of the normative pyroxenes.

At this stage, all of the oxides will have been allocated except for some SiO_2. This is referred to as free silica (since it has not been combined with the other oxides), and appears in the norm as 'quartz'.

The last step in the calculation of a CIPW norm involves conversion of the molecular proportions of the minerals into weight percentages.

Naturally, compositions other than andesitic require modifications in the calculation procedure. In many rocks, for instance, a deficiency of silica results in the appearance of olivine in the norm.

[1] It should be appreciated that many important details have been omitted from this account which is not intended in any sense as a guide to the practical calculation of a norm.

Similarly if alkalis are abundant but there is some shortage of silica, feldspathoids may take the place of some or all of the alkali feldspar.

Discussion of the manipulation of analytical data and of the chemical classification of rocks would be incomplete without reference to the system devised by Paul Niggli[1] which is widely used, particularly by European geologists. Like the CIPW system, this is based on molecular proportions of oxides. Allied oxides are then grouped into four categories: $Al_2O_3 + Cr_2O_3 +$ rare earths $= al$; $(Fe_2O_3 \times 2) + FeO + MgO = fm$; $CaO + SrO + BaO = c$; $Na_2O + K_2O + LiO = alk$.

It is possible to plot the composition of a rock within a tetrahedron defined by these four Niggli values, though in practice triangular diagrams are used. Further subdivisions in the classification are based on values for *si* (the ratio of the molecular proportion of SiO_2 to the combined values *al*, *fm*, *c* and *alk*), *k* (the ratio of K_2O to the total alkalies, *alk*) and *mg* (the ratio of MgO to *fm*).

Norms can be calculated just as easily in Niggli's system as in that of CIPW. The system allows the calculation of normative biotite, hornblende and other complex silicates which are excluded from the CIPW scheme. Thus, with greater flexibility in the choice of compositions of the normative minerals, the whole norm of a rock can be related more realistically to the actual composition. In fact, in theory, if enough trouble is taken, a norm of this kind could be calculated to correspond exactly with a mode. A calculation of this kind would not be used, of course, like a CIPW norm, the function of which is to provide a 'standard mineral' analogue of a *magma's* composition: it would be used when an estimate was required of the actual mineral composition of a *rock* which would result from the crystallization of a magma under specified conditions. In effect the result is a calculated 'mode'. It provides a valuable substitute for a real mode when a rock is to be classified on its mineral proportions and is too fine-grained or otherwise unsuitable for the latter to be measured directly.

As we have stated at the outset, the most appropriate classification for geologists to use under most circumstances is one based on the mineralogical composition of the rocks. Our scheme, in common with most mineralogical classifications, is based primarily on the variation of two characters: the proportions and compositions of the feldspars; and the total silica content of the rock expressed in mineralogical terms, *e.g.* whether or not quartz is present. These two criteria provide, as it were, the vertical and horizontal divisions in a tabular scheme of classification.

[1] *Schweizerische Mineralogische und Petrographische*, **16** (1936), 335. See also Niggli, P., The chemistry of the Keweenawan lavas, *Amer. J. Sci.*, Bowen Volume (1952), 381–412 for discussion of the principles; Burri, C., Petrochemical calculations, *Israel Program for Scientific Translations* (1964).

Silica Percentage as a Factor in Classification

In all igneous rocks, except the minor category of intrusive carbonates, silica is the dominant component. It is natural, therefore, that the proportion of silica to the other components should be regarded as a prime factor in most schemes of igneous rock classification. The most siliceous rocks have come to be known as the Acid Rocks, and those with progressively less silica as Intermediate, Basic and Ultrabasic respectively. This choice of terms is based on a misconception and stems from the time when silicates were regarded as 'salts' of various silicic acids, and rocks containing an abundance of silica were thus regarded as 'acid'. In this book the terms are written with initial capital letters to denote the special meaning with which they have been endowed by petrological usage.

Originally the classification was based on rigid, though arbitrarily defined silica percentages. The value chosen to separate Acid from Intermediate rocks, for example, was 66 per cent SiO_2, a figure based on the silica-content of alkali feldspars. This criterion is reasonable as long as one is dealing with rocks containing a preponderance of alkali feldspar. The problem is, however, that a whole spectrum of feldspars and other minerals is involved, each mineral having a different effect on the total amount of combined silica in the rock. It is worth remembering, for instance, that hornblende and biotite, which are the chief coloured minerals in many Intermediate and Acid rocks, contain relatively low percentages of SiO_2. Igneous biotites commonly contain between about 35 and 38 per cent SiO_2. The influence of the ferro-magnesian silicates is outweighed in many rocks by the anorthite component of plagioclase, containing 43·2 per cent SiO_2.

Obviously in siliceous rocks containing an abundance of these relatively basic minerals, less of the available silica is used up in combination and there is a better chance that free silica will occur than when the alkali feldspars are dominant. Strictly speaking, therefore, to distinguish Acid from Intermediate rocks on the basis of the SiO_2-percentage required for the first appearance of free silica, one would need a sliding scale of values depending on the minerals present, and notably on the proportions of plagioclase and alkali feldspars. A rock containing a somewhat calcic plagioclase as its main feldspar can achieve silica-saturation at a lower level of total SiO_2 than a rock containing predominantly alkali feldspar.

An additional objection to the use of rigid limits of silica percentage is that the necessary chemical data are not readily available to most geologists. Therefore, in recent editions of this book, though the rocks are divided into broadly the same major families, or classes, as previously, the *division has been made on the basis of the proportions of minerals present in the coarse-grained members of each family.*

If the grain-size of a rock is too fine or the rock is composed partly of glass it will not be possible to determine the modal percentage of quartz, and the amount of normative quartz may have to be calculated instead. In fact, with experience, most volcanic rocks—even those of very fine grain—can be identified on inspection and classified by comparison with their coarse-grained equivalents. The precise classification of a doubtful or borderline case is generally only required in a context of petrological research, when chemical data are anyway likely to be readily available.

The Role of Feldspars in Classification

Since early editions of this book we have used the relationship between quartz and different kinds of feldspar as a prime factor in classification. Feldspars are quantitatively important in most igneous rocks and are dominant over other minerals in many types. The different varieties are fairly easily distinguishable and their compositions are determinable in thin sections under the petrological microscope. Basically, we contrast the proportions of alkali feldspar (A) and plagioclase (P) in rocks. There are rocks in which the whole of the feldspar is 'A' and others in which it is exclusively 'P'; but in the majority of rocks both kinds occur together. Here then is an obvious basis of classification.

Alkali feldspars and plagioclase have been described and their interrelationships explained in the Mineralogy section. The former includes the 'pure' end-members, K^+-feldspar and Na^+-feldspar in both 'high' and 'low' varieties together with combinations of the two. 'Plagioclase' in the present context includes all members of the plagioclase family *except* albite and therefore ranges from oligoclase (about An_{10}) to anorthite (An_{90-100}).

Through the years we have consistently divided the full range of feldspar compositions into three equal parts, and have thought in terms of a third and two-thirds of the total feldspar when erecting divisions between three main series. Recently, however, it has been suggested that divisions should be erected at 10, 35, 65 and 90 per cent of either component (A or P). As regards the main divisions at 35 and 65, we can discern no significant advantage of these percentages over our fractions; though we would accept the former if they should be confirmed by international usage.

Any arbitrary division such as this is bound to be a matter of debate. At the present time the whole subject of the mineralogical classification of igneous rocks is being discussed by petrologists from all parts of the World through the medium of the International Geological Congress, stimulated and co-ordinated by A. Streckeisen.[1]

[1] Classification and nomenclature of igneous rocks, *N. Jb. Miner. Abh.*, **107** (1967) 144.

In addition to the divisions based on A/P ratios, further distinctions are made, based upon the proportions of individual alkali-feldspars on the one hand, and the composition of the plagioclase on the other. Details of these further subdivisions are discussed later.

The proportions of the various feldspars can generally be estimated or measured without much difficulty in coarse-grained igneous rocks. Even if a rock is fine grained a petrographer will often be able to assign it to its correct category by using, for example, the evidence of phenocrysts and from experience of other rocks of known composition. There are, however, certain pitfalls of which the beginner should beware. When there are two kinds of feldspar in a rock one variety commonly occurs as conspicuous euhedral crystals, the amount of which one is tempted to over-estimate at the expense of the other variety which may be anhedral, inconspicuously filling the interspaces. It is also easy to over-estimate the relative amount of a generally more calcic core compared with a more sodic rim on zoned crystals of plagioclase: the rim needs to have a thickness of only 0·13 times the diameter of the core to represent an equal volume of material.

If a rock is glassy or too fine grained for the mineral proportions to be assessed by inspection, recourse has to be made to normative values calculated from a chemical analysis. Some authors, including S. J. Shand, advocate the use of normative values in every case, irrespective of grain-size. This procedure certainly has some important advantages: problems arising out of the dual role of albite are eliminated; and the normative values, unlike modal percentages, are unaffected by the degree of crystallization of the rock. In dealing with problems involving petrogenesis and magmas (as distinct from rocks) these advantages may become of paramount importance. Otherwise, as we have already suggested, the fact that a norm is an invisible and indirectly determined quantity renders it unsuitable as the basis of routine classification.

Classification of Quartz-Bearing Rocks

Because most of the igneous rocks containing significant amounts of quartz are relatively poor in mafic constituents they can be classified effectively on the basis of the proportions of their felsic minerals —quartz, alkali feldspar and plagioclase—solely. This is a great convenience since the essential features of the scheme of classification can be represented in simple diagram form. The proportions of the three minerals are plotted with quartz (Q), alkali-feldspar (A) and plagioclase (P) occupying the apices of an equilateral triangle. The value of 'Q' is derived from the amount of modal quartz by dividing this figure by the sum of the total felsic minerals (modal quartz, alkali feldspar and plagioclase) and multiplying by 100.

The QAP triangle, as we shall call it, provides a simple means of plotting the composition of quartz-bearing rocks with a high degree of accuracy, and covering a range from 0 to 100 per cent of each

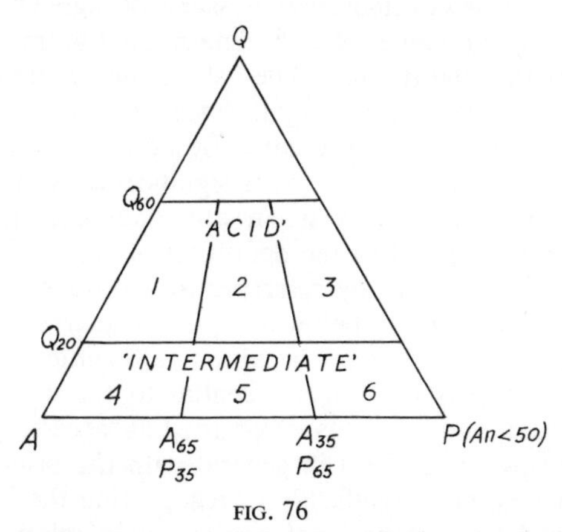

FIG. 76

QAP triangular diagram for the classification of quartz-bearing rocks, showing the fields of composition of the six major families of coarse-grained rocks and their fine-grained equivalents.

of the three constituents named. Up to this point we have been concerned only with the *relative amounts* of quartz, alkali feldspar and plagioclase; but another significant line of variation involving plagioclase has to be taken into account. Not only does the amount of the latter vary from rock to rock, but its composition changes systematically with increasing basicity. This change in composition is not very significant in igneous rocks rich in quartz and alkali feldspars; but it provides a critical means of distinguishing between two major coarse-grained igneous rocks—diorite and gabbro—and their fine-grained equivalents, andesite and basalt, as will be shown in due course. Since it is impracticable to incorporate a fourth variable on a triangular diagram, we limit the use of the QAP triangle to cover only the rocks for which it is really appropriate, *i.e.* Acid rocks and those Intermediate ones—the great majority—which are saturated with SiO_2. In our use of the QAP triangle, P is restricted to plagioclase with an anorthite-content of, at most, An_{50}. This means that gabbros and other Basic rocks which contain plagioclase more calcic than this are excluded from the QAP triangle; these rocks can be classified more appropriately by using other criteria discussed below.

The value of the QAP triangle is demonstrated by the fact that it has been used as the basis of several schemes of classification. Unfortunately all existing schemes differ to a greater or less extent on the nomenclature and numerical limits of the several 'fields' into which the triangle is divided. In the current international discussions on classification to which reference has been made above, however, a considerable measure of agreement has been reached on the Q values which determine the main horizontal boundaries within the triangle. The values chosen are Q_{20} and Q_{60}. The latter is of little practical significance because very few igneous rocks have compositions represented by points above the Q_{60} line. By contrast the Q_{20} line is of great significance. For all the rocks above this boundary quartz is an abundant constituent and its presence is thus essential to the definition of such a rock: this fact is expressed more simply by referring to the quartz as 'essential'. When the composition falls below the Q_{20} line quartz is a subordinate constituent and, indeed, may be absent altogether. Therefore the presence of quartz is no longer essential to the definitions of the rocks concerned, and its status is regarded as 'accessory'.[1]

We are happy to conform in general with the principle of using QAP parameters, and specifically in recognizing the importance of the Q_{20} boundary: international agreement in relation to igneous rock classification is a prize worth striving for. Previous editions of this book have evolved through a series of steps towards this situation. Firstly the rigid chemical distinction between Acid and Intermediate rocks at 66 per cent SiO_2 was abandoned in favour of a 10 per cent limit of modal quartz. This figure was increased to 20 per cent in the last edition largely because many traditional (and typical) Intermediate rocks contain more than 10 per cent but generally less than 20 per cent of modal or normative quartz.[2]

It is, of course, only a small step from 20 per cent modal quartz to the present choice of 20 'Q'. Even so, the reader should appreciate that the two values *are* different. They would actually coincide for a rock consisting solely of felsic constituents and devoid of coloured minerals; but as the amount of the latter rises so the Q-value increasingly differs from the amount of modal (or normative) quartz. As a guide, a rock consisting of 60 per cent feldspars, 20 per cent quartz and 20 per cent mafic minerals would have a Q-value of 25; and conversely a rock with the same amount of mafic constituents but with Q_{20} would contain only 16 per cent of modal quartz.

The Acid and quartz-bearing Intermediate rocks, separated by the Q_{20} boundary are each divided in our scheme into three categories based on the A/P ratios as discussed in the previous section. Thus

[1] It will be appreciated that the word 'accessory' is being used in a sense completely different from that which relates to the 'accessory minerals' such as zircon, magnetite, apatite, etc. These, unlike quartz, only very rarely achieve rock-forming status.

[2] See, for example, the discussion on the composition of andesites.

the QAP triangle is divided into six 'fields' of equal status, or seven if we include the virtually unoccupied apical portion of the triangle above the Q_{60} line. Independent names have to be found for these fields, and while difficulty has been experienced in finding acceptable names for the three fields covering the Acid coarse-grained rocks, it is generally agreed that the most suitable names for the Intermediate fields are those of familiar and widely accepted rock-types of appropriate composition—syenite, monzonite and diorite. It must be noted, however, that we are concerned at this point only with quartz-bearing syenites: there are several members of the syenite family which cannot be accommodated in the QAP triangle as will be demonstrated in due course. The individual fields are rigidly defined by the proportion of the three felsic components only, so that the definitions of the fields cannot agree exactly with the original definitions of the rock-types whose names have been chosen: disregarding the coloured minerals, for instance, means that some adjustment of the original definitions is necessary in order to fit the requirements of the rigid framework of the classification. Despite this difficulty, it is considered much better to adapt established names rather than to create new ones.

Several schemes of classification involve division of the QAP triangle into many more than our six fields. As one might expect, such elaboration leads to difficulties in nomenclature which are discussed in later sections of the book. All that needs to be said at this point is that we believe it is best to keep the primary divisions of the classification as simple as possible, and to adopt a single name to cover all members of the family of rocks within each field. Then, if for any reason more precise definition is required for particular groups of rocks, this can be achieved by division into sub-fields.

We now turn our attention away from the quartz-rich rocks towards the changes which have to be taken into account in classifying the quartz-free Intermediate and more Basic rocks—in effect those rocks which lie at the base of, or outside the QAP triangle.

These changes are of many kinds so that whereas it was possible to 'translate' in a simple way from the original SiO_2 percentage to Q-values in defining the boundary between Acid and Intermediate rocks, no comparable simplicity exists for defining the Intermediate/Basic and Basic/Ultrabasic boundaries in mineralogical terms. The terms Intermediate, Basic and Ultrabasic are retained because they are useful in a qualitative sense in bringing together rocks of broadly comparable SiO_2 percentages; but for quantitative purposes in effect the critical units are the individual rock families, each defined in terms of its limiting mineral composition. As the SiO_2-percentage falls, various mineralogical criteria come into play, depending upon whether one is dealing with alkali-rich rocks or those towards the calcic end of the compositional range. These mineral variations are of three main kinds:

1. The incoming of feldspathoids in the alkali-rich series:
2. increasing anorthite-content of plagioclase in the Calc-Alkali series;
3. variation in the kinds and proportions of coloured minerals.

Naturally, this last kind of variation affects all rock series to some extent, regardless of the nature of the feldspars present; but in practice the coloured minerals tend to play a more important role in classification in the calc-alkali rocks of a Basic or Ultrabasic character.

Each of these three factors is discussed in principle in the following paragraphs; for practical details the reader should turn to the chapters devoted to systematic petrography.

Silica Saturation and the Role of Feldspathoids

In all three families of Intermediate rocks—syenites, monzonites and diorites—feldspars are normally the dominant constituents. As we have seen, if the silica percentage is high enough, the rock is over-saturated with silica and quartz occurs in addition to feldspars. However, in the case of syenites, and to a less extent, some monzonites, if the amount of available silica falls below the amount required for the formation of alkali-feldspars, a proportion of the alkalies and aluminium may go to form one or other of the feldspathoids as explained in the Mineralogy section. The degree of undersaturation is very variable, and in extreme cases may result in the complete elimination of alkali-feldspars. Although feldspathoids occur in a great variety of rocks of different chemical compositions, it is in defining the limits of the syenite family that under-saturation in regard to the felsic constituents becomes most significant, and the details are therefore discussed in the chapter on syenites.

The Role of Anorthite-Content of Plagioclase

In the rock-types containing plagioclase as the dominant feldspar (granodiorite, diorite, gabbro) variation in mineral composition with falling silica percentage takes a completely different form from that obtaining in the alkali-rich rocks. In the plagioclase-rich rocks there are no analogues to the feldspathoids to substitute for plagioclase: instead the lower values of total SiO_2 are reflected in changes in the composition of the plagioclase. Silica-poor anorthite increases at the expense of the more siliceous albite molecule in the plagioclase.

In granodiorite the latter lies within the range oligoclase—andesine; diorites and their fine-grained equivalents, andesites, typically contain andesine; while the plagioclase in gabbros and basalts has an average composition of about An_{65}, with a range covering labradorite to bytownite. Pure anorthite occurs only rarely in some Basic and Ultrabasic rocks. In view of this pattern of variation of plagioclase composition it is natural that a value of An_{50} should have been chosen to mark the boundary between Intermediate and Basic categories, represented by the diorite and gabbro families respectively. Although this criterion serves adequately in the great majority of cases, for instance in distinguishing between typical andesites and typical basalts, it is not the only, or indeed the best, criterion to use in all cases. There are for instance some lavas which form parts of the suites of differentiates in oceanic volcanoes that contain andesine or even oligoclase—that is, 'andesitic' plagioclase—but in other respects are so basalt-like that they are classified as basalts rather than andesites. The generally Basic character of these rocks, known as hawaiites and mugearites, is determined in this case by the relative abundance and nature of the coloured minerals rather than by the composition of the plagioclase.

Colour Index and the Role of Mafic Constituents

In Acid igneous rocks mafic constituents are sparse and are therefore of little consequence in classification. They are more important in Intermediate rocks; but it is in the Basic category that they play an important role—several leading rock-types are defined in terms of the kinds and proportions of the mafic minerals they contain.

The total percentage of mafic constituents in a rock is its **Colour Index (M).** If we think in terms of *all* coarse-grained igneous rocks set out as it were on display, it would be possible to separate them on the basis of 'M' into three groups, respectively light, medium and dark in colour. Three comparable terms embodying Greek roots are leucocratic, mesotype[1] and melanocratic, and it would be logical to allot to each a range of M-values. Thus leucocratic would cover the range 0 to 33, mesotype, 34 to 66 and melanocratic 67 to 100.

Rather than allotting a specific range of M-values to a particular rock-type we believe that the above terms serve a more useful purpose if they are employed in a comparative sense. Granites normally have Colour Indices ranging between 5 and 15. A granite with $M = 25$ although falling within the leucocratic range would, among other granites, be outstandingly dark-coloured, and it would be appropriate to describe it as melanocratic, or (to save a few

[1] The middle term might well have been 'mesocratic', but the roots make this an impossible term: it has been condemned as a 'linguistic miscarriage'. Nevertheless, it is still widely used.

syllables) melagranite. By comparison, a gabbro with precisely the same 'M' would be, among other gabbros, comparatively light-coloured and might be termed leucocratic or leucogabbro; but a diorite with the same 'M' would be average or mesotype. Therefore rocks with exactly the same M-value may correctly be termed leucocratic, mesotype or melanocratic *by comparison with others of their kind.*

By using the contractions leuco- and mela- as prefixes to a particular rock-name greater precision is gained and one rock-name serves instead of three. This course is practicable for those rock-types which consist of a felsic mineral, say nepheline, associated with a mafic constituent, say pyroxene in varying proportions. The ratio of the former to the latter varies actually from essentially 'pure' nepheline-rock to 'pure' pyroxene rock. The rock-type composed of both components in approximately equal amounts is ijolite. Most known ijolites are notably variable rocks: some specimens are pyroxene-rich, others are nepheline-rich. There is a good deal to be said for distinguishing the former as mela-ijolites and the latter as leuco-ijolites rather than using separate specific names and hence obscuring the gradational relationship which exists between them. The reader should compare the petrographic descriptions of ijolites with those of diorites and gabbros to see how the principle is applied in different cases.

It should be noted that some petrologists attach more significance to Colour Index than we do and use it as a basis for distinguishing, say, basalt from andesite; but this gives rise to serious inconsistencies: lavas conforming to our definition of basalt are classified as andesites purely because they have what is regarded as an andesitic Colour Index. A more acceptable name for these rocks would be leucobasalt.

Colour Index is of little practical use in rock classification, except to delimit the group of rocks deficient (to vanishing point) in felsic components but correspondingly rich in mafics, and appropriately termed Ultramafites. Most of these are feldspar-free, and several are monomineralic. A reasonable value of M for this purpose is 90.

Although it is relatively easy to measure the Colour Index of a rock on a smooth surface of a hand specimen—provided it is sufficiently coarse-grained, it becomes increasingly difficult as grain-size diminishes and a point is reached where direct measurement becomes impracticable. However, even if a rock is fine-grained or glassy, its *normative* Colour Index may be used: this is 100 minus the sum of *q*, *or*, *ab*, *an* and *ne*.

Grain size (Degree of Crystallization)

Among other factors that have been employed in rock classification is mode of occurrence, and at one time it was normal practice to divide rocks of *the same composition* into three categories, with a different name for each, according to mode of occurrence. Three collective terms—Plutonic, Hypabyssal and Volcanic—were, and still are, used for three major categories into which all igneous rocks are grouped. For many years we have maintained that these terms are unsatisfactory when applied to igneous *rocks* because they are incapable of scientific definition; they are not mutually exclusive. In particular, it is known that rocks of basaltic composition and texture are not restricted to lava flows, *i.e.* to the volcanic mode of occurrence, but occur also in a multitude of thin dykes which are hypabyssal in mode of occurrence. Most petrologists agree, we believe, that it would be completely unrealistic to apply different names to identical rocks: these two rocks are, beyond question, basalts. We believe that it is impracticable to restrict a given rock-type name to one mode of occurrence: *a rock is distinctive and identifiable in virtue of what it is—not of where it came from.*

If a large collection of basaltic rocks from all three modes of occurrence were to be laid out on display, they would be seen to show continuous variation in degree of crystallinity from extremely coarse to extremely fine-grain (aphanitic). It would be practicable to arrange them in order of degree of crystallinity and to group them arbitrarily into three categories to which the terms Coarse, Medium and Fine could be applied. This is the scheme which we have advocated through several editions of this book, using grain-size as the factor in classification instead of mode of occurrence.[1] For general purposes a rock falls in the fine-grained category if the individual components are too small to be seen with the unaided eye. This does not apply, of course, to any large, first generation crystals ('phenocrysts') which may be present, but only to the matrix ('groundmass') in which they may be embedded. Similarly if the groundmass (or the whole rock in the absence of phenocrysts) consists of grains large enough to be seen, but too small for identification, it probably belongs to the medium grain-size category. These are admittedly only rough-and-ready tests. More precise classification involves measurement of grain size, preferably on thin sections, using arbitrarily chosen numerical limits. The limits in use were chosen by a British Association Committee on Petrographic Nomenclature to correspond as closely as possible with those exhibited in nature by rocks occurring as (1) lava flows, (2) dykes and sills and (3) larger rock bodies of so-called plutonic type.

Apart from variation in grain-size (which is of prime importance) textural variations are of minor significance only: they should not be made the excuse for introducing new rock-names—there are

[1] Wells, A. K., The grain-size factor in classification of igneous rocks, *Geol. Mag.* **75** (1938), 471.

more than enough already. A given mineral assemblage in each of the grain-size groups requires a distinctive name; but textural varieties are adequately covered by using appropriate qualifiers.

The Nomenclature of Igneous Rocks

Some names currently in use are of considerable antiquity—for example, 'syenite' and 'basalt'. In choosing these early names only the most obvious physical features could be employed: thus 'rhyolite' —the rock which flowed; 'trachyte'—rough to the feel, and 'phonolite'—sonorous when struck with the hammer, were introduced and still find their place in our nomenclature.

The advent of the petrological microscope gave impetus to the search for new rock-types and at the same time revolutionized the study of rocks by revealing the details of their mineral composition and textures. Consequently the accumulation of petrographic data at this time was rapid and largely haphazard. The descriptions, judged by modern standards, were often incomplete and definitions overlapped.

During this period, which lasted for several decades, a large number of hitherto unknown rock-types were brought to light by the classic studies of Brögger in Scandinavia, Arthur Holmes in parts of Africa, Lacroix in Madagascar and many others. In general the only 'principle' employed when choosing names for the newly-discovered types was to use the name of the type locality with the uniform termination 'ite' added. The name of the type locality might be that of a mountain chain, a town, village or even a farm, quite regardless of the status of the locality or of the rock itself. Consequently petrographic literature is over-loaded with such names, many of which have an outlandish look to the Anglo-Saxon eye; some are unpronouncable and many defy one's efforts to locate them by gazeteer or atlas.

We base definitions of igneous rock-types on distinctive mineral assemblages using the smallest number of constituent minerals practicable. Thus, troctolite is a two-mineral assemblage: it consists essentially of approximately equal amounts of olivine and calcic plagioclase. The addition of a third mineral say, hypersthene, in significant, but subordinate amounts, is covered by a two-part name, 'hypersthene-troctolite' (with a hyphen). It should be noted that hypersthene in the above example is a *non-essential* constituent. If it increased in amount to the point of becoming essential, we would have a three-mineral assemblage for which a name different from troctolite is required: in this instance it is olivine-norite.

A rock-type is defined in terms of its mineral assemblage; but many rock names consist of two parts—a rock name preceded by the name of a mineral, the two being joined by a hyphen, for example, 'mica-syenite' and 'quartz-diorite'. Syenite is defined, and its position within the classification framework determined, by the proportions of its felsic constituents only: possible mafic constituents are not named in the definition; but if a specific syenite contains mica (biotite) as the chief, or possibly the only dark mineral present, the obvious name to employ is mica-syenite. This is a commonsense usage, and it is correct.

Compound rock-names are in use and may cause difficulty as they are not always used in the same sense. Probably the first such name to come into common use was 'trachyandesite'. Trachyte and andesite are lavas characterized by predominance of alkali feldspar in the former and of plagioclase in the latter. Trachyandesite is neither trachyte nor andesite but is a rock-type intermediate in composition—and in its position in the classification framework—between the two. This particular example sets the pattern for the use of all such compound rock-names. A rock bearing a compound name must lie in horizontal or vertical sequence between the two types bearing names which are compounded in its own. It follows that a compound name should not be applied to the end-member of a sequence of rock-types.

It might well be thought that, of all rocks, those consisting essentially of one mineral only would be the easiest to name; but this is not the case. It would simplify nomenclature considerably if petrologists would agree that all **monomineralic** rocks should bear the name of the mineral of which they are composed, with '-ite' added. Such names would be self-explanatory. Examples which conform to this principle are hornblendite, pyroxenite, hypersthenite and albitite; but other exactly comparable terms are blatant misnomers. Thus, amphibolite, by analogy with pyroxenite, might be expected to be a monomineralic rock consisting essentially of amphiboles; but actually it is neither of these things: it is not even an igneous rock—it is metamorphic, though similar to diorite in mineral composition. The worst offenders, however, are to be found among the names given to certain nepheline- and leucite-bearing lavas. These are discussed more fully in a later chapter; but we may note that, far from being monomineralic, some so-called nephelinites may contain up to 90 per cent mafic minerals.

CHAPTER II

ROCKS CONTAINING ESSENTIAL QUARTZ: THE 'ACID' IGNEOUS ROCKS

Introduction

THE Acid igneous rocks are defined as containing more than 20 per cent quartz among the felsic constituents. This figure may seem high in relation to the original concept of an 'acid' rock as being one with just sufficient silica to ensure the presence of some free quartz. The choice of Q_{20} is based, in fact, on geological experience of the properties of typical Acid and Intermediate rocks viewed in the light of experimental research and knowledge of the relevant petrogenesis. The crystallization of granite, for instance is governed essentially by the thermal trough in the liquidus surface in the system SiO_2-Or-Ab discussed on p. 174. Phase relationships within this system ensure that magmatic granites normally contain about 30 to 40 per cent of normative quartz. Any rock which is otherwise similar to granite but contains significantly less quartz is likely therefore to have originated under some other conditions, and one is justified, therefore, both on petrogenetic and petrographic grounds, in distinguishing it as a variety of syenite.

We divide the Acid rocks into three main families based on the alkali-feldspar/plagioclase ratios, and in each family coarse-, medium- and fine-grained types are distinguished, as shown in the accompanying Table.

Grain size	Alkali-feldspar > $\frac{2}{3}$	Alkali-feldspar and plagioclase each between $\frac{1}{3}$ and $\frac{2}{3}$	Plagioclase > $\frac{2}{3}$
Coarse	Alkali-granites	Adamellites	Granodiorites
Medium	Alkali-Microgranites	Micro-Adamellites	Micro-Granodiorites
Fine	Alkali-rhyolites	Rhyo-dacites or toscanites	Dacites

GRANITES, ADAMELLITES AND GRANODIORITES

Unfortunately the choice of names is still controversial. 'Granite' is a name which has been in use by geologists and laymen alike for a very long time, and naturally it has been used in several different senses. In a general geological context the term 'granite' may relate to almost any coarse-grained Acid rock; but by petrologists it is generally used in a more restricted sense, mostly for such rocks relatively rich in alkali-feldspars and correspondingly poor in plagioclase. Thus in Streckeisen's classification 'granite' extends across most (but not the whole) of the fields we have designated alkali-granite and adamellite. It was largely for this reason that we thought it necessary to add the prefix 'alkali-' for our more restricted use of 'granite' in previous editions of this book. It would, of course, be much more convenient (as well as consistent with the practices we recommend in relation to rock-nomenclature) if the prefix could be dropped. It is possible that, if a symmetrical three-fold division of the Acid igneous rocks were to be widely accepted, plain 'granite' without any qualification would be recognized as the most suitable name to stand in comparison with adamellite and granodiorite. This is what we recommend and hope. We propose to apply the term granite to the whole field; then, if the term 'alkali-granite' is retained it would apply to the appropriate sub-division of the whole field. The situation is complicated by the fact that various authorities have used the name 'alkali-granite' for a separate field in parallel with, but not a part of, their 'granite' field. This is the position in schemes of classification based on a primary five-fold, rather than three-fold division on the basis of the A/P ratios as discussed on p. 193.

Various limiting values of 'A' in relation to 'alkali-granite' have been suggested: Gendler favours 80, Niggli and Trögger 87·5, Nockolds and Streckeisen 90 and Johannsen 95. Actually 'alkali-granite' used in this sense is inherently undesirable, because preponderance of alkali-feldspar is a characteristic of *all* granites. This perhaps rather pedantic difficulty could be overcome by terming granites lying between the A_{100} and A_{90} boundaries as 'alkali-rich granites' or more specifically 'K-rich' or 'Na-rich' as the case may be. As an alternative it might be possible to promote the name of a specific rock-type ('alaskite' has been suggested) to cover the field in question. Na-rich granites (sodic granites) are particularly distinctive, not so much on account of the felsic constituents, albite or albite-rich antiperthite contained in these rocks, but of the mafic minerals which may include Na-rich pyroxenes and/or amphiboles. There is ample justification, therefore, for considering sodic granites separately in the account which follows.

Through several editions of this textbook we have used the name adamellite for the middle member of the coarse-grained Acid igneous rocks. Unfortunately, the name has not caught on in Germany, France and Italy; and alternative names are in use elsewhere. Thus 'adamellite' as we use the term is 'quartz-monzonite' to Tröger and most American petrologists; it is 'normal granite' to Streckeisen and petrologists in the U.S.S.R. As there are serious objections to these alternative names, we propose to continue using 'adamellite'.

Finally, we turn to the remaining major field which is termed granodiorite by all the specialists in petrographic nomenclature whose names have been mentioned in this discussion, so it should be above reproach. It appears to be internationally acceptable and it is so firmly established that it would be unwise to attempt to replace it by a more logical name consistent with the principles we are trying to establish. Strictly speaking the compound name 'granodiorite' is a misnomer: it implies that this rock-type occupies a position in the classification framework between granite and diorite. It does not: it is the most highly silicated member of the continuous series diorite—quartz-diorite—granodiorite.

(1) Alkali Granites

Potassic and Sodi-potassic Granites

Arising out of their experimental studies of the system orthoclase, albite, quartz and water, Tuttle and Bowen have described the relationships between the feldspars in alkali-granites, and have suggested that significant differences in cooling history are indicated by these relationships. Certain granites contain only one kind of feldspar, an orthoclase-albite intergrowth. These one-feldspar granites[1] are probably restricted to high level, sub-volcanic environments, where cooling would be relatively rapid, and therefore the degree of unmixing of the two components in the perthitic intergrowths would be slight. The component minerals exhibit high-temperature optics. Certain British Tertiary granites in the Western Isles of Scotland are of this type and are accepted as being magmatic.

By contrast, other alkali-granites contain two or more kinds of feldspar: albite (not intergrown), orthoclase or (more commonly) microcline may accompany one of the Or-Ab intergrowths. The significance of these differences is discussed more fully below, when the stability relationships are explained. Meantime it will suffice to point out that, provided the feldspars were initially high-temperature, homogeneous types, it is obvious that unmixing has gone much farther in two-feldspar, than in one-feldspar granites, presumably

[1] One-feldspar granites are the '*hypersolvus granites*'; and two-feldspar granites the '*subsolvus granites*' of Tuttle and Bowen.

because of the less steep temperature-gradient with a correspondingly long cooling period.

Quartz averages perhaps 30 per cent of the whole, and varies in its relationship to the feldspars. It is invariably the low-temperature (α) form and builds irregular shaped composite grains, sometimes interstitial to the feldspar, sometimes lobed into it in a manner suggesting replacement (Fig. 77). In the more potassic varieties containing large microclines, some quartz forms relatively small rounded inclusions in the latter; but some is moulded upon the microcline. Finally, quartz frequently participates in the formation of the delicate intergrowth occurring along plagioclase-microcline boundaries, and known as myrmekite (Fig. 74).

Inclusions in the quartz are ubiquitous, and in thin sections are seen to consist of trains of minute bubbles sealed-in in healed fractures in the crystals. Minute acicular crystals, identified (by analogy with more robust, megascopic specimens) as rutile, may occur in considerable numbers—but they have to be searched for in the thin sections. As these rutiles are restricted to the quartz grains, it may be inferred that their substance was originally held in solid solution in the quartz, but separated from the latter by unmixing at the appropriate temperature. Such rutile needles indicate that the quartz in which they occur must have originated as the high-temperature form and thus provide valuable evidence regarding temperatures of crystallization of the granites concerned.

Apart from rutile a little iron-ore may occur, and may well be the only accessory visible in a given thin section; but by crushing rock-samples and using the 'heavy-mineral technique' of concentration and separation, a surprisingly large variety of the rarer accessories may be obtained. Thus it may be noted that the biotite may be crowded with pleochroic haloes formed by irradiation from such radio-active minerals as zircon and xenotime.

All true alkali-granites are relatively deficient in mafic minerals, some more so than others. These are distinguished as **leucogranites**—they are very light-coloured with scattered grains of dark mica. A facies of the Dartmoor granite from Wittabarrow, Devonshire, consists of alkali-feldspar and quartz making up 97 per cent of the rock, with only 3 per cent of mafic and accessory minerals.

Micas are the only mafic minerals commonly seen in these granites and include biotite and/or muscovite often closely associated. The biotite is often partially altered to chlorite.

Examples of sodi-potassic granites are not uncommon among the Armorican granites of south-western England and the Channel Islands (Fig. 77).

It is certainly no coincidence that the alkali-granites seem to be particularly prone to alteration by late-stage or pneumatolytic alteration-processes, as described below. Incipient alteration of these types may cause small flecks of sericite or paragonite to form, often along cleavages, while kaolinization may cause clouding and ultimately opacity of the feldspars.

A distinctive type, of somewhat doubtful affinities, is **charnockite,**

FIG. 77

Alkali-leucogranite, Mount Mado, Jersey, C.I.

Reproduced by permission of the Council of the Geologists' Association, London.

The minerals shown are quartz and two kinds of alkali-feldspar, microperthite and water-clear albite: the rock is a two-feldspar leucogranite. To show the feldspar relationships to advantage the feldspars are drawn as between crossed polarizers, but the quartz is left clear.

described from a locality in Madras. The type-rock contains quartz 40, microcline 48, oligoclase 6, hypersthene 3, biotite 1, and magnetite 2 per cent. In hand specimens the quartz appears bluish, while in thin section it is seen to be charged with extraordinary numbers of minute acicular crystals, identified as rutile, occurring in hundreds of thousands per cubic centimetre. Apart from this, the outstanding feature is the occurrence of distinctly pleochroic hypersthene. This links charnockite with a series of hypersthene-bearing rocks of varying composition—the so-called Charnockite Series,[1] which spans the whole range of composition, from Acid to Ultrabasic.

[1] Holland, T. H., *Mem. Geol. Surv., India*, **28**, Part II (1900), 162; Rao B. Rama, *Bull. Mysore Geol. Dept.*, no. 18 (1945). Howie, R. A., *Trans. Edin. Roy. Soc.*, **62** (1955), 98.

A constant feature is the presence of unusually pleochroic orthopyroxene, of iron-rich type, associated with smaller amounts of augite (rarely), a somewhat distinctive hornblende, or biotite. The most distinctive accessory, not invariably present, is garnet, of pyrope-almandine type.

Among the rocks broadly included in the term charnockite are varieties characterized by a preponderance of microcline—these are the true charnockites, very closely matching the specimens originally described by Holland (1900) from the type locality. In addition, however, others contain two feldspars, a K-type and a plagioclase in the proportions which would place them among the adamellites of our general classification. They are hypersthene-adamellites, showing the distinctive mineralogical and textural features which link them with the Charnockite Series, and are therefore distinguished as 'charnockitic adamellites.' Finally, in other specimens plagioclase within the andesine range (An_{35}) is dominant over microcline, and the rocks fall among the granodiorites. Charnockitic granodiorites have been given the name **'enderbite',** after Enderby Land, the type-locality in Antarctica.[1]

Charnockites are found only in certain Pre-Cambrian shield areas. Some of the better known occurrences include peninsula India and Ceylon, areas of Basement rocks in Africa, *e.g.* the Belgian Congo, Uganda, Natal, Central Sahara and Madagascar; the Fennoscandian Shield; the Scourie area of Lewisian rocks in Scotland; and shield areas of the Ukraine and Siberia. In America charnockites appear to be more restricted. In these areas erosion has uncovered rocks formed originally under conditions of high temperature and particularly high pressure characteristic of the granulite facies of metamorphism. The mineral assemblages of charnockites are those of metamorphic rocks: the association of hypersthene and K-feldspar, for example, is essentially the high temperature and pressure equivalent of some of the biotite that occurs in normal granites of comparable composition. If charnockites are accepted as the metamorphic rocks which they now undoubtedly are, it becomes somewhat academic to decide whether they were originally igneous and intruded as magma. Because all rocks undergo plastic deformation under conditions of deep-seated metamorphism, there is likely to be a large measure of tectonic concordance between the charnockite bodies and the other associated metamorphic rocks. Transgressive relationships which would help to establish original magmatic intrusion are often inconspicuous or absent. Of course, even a transgressive relationship would not prove that the rock crystallized directly to charnockite from magma: intrusion may have

[1] Tilley, C. E., Enderbite, a new member of the Charnockite Series, *Geol. Mag.*, **73** (1936), 312–16.

occurred under conditions of lower crustal pressures giving initially a more normal igneous rock. This appears to have been the case, for example, in some charnockite occurrences in the Enisey range in Siberia in which vestiges of the mineralogy and texture of original Basic igneous rocks have survived. Probably the most convincing evidence of the ultimate igneous origin of charnockitic rocks is provided by their chemical compositions. A typical charnockite series gives a variation diagram ranging from Basic to Acid compositions which is broadly comparable with that of many volcanic suites.

Sodic Granites

True sodic granites are rare, but those which do occur are very distinctive rocks, particularly as regards their coloured silicates. Typical examples occur in ring-complexes in Nigeria[1] and the Sudan. Some of the rock-types building these complexes are ordinary biotite-granites; but of much greater interest are associated sodic rock-bodies of plutonic, extrusive and hypabyssal types, characterized by aegirine and/or riebeckite. The distinctiveness of the mineral suite is, of course, merely the expression of the chemical composition of the magma-fraction from which it was formed, and which has been shown by analysis to have been relatively rich in sodium but poor in Ca, Al and Mg. Deficiency in the first two affected principally the feldspar association in the rocks: the anorthite content is exceptionally low, the plagioclase being nearly pure albite. Further, among coloured silicates the more normal biotite and hornblende were ruled out, their places being occupied by non-aluminous and calcium-free aegirine and riebeckite, sometimes alone, but often in the closest reaction relationship, the latter being usually external to the former. One other special feature of these rocks may be noted: they are rich in an assortment of the rarer accessory minerals, including some of considerable economic importance. A high content of fluorine finds expression in the presence not only of fluorite, but also of cryolite, the corresponding fluoride of sodium, occurring in exploitable amounts in one known riebeckite-granite. Thorite, monazite and xenotime occur, as well as radio-active pyrochlore, visible even in hand-specimens as small honey-coloured octahedra, and valuable as a source of the element niobium. Astrophyllite, seen in thin sections as bright yellow, micaceous-looking aggregates, locally becomes so important in the marginal facies of some riebeckite-granites as to rank as an essential constituent. Nearer to hand 'peralkaline' granites build the islet

[1] Jacobson, R. R. E., Macleod, W. N., and Black, R., Ring-complexes in the younger granite province of northern Nigeria, *Mem. no.* 1, *Geol. Soc. Lond.* (1958).

Analyses of Alkali-Granites

	1 Charnockite	2 Sodic leuco-granite, Mountsorrel	3 Riebeckite-granite	4 Aegirine-granite, Rockall	5 Potassic leucogranite, Dartmoor	6 Potassic granite, Dartmoor	7 Potassi-sodic granite, Dartmoor	8 Alkali-granite Skye
SiO_2	70·65	76·70	76·25	70·31	73·16	73·66	71·69	69·62
Al_2O_3	15·09	12·58	10·86	7·53	13·95	13·81	14·03	13·91
Fe_2O_3	0·80	0·10	1·23	8·32	0·03	0·21	0·57	1·18
FeO	1·53	2·09	0·76	2·44	0·47	1·51	1·93	3·01
MgO	0·53	0·65	0·18	0·02	tr.	0·45	0·66	0·45
CaO	2·66	1·10	0·37	0·35	0·43	0·67	1·49	1·73
Na_2O	2·99	4·90	4·68	5·26	2·57	2·89	3·03	4·27
K_2O	4·69	0·52	4·65	4·19	8·16	5·02	4·59	4·92
H_2O	0·65	0·85	0·50	0·43	0·64	1·66	1·76	0·65
TiO_2	0·46	0·20	0·11	0·26	0·04	0·16	0·33	0·49
Rest	0·10	0·14	0·36	0·54	0·17	0·31	0·29	0·17
	100·15	99·83	99·95	99·65	99·62	100·35	100·37	100·40

1. Charnockite, Madras, India (Anal. J. H. Scoon), R. A. Howie, *Trans. Edin Roy. Soc.*, **62** (1955), 98.
2. Sodic leucogranite, marginal facies of Mountsorrel granodiorite, Charnwood Forest, Leicestershire (Anal. W. H. Herdsman), *Geol. Mag.* (1934), 1.
3. Riebeckite-aegirine granite, Nigeria (Anal. R. O. Roberts), *Geol. Soc. Lond. Mem.*, I (1958), 17.
4. Aegirine-granite, Rockall, P. A. Sabine, *Bull. Geol. Surv. G.B.*, **16** (1960), 156–78.
5. Potassic leucogranite ('aplogranite'), Wittabarrow, Dartmoor (Anal. H. F. Harwood), *Min. Mag.*, **20** (1923), 41.
6. Potassic granite, Haytor East Quarry, Dartmoor (Anal. H. F. Harwood, *op.. cit.*).
7. Potassic-sodic granite, Saddle Tor, Dartmoor (Anal. H. F. Harwood, *op. cit.*).
8. Alkali granite, Glamaig, Skye (Anal. E. A. Vincent), L. R. Wager, *et al.*, *Phil. Trans. Roy. Soc.* (A), **257**, 273–307.

of Rockall[1] in the North Atlantic, the dominant type being one carrying both aegirine and riebeckite. A melanocratic facies has been called **rockallite,** in which the colour-index is very high for granitic rocks—39; the feldspar is albite, 23, while quartz totals 38 per cent.

As defined above, sodic granites are not necessarily excessively alkaline: with rather more Al and less Fe, a type occurs which is no less strongly sodic, but is more ordinary in its coloured mineral content. In such rocks the whole of the Na is locked up in the feldspar, which, by definition, may be albite, but is commonly either microperthite, antiperthite, or cryptoperthite. These are the dominant components, associated, of course, with the requisite amount of quartz, with biotite and the usual accessories.

(2) ADAMELLITES

In this category are included those granitic rocks in which plagioclase accompanies an alkali feldspar in approximately equal amounts: neither the alkali feldspar nor the plagioclase should exceed two-thirds of the total feldspar present. Thus, compared with the alkali-granites, adamellites are distinctive through the increasing importance of Ca-ions in the feldspar. Generally the plagioclase lies within the oligoclase range, but may be andesine. The essential quartz, alkali-feldspar and plagioclase are accompanied by biotite in some adamellites, but by biotite and common hornblende in others, while, as noted above (p. 209), some of the rocks included in the Charnockite Series are charnockitic adamellites. They contain essential quartz, K-feldspar and plagioclase in approximately equal quantities, and pleochroic hypersthene. In the definition of adamellite no limitation is placed upon the content of dark minerals; but as a rule the colour-index is rather higher than in alkali-granite, though lower than in granodiorites.

The name **adamellite** was originally suggested for a type now distinguished as tonalite, but was redefined by Brögger (1895) in substantially the sense in which we use it here. The type locality is the Adamello Complex in the Tyrol; but adamellite is a widely distributed type, though often referred to merely as biotite-granite or biotite-hornblende-granite. A well-known British example occurs at Shap Fell[2] in Westmorland. Part of the complex consists of a particularly handsome rock characterized by numbers of large pinkish phenocrysts of orthoclase, embedded in a granular aggregate

[1] Sabine, P. A., The geology of Rockall, North Atlantic, *Bull. Geol. Surv. G.B.*, **16** (1960), 156–78.
[2] Grantham, D. R., Petrology of the Shap granite, *Proc. Geol. Assoc.*, **39** (1928), 299.

of white oligoclase, quartz and biotite. Micrometric analysis[1] shows this adamellite to contain approximately quartz 24, orthoclase 36, oligoclase 34 and biotite 6 per cent.

(3) GRANODIORITES

Of all the coarse-grained, quartz-rich rocks, granodiorites are quantitatively the most important: indeed they are far more widespread than all the coarse-grained members of the Intermediate and Basic Clans combined. In existing records, however, care must be taken to see the exact sense in which the name is being used—usage is varied, and there is a good deal of confusion as between granodiorite, tonalite and quartz-mica-diorite. In all of these types plagioclase is dominant, to the extent of at least two-thirds of the total feldspar. While granodiorite is quartz-rich, however, tonalite and quartz-mica-diorite contain quartz as an *accessory* only, not as an essential component.

Granodiorites, then, are coarse-grained igneous rocks containing essential quartz with plagioclase dominant, though alkali-feldspar may occur, but must not exceed one-third of the total feldspar content. These felsic minerals are accompanied by a varying proportion of coloured silicates and accessories, of which biotite and hornblende are almost constantly present in the former, and sphene, apatite and magnetite in the latter category.

This is almost exactly the original sense in which the name was first used (Becker, 1892, in conjunction with H. W. Turner[2] and W. Lindgren) as applied to certain rocks of the Sierra Nevada range. Thus defined, granodiorite transgresses the silica-percentage boundary (at 66 per cent) formerly used to separate the Acid from the Intermediate igneous rocks: there are as many granodiorites with less than 66 per cent as there are with more than this amount of total silica. The average is 66 to 67 per cent, and of this about 22 per cent is quartz. There is a perfect gradation into tonalite, which differs only in the subordination of quartz.

Granodiorites in which alkali-feldspar is completely (or almost completely) suppressed, have been termed **trondhjemites** (Goldschmidt, 1916), and consist of plagioclase of the appropriate range of composition for granodioritic rocks—oligoclase to andesine—together with quartz and small quantities of biotite, sometimes proxied by hornblende or pyroxene. With an average SiO_2 percentage of over 70, and 20 to 30 per cent of quartz, it is appropriate to include them as orthoclase-free granodiorites, which in mineral composition are very close to quartz-rich tonalities.

[1] Holmes, A., *Petrographic Methods and Calculations* (1921), 594–9.
[2] The rocks of the Sierra Nevada, 14*th Ann. Rep. U.S. Geol. Surv.* (1894), 478 and 482.

It is in the enormous 'granitic' batholithic complexes of the mountain ranges of western North America that granodiorite, together with the closely similar tonalite, really 'holds the stage'. The two together have been proved by detailed mapping to occupy the greater part of the surface area of the southern California batholith—3,500 out of 4,000 square miles mapped.[1]

In Britain granodiorites are far less spectacular, of course, but nevertheless are well represented, with the main Donegal granodiorite at the head of the list, occupying an area of 30 by 5 miles.[2]

Many loosely named 'hornblende-biotite granites' are granodiorites. As an example we may quote the Mountsorrel intrusion from Charnwood Forest, Leicestershire. The major part of this intrusion is granodiorite containing, on average, quartz 22·6, alkali-feldspar 19·7, plagioclase 46·8, biotite 5·8, hornblende 2·9, and magnetite 2·2 per cent. Similar types occur also among the Caledonian and Devonian granitic complexes of the Highlands and Southern Uplands of Scotland, as in the Moor of Rannoch and Ben Cruachan complexes. The average composition of granodiorites from the Garabal Hill–Glen Fyne complex[3] is all but identical with the Mountsorrel rock quoted above, but there are striking textural differences: the Scottish type includes strongly porphyritic varieties, in which the phenocrysts consist of alkali-feldspar (microline-microperthite) while plagioclase forms subhedral grains, usually zoned, and associated with quartz, hornblende, mica and prominent pale brown sphene.

THE TEXTURES OF GRANITES, ADAMELLITES AND GRANODIORITES

The term 'granitic' used in a textural sense implies no more than that the rock concerned is granular, *i.e.* like a granite. Typically, none of the major constituents is bounded by crystal faces: their shapes have been determined by mutual interference during growth. This most typical of the wide range of granitic textures is sometimes termed 'xenomorphic granular'. It sometimes happens that the feldspars tend to exhibit crystal faces, though the quartz is again anhedral. For this variant, the term 'hypidiomorphic granular' is used.

The handsomest granites are undoubtedly the **porphyritic** types in which phenocrysts of white, grey or red feldspar occur embedded

[1] Larsen, E. S., Batholithic and associated rocks of Corona . . . California, *Mem. Geol. Surv. Amer.*, **29** (1948).

[2] Pitcher, W. S., *et al.*, The main Donegal granite, *Quart. J. Geol. Soc.*, **114** (1958), 259–305.

[3] Nockolds, S. R., The Garabal Hill–Glen Fyne igneous complex, *Quart. J. Geol. Soc.*, **96** (1940), 451.

in a groundmass that may be identical in texture with an ordinary aphyric granite. The phenocrysts may be strongly zoned, and attain in some cases among the West of England granites to 7 inches by 5 inches, measured on the side-pinacoid faces. They frequently lie in parallel orientation and thus provide significant data in structural studies of granitic rock-bodies. Instances of such granites have already been mentioned in the foregoing account. In Britain the Shap adamellite and the porphyritic variety of the Garabal Hill–Glen Fyne granodiorite are good examples; but striking specimens are obtained from the 'giant granite' from Dartmoor.

The great majority of granites may be referred to one or other of these two main textural types. There remain for consideration certain varieties which call for more detailed consideration. **Orbicular granites**[1] contain 'orbs' of various sizes—they may measure several inches in diameter—embedded in a matrix of normal granitic texture. The 'orbs' are, in fact, variously shaped, and they exhibit extraordinary rhythmic banding. Each of the orbs contains a core, which in some instances is of the same composition as the matrix in which they are embedded, but in other cases it is completely different. Again, although the texture of the matrix is sometimes normal granitic, it may show a continuation of the successional crystallization of the component minerals, in the sense that the interspaces between the orbs are filled largely with feldspar, but with mica (biotite) in the centre, representing the last mineral to crystallize. Most specimens so far described occur in Finland,[2] and some of them are objects of outstanding petrological interest. Much has been written concerning the significance of orbicular structure. Nuclei appear to be essential, and may be either xenoliths of wall-rock or cognate xenoliths of granite, *i.e.* material of early consolidation broken up by, and incorporated in, later granitic magma. The magma may have been highly viscous, and diffusion consequently slow. This would favour successional rhythmic crystallization around the nuclei.

The **rapakivi texture** also was originally described from Finnish granites. In typical specimens large flesh-coloured potassic feldspars form rounded crystals a few centimetres in diameter, and are mantled with white sodic plagioclase, in some cases rhythmically zoned with orthoclase. These feldspars are embedded in a matrix of normal texture, but consisting chiefly of quartz and coloured minerals. Recently the term 'rapakivi' has been applied to granites and porphyritic microgranites which contain phenocrysts of reddish

[1] Johanssen, A., *Descriptive Petrography of the Igneous Rocks*, vol. iii (1937), 248. See particularly the photographs of a range of different types of orbicular granites.

[2] See Sederholm, J. J., On orbicular granites . . . , *Bull. Com. Géol. Finlande*, No. 83 (1928).

orthoclase with narrow mantles of whitish sodic feldspar; for example, some facies of the Shap, Dartmoor and Jersey granites and associated minor intrusions. It is customary in these days to regard phenocrysts with suspicion—the possibility that they originated elsewhere than in the magma represented by the matrix in which they are now embedded must be given due consideration. With the Finnish rapakivi granites this may well have been the case, for the rounded form of the 'phenocrysts' suggests, but does not prove, magmatic corrosion. Growing crystals develop plane surfaces only if surface-tension conditions are appropriate, otherwise curved surfaces develop. Therefore a rounded shape may be original. Genuine rapakivi granites are orthoclase-rich, therefore orthoclase crystallizes first, and with slow cooling large phenocrysts may be built up. With falling temperature quartz begins to crystallize, and subsequently all three components—orthoclase, quartz and sodic plagioclase—may crystallize simultaneously. Some of the plagioclase would be deposited round the orthoclase nuclei, but some of it might form independent crystals, particularly if crystallization was accelerated. It has been suggested that continuous build-up of vapour-pressure may have led to the ultimate fracturing of the 'roof' above the granite, allowing the water vapour to escape, with the consequence of the setting up of many new centres of crystallization. Much of this cannot be proved, of course, but here is a suggested mode of origin which would account for the distinctive features of these interesting rocks.

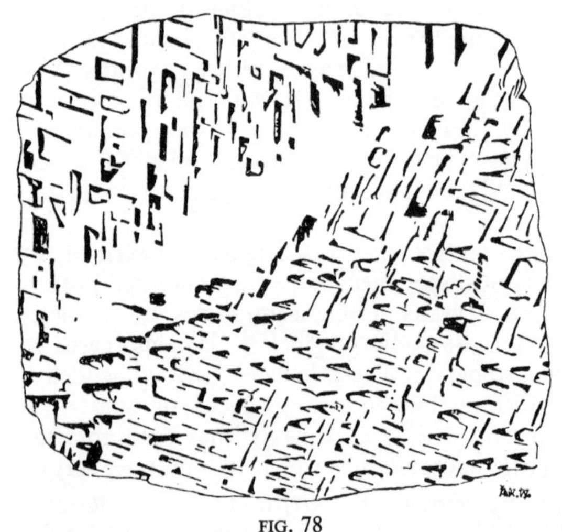

FIG. 78

Sketch of hand-specimen of runite (graphic granite). Microcline plain; quartz hieroglyphs in black, showing shapes controlled by three directions in the host-mineral. *Natural size.*

The **graphic** or **runic texture** is one of the most distinctive shown by granitic rocks, and is particularly characteristic of the pegmatitic facies. Both terms have reference to the marked resemblance of the small quartz 'hieroglyphs' to runic characters, which show up clearly against the background microcline in which they are embedded (Fig. 78). This texture is considered more fully under the heading 'pegmatites' below.

Finally, some granites contain small irregular cavities lined with well-terminated crystals of the normal constituents of the rock, accompanied by some of the rarer accessories. Such granites are said to be **miarolitic.** In Britain the granites of the Mourne Mountains in Ireland and Lundy Island off the Somerset coast provide typical specimens. Smoky quartzes, well-terminated feldspars, 'books' of white mica as well as rarer apatite and topaz crystals occur in the manner described. Again there is a link with pegmatites, for the most likely cause of the phenomenon is the existence in the magma of 'pockets' of gas which included the fluorine sealed up in the white mica, topaz and apatite.

GRANITE-PEGMATITES AND APLITES

The rocks we have studied so far represent the major portion of the granite magma: the final residuum of large intrusions is naturally rich in the fugitive constituents of the magma, and by the freezing of this residuum rocks are produced which differ markedly in both mineral composition and texture from the normal granite. In texture they fall into two contrasted types: the first of relatively coarse grain; the second relatively fine. The former are the granite-pegmatites, and the latter granite-aplites.

Granite Pegmatites

Few rocks are so well endowed with the power of capturing the imagination as the pegmatites. They are extremely coarse-grained and contain not only the world's largest crystals, but also the choicest mineral specimens. Further, some pegmatites are of considerable economic importance, since they are formed by the crystallization of residual solutions in which there is often a marked concentration of rare elements.

The place that the pegmatite fluids hold in the cooling history of a granite mass has already been mentioned (p. 164). That the main part of the crystallization is completed before the formation of the pegmatitic residuum is obvious from the relationship between pegmatitic bodies and the main granite masses with which they are associated. Most pegmatites occur in the form of irregular segregations, veins or small dykes or sills, which are especially abundant in the marginal parts of the parent granite, or in the country rock surrounding it, though some pegmatite veins occur at a distance from the main intrusion, and appear to have no visible connection with it.[1]

Pegmatite dykes do not, however, form an invariable part of all exposed granite masses: for although most of the latter contain occasional ultra-coarse-grained schlieren or clots, some are associated with a profusion of pegmatitic veins and dykes. The Pre-Cambrian granites seem particularly well endowed in this respect; but pegmatites are very poorly represented in association with many granites of later periods, for example the great granitic batholiths of the Andes. In many instances the pegmatite dykes run conformably with the strike of the country rock, and in such abundance as to simulate a dyke-swarm.[2] Pegmatites are characteristic of regions of compression, not of tension, as is the case with true dykes. Further, unlike true dykes, pegmatites are normally quite short, irregular in form, and rarely parallel-sided. The extent to which pegmatites are formed by in-situ replacement remains an open question.

Mineral Composition

Alkali-feldspar is the dominant constituent of all granitic pegmatites. In those classified as **simple pegmatites** it is generally microcline-microperthite, associated with quartz and white mica. Among the abundant and varied accessory minerals are those normally occurring in granite, together with others which are often regarded as 'pneumatolytic', such as tourmaline, topaz and other fluorine-bearing minerals and various ores, among which cassiterite is one of the most important.

Less commonly pegmatites consist of a much more varied mineral assemblage. In these **complex pegmatites** albite, or clevelandite, is an important constituent, often exceeding microcline in amount, and is accompanied by a suite of lithium-bearing minerals including red and green parti-coloured tourmalines, spodumene and lithia-micas, also beryl and minerals containing niobium, tantalum and other rare elements. The special interest attaching to the complex pegmatites is discussed below, under the heading 'paragenesis'.

Textural Features

One of the most characteristic features of granite-pegmatites is the development of the graphic or runic texture, resulting from the close intergrowth of feldspar—generally microcline-microperthite and

[1] Anderson, Olaf, Discussion of certain phases of the genesis of pegmatite, *Norsk. Geol. Tiddskr.*, **12** (1931), 1.

[2] See *e.g.* the account given by Gevers, T. W., and Frommurze, H. T., The tin-bearing pegmatites of the Erongo area, S.W. Africa, *Trans. Geol. Soc. S. Africa*, **32** (1929), 11.

quartz. It was to this structure that Haüy originally applied the term 'pegmatite,' although in later years this has given place to 'graphic' or 'runic'.[1]

Graphic granite presents many striking features and as many problems. In the most regular examples the quartz takes the form of sub-parallel, elongated prisms which pass through the feldspar, and present the characteristic hexagonal cross-sections, often flattened and distorted in the manner of ordinary quartz crystals. Fersmann has shown that the vertical axes of the quartz crystals are so inclined as to make an angle of approximately 70 degrees with the *c*-axis of the host feldspar. However, other investigators have been unable to confirm this regularity suggested by 'Fersmann's Law'. Apart from their attitude, it has been found that the proportion of quartz to feldspar is fairly constant, at about 30 to 70 per cent. These various features strongly suggest that the structure results from the crystallization of a eutectic or cotectic mixture of the two components. Despite this, however, origin by replacement of the feldspar by quartz has been urged by certain authors.[2]

The most striking textural feature of pegmatites is their extraordinary coarseness of grain. Impressive examples are afforded by a beryl crystal 19 feet, and a spodumene 47 feet long, associated with microcline-perthites several feet through, discovered in a pegmatite at Keystone, in the Black Hills, South Dakota. The growth of such large crystals must result primarily from the very low viscosity of the pegmatitic fluid, caused by the presence in it of abundant fluorine, magmatic water and other substances of low atomic and molecular weights. T. Quirke and H. Kremers[3] have suggested that the movement of fluid of constantly varying concentration and temperature through the interstices of the crystallizing pegmatite may be an important contributory factor in building up large crystals: a constantly replenished mother liquor could, in this way, largely surround each growing crystal. Occasionally, however, details of crystallization indicate that once a pegmatitic residuum has accumulated, it has remained static until crystallization has been completed. The Cornish pegmatite illustrated in Fig. 79 shows the growth of crystals, including thin acicular tourmalines, perpendicular to the roof of the sill-like intrusions from which the specimen was collected, and it is difficult to imagine the growth of such crystals in anything other than an undisturbed medium.

The *paragenesis* or order of crystallization of the several consti-

[1] Haüy's original term has survived, however, in 'micropegmatite', which is a comparable structure, though on a finer scale, seen in the interstitial quartz-feldspar intergrowth in certain quartz-gabbros and quartz-dolerites.

[2] Schaller, W., Mineral replacements in pegmatites, *Amer. Min.*, **12** (1927), 59; also Wahlstrom, E., Graphic granite, *Amer. Min.*, **24** (1939), 681.

[3] Pegmatite crystallization, *Amer. Min.*, **28** (1943), 571.

tuents has been worked out in full detail for a large number of pegmatites. It is usually found that intergrown quartz and alkali-feldspar (graphic granite) head the list. The subsequent course of

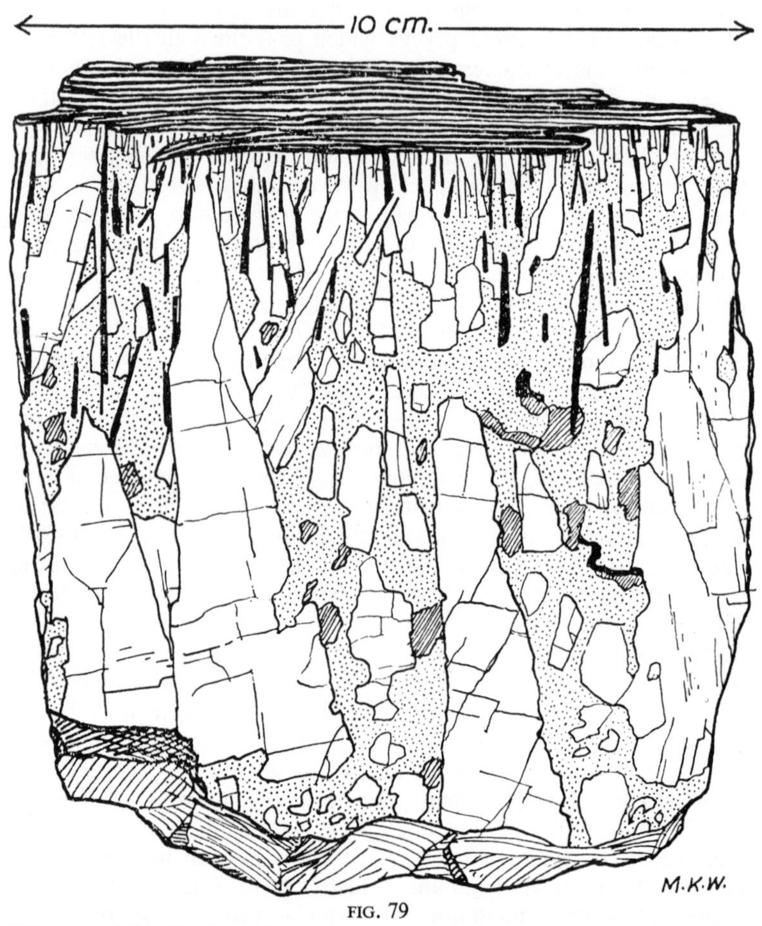

FIG. 79

Diagram of block of granite-pegmatite showing the arrangement of crystals developed perpendicular to the roof of the sill, Porthleven, Cornwall. Tourmaline crystals, black; mica, shaded; quartz, stippled; and perthite, blank. Roof rock is tourmalinized shale ('killas').

mineral deposition appears to be controlled by the gradual passage of solutions of varying composition through the 'pegmatitic thoroughfare'. These from time to time are liable to effect changes in composition and textural relationships between the original components. In this way the albite (clevelandite) of the complex pegmatites is produced at the expense of the earlier microcline-microperthite.[1] Striking replacement textures between these two minerals are so frequently observed that the mechanism of this type of albitization is beyond question.

When the migrating solutions dissolve previously formed minerals

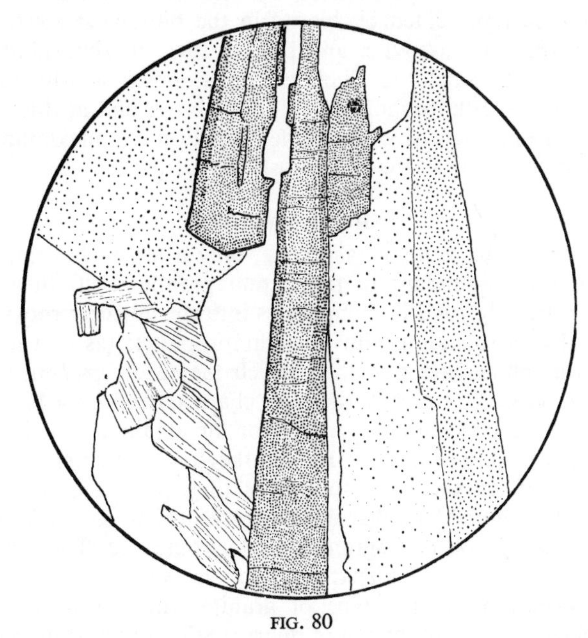

FIG. 80

Granite-pegmatite from pegmatite-aplite sill, an offshoot of the Godolphin granite, Porthleven, Cornwall. From the upper part of the specimen shown in Fig. 79. Zoned tourmalines (*heavy stipple*); orthoclase (*light stipple*); muscovite (*lined*); quartz (*blank*).

faster than they deposit new ones, open cavities or vugs may result. These are often lined with fine crystals of smoky quartz, adularia, clevelandite, etc., and with hydrothermal incrustations. The quartz crystals in these cavities are in some cases continuous with the quartz rods in surrounding graphic granite. The quartz crystals which grew freely into the cavities are of the α-type, while the intergrown rods show signs of having inverted from the β-quartz type. In other words, during the growth of the quartz rods towards the vugs, the temperature fell below the inversion point at which high- is converted into low-temperature quartz: *i.e.* at about 575° C. Thus for once we can make use of a reasonably accurate point on a geological thermometer, and state that pegmatite containing graphic

[1] Schaller, W., The genesis of lithium pegmatites, *Amer. J. Sci.*, **10** (1925), 269.

granite normally commences to crystallize not far above 575° C, say at 600° C.

One of the most distinctive mass-characters of many pegmatites is a more or less pronounced zonal structure within the rock-body, shown most obviously by textural differences, and on closer examination by mineralogical differences between successive zones. Thus mica may be represented by biotite in the outermost zone, giving place inwards to muscovite, and this, in turn, to zinnwaldite, and in the centre, to lepidolite. Such zoning is consistent with the idea of the 'pegmatite-thoroughfare' outlined above, and in this specific case indicates progressive enrichment in lithium, maximum concentration occurring in the innermost, latest fraction.

Granite-Aplites

These rocks occur as veins, as a rule, only a few inches thick, and although most abundant in the parent granite itself, they sometimes penetrate beyond its boundaries into the adjacent rocks. They are found in association with mica-lamprophyres (as rich in mafic minerals as aplites are poor), to which they are *complementary* in Brögger's sense. Chemically they are characterized by a high silica content and a considerable proportion of alkalies, with iron and magnesium in subordination. To this they owe their light colour—white to buff. In the hand-specimens they present a remarkably even and fine-grained saccharoidal texture, which, under the microscope, is seen to be microgranitic or micrographic. The dominant constituent is usually feldspar—a potassic, sodic or lime-bearing variety, according to the type of granite with which the aplites are associated. The proportion of quartz varies, but in some varieties this mineral predominates, indicating a passage to quartz veins.

Relation of Aplite to Pegmatite

Aplites and pegmatites both occur as veins, dykes or sills in granites or the nearby country rock. Further, both types may be seen in the closest association in composite intrusions which may be pegmatitic in their marginal parts, but aplitic in the centre. Occasionally the two may be interbanded; or a vein dominantly aplitic may contain irregular pockets of pegmatite. This intimate association demonstrates that both rock-types represent granitic residua and that both have been generated in essentially the same way. There has been much speculation as to the cause of the striking difference in grain-size between these two residua. Probably aplites represent a 'dry' and poorly fluxed fraction of the residual granitic magma, while the fugitive constituents were concentrated in the pegmatitic fraction. Doubtless the latter would be far more mobile than the former. The truth of this supposition is established at Porthleven, Cornwall, where pegmatite occurs in contact with the *under-side* of sedimentary xenoliths suspended in granitic (largely aplitic) sills. Such local pegmatite has obviously originated by the arrest of the upward-streaming volatiles.

There is no doubt that many pegmatites and aplites have originated in the manner outlined above; but in some cases a metasomatic origin is indicated by the field-relations of the rocks. If it is conceded that some granitic-looking rocks have originated by replacement, there are even stronger grounds for believing that the same is true of some pegmatites.[1]

PNEUMATOLYSIS

Following the final consolidation of the magma, the fugitive constituents are released, and escaping through joints and other fissures may effect striking changes in the mineral composition of the parent rock. These effects are covered by the term pneumatolysis, which implies that the fugitives are in a gaseous state. This is probably true in some cases, but not in others. Whether they exist as a true gas phase or not, however, they behave essentially as chemically active solutions. Pneumatolytic modifications are most strikingly displayed by granitic rocks (though they are not restricted to the latter), and include (*a*) greisening and (*b*) tourmalinization. Kaolinization is sometimes included, but is better regarded as hydrothermal alteration.

(*a*) **Greisening.**—Greisen, composed essentially of white mica and quartz, is one of the most distinctive products of this type of activity, and appears to have three different modes of occurrence, depending upon the degree of consolidation and fissuring of the parent rock.

(1) Most commonly greisen is a marginal modification of granite adjacent to quartz- and mineral-veins. The alteration is very localized: normally it extends for a distance of a few inches only from the contact. Usually there is convincing evidence of replacement, such as the pseudomorphing of feldspar crystals by aggregates of white mica. The latter is often a variety containing lithium and fluorine, such as zinnwaldite. Other fluorine-bearing minerals, notably fluorite and topaz, are also commonly present. Purple patches of fluorite give colour to an otherwise light grey or white rock. Topaz is an almost constant accessory, and in some specimens becomes an important constituent, even exceeding the white mica in amount. The end-product of this line of variation is a topaz-quartz rock, the former making up perhaps nine-tenths of the rock, to which the name 'topazfels' is commonly, though incorrectly, applied. Granites vary widely in the extent to which they display pneumatolytic effects, and this type of greisening is developed

[1] See King, B. C., *J. Geol.*, **46** (1948), 459.

extensively in only a few areas, such as the tin-mining districts in the Erzgebirge in Saxony, in Cornwall and in northern Nigeria.

(2) In other occurrences the volatile-rich residuum does not alter the parent granite, but gives rise to veins and thin dykes of white-mica, quartz rock which crystallizes in fissures in the granite. Such occurrences are fundamentally different in nature and origin from the demonstrably metasomatic greisens noted above. They are just as definitely primary igneous rocks as aplites and pegmatites. Thus if the name 'greisen' implies a metasomatic origin as well as a particular mineral assemblage, another name is needed for these rocks. Spurr[1] suggested the name 'esmeraldite' for primary quartz white-mica rocks. A better alternative is to apply the term 'greisen' to all such rocks, and to indicate the mode of origin by appropriate qualifiers.

(3) Large bodies of greisen sometimes occur as apophyses and marginal facies of granite masses. They are not related to jointing or fissuring in any way, and are obviously different from the greisens considered under (1) and (2) above. A typical example is afforded by the outcrop at Grainsgill, just north of the main part of the Skiddaw granite, Cumberland. The chemical gains and losses of the greisen as compared with the main granite are different in this case from those which usually apply.[2] There is usually a considerable increase in H_2O, F and probably Al, while Na may decrease to zero. Other components may show small but haphazard changes. When the analyses of Skiddaw granite and Grainsgill greisen are compared, there is found to be an increase in SiO_2 in the latter, which is too great to be accounted for by hydrothermal alteration.[3] A. Harker[4] suggested that the greisen crystallized from a particularly acid magma-fraction which was separated from the main granite and driven northwards by filter-press action. As the greisen yields evidence of replacement of the original feldspars, however, it is probable that two processes have been involved: firstly differentiation on the lines suggested by Harker, and secondly, deuteric alteration by an active residuum. Retention of the late-stage solutions in such large volumes must depend on such geological 'accidents' as the shape of the differentiated body—ideally an apophysis of the main mass—and the degree of fissuring of the surrounding rocks, which must affect the escape of the volatiles.

[1] The S. Klondyke District, Esmeralda County, Nevada, *Econ. Geol.*, **1** (1906), 382.

[2] A valuable summary of the gains and losses involved in greisening is given by G. J. Williams; A granite-schist contact in Stewart Island, New Zealand, *Quart. J. Geol. Soc.*, **90** (1934), 348.

[3] Hitchen, C. S., The Skiddaw granite and its residual products, *Quart. J. Geol. Soc.*, **90** (1934), 158.

[4] Carrock Fell granophyre and Grainsgill greisen, *Quart. J. Geol. Soc.*, **51** (1895), 143.

This three-fold division of greisens and greisen-like rocks into (1) primary, (2) metasomatic and probably pneumatolytic and (3) deuteric, is paralleled among granites affected by tourmalinization.

(*b*) **Tourmalinization.**—Tourmaline, like muscovite, appears as a normal constituent of some granites of the more acid and alkali-rich types, as in the Carnmenellis mass in Cornwall. A particular concentration of boron produces a brown iron-rich tourmaline in place of biotite, so that the two minerals are very seldom seen together. With increasing flux-concentration tourmaline increases at the expense of other constituents of the granite; in particular, the feldspar is progressively eliminated. **Luxullianite**[1] represents an arrested stage of tourmalinization, in which some brick-red feldspar has survived, though the outlines of the crystals are much corroded. Between them quartz and finely acicular black tourmaline are much in evidence (Fig. 81).

Under the microscope luxullianite is a beautiful rock. The black tourmaline (schorl) occurs in the form of delicate needle-like crystals radially disposed and often clustered around corroded relics of earlier 'massive' brown tourmaline. The tourmaline needles freely penetrate into secondary quartz, a mosaic of which forms the general background of the sections. The feldspars have been deeply embayed and show progressive replacement chiefly by quartz.

A second stage of tourmalinization is represented by rare tourmaline-quartz rocks in which none of the original components save quartz has survived; but although completely replaced, the shapes of the feldspars can still be detected, though their substance is chiefly quartz-mosaic.

In other instances tourmaline occurs in imperfect stouter prisms of random orientation, embedded in coarse quartz mosaic, giving a black-and-white rock of distinctive appearance. A well-known example of this type is that of Roche Rock, Cornwall, which forms an isolated outcrop with no visible connection with the nearby main granite intrusion. There is, indeed, no evidence, microscopic or otherwise, that this rock ever was a normal granite: it may well represent a magma-fraction drastically enriched in 'fugitives' and analogous with the 'primary greisen' noted above. Apart from such occurrences, however, the same mineral assemblage, sometimes with cassiterite in addition, occurs in the form of primary veins.

Finally, it may be noted that the activities of the fugitive constituents are not limited by the boundaries of the parent granite, for the surrounding rocks are sometimes extensively tourmalinized.

(*c*) **Kaolinization.**[2]—The partial alteration of feldspars into an

[1] Wells, M. K., A contribution to the study of luxullianite, *Min. Mag.*, **27** (1946), 186.

[2] Handbook to the collection of kaolin, etc., *Mem. Geol. Surv.*, (1914).

exceedingly fine aggregate of flaky minerals of the kaolinite type is a ubiquitous phenomenon in granitic rocks. Such kaolinite is usually accompanied by sericite, and together these two minerals are

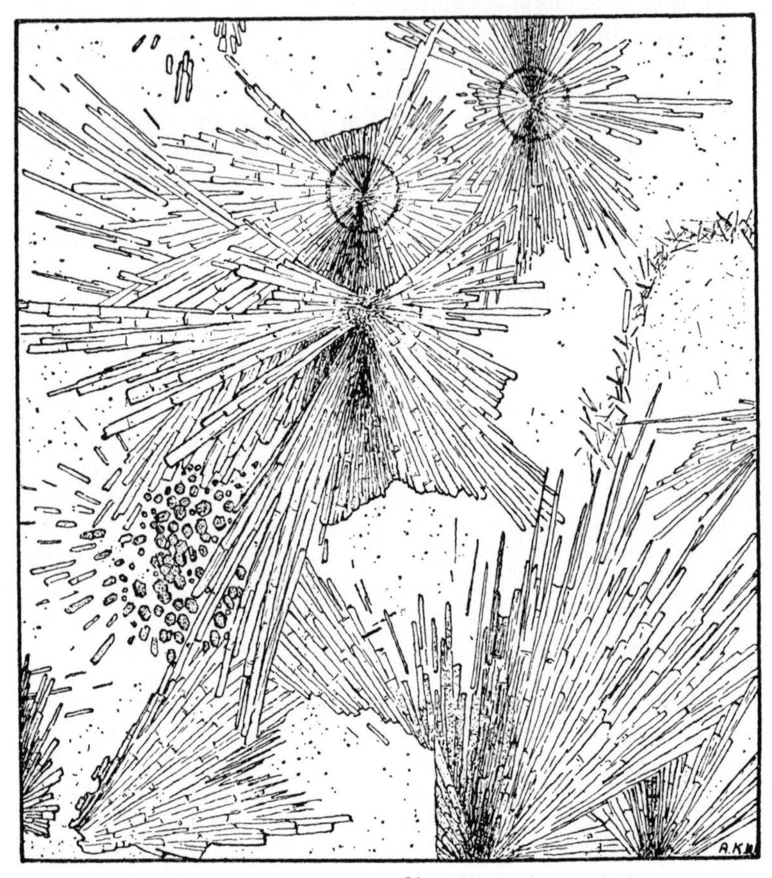

FIG. 81

Section of luxullianite (the type rock), Luxulyan, Cornwall.

Tourmaline forms radial aggregates. Some of the quartz crystallized before the tourmaline and some is secondary after feldspar. The boundary of a crystal of the latter, completely silicified, is outlined with acicular tourmaline on the right, near the top of the section. Tourmaline crystals reach a length of 2 mm.

the chief cause of the opacity of orthoclase in the hand-specimen and its cloudiness in thin section. Occasionally kaolinization is so complete that the whole rock is reduced to a thoroughly rotted condition, becoming friable and so soft that the rock can be easily

FIG. 82.—'Chinastone', St Austell, Cornwall, England.

Orthoclase strongly kaolinized; albite, fresh and in part occurring in regular orientation within orthoclase. Quartz, irregular, with trains of inclusions; also forming a few 'hieroglyphs' in orthoclase. White mica in large irregular masses with tabular inclusions of fluorite. Also small sheaves of sericite. Fluorite, irregular masses showing strong *negative* relief and octahedral cleavage. Topaz, top left below white mica, shows moderate relief and traces of (001) cleavage. (*A.K.W.*)

dug away, or even removed by a high pressure jet of water. In this condition the rock is of great economic importance as **china clay,** perhaps the most valuable raw material in the ceramic industry. Of the original minerals of the granite, only the quartz survives kaolinization. Large masses of the St Austell granite[1] in Cornwall have been modified in this way, giving china clay in association with other late-stage alteration products such as quartz-tourmaline rock and **chinastone.** The latter rock is comparable with luxullianite in a sense, as it represents an arrested stage of alteration—this time of kaolinization—and exhibits a wide variety of replacement textures. Two feldspars are present: strongly kaolinized orthoclase, often microperthitic, and relatively fresh euhedral albite. The orthoclase may be extensively veined by secondary quartz and by fluorite. The latter is an essential component in more than the usual petrographical sense, for it gives the rock its fluxing properties in the glazing of china. It occurs in a variety of forms, either replacing feldspar or penetrating along the cleavage planes of muscovite as shown in Fig. 82 (top right and left).

The chief agent of kaolinization is probably high-temperature H_2O which attacks the orthoclase ($KAlSi_3O_8$), etc., to form kaolinite, $Al_2Si_2O_5(OH)_4$, and releases a certain amount of potassium which may combine with CO_2 to form a carbonate and thus be removed as a soluble component; or the potassium may combine with other components to form sericite which is so commonly associated with the kaolinite. Carbon-dioxide may thus play a part in these reactions, but it is uncertain how important a part.

MICROGRANITES, MICRO-ADAMELLITES AND MICRO-GRANODIORITES

In view of the practical difficulty met in estimating the relative amounts of potassic and sodic feldspars (except by calculation of the norm), it is necessary to be satisfied with generic, rather than specific, identification. Actually there is little point in trying to specify that a given microgranite is either potassic or sodic—most are potassi-sodic or sodi-potassic. The feature that attracts more attention than the composition of the feldspar is the nature of the coloured silicate(s) present.

In some microgranites biotite occurs; but it is evident that these dyke-rocks are nearer to rhyolites than to granites so that pyroxene—usually a light coloured diopsidic augite—takes precedence over mica. Hedenbergite, accompanied sometimes by iron-rich olivine, occurs in microgranites from northern Nigeria; while ferrohorton-

[1] Exley, C. S., Magmatic differentiation and alteration in the St Austell granite, *Quart. J. Geol. Soc.*, **114** (1959), 197–230.

olite occurs in a Tertiary granophyre (Meall Dearg, Skye). In more strongly sodic microgranites such NaFe-rich pyroxenes and amphiboles as aegirine and riebeckite occur—they have already been noted as occurring characteristically in granites of sodic types. It is customary in naming these rocks to hyphenate the mafic mineral-name on to the term 'microgranite': thus aegirine-microgranite and riebeckite-microgranite are informative names. The former is still disguised under the term grorudite (Brögger, 1894): from the details of its composition, SiO_2, 76 per cent, orthoclase-microperthite (with large excess of K_2O over Na_2O), aegirine and, of course, quartz, it is evident that grorudite is synonymous with aegirine-microgranite. Nothing is gained by retaining this rock-name.

As regards texture, microgranites grade with diminishing grain-size into rhyolites, and with increasing grain-size into granites—the limits in both directions are arbitrary; the textures are intermediate between, and grade into those displayed by rhyolites and granites respectively. Three different textures claim our attention: (1) porphyritic, (2) aphyric (the clumsy term 'non-porphyritic' is often preferred, unfortunately); and (3) micrographic.

(1) **Porphyritic microgranites, micro-adamellites** and **micro-granodiorites** are megascopically the most distinctive members of the family (Fig. 83). They have been widely termed 'quartz-porphyries' or 'granite-porphyries' though the terms favoured here are less liable to misinterpretation than the other two. The phenocrysts (first generation crystals) are believed to have formed in depth, about relatively few, widely spaced centres of crystallization in an environment where growth was uninhibited and perfect crystals were developed, notably the feldspars which are euhedral and plane-faced. Quartz is the high (β) form, of distinctive dihexagonal bipyramidal shape; the crystals are both fewer and smaller than the feldspars. A handsome British example forms a dyke striking across the beach at Prah Sands in South Cornwall and known as the 'Prah elvan'. It is noteworthy for the fluxional disposition of the abundant feldspar phenocrysts it contains. The matrix of these rocks is visibly granular and is seen in thin section to be a micro-crystalline mosaic of quartz and alkali-feldspar.

(2) **Aphyric microgranites,** etc. resemble the previous rocks in composition and mode of occurrence; but as the qualifier implies, there are no porphyritic crystals: in a sense they are all ground-mass. Texturally and, up to a point, mineralogically too, these rocks are close to granite-aplites but differ by containing appreciably more coloured mineral: aplite represents only a felsic residue from granite, but an aphyric microgranite represents *all* the granite, mafic as well as felsic components.

(3) The most interesting texture is the **micrographic** which has

several features in common with the graphic, considered above. 'Micrographic microgranite' is a ponderous term and **granophyre** (Rosenbusch, 1872) is a more acceptable name for this group of

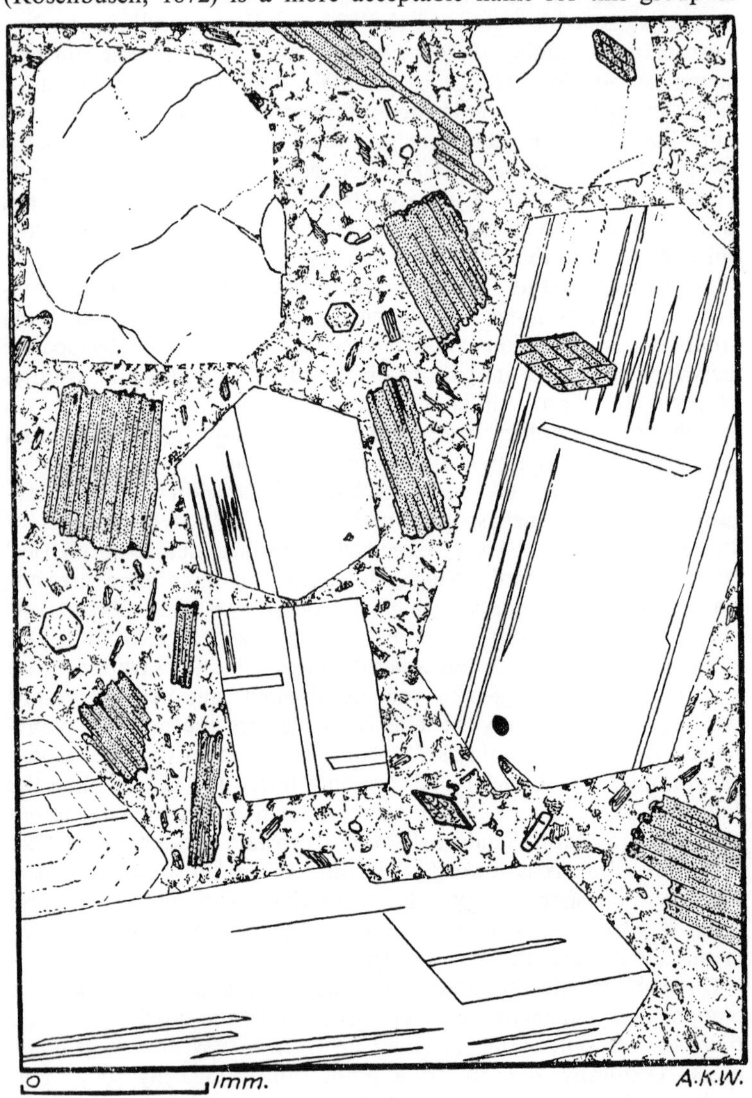

FIG. 83

Porphyritic micro-granodiorite, south of North Bend, British Columbia.

Phenocrysts of quartz, zoned plagioclase, common hornblende and biotite are embedded in a microcrystalline groundmass of the same minerals. Apatite and sphene are also shown. The twinning is indicated in the plagioclase.

rocks. Most granophyres are porphyritic and, in general, are close to porphyritic microgranites: in fact it is only the groundmass texture which distinguishes them. This consists of intergrown quartz and alkali-feldspar, the former occurring as isolated 'hieroglyphs' like those seen in the coarser-grained graphic granites and pegmatites, or sometimes as 'fibres' which have grown away at right angles to the faces of the feldspar phenocrysts and which expand in girth distally. The hieroglyphs or fibres are in parallel optical orientation within each area of micropegmatite and therefore extinguish simultaneously (Fig. 84). The micropegmatite patches may be roughly rounded, and within each, the quartz hieroglyphs are arranged centrically. This type of structure grades into the spherulitic as the quartz individuals become finer, thinner and more regular in their radial disposition.

With regard to their nature and origin all granophyres were at one time accepted as being magmatic rocks of essentially granitic composition; but certain granophyres are now believed to be metasomatic in origin. The evidence seems to be convincing in the case of certain granophyres which are marginal to Karroo dolerites, and which appear to have been formed by the metasomatism of the sediments into which the doleritic magma was intruded.[1] There is nothing inherently improbable in this; but we think it is unwise to conclude on this evidence that all granophyres have originated in the same way—a view that is apparently held in some quarters.[2] It may be appropriate, therefore, to restate the case for believing that most granophyres were formed by direct crystallization of a melt, and must therefore be regarded as magmatic. The evidence concerns the composition and textures of the rocks, their field relations and affinities with other rocks, notably certain rhyolites. First as regards mineral composition: the proportion of quartz to alkali-feldspar corresponds to the lowest-temperature range of compositions established experimentally for the synthetic 'granite' system, quartz–orthoclase–albite. Characteristic mafic minerals include pyroxenes and olivines of magmatic types, including hedenbergite ($Ca_{42}Mg_7Fe_{51}$) and ferrohortonolite (Fa_{85}) occurring in the Meall Dearg granophyre in Skye: the same minerals are found in high-level late fractions forming part of the Skaergaard complex, also in rhyolitic glasses including pitchstones. As regards texture, the fact that it can be closely matched among metallurgical textures produced during crystallization of eutectic mixtures has long been regarded as significant. So, to us, is the fact that the textures observ-

[1] Walker, F., and Poldervaart, A., Karroo dolerites of the Union of S. Africa, *Bull. Geol. Soc. S. Africa*, **60** (1949), 591.

[2] See *e.g.* Holmes, A., in discussion on Hughes, C. J., The Southern Mountains igneous complex, Isle of Rhum, *Quart. J. Geol. Soc.*, **116** (1960), 111.

FIG. 84

Aphyric granophyre, Fjardardal, Iceland.

Orthoclase is shown stippled, quartz black and white.

able in the mesostasis of quartz-dolerites are identical with those seen in some granophyres. The textures in the quartz-dolerites are magmatic and it would be extraordinary if identical textures in granophyres were produced in some other way.

Finally, granophyres occur in the so-called volcanic association. In composition they closely match certain rhyolites; they have the same mode of occurrence as various microgranites and pitchstones; they show the same close association with Basic (basaltic) rocks in various Acid/Basic associations, including composite dykes and sills, while the recently described granophyre pipes occurring in doleritic and gabbroic rocks on Slieve Gullion and elsewhere are particularly significant in this respect.[1]

We would emphasize one point. We are satisfied from the above evidence that the majority of granophyres are magmatic. How the magma was formed is another matter. Selective melting of material of the appropriate composition by basic magma at a temperature of 1,100 to 1,200° C is well within the realms of possibility; and we would class arkose, rhyolitic 'ash' and an older granodiorite among the materials 'of appropriate composition'.

One of the most recently described granophyres is that which forms the scarp feature of Cader Idris in Merionethshire, North Wales.[2] Its chilled facies is indistinguishable from certain rhyolitic lavas with which it is closely associated. It is interesting to note that this granophyre has *caused* metasomatism and mobilization of the sediments into which it was intruded, under a cover perhaps 1,500 feet thick. This illustrates how difficult it may be to sort out cause and effect.

Microgranites are widely distributed. Types corresponding in composition with the alkali-granites are associated with, and occur as offshoots from, major intrusions of this nature. Thus the coarser-grained dyke-rocks, locally termed 'elvans', found in the near neighbourhood of, or actually cutting, the Cornish and Dartmoor granites, are potassic or sodi-potassic microgranites. Those with adamellitic and granodioritic affinities tend to occur in association with, for example, the Caledonian 'granites' of these types in Scotland. Many of the minor intrusions in the English Lake District are microgranites: the St John's Vale intrusion, characterized by phenocrysts of orthoclase, quartz and small red garnets, is a porphyritic potassic microgranite. Among types which are correspondingly sodic we may note **riebeckite microgranite** which occurs at Ailsa Craig in the Firth of Clyde, on Holy Island off the coast

[1] Elwell, R. D., *et al.*, Granophyre and hybrid pipes . . . of Slieve Gullion, *J. Geol.*, **66** (1958), 57–71.

[2] Davies, R. G., The Cader Idris Granophyre and its associated rocks, *Quart. J. Geol. Soc.*, **115** (1959), 189–216.

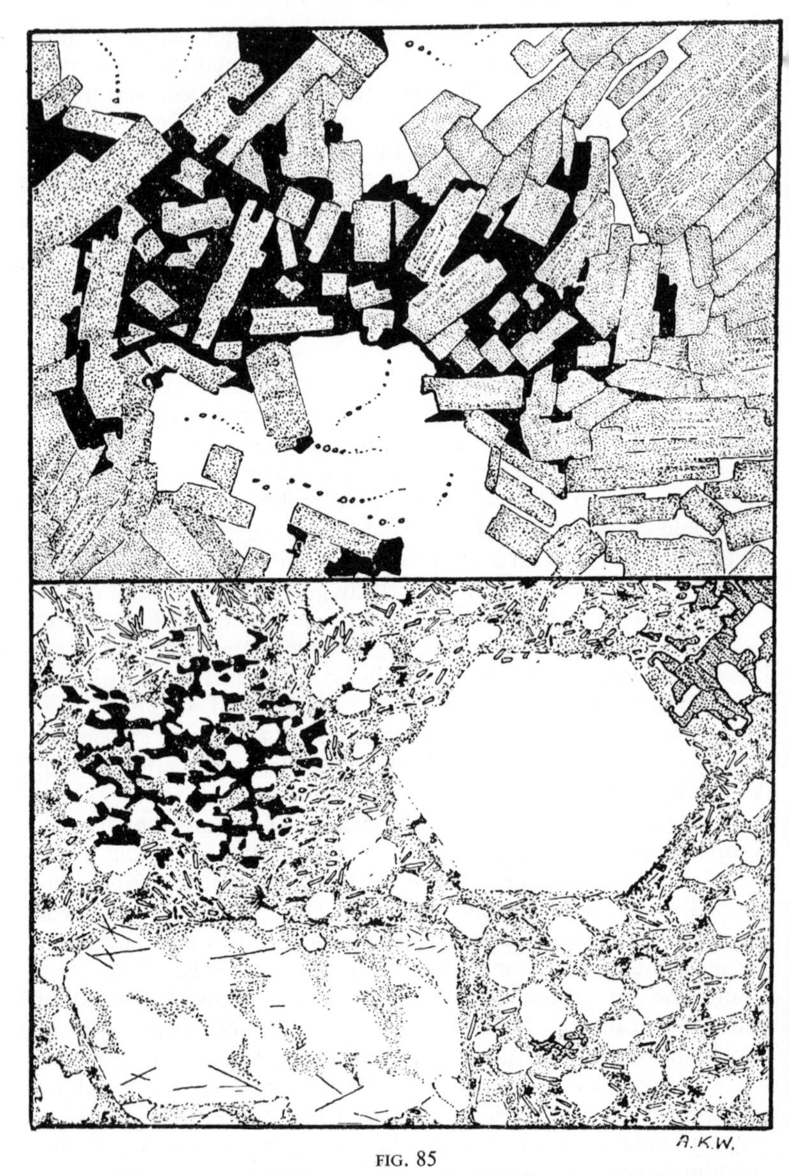

FIG. 85

FIG. 85 (*above*).—Riebeckite microgranite, Ailsa Craig, Firth of Cyde, Scotland. Alkali-feldspar embedded in poikilitic riebeckite (shown black) and quartz.

FIG. 86 (*below*).—Riebeckite-acmite-microgranite, Mynydd Mawr, Caernarvonshire. Phenocrysts of β-quartz and alkali-feldspar in microcrystalline groundmass, with acicular prisms of acmite and micropoikilitic riebeckite, top right and near top left—black (in position of maximum absorption).

of Arran and at Mynydd Mawr in North Wales (Figs. 85 and 86). Although closely similar in mineral composition, the two examples figured differ in textural detail. The micropoikilitic habit of the riebeckite, sometimes giving it a 'mossy' appearance, is characteristic.

It may be noted that the Ailsa Craig rock has been called 'riebeckite-aplite'. But the term 'aplite' implies not only a peculiar mineral composition and a distinctive texture, but a special mode of origin. Unless there are strong grounds for believing that the rock in question has originated in the manner described above under 'aplites' the term aplite should definitely not be used.

Aegirine-microgranite occurs among the dyke rocks connected with the alkali complexes in Sutherlandshire, for example, Cnoc-na-Sroine in Assynt.

One of the best known British examples of granophyre is that which occurs at Carrock Fell: the intergrowth in this case is relatively coarse. The rock forming the Armboth dyke in the Lake District is a handsome porphyritic variant with phenocrysts of quartz, bright red orthoclase and small garnets set in a dun-coloured graphic groundmass. The prominent Fort Regent mass lying on the outskirts of St Helier, Jersey, is only in part medium-grained, but is beautifully micrographic in places.

RHYOLITES, TOSCANITES AND DACITES

The classification of fine-grained igneous rocks must follow the plan adopted for their coarse-grained equivalents. Therefore, using the QAP triangle as a basis we recognize six fields bearing the names of leading rock-types of appropriate composition. It is essential to give independent names to each: they are all self-contained, of the same status and mutually exclusive. It is agreed among petrologists that rhyolite is the fine-grained ('volcanic') equivalent of granite, trachyte of syenite, andesite of diorite and dacite of granodiorite. When these names are inserted into the QAP triangle two fields remain to be named—those corresponding to adamellite and monzonite. There are alternative names for both. For the fine-grained equivalents of adamellite the compound name rhyodacite is taxonomically correct and self-explanatory. However, such names are liable to misinterpretation, so we prefer the synonym, toscanite. Similarly, and for the same reason we choose latite in preference to trachyandesite.

Rhyolites (sensu stricto), Alkali Rhyolites

If for the sake of simplicity one collective name is required to include all the Acid igneous rocks of fine grain, it must be rhyolite, which thus has the same status as 'granite' for coarse-, and 'microgranite' for the medium-grained rocks. This is one of the older rock names, and was used by von Richthofen (1860) on account of the flow banding frequently exhibited by these rocks. Liparite (Roth 1860), from well-known occurrences in the Lipari Islands, is synonymous with rhyolite.

The upper limit of grain-size is 0·05 mm, this having reference to the diameters of quartz or feldspar grains in the groundmass. Grains of this size are, of course, irresolvable with the naked eye, so for practical purposes the dividing line between microgranites and rhyolites can be placed at the limit of unaided vision for the groundmass components. No limit of size is laid down concerning the phenocrysts which are present in many specimens.

In our scheme of classification rhyolites are analogous to the granites (*sensu stricto*) and microgranites, and like them, may be further subdivided on the basis of the proportions of K^+ and Na^+. It has been customary to distinguish potassic and sodic categories; but these terms are applicable only to rhyolites of extreme composition: the majority contain both K^+ and Na^+ in widely varying proportions, perhaps on average with a slight bias in favour of K^+.

Potassic rhyolites.—In these rocks the dominant feldspar is the high-temperature form, sanidine, often in the form of glassy clear phenocrysts and/or microlites or granules in the groundmass. The sanidine often contains a considerable proportion of the albite molecule, either in solid solution in apparently homogeneous crystals or partially exsolved in crypto-perthitic intergrowths. Occasionally the latter display a distinctive blue chatoyancy like that shown by moonstone.

In all rhyolites free silica may be present not only as inverted β-quartz,[1] but also as tridymite and even cristobalite. Of these, β-quartz is by far the commonest, and the only obvious form of silica even under expert examination. It occurs as well-formed bipyramids, which, however, may show any degree of magmatic corrosion, and in extreme cases may be reduced to shapeless wrecks of the original crystals (Fig. 87). By comparison with granite one would expect the micas and hornblende to be the most commonly occurring coloured silicates; but actually although biotite is common, hornblende is less common in rhyolites than pyroxene. This is consistent with the higher temperature at which consolidation took place in the case of rhyolite, and the boiling off, instead of retention, of volatiles in lava flows doubtless also favoured the crystallization

[1] Readers may be reminded that all quartz in igneous rocks, in rhyolites as well as in granites, is 'low' quartz. That occurring in the rocks under discussion was precipitated as 'high' quartz but subsequently inverted into 'low' quartz during cooling.

of the anhydrous pyroxene. The biotite is often very strongly coloured, so that basal sections appear almost black. The pyroxene is often nearly colourless—actually a very watery light green—in thin section, and is a diopsidic augite.

Among the sodi-potassic rhyolites special mention may be made

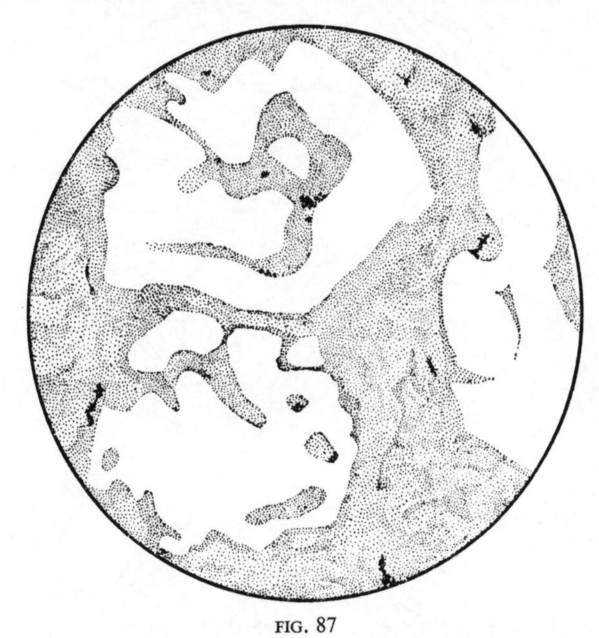

FIG. 87
Devitrified porphyritic pitchstone, Petit Portelet, Jersey, C.I.
Highly corroded phenocrysts of quartz in 'felsitic' (originally glassy) groundmass.

of the Si-rich members of the pantellerite suite (see p. 293). These rocks, termed **quartz-pantellerites** are characterized by phenocrysts of β-quartz and anorthoclase near to $Or_{50}Ab_{50}$ in composition, accompanied by a wide range of coloured minerals notably rich in iron, including hedenbergite, fayalite and various strongly coloured Na- and Fe-rich amphiboles. The presence of iron-rich silicates is to be expected in many rhyolites from their general position as end-members of differentiation sequences resulting from fractional crystallization.

Sodic Rhyolites.—Like sodic granites these rocks may be very distinctive, largely by reason of the coloured silicates they contain, which may be the same as those already described in the corresponding granites and microgranites: *viz.* aegirine among the pyroxenes and riebeckite among the amphiboles being characteristic (Fig. 88). The coloured silicates are accompanied by quartz, of

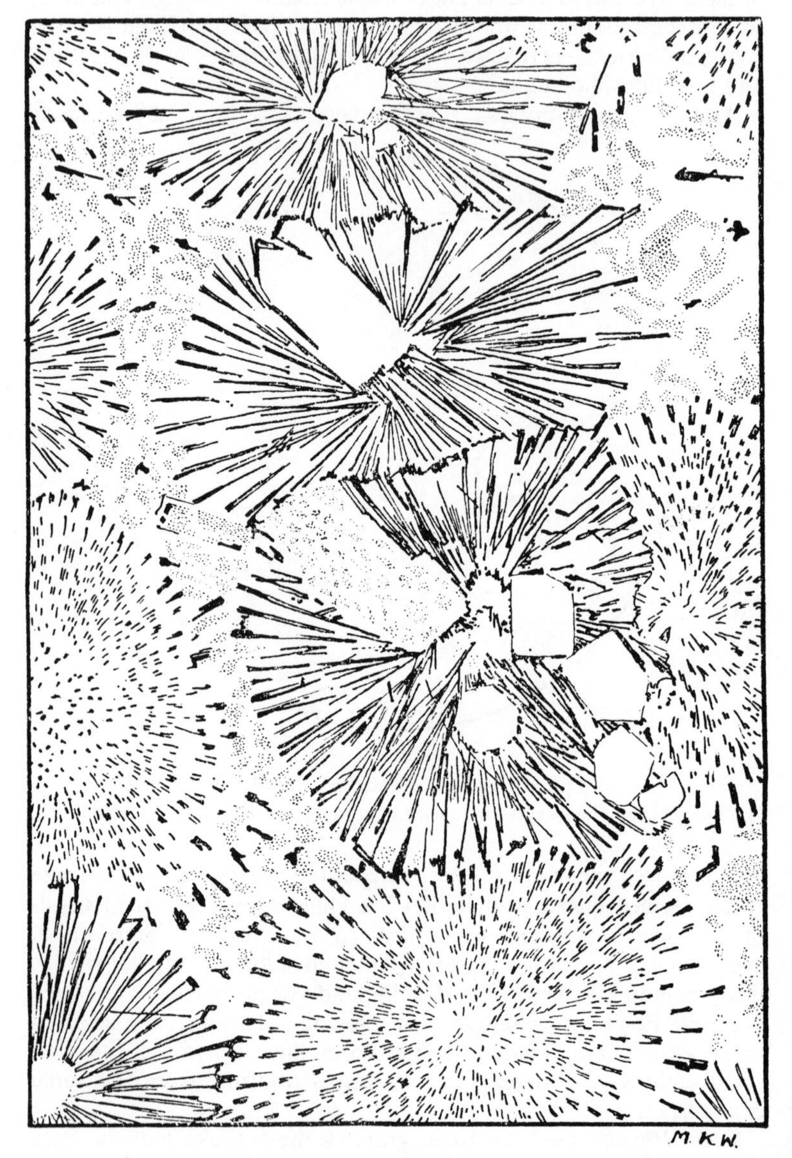

FIG. 88

Spherulitic riebeckite-rhyolite, between Rishi and Ayaka Leru, northern Nigeria.

Spherulites of riebeckite enclosing phenocrysts of 'high' quartz and sodipotassic feldspar in a devitrified (felsitic) matrix.

course, and ideally by albite, though in very many instances the feldspar, as in the corresponding coarse-grained rocks, is sodi-potassic, often recorded as anorthoclase which would unmix, presumably to antiperthite, in a plutonic environment.

From the facts of their mineral composition such riebeckite- and

ANALYSES OF ALKALI-RHYOLITES

	1 Potassic rhyolite	2 Pitchstone Arran	3 Sodi-potassic rhyolite obsidian.	4 Sodic-rhyolite (Pantellerite)
SiO_2	73·76	73·20	76·78	68·63
Al_2O_3	11·98	10·75	12·09	10·30
Fe_2O_3	1·14	0·95	0·56	5·60
FeO	2·40	1·02	0·81	2·61
MgO	0·76	0·15	0·10	0·37
CaO	0·32	0·76	0·57	1·07
Na_2O	0·53	3·78	3·79	6·14
K_2O	7·38	4·20	4·93	4·17
H_2O	1·75	4·70	0·20	0·53
TiO_2	0·12	0·16	0·08	0·35
Other constituents	0·34	0·61	0·38	0·23
	100·48	100·28	100·29	100·00

1. Potassic rhyolite, Cwm Caregog, Snowdon (Anal. R. J. C. Fabry), *Quart. J. Geol. Soc.* (1927), 368.
2. Sodi-potassic pitchstone, Arran, Scotland (Anal. E. G. Radley), Geol. of Arran, *Mem. Geol. Surv.* (1928).
3. Obsidian, Obsidian Cliff, Yellowstone National Park, U.S.A. (Anal. E. S. Shepherd), *Am. J. Sci.*, 35A (1938), 329.
4. Sodic-rhyolite (Pantelleria type). Average of twelve pantellerites quoted from Daly (1933), p. 10.

aegirine-rhyolites are evidently comagmatic with the corresponding granites and dyke-rocks with which they are closely associated in the field, for example in the alkali-complexes in Nigeria (Fig. 88). The type illustrated is a very unusual one as the mineral involved in the spherulites is riebeckite which adventitiously encloses small phenocrysts of sodi-potassic feldspar and inverted β-quartz. The groundmass is a devitrified glass, represented by a felsitic aggregate of quartz and feldspar. The name **'quartz-keratophyre'** has been used in several senses, but the practice in this country is to apply it to sodic rhyolites characterized by the presence of albite or albite-rich feldspar, associated with a little coloured silicate, the original nature of which has become obscure through alteration: it is represented by uninformative patches of chlorite. Lavas and texturally similar minor intrusives of this type occur among the Ordovician and Devonian eruptive rocks of North Wales and Devon and Cornwall respectively. Quartz-keratophyre is a type of soda-rhyolite, and a member of the spilitic suite.

Toscanites (Rhyodacites)

The rocks in the fine-grain category corresponding with adamellites are known by two names: the older is toscanite (Washington, 1897) derived from the Italian name for Tuscany, whence the original rocks so named were described. The alternative name is more informative: **rhyodacite** conveys the correct impression of the essential feature of these rocks—the sharing of the characteristics of rhyolites and dacites. Thus in rhyodacites two kinds of feldspar, an alkali type characteristic of rhyolites, and plagioclase, generally oligoclase-andesine, are of equal status. The term is comparable with trachyandesite, which occupies an analogous position among the 'Intermediate' lavas. Most rhyodacites are porphyritic and all three felsic constituents may occur as phenocrysts, though commonly the latter include inverted β-quartz and plagioclase, while the alkali-feldspar, sanidine, forms the major part of the groundmass.

Dacites

Dacites are important members of the Calc-Alkali (or orogenic) Suite and occur in close association with andesites into which they grade with decreasing silica. Especially in America the distinction between andesite and dacite is based on silica percentage and on this basis dacites contain between 63 and 68 per cent SiO_2. The colour index is low compared with andesites; but the chief means of distinction lies in the role of quartz. The high percentage of normative quartz in dacites leads to the early precipitation of modal quartz as phenocrysts. It would simplify nomenclature if the presence of the latter were made an essential part of the definition of dacite. There is no doubt that these rocks are often misnamed as andesites or quartz-andesites.

On average dacites have a somewhat lower silica percentage than rhyolites as might be expected on account of the higher ratio of alkali-feldspar to plagioclase diagnostic of the latter as compared with the former. In some classifications a silica percentage of 68 is used to separate dacite from rhyolite; but the fundamental difference between the two is not silica percentage but the ratio of alkali-feldspar to plagioclase.

Rocks Containing Essential Quartz

The recognition of a distinct category of rocks within the dacite field having a value of P greater than 90 raises a problem of nomenclature. In our view these rocks are simply plagioclase-rich dacites—they need no special name; but they have been, and are being distinguished as 'quartz-andesite'. This term, properly used, can only logically refer to a rock, *in the andesite field*, containing an appreciable amount of quartz, which, by definition, must not exceed the Q-value of 20.

It is difficult to know just how important dacites are in the volcanic fields in which they occur because of loose terminology: so many have been recorded as andesites and quartz-andesites, while others have been identified as ignimbrites. However, there is no doubt as to their very wide distribution in orogenic belts, and in some of these they occur in enormous volumes.

Textural Features of Acid Lavas

As explained more fully below, many rocks which in regard to chemical and mineral composition qualify as rhyolites, rhyodacites (toscanites) and dacites are ignimbritic in character; and since one cannot intelligently discuss the nature of ignimbrites without reference to their critical textural features, it follows that much of the information which might be included under the present heading must be deferred. It remains to describe the textures of the very fine-grained and glassy Acid igneous rocks which occur not only as conventional lava-flows but also as near-surface (sub-volcanic) dykes and other minor intrusions.

Due to the high viscosity developed by Acid melts as they near the surface and lose some of their dissolved volatiles, many rhyolites tend to be vitreous. The proportion of natural glass to crystalline material in what we may conveniently term the Acid lavas (though including identical rocks occurring in minor intrusions) is infinitely variable. A specimen of obsidian or of pumice may be wholly glass; but these terms indicate the physical condition of the glass, not its composition, so that they should not be used without a suitable

(*a*) Margarites.

(*b*) Globulites.

(*c*) Trichites.

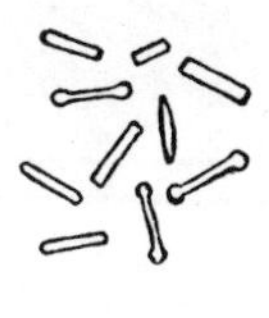
(*d*) Belonites.

FIG. 89

Some types of crystallites.

rock-name qualifier, *e.g.* rhyolitic obsidian or dacitic pumice. In the absence of phenocrysts it is impossible to distinguish between obsidian or pumice of rhyolitic, rhyodacitic or dacitic composition unless chemical data are available.

At its best **obsidian** is a pure natural glass of rhyolitic (or granitic) composition, black in colour, naturally vitreous in appearance, and

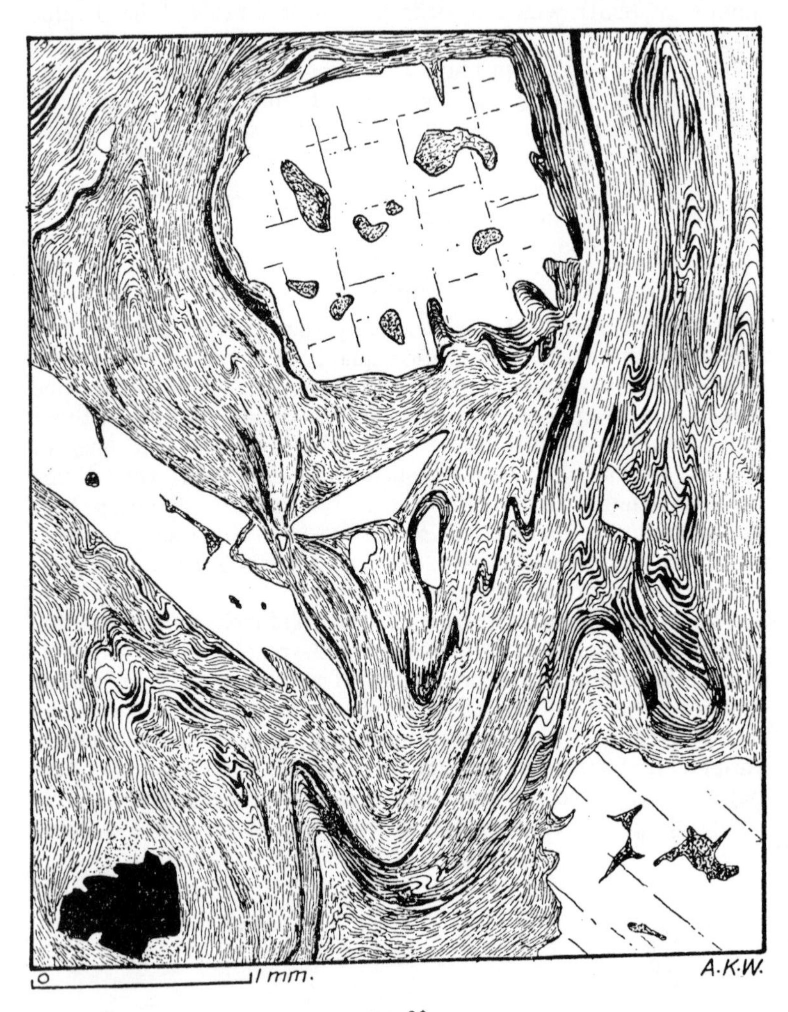

FIG. 90

Porphyritic pitchstone, Sgùrr of Eigg, Hebrides.

The corroded phenocrysts are of β-quartz and alkali-feldspar ($Or_{42}Ab_{52}An_6$): they have been chipped and the fragments scattered by the flow, indicated by the banding.

usually exhibiting a striking conchoidal fracture. Obsidian may be easily trimmed up to any desired shape—to a point or sharp cutting edge, for example; therefore this rock was largely used by primitive peoples in the manufacture of stone implements and weapons. Although obsidian is the commonest of the natural glasses, it is not widely distributed, the occurrences at Obsidian Cliff in the Yellowstone Park, at Mount Hecla in Iceland, and in the Lipari Isles, being the best known. In thin section obsidian is colourless and isotropic. It is rarely completely devoid of crystalline material, which may occur as minute scattered crystallites (Fig. 89), or spherulites, either isolated, in irregular groups or trains, or in definite bands (spherulitic obsidian).

Although the proportion of glass in **pitchstone** is very high, there is much more crystalline material present than in obsidian. Consequently pitchstones are duller in appearance, they often have an irregular hackly fracture, while the lustre tends to be resinous rather than vitreous. The crystalline material may take the form of phenocrysts of any of the minerals appropriate to rhyolites—β-quartz, sanidine, oligoclase, light green pyroxene are all common—or as microlites. The latter have been identified in some cases as feldspar, in others as pyroxene, presumably of the same type as that occurring as phenocrysts. The microlites may appear rod-like in thin section, but in some cases they form stellate or feathery groups (see Fig. 92), or may resemble fern fronds. In these glassy rhyolites of both

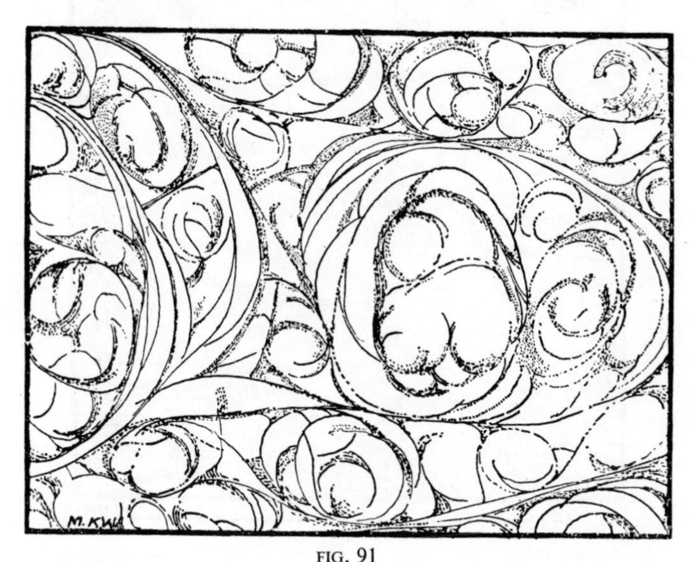

FIG. 91
Perlitic obsidian.

types, flow structure is common: it is shown sometimes by the parallel alignment of the microlites which may sweep round phenocrysts like eddies in a stream, and convey a vivid impression of the viscous flow of the magma. In other cases the glass itself is colour-banded (Fig. 90), or layers of vitreous and non-vitreous lava may alternate, doubtless resulting from a degree of heterogeneity in the lava. Perlitic structure, resulting from tension set up by contraction during cooling, occurs in both obsidian and pitchstone (Fig. 91). These two terms are used rather loosely: in the opinion of some, the distinction between them should rest on the different degree of crystallinity, as noted above; but in the opinion of others, the difference in water-content is more significant. In obsidian the amount is small—usually less than 1 per cent—while in pitchstone it may rise as high as 10 per cent. Johanssen has advocated drawing the line at 4 per cent of water, but although this would make for accuracy, it is a drawback being unable to name a specimen until an analysis is available.

In Britain famous examples of pitchstones occur among the Tertiary intrusives in the Hebrides, in Arran and Eigg. The Arran pitchstones are well known on account of the interesting textures they exhibit (Fig. 92). Further, some contain iron-olivine, fayalite,

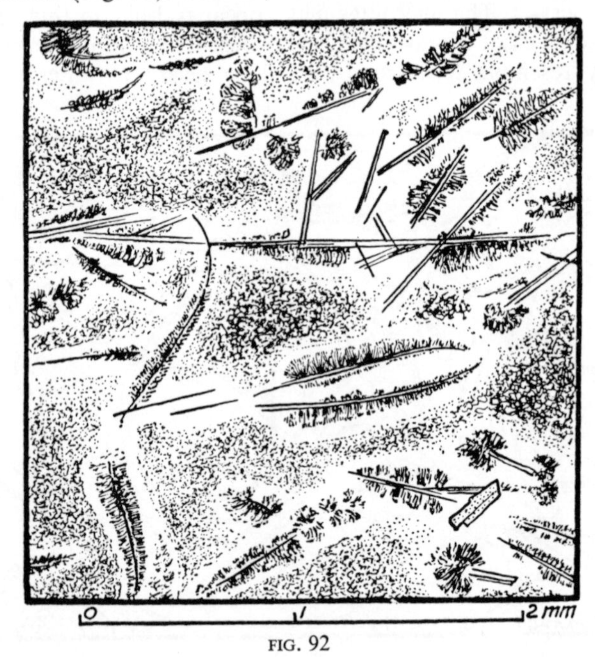

FIG. 92

Pitchstone, Corriegills, Arran, with pyroxene crystallites surrounded by zones of clear glass.

in addition to ferro-augite and orthopyroxene, feldspars and quartz phenocrysts (Fig. 93).

It was mentioned incidentally above that rhyolites may be

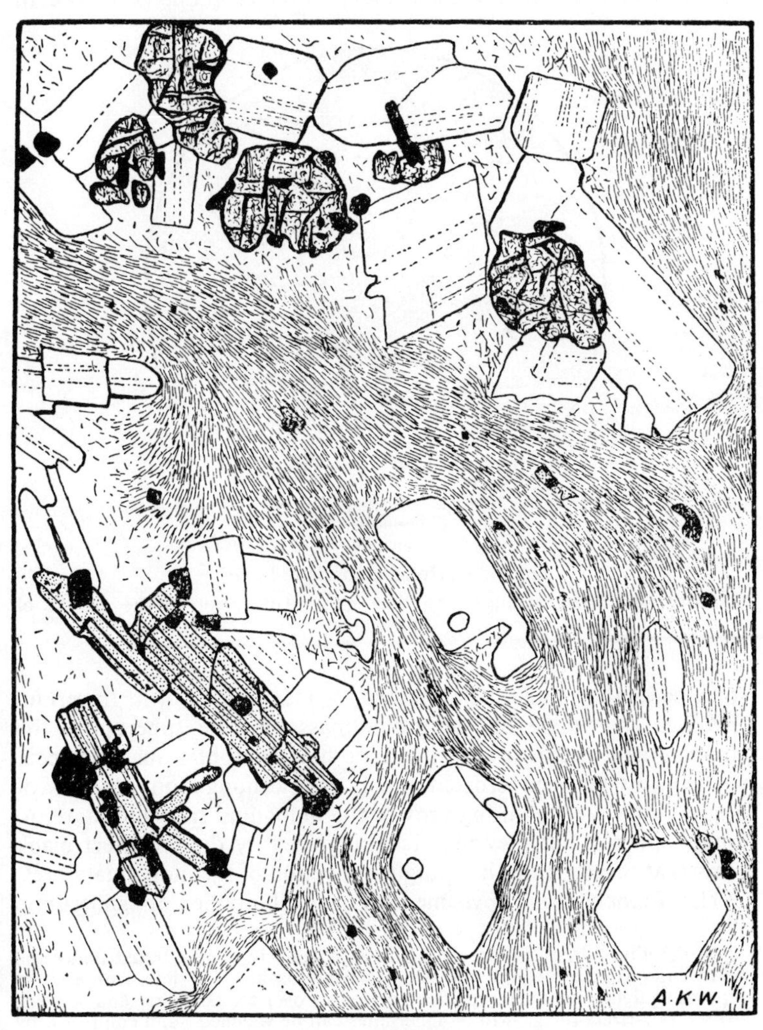

FIG. 93

Fayalite-pitchstone, Glen Shurig, Arran. Fayalite phenocrysts occur in the upper half of the section, a greenish ferroaugite in the lower half. Iron ore, plagioclase (An_{27}) and β-quartz are also present, the latter giving typical hexagonal basal section. The fluxion structure shown by groundmass microlites ('spiculites') is noteworthy.

spherulitic.[1] Ideally the **spherulites** consist of near-spheres, often about the size of a pea, but varying from smaller than a pin's head to larger than a man's fist. In the case of the smaller ones a section shows the spherulites to consist of radially disposed fibrous crystals, which are not easily identified under the microscope (Fig. 94). In

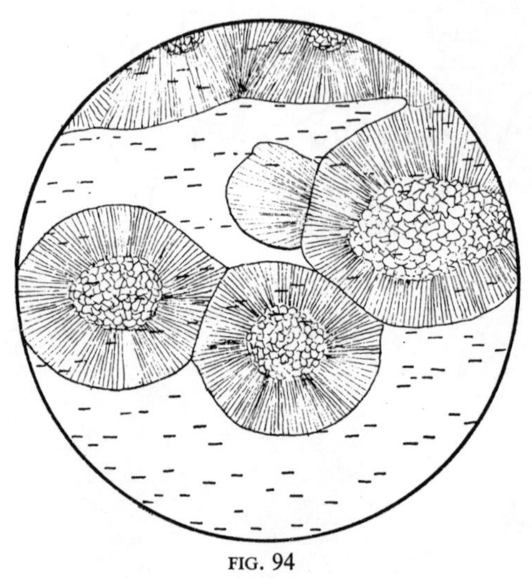

FIG. 94

Spherulitic rhyolite-obsidian, Lipari.

Showing 'spiculites' arranged by flow movements prior to the growth of spherulites.

some cases they contain cristobalite, or may consist of it. In certain rhyolites the whole of the groundmass may be spherulitic: the characteristic radial arrangement of fine microlites may be seen, particularly between crossed polarizers, although on account of mutual interference during growth the individual spherulites are no longer spheres, but they may be recognized by the imperfect black cross that results from the straight extinction of the individual fibres.

The French name 'pyromerides' is still applied to spherulitic

[1] Spherulitic crystallization can be most effectively demonstrated under the microscope using a pinch of cholesterol acetate on a 3 × 1 slide, with a cover slip resting lightly on it. The powder melts easily over a small flame, and between crossed polarizers the growth of spherulites can be watched—and controlled by blowing cold air onto the slide.

Often a better impression of the internal structure of one of these glassy rocks may be derived from the examination of a *thick* section, smoothed on one side which is then cemented to a glass slip and examined in strong surface illumination, preferably using a binocular microscope (not necessarily petrological). It is possible to see below the surface and obtain, within limits, a three-dimensional impression of the structure.

rhyolites of the north coast of Jersey, Channel Isles, which are perhaps unique. The spherulites often show a strongly developed concentric structure, superimposed upon the more normal radial arrangement. In this respect they are allied to the so-called *lithophysae*—stone bubbles which superficially resemble spherulites but are hollow inside. Lithophysae may reach several centimetres in diameter.

With slower cooling rhyolitic magma will not congeal as glass, but will become wholly crystalline. There may be difficulty in deciding whether this holocrystalline condition is original or the result of devitrification; but in the absence of relics of the glassy state, referred to above, and if the individual grains are clearly defined, it is almost certainly an original structure. As noted above, rhyolites may be porphyritic to any degree: the phenocrysts may be minute and widely scattered, or they may be relatively large and so closely packed as almost to exclude the groundmass. The name 'nevadite' was applied by von Richthofen to a rock of the latter type. It is best, however, to introduce new names for rocks with distinctive mineral composition, rather than some mere textural peculiarity, which can always be covered by appropriate qualifiers. Richness in phenocrysts is not an attribute of one kind of rhyolite, however, and while one lava described as nevadite is a phenocryst-rich potassic rhyolite (with porphyritic sanidine and quartz), another has been called by the same name on account of the abundance of phenocrysts, but as they include oligoclase, biotite and hornblende in addition to sanidine and quartz, we are dealing with a rock in a different category altogether: it is porphyritic dacite.

NATURE AND MODE OF ORIGIN OF IGNIMBRITIC ACID IGNEOUS ROCKS

Because of their high viscosity and large content of volatiles, siliceous magmas tend to be erupted with explosive violence, especially when erupted from a constrictive vent. In these circumstances the lava column is completely pulverized, some of it is thrown high into the air, dispersed by wind, chilled in transit and is ultimately deposited as an air-fall ash. The remainder, it may be a large part or, in extreme cases, the whole of the erupted material issues from the vent as a suspension of effervescing lava fragments in white-hot, turbulent volcanic gases which may be rapidly dispersed over a wide area. The velocity at the point of emission was estimated at 74 m.p.h. for the disastrous eruption of Mt Pelée in Martinique in 1902, and 100 m.p.h. at Mt Lamington. This part of the *nuée ardente* moves as a dense cloud close to the ground, and the fragmental lava settles, drowning the topography and forming an extensive, flat-topped sheet now commonly referred to as an ash-flow, tuff-lava or ignimbrite. The extraordinary mobility of the latter results from the fact that the comminuted lava fragments are cushioned by hot volcanic gases[1] so effectively that friction is eliminated: therefore the material moves as freely as a liquid. Much of the 'ash'—typically more than 70 per cent of it—consists of comminuted pumice and glass fragments ('shards') from 4 mm down to volcanic dust. The pumice was not in the physical condition with which we are familiar, but was strongly vesiculating magma-froth reminiscent of the foam on top of a saucepan of boiling milk. The exsolution of gases was actively taking place, the vesicles were still expanding, and their walls were still plastic when the froth was pulverized. The disintegration of this pumice-in-the-making gave rise to shards of distinctive shapes ranging from nearly complete glass bubbles, thin fragments of their walls and larger pieces of glass between a number of adjoining bubbles (Fig. 95). The identification of these shards is essential to the identification of an ignimbrite. The extent to which they remain identifiable depends on the load to which they were subjected before the ultimate freezing of the flow. Ignimbrites generally contain larger pieces of pumice up to block-size. These sometimes still look like pumice; but often they were highly plastic when incorporated in the flow, and were flattened to such a degree that the vesicles were completely eliminated.

The temperature of rhyolitic magma at the point of emission may be as high as 900° C. The nature of the dense, particle-charged gas cloud travelling at great speed over the ground ensures a minimal loss of heat, so that the glass shards and pumice fragments, even if transported for considerable distances must be at near magmatic temperatures when they come to rest, and welding is inevitable. It has been shown experimentally that the welding temperature of dry pumice is 775 to 900° C, but may be as low as 600° C in the presence of water vapour and under loads approximating to those obtaining under natural conditions. Other things being equal, welding is best developed under the following conditions: in thick, rather than thin, flows; in the central parts of the flow where heat is retained as a consequence of the insulation provided by the sub-jacent and superincumbent 'ash'; towards the source, rather than the distal parts of the flow, so that the welded zone tends to thicken towards the source of the eruption.

Welding is effected along surfaces of contact between the shards, forming thin sheaths of glass which may be distinctive in thin

[1] Tazieff, H. (*Bull. Volc.* **34** (1970) 1) has shown that CO_2 can be a very important constituent of high-temperature volcanic gases, and has suggested that, because of its relatively high density, CO_2 would be more efficient than H_2O as a transporting agent for ignimbrite flows.

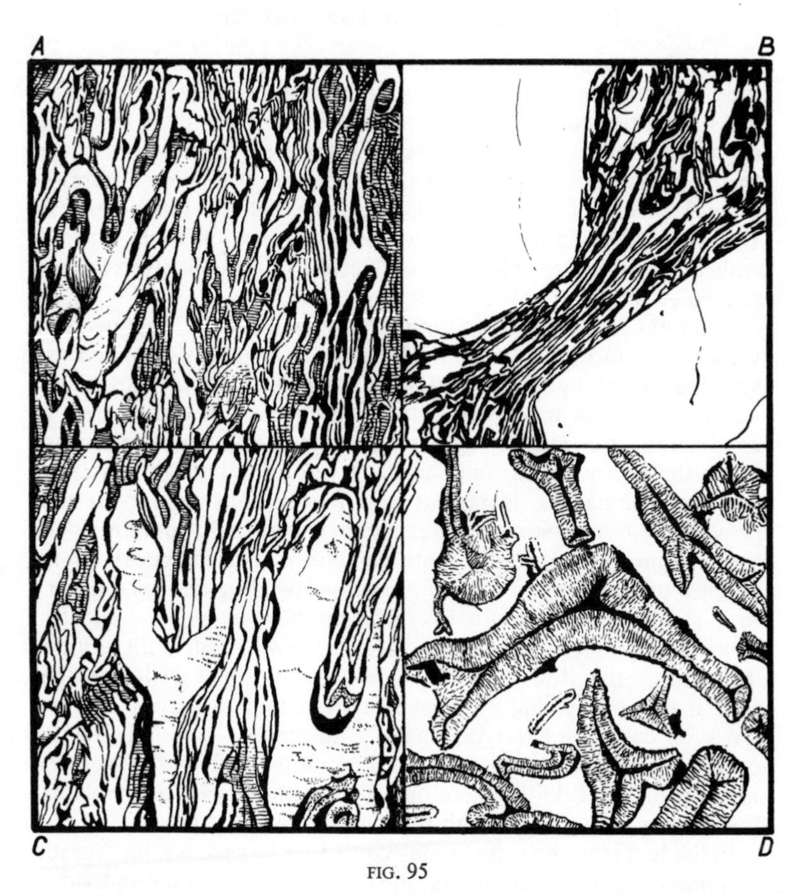

FIG. 95

Thin sections of ignimbrites: (A) from Central Oregon, (B) from Yellowstone Park, Wyo., (C) from S. E. Idaho and (D) from Valles Mts., N. Mexico. A. C. and D based on photomicrographs in 'Ash-flow tuffs . . .' by Ross, C. R., and Smith, R. L., *Geol. Soc. Amer. Prof. Paper*, **366** (1961), Pl. 68, 70 and 74. (B) based on photomicrograph in 'Welded tuffs and flows in the . . . Yellowstone Park, Wyoming' by Boyd, F. R., *Bull. Geol. Soc. Amer.*, **72** (1961), Pl. 5. (*Drawings by A. K. W.*)

In (A) and (C), particularly the latter, 'shards' of two sizes occur—fragments of thick-walled pumice are embedded in a matrix of thin-walled shards, some of extremely complicated shape, which are flattened against and moulded upon the former, *e.g.* around the H-shaped shard on the R-hand side of the section. In (B) thin-walled pumice fragments were squeezed into parallelism between two phenocrysts of β-quartz which had been stripped bare of adhering glass before being incorporated in the ignimbrite. Some of the fragments, top right and bottom left, are probably lying in the plane of the section; but those between the phenocrysts are orientated by the crystal faces of the latter. All three sections give a strong impression of a high degree of plasticity and toughness of the fragments of pumice at the time of incorporation, but prior to welding. Section D, a devitrified ignimbrite, shows the shapes of individual 'shards' which crystallized to form an intimate intergrowth of cristobalite and sanidine, resulting in the development of axiolitic structure (Zirkel). See text.

section by reason of slight differences in colour and in refractive index. The ultimate product of welding is a virtually homogeneous glass—an obsidian—in which it may be extremely difficult to discern any trace of original fragmentation. Further, welding is not restricted to the finer fragments: it may occur also *internally* in compressed pumice, blocks of which may also be converted into black obsidian. The regular orientation of flattened lenticles of such obsidian embedded in lighter coloured 'ash' results in a distinctive structure termed **eutaxitic.**

As might be expected, an ignimbrite flow normally displays some variation from top to bottom. The basal layer is contaminated to a variable extent by extraneous material picked up in transit, and welding is not to be expected as a consequence of loss of heat by conduction; the middle layer tends to be relatively massive and dense through welding; but the top layer, being typically unwelded and less compacted, is porous and poorly consolidated. It may prove difficult to distinguish this latter type of rock from an air-fall ash. Ignimbric flows differ from genuine pyroclastic rocks by the absence of sedimentation features such as bedding and sorting; and as a result of their origin by turbulent flow in a gas cloud, the constituent pumice fragments and blocks show a random distribution throughout the flow unit. It is noteworthy that well-developed columnar jointing, indistinguishable from that displayed by lava-flows of conventional type, may be displayed, especially by the central, more massive parts of an ignimbrite. Differential weathering of these several lithologies may give rise to a terraced or plateau topography, comparable with that formed by flood basalts.

Individual ignimbrite flows may cover very large areas, measurable in some cases in thousands of square kilometres; while individual thicknesses, though averaging 10 m may reach a thickness of 100 m or more. The total volume of ignimbrite flows, augmented probably by some ash deposits, is truly enormous in some regions of rhyolitic volcanicity. Thus in south Sumatra they occupy an area of 30,000 km^2 and have a volume estimated at 2000 km^3. They cover 15,000 km^2 in North Island, New Zealand, including the well-known Lake Taupo volcanic region, while other noteworthy occurrences include the Arequipa region in Peru, Crater Lake, Wyoming, the Yellowstone Park including Obsidian Cliff, the area picturesquely named the Valley of Ten Thousand Smokes in the vicinity of the Katmai volcano, the eruption of which in 1912 serves as the type eruption of the kind which produces ignimbrites.

Much has been said and written about these rocks since 1935, the year of the original definition of ignimbrite by Marshall[1] with

[1] Marshall, P., Acid rocks of the Taupo-Rotorua volcanic district, *Trans. Proc. Roy. Soc. N. Z.*, **64** (1935), 323–66.

reference to New Zealand examples; and of welded tuffs by Mansfield and Ross[1] for comparable rocks in Idaho. Other terms including tuff-lava and ash-flow have been, and are still being used by different writers. The problem is that no single term is adequate to cover all the varied characteristics of these rocks—their lithologies as well as their mode of origin. The word 'ash' for example is appropriate to the extent that it indicates the fragmental character (before welding) of these rocks; but it conveys an entirely wrong impression of their mode of origin. True volcanic ash is the product of violently explosive volcanic eruption of the type represented by that of Krakotoa in 1883 during which a large part of the island disappeared, with fragmentation of a great volume of various kinds of solid rock, and the expulsion of the finest dust particles into the stratosphere where they remained for a considerable time. By contrast, ignimbrites are products of a Katmaian type of eruption in which, to quote Howel Williams,[2] 'the magma frothed quietly to the surface instead of being propelled high above the vent'. The ignimbrite, tuff-lava or ash-flow—call it what you will—is erupted as a turbulent mixture of expanding gases and gas-emitting lava fragments. If the magma was rhyolitic and welding took place after eruption (as is typically the case), the end-product is rhyolitic too and may be indistinguishable from rhyolitic lava—which in the compositional sense it is, of course—erupted in the conventional manner. In fact there may well be a complete transition not only in textural detail but also in mode of eruption, between ignimbrites and conventional lava-flows, with froth-flows the connecting link.[3] These show the field characteristics of normal lava-flows; but under the microscope they display the distinctive textural features of ignimbrites.[4] The differences between the three types of eruption depend apparently on the amount and activity of the volcanic gases available.

It will have been gathered from the foregoing account that the textural features of an ignimbrite may be distinctive and easily identified in thin slice; but in other cases it may prove difficult or impossible to infer the type of eruption involved. Certain features such as flow-banding, the occurrence of spherulites or lithophysae and the development of columnar structure are not safe criteria to use in attempting to distinguish between normal lava-flow rhyolites and rhyolitic ignimbrites: they are common to both, though there may be subtle differences.

[1] Mansfield, G. R., and Ross, C. S., Welded rhyolitic tuffs in southeastern Idaho, *Trans. Amer. Geophys. Union*, **16** (1935), 308–21.

[2] Williams, Howel, Problems and progress in volcanology, *Quart. J. Geol. Soc.*, **109** (1954), 311–22.

[3] Kennedy, G. C., Some aspects of the role of water in rock melts, in *Crust of the Earth*, Geol. Soc. Amer. Special paper no. 62, (1954), 762pp.

[4] Boyd, F. R., Welded tuffs and flows in the Yellowstone Park, Wyoming, *Bull. Geol. Soc. Amer.*, **72** (1961), 406.

Genuine flow-banding in Acid lavas has been referred to and illustrated above. The banding is sinuous and convoluted; and individual bands can be followed continuously for, it may be, several centimetres. The analogous banding in an ignimbritic rhyolite does not involve viscous flow; but results from close-packing, flattening and elongation of unsorted shards and pumice fragments. The banding tends to be more regular and flatter, and close inspection will often show that an individual band is discontinuous and built up of small, often minute fragments, some of which may display the characteristic cuspate and bifurcating shapes of glass shards.

Devitrification

Volcanic glass at normal air temperatures is in a state of unstable equilibrium and there is an inherent tendency for it to crystallize spontaneously. This recrystallization occurs most readily under the influence of a stimulus such as is provided by the presence of trapped volcanic gases or incipient metamorphism affecting more ancient rocks; but the important point is that, given sufficient time, crystallization is inevitable, and all rocks which have retained their original glassy character have been erupted, geologically speaking, in relatively recent times. This deferred spontaneous crystallization is termed devitrification, as a consequence of which a wholly glassy rock such as obsidian may be converted into one which is holocrystalline. Although evidence of their originally glassy nature may be clearly proved by traces of perlitic structure, for example, examination between crossed polarizers shows them to be completely crystalline—often quite coarsely microcrystalline, the whole rock breaking up into a mosaic of ill-defined grey-polarizing areas bearing no relation whatever to the original structure. This is often called felsitic texture, and the rock itself a **felsite.** The term felsite is a very broad one, applied to compact pinkish or grey dyke-rocks satellitic to granite masses. Some felsites are demonstrably devitrified rhyolitic glasses, but in some the crypto-crystalline (not microcrystalline) groundmass may be original. Many examples occur among the lavas of Uriconian age in Shropshire, and of Ordovician age in North and South Wales. Similar phenomena occur among the Tertiary minor intrusions in Arran, Scotland, a familiar example being the 'spherulitic felsite', which, although not illustrated, is fundamentally like Fig. 94 except that, as its name implies, it has been completely devitrified. The development of the existing structure involved (1) crystallization of certain high-melting point accessories; (2) radial crystallization of fibrous aggregates of intergrown feldspar and cristobalite after the lava had come to rest and cooled to the appropriate temperature; and (3) the spontaneous redistribution of the components of the spherulites to form felsitic material. It is to the third stage only that the term devitrification can be properly applied.

Many welded ignimbrites on examination in thin sections are seen to be wholly, though minutely, crystalline. The actual minerals involved cannot be identified under the ordinary microscope, even with high magnification; but with carefully controlled lighting it can be seen that two minerals are involved, one of them with an exceptionally low refractive index. X-ray analysis has shown them to be cristobalite and sanidine. In some ignimbrites each component fragment behaved as an independent unit: crystallization commenced on the margin of the shard and continued until the crystals met at a well-defined middle line. The resulting **axiolitic structure** is very distinctive. It was described and figured long ago by Zirkel, though its significance was not then understood. Axiolitic structure may still be recognized even in Pre-Cambrian ignimbrites, though the original cristobalite has given place to a micro-crystalline aggregate of stable quartz grains.

Under different conditions the minute crystals maintain a general parallelism quite independent of the shapes and dispositions of the shards, and presumably grew at right angles to the general cooling surface. In other ignimbrites radially disposed crystals grew outwards from relatively few and widely spaced centres, forming spherulites which are identical with those occurring in normal flow-rhyolites except that, in some cases, 'ghosts' of original shards may be seen within the boundaries of spherulites.

We would emphasize that in all three cases we are seeing the results of a *primary* crystallization which took place at near magmatic temperatures presumably when the latter had fallen to the eutectic temperature for cristobalite and sanidine, when simultaneous crystallization of these two components occurred. This is not devitrification in the sense explained above; but because cristobalite and sanidine are both metastable, given time, both will recrystallize into quartz and orthoclase.

It may be noted that gas-phase crystallization causes tridymite to form in cavities in ignimbrites, so that three forms of crystalline silica—cristobalite, tridymite and β-quartz may all be present.

DISTRIBUTION AND MODE OF ORIGIN OF ACID VOLCANIC ROCKS

A fact of cardinal significance concerning the distribution of Acid volcanic rocks is their virtual restriction to continental regions, and absence from the oceanic basins. Although they are represented in most if not all the continental volcanic regions and in island arcs, they achieve their maximum development in regions of active orogeny and uplift. Acid volcanic rocks, in this context broadly referred to as rhyolites, occur as the end-members of the Calc-Alkali Suite, in close association with andesites and basaltic andesites, for example in the Andes; though elsewhere the assemblage includes rhyolites and basalts, for example in the Yellowstone Park volcanic region. Both the proportions of the different types of volcanic rocks and the sequence in which they were erupted are matters of great significance in connection with the problem of the origin of rhyolitic magma. The whole question is interwoven with the problem of the origin of granites, discussed in the next chapter, so at this point it is only necessary to refer briefly to some of the main points.

The paucity of rhyolites among the predominant basalts of oceanic suites suggests very strongly that fractional crystallization and differentiation of basaltic magma is the only process involved: the very small amounts of sodic rhyolite associated with trachytes in a few of the oceanic islands such as Samoa, belong to the alkali olivine-basalt suite (see p. 134) and are products of extreme differentiation generally erupted late in the volcanic sequence. By contrast, in many of the calc-alkaline associations, the proportion of basalt to rhyolite is quite inadequate for the rhyolites to have been derived from a basaltic parent magma. In Yellowstone Park volcanic region, for example, it is estimated that basalts constitute less than one per cent of the total volcanic rocks erupted.

The situation regarding the possibility of andesitic magma having been the parent of rhyolitic rocks is very different. Andesites contain on average considerable quantities of normative quartz, while one third of the total feldspar they contain is of the kind required in granites and rhyolites. It follows, therefore, that in composition many andesites are not far removed from their more Acid associates in the Calc-Alkali Suite: in fact the compositions of some siliceous andesites overlap those of many plutonic rocks which, on the basis of their quartz-content, are classed as granodiorites (see p. 213). Variation diagrams clearly demonstrate a genetic linkage between the andesites, dacites and rhyolites occurring in a number of volcanic provinces that have been adequately studied. Unfortunately there is no certain way of distinguishing between the products of magmatic differentiation and of selective melting of rocks in deep parts of the crust. However, consideration of the broad geological factors involved suggests that the latter process must play an important part in the evolution of rhyolitic magmas on a large scale.

One of the strongest indications in this direction is provided by the mode of eruption of the Acid volcanic rocks, particularly ignimbrites. Frequently the source or sources of the associated pyroclastic and ignimbritic flows are difficult to locate; but fortunately in some cases it has been possible to identify the source as a collapsed caldera or some other form of fault-bounded volcano-tectonic depression. The diameters of some of these structures are too great for them to have been formed by any process associated with a simple volcano with, for example, a restricted feeder system. There is, in fact, strong evidence to show that the calderas overlie batholythic intrusions. It is a fascinating thought that, in the opinion of many geologists, the volcanic rocks and associated hot springs and geysers in the Yellowstone Park overlie a batholith that is still cooling—and breathing—only a few kilometres below the surface.

At the time of active volcanism, magma that accumulated towards the roof of the growing batholith must gradually, and probably over a very long period of time, have changed in composition, presumably becoming more siliceous, and certainly enriched in volatile components. With the build-up of gas pressure, the overlying crust would eventually fracture and probably after an initial phase of violent explosion with the consequent ejection of volcanic ash, the gas-swollen and fluidized contents of the magma reservoir

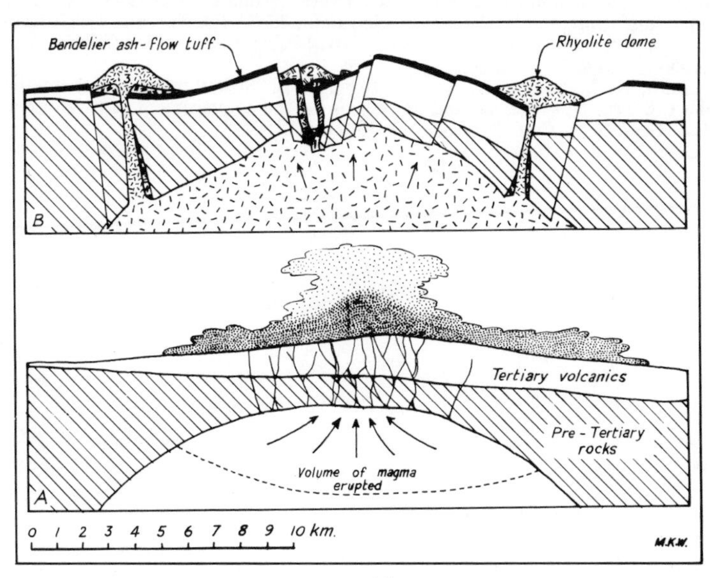

FIG. 96

Schematic sections showing the evolution of the Valles caldera, New Mexico. (A) shows the phase of eruption of the Bandelier ash-flows; and (B) shows the structure after caldera collapse, the eruption of rhyolitic magma in three stages, the last forming the rhyolite domes (3) located on ring-faults which enclose the caldera, and final doming of the floor of the caldera by the underlying intrusion. (*After R. L. Smith. R. A. Bailey and C. S. Ross* (1961).)

would be erupted in large part as ignimbritic flows. The emptying of the magma-chamber would lead to the subsequent caldera collapse.[1] This sequence of events has been shown to have occurred during the formation of the great Valles Caldera and associated ignimbritic flows of the Bandelier Tuff in New Mexico,[2] and there are other examples of a similar type that could be quoted from other parts of the United States, the Oslo province in Norway and elsewhere. The authors of the Valles account also suggest that a good case may be made out for regarding the original type example of cauldron subsidence intrusion at Glencoe in Scotland and the associated granite batholith of the Etive Complex (Fig. 146) as an additional expression of similar phenomena.

The importance of these observations from the point of view of rhyolite petrogenesis lies in the fact that the magma was derived from batholithic intrusions: the latter could scarcely have developed without crustal rocks being involved at some stage of the process, either by melting at crustal levels or by the contamination of magma rising from some deeper source.

[1] For the development of ideas concerning the relationship between caldera formation and ignimbrite eruption see Howel Williams, Calderas and their origin, *Calif. Univ. Publ., Dept. Geol. Sci. Bull.*, **25** (1941), 239, and Doris Reynolds, Calderas and ring complexes *Koninkl. Nederl. Geol. Mijnb. Genootschap*, **16** (1956), 355–98.

[2] Smith, R. L., Bailey, R. A., and Ross, C. S., Structural evolution of the Valles Caldera, New Mexico and its bearing on the emplacement of ring dykes, in *U.S. Geol. Surv. Prof. Paper* 424-D (1961), D145–9.

CHAPTER III

ORIGIN OF THE GRANITIC ROCKS

'GRANITIC rocks' includes not only granites within the whole range of quartz-rich rocks from granodiorites to alkali-granites (true granites), but also their fine-grained equivalents, the rhyolites, and the medium-grained rocks occurring as associated minor intrusions. The lavas and associated dyke-rocks are demonstrably magmatic; so too are the comagmatic granites, and for these assemblages there is no problem of origin: they represent rhyolitic magma cooled under different conditions. The reality of rhyolitic magma is emphasized every time rhyolitic lava is erupted. But many granites are not visibly or even inferentially connected with surface flows, and therefore their magmatic origin is incapable of direct proof. It is with the origin of such granites that we are immediately concerned, and the nature of the evidence is examined in the following pages.

The first step towards building up a reasonable theory of the origin of the granitic rocks must be based upon the observation that they greatly predominate among plutonic rocks; but that among the volcanics the dominant type is basalt. Consideration of the distribution of these two contrasted types had led to the belief that granite can only originate where a granitic shell, the sial, is already in existence. This conclusion follows naturally from the fact that granites are restricted to the continental areas, but are absent from the ocean basins. In the latter only rocks of direct basaltic parentage can be formed. This seems to indicate that, in general, granites are not derived from basaltic magma by differentiation as has been advocated in the past, notably by Bowen. However, a particular variety of basalt termed tholeiite, of very wide distribution in continental regions, is distinguished by containing interstitial patches of residual glass. The latter may contain up to some 70 per cent of SiO_2: it is therefore thoroughly acid, and potentially granitic. If at the appropriate stage of crystallization it could be squeezed out of the basalt and allowed to crystallize independently, it would certainly form granite, microgranite or rhyolite according to the environment in which it cooled. This may well be the mode of origin of the occasional rhyolitic lavas intercalated among the dominant basaltic flows in many volcanic regions, and also of the minor granitic intrusions associated with the more abundant Basic rocks. The quantity of granite originating in this way is small at best, and this mode of origin is quite inadequate to account for the enormous amounts involved, for example, in the orogenic belts in the western Americas.

It is probable that the formation of most granites has involved selective melting of the sial as advocated by Eskola, with the most soluble combinations of minerals melting first. The fact that quartz and alkali-feldspar (orthoclase and albite) in the approximate proportion of 1 : 1 : 1 combine to form the lowest-melting-temperature assemblage of rock-forming minerals is of paramount importance in this connection. Wholesale melting of rocks is termed **anatexis;** while the formation of new magma capable of intrusion at higher levels in the crust is termed **palingenesis,** implying rebirth of a new rock from the melting of a previously existing one. These processes can operate most effectively when the sial is thickened and crustal temperatures are raised in the root-zones of mountain chains, thus accounting for the fact that plutonic rocks of the tonalite-granodiorite-granite association are largely confined to orogenic regions.

Petrologists seldom concern themselves very much with the probable sources of basaltic magmas, being content to assume derivation from the deeper crustal layers or from the underlying mantle. These are safely beyond the range of direct observation, at depths of the order of 20 or 30 km. In the case of granitic rocks, however, deep erosion in shield areas and in the cores of ancient mountain chains should expose or approach the levels where selective fusion of the granitic components might have occurred in the past. These are the great theatres of plutonic metamorphism, and in them even the granitic rocks are more metamorphic than igneous in character. This is so much the case, in fact, that the magmatic origin of 'granites' in general has been questioned. A long and vigorous controversy has ensued between those upholding the magmatic origin of granites ('magmatists') and those regarding granites as essentially metamorphic products of granitization ('transformists'). There is no doubt that extreme and sometimes absurd claims have been made for both hypotheses; but now it is generally realized that the arguments of both schools of thought are valid if applied to granites of the right kinds and in the right environments. Probably no one has done more to bring about this realization than H. H. Read whose writings on the subject, collected in the book *The Granite Controversy*, form a critical commentary upon the whole problem.

Conclusions regarding the origin of any particular granitic rock must take into account the following:

(1) The structure of the granite mass and its relationship to surrounding rocks.
(2) Chemical composition of the granite, particularly by comparison with volcanic rocks of known origin.
(3) Mineralogical and textural features which provide evidence of the conditions of crystallization and cooling.

Each of these factors is examined in turn below, with particular reference to intrusive granites that may reasonably be regarded as essentially igneous rocks: the problems of granitization and the origin of 'metamorphic' granites are mainly reserved for the final section of this chapter. Two points should be emphasized here. As this is a book on igneous petrology, the granites of metamorphic character are treated much more briefly than they deserve. Secondly, the division of granites into two categories is artificial and a matter of convenience only. All granitic rocks belong to a series with gradational properties.

(1) Field Relations and Problems of Granite Emplacement

In the following brief survey the main contrasts between granites at the two ends of the series are emphasized: those originating early in an orogenic cycle at deep levels in the crust are contrasted with granites formed later and generally at higher levels.

Very deep-seated granites are often concordant, with margins parallel to the foliation and/or bedding of the surrounding schists and gneisses. They display a wide diversity of form and size, from diffusely scattered augen measuring perhaps a centimetre or so in diameter, to thick sheet-like or lenticular masses of great lateral extent. Internal foliation is often conspicuous and there are clear indications either that the granites were in existence before folding and metamorphism occurred (*i.e.* the granites were *pre*-tectonic); or that they came into being during folding, thus being *syn-tectonic*. The latter are more commonly discordant than the pre-tectonic granites and they may form irregular veins in 'injection complexes', or be intimately mixed and interlaminated with metamorphic rocks of various kinds, forming the so-called **migmatites**. A uniformly high grade of regional metamorphism frequently extends through the granites and associated rocks of any particular area, indicating crystallization of all the rocks concerned under uniform conditions of temperature and pressure.

High-level granites in orogenic regions are generally much more obviously intrusive, forming plutons which are often of very large size. Contacts are either sharp or else there is a relatively narrow transition zone between the granite and surrounding rocks. The latter often show little or no sign of regional metamorphism, though they may have been folded before intrusion of the granite. Many high-level granites are thus shown to be *post-tectonic.*

One of the most significant differences between deep-seated and high-level granites lies in the aureoles of thermal metamorphism which characteristically surround the latter, but not the former. The disharmony between the energy levels represented by such granites and the metamorphosed rocks into which they are intrusive is most significant. The thermal metamorphism which they have caused is but one of several phenomena which lead to the belief that high-level granites originate in general from the intrusion of hot and relatively fluid material. Whether this is regarded as magma is partly a question of definition and partly dependent upon facts of chemical composition which are discussed below. Granite is believed to be capable of diapiric[1] intrusion even when it is largely crystalline, the mechanism of intrusion being closely comparable with that of rock-salt which possesses the property of plastic flow. Unfortunately even a close study of flow-structures may fail to differentiate this condition from one involving dominantly fluid magma, because so much of the structural character of a granite is determined ultimately by late- or even post-crystallization movements. In view of the gradation that must exist and the impossibility of deciding the degree of fluidity in particular cases, it seems best to the writers to regard all granites in this category as magmatic in the wide sense.

The main problem arising concerns the means whereby the granite mass came to occupy its present position: and it is commonly argued that the **space problem** is less acute if replacement rather than displacement is invoked. Actually there is a space problem in both cases; and intrusion necessitating displacement of country-rock does not involve a greater space problem for magmatic rocks than for rock-salt or carbonatites. In Iran, Poland and elsewhere it is evident that plugs of rock-salt have punched their way through great thicknesses of overlying strata, producing little structural disturbance of the latter, and being driven upwards by their own gravitational potential only. The mechanism of emplacement may not be obvious from the structural evidence; but in the case of salt domes, at least there is no question as to their origin by displacement rather than replacement.

Space may be made for the emplacement of granite in several ways: probably in any specific case a combination of different processes was involved.

Harker long ago argued that magma could shoulder aside country-rock or lift the 'roof' in making space for itself; mechanical displace-

[1] Diapirism is literally a breaking through an envelope of country-rock. This may be achieved by intrusion of either fluid or mobile solid material, and gives rise to cross-cutting relationships.

ment by cauldron subsidence or piston faulting has been already described; while piecemeal stoping is proved to have taken place in many cases where granite is seen to be charged with xenoliths.

The solution of the space problem presented by a particular granite depends upon detailed structural study of the rock mass. The nature of the problem is well illustrated by the Flamenville granite which outcrops on the coast of Brittany.

When first mapped it was thought that relics of original bedding could be traced through the granite as a *ghost stratigraphy* which linked up with the country-rocks on either side. If true, this would have made a most convincing argument for the granite having been formed *in situ* by replacement (*i.e.* granitization). Structural mapping has shown, however, that the country-rocks surrounding the granite have been intensely disturbed by forceful intrusion of the latter; while internal flow layers parallel to the margin suggest that the intrusion is mushroom- or pear-shaped with the widest part above a narrow stem.[1]

The general characters of the main kinds of granitic intrusions have been indicated above in Part II, Chapter 2. Some idea of their variety may be gained from a study of the 'granites' of Donegal, in Northern Ireland (Fig. 97). Four main types of intrusion are represented, each emplaced under different conditions. The earliest intrusion is essentially granodioritic and penetrated the country rock, consisting of Dalradian metamorphic rocks, in an intricate manner. The whole complex is migmatitic. Although the Dalradian rocks occur as inclusions enclosed within, and injected by the granodiorite, they have retained to a great extent their original positions in space. Preservation of such a 'relict stratigraphy' indicates that the intruded rocks have suffered only slight displacement; but the fact that some rotation of xenolithic blocks and some relative displacement can be demonstrated in the field is regarded by Pitcher as evidence of the essentially magmatic character of the granodiorite.[2]

The Rosses granite complex[3] cuts the older granodiorite. It has been formed by successive intrusions of four distinctive types of granite which are bounded by sharp and steep contacts giving polygonal outcrops. The boundaries may be regarded as ring-faults whose attitude has largely been controlled by master joints developed in each previously consolidated granite. The intrusion may be described as 'permitted' *i.e.* resulting from the inflow of magma into

[1] Martin, N. R., The structure of the granite massif of Flamenville, Manche, north-west France, *Quart. J. Geol. Soc.*, **108** (1953), 311–42.

[2] Pitcher, W. S., The migmatitic older granodiorite of Thorr District, Co. Donegal, *Quart. J. Geol. Soc.*, **108** (1953), 413–46.

[3] Pitcher, W. S., The Rosses granitic ring-complex, County Donegal, Eire, *Proc. Geol. Assoc.*, **64** (1953), 153–82.

potential cavities induced by the ring-faulting. In this respect the Rosses complex is in complete contrast to the neighbouring Ardara pluton[1] which is of the same general character as that of Flamenville,

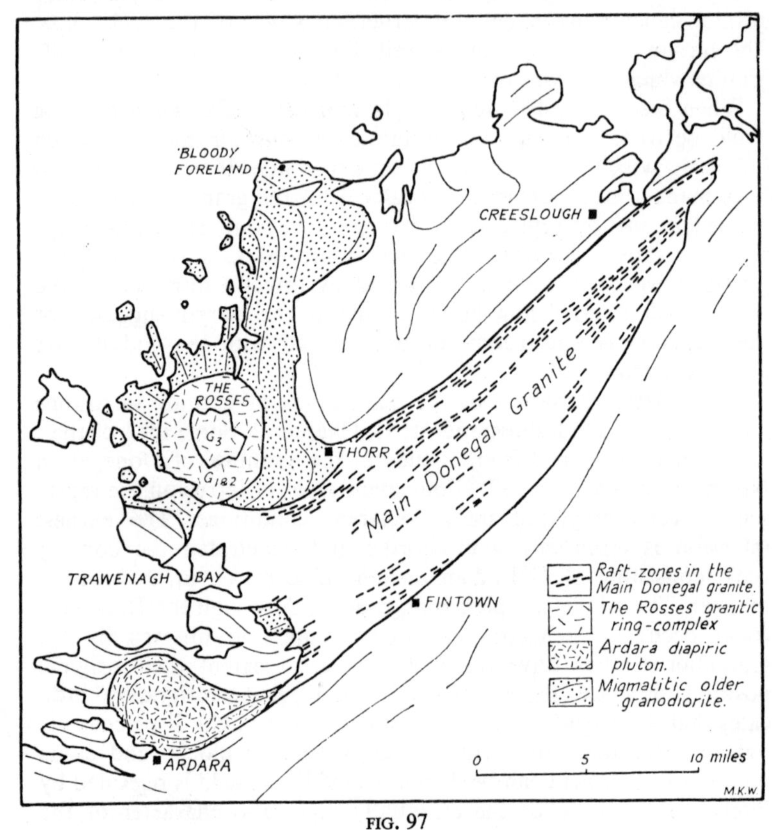

FIG. 97

Diagrammatic map of the Donegal Granite.

(Based on the work of Read, Pitcher and others.)

having been *forcefully* intruded as a diapir. The surrounding rocks have been pierced through and pushed aside in the process. It is interesting to note that this also is a multiple intrusion with a contaminated and more basic facies near the margin, crowded with xenoliths. This is a very common feature of Acid plutons and is displayed by many of the late-Caledonian intrusions in Scotland.

The Main Donegal Granite has the dimensions of a batholith.

[1] Akaad, M. K., The Ardara granitic diapir of County Donegal, Ireland, *Quart. J. Geol. Soc.*, **112** (1956), 263–90.

Much of the interest of this great mass is centred upon the trains of xenolithic inclusions which have been drawn out parallel to the length of the intrusion. The gradual divergence of the trains towards the south-west suggests that they were drawn out in this direction by lateral flow movements. Emplacement of the granite probably involved 'lateral magmatic wedging';[1] but the process must have been immensely complex and long drawn out as evidenced by the structure and metamorphism of the surrounding rocks. These have been subjected to tight folding, and multiple metamorphism which in some respects has a 'regional', rather than local 'contact' appearance.

(2) Factors Involving Chemical Composition

There can be no doubt as to the existence of magma of granitic composition: its reality is demonstrated—often with dramatic effect—every time rhyolitic lava pours out of a hole in the ground during a volcanic eruption.[2]

The similarities of composition of rhyolites, rhyodacites and dacites on the one hand, and of granites, adamellites and granodiorites on the other are so marked in some igneous provinces that they amount to virtual identity. Notable examples are provided by sodic rhyolites and riebeckite-granites in the province of the 'younger granites' of northern Nigeria (see p. 210). Similarly the analysis of Arran pitchstone quoted on p. 239 corresponds almost exactly with that of a granite from the same area. These examples—and many others that could be quoted—come from volcanic regions where the magmatic origin of the intrusive rocks scarcely needs confirmation; it is only when the same chemical similarities can be shown to apply to genuinely plutonic 'granites' that they form really valuable clues as to the magmatic origin of the latter. It is surely significant, therefore, that the chemical variation diagrams for O.R.S. lavas in Scotland apply almost equally well to the plutonic series of diorites, granodiorites and granites of the region;[3] and also that lavas in southern California can be matched in composition with the various intrusions which make up the great batholith of that region.[4]

This last example introduces another significant feature shown by many granitic rocks: within the limits of a single intrusion the rock may be remarkably homogeneous in texture and composition.

[1] Pitcher, W. S., and Read, H. H., The main Donegal granite, *Quart. J. Geol. Soc.*, **114** (1959), 259–305.

[2] See the very readable discussion by M. Walton, The emplacement of granite, *Amer. J. Sci.*, **253** (1955), 1–18.

[3] Nockolds, S. R., The Garabal Hill–Glen Fyne Complex, *Quart. J. Geol. Soc.*, **96** (1940), 451.

[4] Larson, E. S., Jr, Batholith and associated rocks of Corona, Elsinore, and San Luis Rey quadrangles, Southern California, *Geol. Soc. Amer. Mem.*, **29** (1948).

Such homogeneity is to be expected of a magmatic rock, but is difficult to explain for one of this order of magnitude if a metasomatic origin is claimed.[1] These arguments apply to the whole range of granitic rocks defined in a wide sense; but there is one aspect of the composition of alkali-granite (*i.e.* granite *sensu stricto*) which sets it apart from the rest. The proportions of quartz and alkali feldspars in many such granites vary within very narrow limits, and correspond closely to the low melting trough in the ternary system albite-orthoclase-silica (Fig. 73). This, of course, provides strong presumptive evidence that these particular granites originated from melts.

(3) **Mineralogical Evidence Relating to Temperatures of Crystallization**

It has been shown experimentally that the minimum temperatures of the melts in the Or–Ab–SiO_2 system can be drastically lowered under conditions of high water-vapour pressure, from about 950° C for 'dry' melts to about 650° C with a pressure of 4000 kg/cm^2.[2] By making certain assumptions regarding the thermal gradient (believed to be of the order of 30° C for every kilometre increase in depth—but certainly variable from place to place in the crust), and also by assuming that the water-vapour pressure is equivalent to load-pressure, etc., it is possible to predict the levels in the crust where granite should become melted. In view of the number of variable factors involved it would be misleading to quote precise figures; but the experimental results indicate, with a high degree of probability, that melting must commonly occur at depths between about 10 and 20 km.

Very extensive research has been carried out by Tuttle, Bowen and others on the stability of the principal granite minerals, quartz and the alkali feldspars, in an attempt to use the evidence of the structural state of the minerals as indicators of the temperatures of their crystallization. Details relating to the inversion of α- to β-quartz have been described in the Mineralogy Section. Since there is a small but measurable difference between the inversion temperature for quartz that originally crystallized in the high (β) state compared with primary α-quartz, this difference should provide a most valuable clue regarding the temperatures under which granites have been formed. In fact it has been found that the quartz in all the genuinely plutonic granites so far tested has the inversion charac-

[1] See *e.g.* Chayes, F., The finer grained calc-alkaline granites of New England, *J. Geol.*, **60** (1952), 207–54.

[2] Tuttle, O. F., and Bowen, N. L., Origin of granite in the light of experimental studies in the system $NaAlSi_3O_8$-$KAlSi_3O_8$-SiO_2-H_2O, *Geol. Soc. Amer. Mem.*, **74** (1958). This important memoir contains critical discussions on all of the experimental work on synthetic granites and the granite minerals carried out by Bowen and Tuttle and their colleagues.

teristics of the primary low form. This might be taken to imply that *all* really deep-seated granites must have crystallized below about 600° C.[1] However, even the most ardent advocate of low-temperature granitization would scarcely claim this to be the case. It must be concluded, therefore, that during the long and involved history of cooling of any plutonic rock, quartz thoroughly *recrystallizes* and takes on the characteristics of the primary low variety regardless of its original state. This is a disappointing conclusion because the evidence fails just where it is most needed—for the deep-seated granitic rocks. Of course it provides added confirmation of the high temperature origin of many high-level and generally fine-grained granites, many of which contain quartz showing the primary β characteristics.[2]

Much the same situation arises in the case of the alkali feldspars, the stability relations of which have been described above. Briefly, a single phase of alkali feldspar (anorthoclase or sanidine as the case may be) is stable at high temperatures and occurs in Acid lavas and rapidly cooled rocks of the appropriate composition. In a very restricted range of granitic rocks, for which the cooling rates have been somewhat slower, the place of sanidine may be taken by an unmixed perthite. As long as only one kind of alkali feldspar is present it is safe to assume that this must have crystallized originally at temperatures above the level of the solvus (see p. 175).

In most, if not all, plutonic alkali-granites, however, two separate and discrete kinds of feldspar occur, the one dominantly potassic (commonly microcline or microcline microperthite) and the other sodic. This may arise in one of three ways:

(1) Extreme unmixing, particularly if associated with recrystallization, could give rise to two alkali-feldspars from an original single phase precipitated at high temperatures. All evidence of unmixing would be lost in the subsequent recrystallization.

(2) Under conditions of very high water-vapour pressure, the granite liquidus can be depressed to a point where the solvus is intersected (see p. 175). It is theoretically possible for two alkali feldspars to crystallize directly from the melt under these conditions.

(3) The feldspars may have crystallized as sub-solvus phases under metamorphic conditions.

There is no certain way of distinguishing between the three possibilities, although the texture and environment of the granite may provide some indication. If, for example, the granite has a homogeneous composition, and uniform feldspar proportions are

[1] The inversion temperature is $573 \pm 1°$ C at atmospheric pressure; but it rises to somewhat above 600° C under very high pressures.

[2] See *e.g.* Tuttle, O. F., and Keith, M. L., The granite problem: Evidence from quartz and feldspar of a Tertiary granite, *Geol. Mag.*, **91** (1954), 61–72.

maintained throughout, then either of the first two explanations may apply. A heterogeneous distribution of feldspars, however, may well have resulted from granitization: further evidence of the latter may be provided by obvious replacement of K-feldspar by albite or *vice versa.*

(4) Metamorphic Granites and Granitization

Some of the more important characteristics of really deep-seated granites occurring typically in Precambrian shield areas have already been noted. Generally such granites display a harmonious relationship towards surrounding rocks both structurally and as regards metamorphic grade. This fact makes it difficult to assess the true nature of the granites concerned, and three possibilities have to be considered: firstly, the granites may be relics of an earlier generation of rocks of the same composition, *i.e.* earlier granites, rhyolites, etc., which have been recrystallized during the current phase of regional metamorphism. The formation of granites by this simple process—recrystallization of material of appropriate composition—is illustrated by the Bréhat granite in Brittany,[1] which is believed to have resulted from recrystallization of a pile of eruptive rhyolitic rocks. Such recrystallization does not take place in the dry condition in all probability: it is much more likely to happen under the impetus of an advancing wave of granitization.

Secondly, the high-grade metamorphism may have caused the segregation of granitic components from the surrounding rocks by processes relating to selective fusion. Thirdly, the granitic components may have been introduced into the rocks either as an accompaniment to the metamorphism or as a separate event. Whichever of these explanations applies, there is liable to be the most intimate mixing of the granitic and non-granitic rocks and minerals on all possible scales.

J. J. Sederholm[2] has described rocks of this kind as **migmatites.** The granitizing agents were termed by him 'ichors', which possess properties intermediate between aqueous solutions and very much diluted magma, much of it being probably in a gaseous state. Detailed studies of such rocks, which are intermediate between metamorphic rocks and those of igneous aspect, have been carried out in a few areas in Britain, chiefly in the far north among the high-grade schists, gneisses and granulites of the Highlands. By tracing particular bands along the strike in the direction of increasing metamorphism, progressive granitization may be observed, involving the elimination of certain minerals including garnet and

[1] Lafitte, P., *Bull. Serv. Carte Géol., France*, **53** (1955).

[2] On migmatites and associated Precambrian rocks of southwestern Finland, *Bull. Com. Géol., Finlande*, no. 77 (1926), 89.

muscovite, while plagioclase crystals increase in size and become more sodic. The growth of large porphyroblastic feldspars is characteristic. The resulting rocks are appropriately termed **permeation gneisses;** they are metamorphic, in part metasomatic and represent a stage in the formation of rocks of granitic composition and of igneous aspect. Actual penetration of the permeation gneisses by granitic fluids followed, forming **injection gneisses.** In these, distinct interbanding occurs of quartzo-feldspathic layers of granitic composition, with biotite-rich layers representing original pelitic (argillaceous) material. These are typical migmatites. Not infrequently original sedimentary structures are faithfully preserved in the migmatites. With increasing metamorphism the distinction between layers of contrasted composition tends to become less pronounced, the rocks become 'homogenized' by a process of metamorphic diffusion and in extreme cases no trace of the preceding stages outlined above have survived. There can be no doubt, in view of these facts, that migmatites and more advanced granitic gneisses are products of interaction between country-rock and 'juices' (to be noncommittal) derived from granitic magma. The latter was considered by Sederholm and his adherents to be a necessary prerequisite for granitization. In other words, granitization can take place only if and where granitic magma is available.

There has been much speculation regarding the nature of the granitizing medium. According to the evidence of Tuttle and Bowen this should be in the form of a silicate melt provided there is a sufficient concentration of H_2O and the temperature is in the region of 600° C or more. There are many instances, however, where the evidence provided by the mineral associations in the accompanying metamorphic rocks suggests that granitization occurred at appreciably lower temperatures. In such cases relatively dilute aqueous solutions may be responsible (*cf.* the ichor of Sederholm). As an alternative it has been suggested that granitization may be brought about by the diffusion of ions through crystal lattices and particularly along crystal boundaries, this being in the 'dry' state and without the intervention of a liquid phase.[1] The migration of ions is prompted by rising temperature and pressure combined: the activity of the ions depends upon the energy level. Detailed geochemical researches on contact phenomena have demonstrated that the various elements involved in the metasomatism tend to migrate in a certain order. Where a particular element or combination of elements has accumulated at the limits of diffusion within the metasomatized rock, a zone distinctive in mineral composition results, and has been termed a 'front'. It has been shown that in some instances, at least, a **basic front**, involving

[1] Backland, H., The granitization problem, *Geol. Mag.*, **83** (1946), 110.

Ca, Mg and Fe is first produced and moves in advance of the acid constituents.

The most enthusiastic exponent of the concept of metasomatic fronts in this country is D. L. Reynolds who has provided several detailed studies of contact phenomena shown by the Caldedonian and Tertiary intrusions in Northern Ireland. Various references are given in the last part of this book. The 'basic front' phenomena are well displayed in the dark zone rich in mafic minerals frequently developed at the contact between a basic xenolith and the magma in which it is enveloped. Indeed, basic and ultrabasic xenoliths of some observers are interpreted as being due to basic fronts advancing inwards during the course of the granitization of sedimentary rock.[1]

The well-known porphyroblastic feldspars occurring in the contact zone surrounding granites are in the nature of a test case as regards granitization. A finely displayed contact between Malmesbury Shale and the granite at Sea Point just outside Cape Town provides critical evidence.[2] Here the shale has obviously been extensively migmatized, and impregnated by material of granite origin, the physical and chemical nature of which is not important for the moment. The darker, more shaly portions of the mixed rock are charged with large euhedral crystals of alkali feldspar, and all the evidence suggests that they have grown *in situ* within the modified country-rock as **porphyroblasts** formed by metasomatism. Feldspar porphyroblasts of identical appearance are not infrequent in fine-grained basic xenoliths enclosed in many granites such as Peterhead, Shap, and the Jersey granites, and they convey the impression that with a slightly higher degree of alteration of the matrix in which the porphyroblasts are enclosed, the resultant rock would be close to a porphyritic granite in at least appearance.

In consequence of this many geologists have come to regard the growth of feldspar porphyroblasts in xenoliths, and in country-rock neighbouring granites, as representing an advance guard of granitization. It is important to realize that the porphyroblasts are frequently indistinguishable from feldspar phenocrysts in associated granites: hypotheses designed to account for the former have quite naturally, therefore, been applied to the latter also. Reasoning along these lines has led to the idea that the 'phenocrysts' in some granites—even in granites of magmatic appearance—have arisen by granitization of essentially solid rock. It is scarcely necessary to point out that this is completely contrary to experience gained from a study of

[1] See *e.g.* Reynolds, D. L., The geochemical changes leading to granitization, *Quart. J. Geol. Soc.*, **102** (1946), 389.

[2] Mathias, M., and Walker, F., On a Granite-slate contact at Sea Point, Cape Town, *Quart. J. Geol. Soc.*, **102** (1946), 499.

volcanic rocks in which phenocrysts represent early precipitates from the magma.

The explanation given above for the large feldspar crystals in apparently intrusive and magmatic granites is too inconsistent to be accepted without very serious reservations. We prefer to accept these large crystals at their face value, that is, as phenocrysts which may have had long and complex histories of crystallization. Thus, although they probably commenced growth at an early stage in the cooling of the magma, they doubtless suffered continuous modification by reaction with residual fluids. This would be consistent with the plutonic environment. Any residual magma would be somewhat enriched in volatile constituents and would permeate xenoliths and adjacent country-rock, at least for a limited distance from the granite. This fluid would, as it were, carry the conditions of physico-chemical equilibrium pertaining to the granite across the contact and into the country-rock. Provided that the composition of the latter were suitable, then similar mineral phases could grow in equilibrium with the magmatic residuum on either side of a granite/country-rock contact.

The significance of feldspathization in the contact aureoles of granites has been discussed in some detail because it illustrates how widely different interpretations can be put on even a minor (though critical) part of the evidence in the granitization controversy.

CHAPTER IV

SYENITES AND RELATED ALKALI-RICH ROCKS

General Introduction to Intermediate Rocks

The syenite family is the first of three which constitute the Intermediate category. The latter were at one time defined to include all igneous rocks containing between 55 and 66 per cent of silica; but, as already explained, this is not a satisfactory basis of classification. The Intermediate rocks are those with compositions lying between the Q_0 and Q_{20} parallels in the QAP triangle, together with their under-saturated equivalents. The three families are of equal status and occupy fields 4, 5 and 6 in Fig. 76 and, are separated from one another by the field boundaries passing through the points $A_{65}P_{35}$ and $A_{35}P_{65}$.

The three families are:

Syenites, with alkali-feldspars dominant to the extent of at least 65 per cent of the total feldspar;

Monzonites (= syenodiorites), with approximately equal amounts of alkali- feldspar and plagioclase ($An < _{50}$);

Diorites, with plagioclase ($An_{<50}$) amounting to 65 per cent or more of the total feldspar.

SYENITES

Classification and Nomenclature

The minimum specification of syenite, stripped of all non-essentials and excluding for the moment the feldspathoid-bearing varieties, is a coarse-grained assemblage of alkali-feldspar with usually a small amount of unspecified mafic constituents and accessory minerals. A limited amount of both quartz and plagioclase is permissible; but the field boundaries referred to above impose strict limitations on both. No rock with a composition falling outside the syenite field can bear the name 'syenite'; and all rocks within it must be named 'syenite' with suitable qualifiers as necessary.

The syenite family comprises three categories of rocks which differ in their degree of silica-saturation and grain-size:

1. Over-Saturated. Quartz-syenites: rocks of syenitic aspect in

hand specimens; syenitic composition confirmed by examination of thin sections, but containing quartz in an accessory capacity.

2. Saturated, containing neither quartz nor feldspathoid. It is doubtful if any large rock-body is of this type: most syenites are either over- or under-saturated.

3. Under-Saturated, differing fundamentally from the above by containing nepheline or other feldspathoid proxying for alkali-feldspar to any extent.

Each of the categories in 1, 2 and 3 above may be further subdivided on the basis of the kind of feldspar present: it may be

(*a*) potassic (orthoclase or microcline);
(*b*) sodipotassic (perthites); or
(*c*) sodic (albite).

Perhaps largely because of their relative rarity a curious mystique has come to be attached to the occurrence of feldspathoids in igneous rocks. There has been a tendency to regard the presence of, say, even a minute amount of nepheline in a syenite as of greater significance than a comparable amount of quartz. Strictly speaking, the one is no more significant than the other: each indicates that a phase boundary has been crossed. The name 'nepheline-syenite' is commonly applied, therefore, to a rock of syenitic aspect in which perhaps only a minute amount of nepheline has been discovered in a thin-section, though a rock containing a comparable amount of quartz would not necessarily be distinguished as 'quartz-syenite'. Strictly speaking, of course, any departure in either direction from the condition of exact saturation should be indicated by using the relevant mineral qualifier, (Fig. 98). This means that just-saturated syenites—**orthosyenites**—would be confined in the scheme of classification to the hypothetical base-line separating the QAP and FAP triangles.[1] In practice it is necessary to allow a certain amount of latitude in the definition of orthosyenite: authorities who use the QAP triangle have agreed on Q_5 as a suitable limit, and we suggest that the comparable value, F_5, should be used for the limiting proportion of feldspathoid.

It would be very convenient if we could extend the comparison between the two triangles further and use F_{20} as the limiting proportion of feldspathoid for any rock within the syenite field. In a sense we are free to do so because the choice is entirely arbitrary. However, there is one major difference between the occurrence of quartz and of feldspathoid which must be taken into account. Whereas the physico-chemical conditions set a natural limit to the

[1] We do not illustrate the FAP triangle, as any internal boundaries dividing it into specific 'fields' are still *sub judice*; but the reader will appreciate that it is analogous to the QAP triangle with which it is in mirror-image relationship across the common AP base-line.

proportion of quartz which normally occurs in igneous rocks, there is no such restraint on the proportion of feldspathoid. In intrusive alkali complexes it is quite normal for rock-types to occur in which

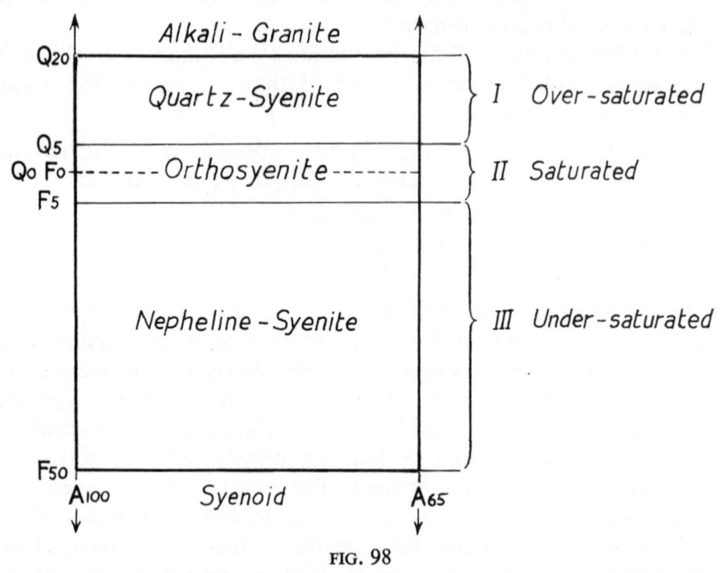

FIG. 98

Classification of the rocks of the Syenite Family, based on the degree of silica saturation.

feldspathoids are in excess of feldspars, or even contain the former to the exclusion of the latter. In view of this fact, we suggest that the natural limit for feldspathoidal rocks in the syenite field lies at the point where the proportion of feldspathoid balances that of feldspar. Beyond this limit feldspathoidal syenites give place to **syenoids** which, as the name implies, have strong affinities with syenites, but lack the essential dominance of alkali feldspar which is the distinguishing characteristic of all true syenites.

1 AND 2. OVER-SATURATED AND SATURATED SYENITES

These are dealt with together for the reason stated above. Varieties containing (*a*) potassic feldspar, (*b*) sodipotassic feldspar, or (*c*) sodic feldspar may be distinguished.

It is of historic interest perhaps to refer to the rock from Syene (Aswan) in Egypt, from which Pliny derived the word syenite. Actually this rock is a particularly handsome hornblende-*granite*. The rock selected as the type by A. G. Werner comes from Plauen near Dresden and is widely known as the **Dresden syenite**, and countless generations since Werner's day have been taught to regard it as representative of its kind. It is an over-saturated potassic syenite containing considerable amounts of oligoclase—in some facies too much (up to 37 per cent of total feldspar) which places these varieties among the monzonites. In hand-specimens dull reddish, fluxionally arranged feldspars are obvious, the coloured silicate is identified as common hornblende on examining thin sections, while megascopic sphene, together with apatite and magnetite, are the usual accessories. In hand-specimens the feldspars are all uniformly brick-red, and this doubtless deceived early observers into thinking that only one kind of feldspar was present. Examination of thin sections shows that there are two kinds, orthoclase and oligoclase. Both are of the same crystal habit. It is rather easy, therefore, to over-estimate the amount of orthoclase in a section of Dresden syenite: the oligoclase is liable to be overlooked. It occurs as complex twins (Carlsbad-Albite) and by reason of the straight extinction of plagioclase of this composition the simple twinning is much more obvious than the lamellar Albite type.

Tyrrell[1] has described from Spitzbergen a syenite of the same general variety, but much nearer to the ideal. Quartz is present, but in insignificant amounts (3·3 per cent); microcline makes up 62 per cent of the rock, with oligoclase only 4 per cent; while biotite and augite in equal amounts make up the remaining quarter.

Equally typical syenites occur in alkali complexes in South, central and eastern Africa. One distinctive type contains a little interstitial green pyroxene, but the rest consists of platy, fluxionally arranged alkali-feldspars, dull red in the hand specimen, and giving the impression that every crystal is a Carlsbad twin.

If **albitites** are really what the name implies, they conform ideally to an important syenitic type, the saturated sodic syenites. Albitite (H. W. Turner, 1896) was first recorded from Plumas Co., California; later records include Cape Willoughby, S. Australia, the Shetland Isles, Scotland,[2] and Kotaki, Japan. Albitite is virtually monomineralic. It contains no free quartz although the SiO_2 percentage may be as high as 68. Over-saturated albitites (**quartz-albitites**) occur in dykes in the orogenic zone in Japan, forming part of a dyke-phase associated with ultrabasic serpentinites, and including also albitites, trondjemites and other highly feldspathic rock-types. Albitite appears to be closely associated with serpentinites: 'lenses' of the former occur embedded in the latter, but sheathed in jadeite. It will be remembered that jadeite is $NaAlSi_2O_6$, while albite is $NaAlSi_3O_8$, so that the jadeite shell probably represents desilicated albite.

[1] Tyrrell, G. W., *Trans. Roy. Soc. Edin.*, **53** (1922), 225.
[2] Walker, F., An albitite from Ve Skerries, Shetland Isles, *Min. Mag.*, **23** (1932), 239.

As might be expected, syenites in which the feldspars are of less extreme composition are much commoner than those containing albite or orthoclase. The feldspars are sodi-potassic—occurring either as discrete crystal grains of albite *and* orthoclase or microcline, or intergrown as perthite or antiperthite. Examples are found in both the over-saturated and saturated categories. Over-saturated sodipotassic syenites are well represented by **nordmarkite** (W. C. Brögger, 1890), described originally from a locality near Oslo, Norway. The dominant component is microperthite accompanied by rather much quartz, which places some specimens of nordmarkite on or over the borderline between granites and syenites. Subordinate coloured silicates include aegirine, riebeckite or arfvedsonite, either alone or in company. If desired, therefore, distinction may be made between, say, aegirine-nordmarkite and riebeckite-nordmarkite. The nature of the coloured silicate(s) may best be indicated in this way. In Britain nordmarkite forms the upper part of the stratified laccolith forming Cnoc na Sroine in Assynt; it also occurs (with pulaskite) in the Ben Loyal Complex[1] and the Loch Ailsh Complex in Sutherlandshire. The modal analysis of the Ben Loyal nordmarkite is: 77 per cent feldspar (albite 47, orthoclase 30, therefore antiperthite) quartz 12, amphibole 9 with accessory sphene and magnetite.[2]

Perthosite (J. Phemister, 1926) described from a Scottish type-locality is a sodipotassic leucosyenite with perthitic feldspars and only 5 per cent of mafic and accessory constituents. Even with 65 per cent silica the rock is quartz-free; it is exceptionally rich in alkalis, with Na_2O, 7·4, in excess of K_2O, 5·2 per cent.

Pulaskite is a very similar type, described from Pulaski Co., Arkansas, by J. F. Williams (1890); but much nearer home it forms the greater part of the Ben Loyal Complex in Sutherland, northern Scotland. This syenite also is markedly leucocratic, but with antiperthitic feldspars making up almost the whole of the rock, with a little aegirine and accessories.

A most distinctive type, both as regards megasocopic and microscopic characters, is the very handsome Norwegian rock, widely used for ornamental purposes, and termed **larvikite** (or **laurvigite**) by Brögger (1890). The type is characterized by its coarse grain and distinctive feldspars, which, especially on polished surfaces, exhibit a beautiful blue schillerization. They were referred to as anorthoclase and were regarded as a cryptoperthitic intergrowth of orthoclase and oligoclase. In this respect they are akin to the feldspars in the lava-type, kenyte. Recently, however, it has been claimed that

[1] Read, H. H., in *Geology of Central Sutherlandshire*, *Mem. Geol. Surv., Scotland* (1931), 174.

[2] Knorring, O. von, and Dearnley, R., Note on nordmarkite . . ., *Min. Mag.*, **32** (1960), 389.

FIG. 99. Larvikite, Larvik, Norway.

Schillerized titanaugite (*stippled*); olivines with strong surface relief (*irregular stipple*); titanomagnetite, small early octahedra and irregular patches closely associated with lepidomelane; feldspar, fresh (see text); nepheline scarce (*near top centre and l.h. end of titanaugite*); apatite abundant in mafic clots.

instead of a single feldspar-phase, there are *two* present—oligoclase and alkali-feldspar. The former is largely antiperthitic and it is exceedingly difficult to identify as it rarely shows twinning. Actually the plagioclase member is intricately twinned on the Albite Law, but on so fine a scale that the lamellae are invisible under ordinary conditions. To complicate the issue still further, the twinning imparts a monoclinic symmetry to the optical orientation of the crystals. This elusive internal structure finds outward expression in the development of a distinctive crystal habit under favourable conditions, well exemplified in the rhomb-porphyries described below. Larvikite has always been regarded as a syenite, but in view of these facts, its retention in this family is questionable: some larvikites appear to be monzonitic in composition: the *substance* of oligoclase is present in the requisite proportion, whether it is intimately intergrown with the alkali-feldspar, or occurs in discrete crystals.[1] The specimen illustrated in Fig. 99 is *apparently* syenitic, however, and so far as superficial characters are concerned, it would be out of place among the monzonites. Larvikite grades into **lardalite,** in effect a nepheline-larvikite, but with albite in place of oligoclase.

On account of the dark colour of the strongly schillerized feldspars the mafic minerals are not conspicuous in larvikite specimens, but are exceptionally striking in thin sections. They form clots from which feldspars are excluded; the assemblage is not one which would be regarded as typically syenitic but would appear more 'at home' in a gabbroic rock. The clots include schillerized titanaugite, iron-rich olivine, abundant apatites of large size and perfect shape, irregular masses of iron-ore from which crystals of lepidomelane tend to radiate. Late-formed finger-like outgrowths of pyroxene penetrate outwards into the surrounding feldspars. A second iron-ore occurs as a normal accessory in the form of octahedra of titano-magnetite embedded usually in titanaugite. It may well be that these extraordinary clots are of xenolithic origin.

Of the rock-types described above, albitite and perthosite are notably leucocratic, while nordmarkite and the 'Dresden syenite' are mesocratic. Potassic melasyenites, as noted above, are rich in mafic minerals. For saturated potassic melasyenites the most appropriate existing name is **shonkinite** (Weed and Pirsson, 1895), from the well-known Shonkin Sag laccolith in the Bearpaw Mountains, Montana.[2]

[1] Barth, T. F. W., The igneous rock complex of the Oslo region; ii, *Systematic Petrography of the Plutonic Rocks*, *Oslo* (1945), p. 76; but *cf.* Oftedahl, C., *ibid.*, ix, *The Feldspars* (1948), 71.

[2] Barksdale, J. D., The Shonkin Sag laccolith, *Amer. J. Sci.*, **33** (1937), 321–59; also Osborne, F. F., and Roberts, E. J., Differentiation in the Shonkin Sag laccolith, Montana, *Amer. J. Sci.*, **22** (1931), 331–53.

The original shonkinite contains olivine; but for the purposes of classification it seems best to ignore this in the definition and to prefix the rock-name with 'olivine-' when that mineral occurs: thus the type rock is olivine-shonkinite. Similarly some facies of the type rock contain a little nepheline. So long as the latter is merely an accessory, this variety may be appropriately termed nepheline-shonkinite; but if nepheline is of essential status, this name is no longer appropriate. The mineral assemblage orthoclase, nepheline, mafic minerals defines malignite, described below.

UNDER-SATURATED SYENITES AND 'SYENOIDS'

Within this important category mineral variation involves the kind(s) of feldspathoid present, the kind(s) of feldspar associated with the latter, the relative proportions of the one to the other, and the role of mafic constituents. The latter involves not only the kinds, but also the amounts of mafic minerals occurring. There are thus five variables, and ample scope, were we so inclined, for defining a large number of specific types. As a matter of interest there are actually more named types of nepheline-syenite than are found in any other category: we believe that many of these are superfluous in the sense that they differ only from better known rock-types in insignificant points of detail, which are adequately covered by appropriate qualifiers as indicated below.

Theoretically it would appear possible to distinguish two series of under-saturated syenites, one potassic, the other sodic; but actually this is impracticable. The significant minerals in this category are the feldspathoids, and therefore due weight is given to this fact by recognizing: (*a*) pseudoleucite-bearing types; and (*b*) nephelinic types. The former are exceedingly rare: the great majority of feldspathoidal syenites are nephelinic. The several mineral associations and the names of the corresponding rock-types are tabulated below:

(*a*) with pseudoleucite accompanied by orthoclase: borolanite
(*a'*) with pseudoleucite only—'feldspar-free' types: fergusite, missourite[1]
(*b*) with nepheline accompanied by potassic feldspar: malignite
with nepheline accompanied by perthitic feldspars: foyaite
with nepheline accompanied by separate K^+ feldspar and albite: litchfieldite

[1] There is a difficulty of terminology here, amounting almost to a contradiction in terms. Pseudoleucite consists largely of orthoclase; but these rocks are feldspar-free in the sense that they consist of *pseudoleucite* and mafic minerals only.

with nepheline accompanied by albite: mariupolite and monmouthite

(*b'*) feldspar-free syenoids with nepheline: the ijolite series; urtite, ijolite and melteigite

In due course it will be shown that leucite-bearing lavas are much commoner than coarse-grained rocks of the same composition on account of the chemical instability of leucite under deep-seated conditions; and there is a closer and more complete similarity between the various nephelinic lavas and the leucitic analogues.

The essential facts of the mineral composition of the main feldspathoidal syenites are indicated in the above list. Certain details of interest or significance are noted in the following brief account of individual types.

Types containing Pseudoleucite

Borolanite is a most interesting rock although its status is uncertain. It is one of the few British type-rocks and it is certainly distinctive enough to justify a specific name. The type area is the Loch Borolan Complex in Assynt, in the north-west Highlands of Scotland. It was originally described by Sir J. J. Teall who was impressed by the resemblance of certain prominent whitish crystals or aggregates to pseudoleucites.[1] Superficially the resemblance is close in some facies of the rock, but not in others. Thin sections show the whitish aggregates to consist of orthoclase associated with nepheline, usually decomposed. The abundant mafic constituents include melanite garnet, rich in titanium and associated with much sphene, pyroxene, greenish biotite, with purple fluorite as an unexpected accessory. The rock also contains polygonal areas said to be zeolitic aggregates after sodalite.

The felsic mineral association, orthoclase with nepheline, places borolanite among the nepheline-syenites, in this respect being close to malignite apart from the fact that it is rich in melanite. The systematic position of the rock depends, of course, upon whether it contains pseudoleucite or not. The uncertainty arises from the fact that borolanite occurs in a belt of country much affected by overthrust faulting; and Shand in a re-examination of the Loch Borolan Complex was led to believe that Teall's pseudoleucites were orthoclases which had been mechanically rounded during the earth-movements.[2]

With increasing melanite, borolanite grades into a type of melasyenite particularly rich in garnet, to which Shand originally gave

[1] *Trans. Roy. Soc. Edin.*, **37** (1892), 163.

[2] Shand, S. J., On borolanite and its associates in Assynt, *Trans. Roy. Soc. Edin.* (1909–10), 202–15 and 376–416.

the name ledmorite, from the nearby Ledmore River; but he subsequently withdrew the name in favour of melanite-malignite. The complex also contains in its basal parts a pyroxene-rich facies reminiscent of the corresponding ultramafic associate of melteigite described below.

There is no such uncertainty concerning **fergusite** (Pirsson, 1905) described originally from the Highwood Mountains, Montana. It is chemically similar to shonkinite but is under-saturated, with leucite (or pseudoleucite) taking the place of orthoclase. In the type rock pseudoleucite, consisting of the usual association of orthoclase with nepheline, makes up more than half the rock (65 per cent); aegirine-augite is the mafic constituent and magnetite the chief accessory. A stock of fergusite has more recently been described from Tashkent in the U.S.S.R., while of greater interest are certain occurrences in the Roman volcanic province. Here fergusite containing *fresh* leucite is represented by ejected blocks only, associated with italite (see p. 89). Only two minerals are present—leucite and pyroxene, with an unidentified whitish material. Evidently the rock is not of deep-seated origin and by reason of its mode of occurrence the expected alteration into pseudoleucite has not occurred. Chemically, as might be expected, outstanding features are low silica percentage (about 48 per cent) and high K_2O (9·5).

Missourite (Weed and Pirsson, 1896) also comes from Highwood Mountains and is closely related to fergusite, but is more melanocratic and contains olivine. As compared with fergusite, missourite contains only 15 per cent pseudoleucite, about half the rock is augite, while olivine is also abundant. Thus missourite is olivine-melafergusite—the name missourite is not essential. Among the fine-grained leucitic lavas one type, olivine-leucitite is well known and is mineralogically the fine-grained equivalent of olivine-melafergusite.

Nephelinic Syenites and Syenoids

Malignite (Lawson, 1896) takes its name from the type occurrence on the Maligne River in Ontario. The significant felsic mineral association is nepheline and orthoclase, in about equal amounts in the type rock. Clinopyroxene makes up about half the rock, while biotite, apatite and sphene also occur. With diminishing nepheline, malignite grades into shonkinite.

A malignite of particularly striking appearance is illustrated in Fig. 100. This is a well-known rock in teaching collections on account of the variety of minerals it contains. Two feldspathoids, fresh nepheline and euhedral crystals of a member of the hauyne-nosean group (nearer to nosean than to hauyne) together with interstitial poikilitic orthoclase are the felsic constituents. Euhedral zoned

FIG. 100. Olivine-malignite, Katzenbuckel, Odenwald.

All the component minerals (except orthoclase) are euhedral: aegirine-augite with more sodic outer zones (*stippled*); olivine, three small phenocrysts marginally altered to serpentine; nepheline, perfectly euthedral, well cleaved, with the cleavage emphasized by incipient marginal alteration; nosean/hauyne rhombdodecahedral, zoned, (*fine irregular stipple*); orthoclase, poikilitic towards all other components. Apatite usually cored; magnetite octahedra, sometimes in parallel growth.

aegirine-augites with deeper coloured, more sodic outer zones accompany fewer and smaller olivines, which are partly sepentinized. Prominent among the accessories are magnetite in unusually perfect octahedral crystals, and abundant apatites varying widely in size. Many of the larger apatites are distinctly cored. It will be realized that in squeezing this distinctive rock into the classification as a variety of malignite we are doing it less than justice. We know of no other rock in which the association of potassic feldspar (orthoclase) and sodic feldspathoid (nepheline) is so strikingly shown. The rock is malignite to that extent: but it should be distinguished as an olivine-hauyne-malignite.

Another very distinctive rock-type may be included in the malignite category, though Shand termed it a foyaite (see below): it contains microcline instead of orthoclase together with nepheline, mafic minerals and accessories. It forms a part of the Pilaansberg Complex in South Africa. The microcline is an apple-green variety resembling the well-known teaching material from Pike's Peak, Colorado. This feldspar contains barium in significant amounts.

Of the undersaturated syenites in which the feldspar is sodipotassic the type termed **foyaite** (Blum, 1861) is by far the most important, and is probably the most widely distributed of all the feldspathoidal syenitic rocks. Since the original discovery at Foya Peak, Serra de Monchique, Portugal, foyaites have been discovered in many alkali-rich complexes, notably in several in South Africa, described by Shand. These foyaites appear to conform to a type in which roughly one-quarter of the rock consists of nepheline, sodipotassic feldspar averages about 60 per cent, sodic pyroxene about 10, and sphene and other accessories 5 to 6 per cent. The alkali feldspar forms flat tabular white crystals contrasting with the red-weathering nepheline and the black lustrous prisms of aegirine. The habit of the nepheline varies: in some cases it is interstitial to the feldspars, in others it is euhedral, forming the characteristic stumpy hexagonal prisms, the identification of which, in hand-specimens, is easy by reason of the quarter-inch square and six-sided sections visible on the weathered surface. Occasionally significant accessories occur in a foyaitic mineral assemblage. Thus in the Pilaansberg ring-complex in the Transvaal Shand records a white sodalite-foyaite which 'glitters in the sunlight like marble'. This is transitional into sodalite-syenite, considered below.

Litchfieldite (Bayley, 1892) resembles a leucofoyaite in bulk composition, but differs in that it is a *two-feldspar* instead of a one-feldspar nepheline-syenite. Albite is dominant, but is accompanied by microcline. In the mass the rock is whitish, with scattered black lepidomelane crystals and lemon-yellow patches of cancrinite, while sodalite is also present. The type-rock is a cancrinite-sodalite leuco-litchfieldite.

The most strongly sodic syenites consist of nepheline associated with albite and mafic minerals; and two named varieties may be noted. **Mariupolite** (Morozewicz, 1902) contains albite, about three-quarters of the rock, nepheline 13 per cent, aegirine 7·5 and lepidomelane 4 per cent. The important feature is the minor role of nepheline. We consider that mariupolite is sufficiently well defined in terms of the feldspar-feldspathoid relationship only: the type-rock contains aegirine as the chief mafic mineral; but there is room in the scheme for, say, riebeckite-mariupolite as well as for aegirine-mariupolite, and this may be regarded as a kind of test case. Mariupolite *could* be defined by the mineral assemblage albite (in large excess), nepheline *and aegirine*. If this is done a new name must of necessity be found to cover those mariupolite-like rocks which contain a mafic mineral other than aegirine. The necessity for introducing new names does not arise if it is agreed that mariupolite is defined by the feldspar-feldspathoid relationship *alone*, without specifying the mafic minerals which may be present in different varieties: they are most satisfactorily covered by prefixing the name (or names) of the mafic constitutents to the general rock-name mariupolite, as above.

By contrast, the second essential assemblage involves the same two minerals but with roles reversed: nepheline predominant, albite accessory. This assemblage is represented by **monmouthite** (Adams, 1904), named from Monmouth Co., Ontario. Nepheline makes up more than three-quarters of the rock. In the type rock an amphibole is the mafic constituent; but aegirine-monmouthite has also been described.

The mineral which most commonly acts as proxy for nepheline is the allied silicate, sodalite. As noted above, sodalite may occur, in a very subordinate capacity, in such types as foyaite. As soon as it attains to the status of an essential component, a new name is required. **Ditróite** (Zirkel, 1866) from Ditró in Transylvania, is such a rock. In it nepheline may be as abundant as feldspar (microcline-microperthite in the type-rock), and is accompanied by both sodalite and cancrinite. The latter is apparently secondary after nepheline. While a green biotite occurs in some specimens from the type-area, aegirine-augite rimmed with very dark green arfvedsonite is more characteristic. Ditróite is a sodalite-bearing nepheline-syenite. Obviously with the roles of the two feldspathoids reversed, the rock would be nepheline-bearing sodalite-syenite. True **sodalite-syenite,** containing no nepheline, lies at the end of this line of variation, and is represented by a rock to which this name was originally given, occurring at Julianehaab in Greenland. Sodalite is not particularly abundant in it, amounting to some 8 per cent, but is the only feldspathoid present. Much of it is interstitial to lath-shaped sections of microcline-microperthite, but some of it occurs in the form of euhedral rhombdodecahedrons embedded in the feldspar. It is thus obviously primary. Similar sodalite-syenites have been described from Liushe in the (late) Belgian Congo, and from Nyasaland. Sodalite-rich syenites are rare; but an example from St Hilaire Mountains, Quebec, contains up to 70 per cent; while a facies of the sodalite-syenite in Greenland grades into an almost pure sodalite rock, of an intense blue colour.

It is appropriate to consider **analcite-syenite** next, if for no other reason than the close similarity between sodalite and analcite. Both are Cubic, colourless and have low refractive indices, and there is no easy optical method of differentiating between them: a micro-chemical test is the only certain means of distinction. 'Analcite-syenite' used without qualifiers implies the association of analcite with alkali-feldspars and coloured minerals. G. W. Tyrrell[1] has described a coarse-grained analcite-syenite from an Ayrshire locality. The dominant mineral is anorthoclase rimmed with orthoclase, but it is accompanied by some plagioclase. The dark minerals are titan-augite rimmed with aegirine, and barkevikite. All these minerals are embedded in an abundant matrix of analcite. In this locality the analcite-syenite forms the upper part of a composite sill, the lower part of which is a basic analcite-gabbro (crinanite). This suggests that analcite-syenite is a differentiate of a Basic magma carrying the constituents of analcite. The occurrence of thin veins of the former in a differentiated teschenite sill at Hallcraig in Fifeshire[2] supports this hypothesis, as teschenite is closely similar to crinanite.

Nepheline-syenite may contain cancrinite. In many instances its relations to the nepheline indicate that it is an alteration product of the latter; but in other cases it is not only more abundant, but it has the status of an essential component, and in appearance seems to be primary. By analogy with other types considered above, such rocks are cancrinite-nepheline-syenites or **cancrinite-syenites,** depending upon the amount of cancrinite in the rock.

Nepheline-syenites and more particularly the pegmatitic facies are noteworthy on account of the wide variety of accessories they contain. Many of these are titanium and zirconium silicates. One of the more striking is eudialyte, which in some rocks is obvious in the hand-specimen, but which rarely ranks as an essential component. **Eudialyte-nepheline-syenite,** named chibinite (Ramsay, 1894) from a

[1] *Quart. J. Geol. Soc.*, **84** (1928), 540.
[2] Campbell, R., The Braefoot outer sill, Fife, *Trans. Geol. Soc. Edin.*, **13** (1933), 148.

locality in the Kola Peninsula in the U.S.S.R. is a very striking-looking rock in the hand-specimen: it consists of black aegirine prisms and bright cherry-red eudialyte[1] in a matrix of nepheline and microcline-microperthite.

Finally, reference may be made to another rare type—**corundum-syenite**—which occurs in the Haliburton—Bancroft area in Ontario[2] and in the Ural Mountains. The rock from the latter locality contains no nepheline, but it falls in the category of under-saturated syenites in virtue of the presence of corundum. This mineral makes up nearly a quarter of the rock, most of the rest being microperthite, the third component being biotite in very small amount. The Canadian specimens on the other hand are corundum-bearing nepheline-syenites, often of pegmatitic facies. The origin of the corundum-bearing syenites is discussed later.

Types of nephelinic syenoids containing no feldspar are completely under-saturated so far as felsic constituents are concerned. The only variability involves the kinds and proportions of mafic minerals. Three rock-types of wide distribution and closely associated in the field fall into this category. Their essential relationships are shown in the appended table. As the constituent minerals are essentially the same in all three, and only the proportions vary, it is obviously possible to use only one rock-name, that of the central, 'average' type, ijolite, distinguishing the melanocratic variety as **mela-ijolite,** and the leucocratic variety as **leuco-ijolite.** Pulfrey has suggested dividing the series rigidly at the points in the colour-index scale as shown by the figures on the right:—

	Colour Index
nepheline-rock	
	10
urtite—leuco-ijolite	
	30
ijolite—ijolite	
	70
melteigite—mela-ijolite	
	90
pyroxenite	

Urtite (Ramsay, 1894), a whitish rock of syenitic aspect, is named from the type-locality Lujaur-Urt in the Kola Peninsula, U.S.S.R. Nepheline makes up 85 per cent of the rock, the rest consisting of

[1] This is a very different colour from the red altered nepheline seen in many foyaites: it is a 'raw' deep pink. Further, the eudialyte has a vitreous lustre: the altered nepheline is dull.

[2] Adams, F. D., and Barlow, A. E., Geology of the Haliburton–Bancroft area, Ontario, *Geol. Surv. Mem. Canada*, no. 6 (1910).

mafic minerals, in most instances (though not exclusively) aegirine. Although the colour index is so low (only 15), the silica percentage is very low (45 per cent), while, as might be expected, both alumina

FIG. 101

Analcite-ijolite, Neudorfel, Schreckenstein, Austria.

The chief components are titanaugite and nepheline, both in well-formed crystals, especially the latter. Analcite is interstitial to these, and is partly altered to a zeolitic aggregate. Apatite and iron-ore are prominent accessories. The nephelines are traversed by canal-like fractures, while the cleavage is accentuated by incipient alteration.

(30 per cent) and alkalis (20 per cent), chiefly soda, are both exceptionally high. In the type rock the mafic mineral is again aegirine; but as defined above, it might be any of the soda-iron amphiboles or pyroxenes.

A small enough specimen of urtite may be pure nepheline rock, and in the mass the name which should be applied to a rock of this composition is nephelinite: no other is really appropriate. Unhappily this name is widely used for the nephelinic lavas referred to below under 'ijolite'.

Ijolite (Ramsay and Burghell, 1891), originally named after a locality in Finland, consists of nepheline making up approximately half the rock, the rest consisting of unspecified coloured silicates and accessories. The type-rock contains aegirine; but an ijolite described from Songo, Sierra Leone, contains a pyroxene within the diopside-hedenbergite range.[1] Ijolite is much more widely distributed than most of the types so far noted: it occurs in association with nepheline-syenites (usually foyaites) in many alkali complexes, for example that at Spitzkop, Sekukuniland, where it is represented by a handsome coarse-grained rock in which black prismatic aegirines are embedded in a matrix of dull red nepheline with a little bright yellow cancrinite. It may be noted that certain important very basic feldspathoidal lavas consist of the ijolitic mineral assemblage: there can be little doubt that they are comagmatic. With decreasing mafic minerals ijolite grades into urtite, and in the opposite sense into **melteigite** (Brögger, 1921). In the type-rock nepheline makes up about a quarter of the whole rock, the rest being mafics and accessories. Pyroxene (aegirine or aegirine-augite) is strongly dominant: the rock is exceedingly melanocratic and is at the opposite end of the scale to urtite, the two being complementary.

MICROSYENITES

Microsyenites are the medium-grained equivalents of syenites on the one hand and trachytes. leucitophyres and phonolites on the other. They are not common rocks, and in this respect are less important than the other grain-size groups. Several alternative names have been applied to them in the past, some of general significance, others with more specific meaning. As most of these rocks are porphyritic, with more or less prominent phenocrysts of orthoclase, they have been collectively grouped as 'porphyries' (as distinct from 'quartz-porphyries', which are Acid rocks of the same status, but of granite composition, of course). In a former edition of this book we used 'microsyenite' for aphyric types, and

[1] Baker, C. O., Marmo, V., and Wells, M. K., The ijolites at Songo, Sierra Leone, *Col. Geol. Min. Res.*, **6** (1956), 407–15.

'syenite-porphyry' for those with phenocrysts. It seems better to call all these rocks, regardless of texture, by one name, and, by analogy with the granitic rocks, **microsyenite** seems the obvious choice. The two textural variants are then 'aphyric microsyenites' and 'porphyritic microsyenites', the meanings of which are self-evident.

As regards mode of occurrence, these rocks are intrusive; but the mere fact that a syenite rock occurs as a sill or dyke does not automatically entitle it to the name microsyenite. The latter should only be used if the rock is of medium grain; if of fine grain, it must be termed trachyte, if preferred, with the prefix 'intrusive' to remove any possible doubt.

Certain of these rocks are definitely potassic, and differ from the corresponding syenitic and trachytic types only in grain-size. **Potassic microsyenites** often have a distinctive texture referred to as orthophyric, in effect a coarser trachytic texture resulting from the close packing of short stout prisms of orthoclase. Orthoclase is the dominant mineral, often occurring in two generations, and accompanied by hornblende, biotite, and, in the more acid examples, by a little interstitial quartz. A particularly good example of a porphyritic potassic microsyenite occurs at Goodsprings, Nevada, and is noteworthy for the beautiful phenocrysts of reddish orthoclase—simple crystals or Carlsbad twins—for which it is famous. Orthoclase crystals from this rock are to be found in all good teaching collections. The groundmass also consists chiefly of orthoclase, and the amount of coloured minerals is small, as in most syenitic rocks.

If the magma from which these potassic microsyenites were formed became desilicated, leucite would crystallize at high temperature, but under intrusive conditions it would invert into pseudoleucite with falling temperature, and the rock would contain pseudoleucite and orthoclase (the proportions depending upon the degree of under-saturation achieved), accompanied by coloured silicates of the appropriate composition. Such a rock would be exactly analogous with pseudoleucite-syenite on the one hand and leucitophyre on the other and forms the connecting link between these two types. It is equally rare, but examples occur in the alkali-complex at Magnet Cove, Arkansas, and others have recently been noted, under the names 'leucite-porphyry' and 'pseudoleucite-porphyry', associated with 'orthophyres' and potassic trachytes in an Eocene volcanic area in the Pambak Mountains, Armenia.

Sodic and sodipotassic microsyenites compare closely with corresponding coarse-grained rocks, with which they are intimately associated in the field, either as marginal phases or as dykes and sills. Provided the general principle of naming such relatively finer-grained facies of distinctive plutonites after the latter as explained above (p. 287) is accepted, a clearer picture is evoked by using such specific names as microlarvikite, microfoyaite and porphyritic micropulaskite, for example, and further description is scarcely necessary. Unfortunately even today names are applied to such rocks

FIG. 102

Microijolite. Spitzkop, Sekukuniland, S. Africa.

Nepheline, the most abundant constituent, clear, but some grains zoned and with a pronounced basal parting (*e.g. top right*). Aegirine, irregular (*fine stippling*); apatite of very unusual habit (*top left*), growing simultaneously with aegirine and nepheline (note lobed contacts). One small ovoid grain of calcite with a reaction rim of cancrinite ('nepheline-carbonate'), (*bottom right*). One grain of iron ore.

which give no indication whatever of their relationship to other named parental types. Thus in a recent revision of the classic Oslo petrographic province, the relatively finer grained larvikites are called akerites.

The Oslo district is also famous for its **'rhomb-porphyries'**[1] so named by von Buch (1870) on account of the distinctive shape of the cross-sections of the feldspar phenocrysts. These rocks occur as lava-flows and dykes, and, although some are definitely fine-grained, most measured specimens lie just over the boundary between fine and medium, and are therefore appropriately considered here. Further, the composition of the rocks varies widely: some are monzonitic; but others, with a very low content of An (less than 1 per cent in extreme cases), are clearly microsyenitic. The phenocrysts are antiperthitic, the dominant partner being oligoclase or andesine, rimmed with and veined by 'anorthoclase' to the extent of 30 to 50 per cent of the whole. The groundmass feldspars are microlites of anorthoclase, approximately $Ab_{52}Or_{45}An_3$. In mineral composition, therefore, rhomb-porphyries must closely resemble larvikites, with which they are presumably consanguineous. We consider 'porphyritic **microlarvikite**' to be an appropriate name for them.[2]

Under-saturated types of microsyenites are well exemplified by the **microfoyaite** type, which is the mineralogical equivalent of foyaite, in the medium grain-size group. The essential components are therefore alkali-feldspar and nepheline associated with such minerals as aegirine, aegirine-augite or riebeckite. These mineral names may be prefixed to the rock-name to give added precision. Riebeckite-microfoyaite grades into the commonest type of phonolite as the grain becomes finer. Shand employed 'microfoyaite' to distinguish certain fine-grained facies of foyaites occurring in the alkali-complexes of Spitzkop and Pilaansberg in South Africa. This precedent may well be followed without rigid adherence to a fixed limit of grain size: the thing that matters is that there is an appreciable difference in texture between the parent rock and the facies which it is desired to indicate as something different.

SYENITE APLITES AND PEGMATITES

Syenites are rare rocks, of very limited distribution. It follows therefore that syenitic rocks analogous with the granite-pegmatites

[1] Oftedahl, C., Studies on the igneous rock complex of the Oslo region. vi. On Akerites, . . . and Rhomb-Porphyries, *Skr. utg. av Det Norske Vidensk.-Akad. Oslo.*, no. 1 (1946), 37.

[2] Specimens of several types of rhomb-porphyries were brought to this country by the North Sea ice and are spread along the Yorkshire coast as erratics familiar to glaciologists.

already considered, will be very uncommon. They do occur, however, particularly in southern Norway where they are important on account of the rare minerals some of them contain.

In regard to syenite-aplites, bearing in mind the nature of the syenites, and the improbability of there being extensive bodies of syenitic magma, it follows that the alkali-rich residua, comparable with granitic aplites, must be exceedingly rare. Further, as syenites themselves are notably leucocratic rocks, it must prove very difficult to be certain that a given rock, suspected of being a syenite-aplite, is really such, and not merely a fine-grained facies of a normal syenite. The **bostonites,** named by Rosenbusch (1882) from Boston, Mass., are fine-grained rocks, occurring as dykes, and closely resembling trachytes in appearance. Mineralogically and texturally these rocks are leucocratic microsyenites or intrusive trachytes as the case may be. If the relations of these dyke-rocks to the parent mass are the same as those obtaining between granite-aplite and granite, then they are syenite-aplites; and if they match the type bostonites in the details of their mineral contents and texture, then they are syenite-aplites of the Boston type, or bostonites.

THE TRACHYTES AND RELATED TYPES

The fine-grained members of the alkali series of rocks of Intermediate composition are the trachytes, which thus correspond in mineral content with the syenites. The name trachyte was first applied to all volcanic rocks that, on account of vesicularity, were rough to the touch. Later the term was restricted, first to lavas of Intermediate composition, and later to those containing dominant alkali-feldspar.

As a consequence of high viscosity trachytic magma is rarely represented by lava-flows of conventional type; but when a true trachytic lava-flow is encountered it is seen to cover only a small area near the centre of eruption and to build a lava dome or tholeoid, in contrast to the extensive thin sheets characteristic of basaltic flows. Trachytic shield volcanoes do occur, for example in the Kenya Rift Valley region and are seen to consist largely of sheets of pyroclasts and flows some of which must have originated in a similar manner to the rhyolitic ignimbrites described above in the previous chapter. Trachytic obsidian and pumice are both widely distributed. Trachytes identical with those occurring as lava-flows occur also as minor intrusions, notably as dykes, sometimes seen to have acted as feeders to the surface flows. Trachytes are readily divided into three groups on the silica-saturation principle: (*a*) those containing free silica—the quartz-trachytes; (*b*) those which are exactly saturated, containing neither quartz nor feldspathoid—the ortho-trachytes; and (*c*) those which contain an unsaturated mineral of felsic type, including nepheline, leucite and other feldspathoids.

(*a*) Quartz-trachytes and (*b*) Orthotrachytes

Repetition is prevented by describing these two categories together: the sole difference is the occurrence of accessory quartz in the former. In theory, the same degree of latitude is allowed as for the syenites: quartz may occur up to 20 per cent of the felsic minerals. It never occurs as phenocrysts, but is restricted to the groundmass, where it may occur as interstitial grains lying between the feldspar microlites, or it may form micropoikilitic patches, optically continuous over an area which may include parts of a number of feldspar microlites: in other words, the quartz forms little 'lakes' in the slide into which the ends of a number of microlites of feldspar may penetrate. Thus in so far as the mere names of the component minerals are concerned, there is nothing to choose between quartz-trachyte and rhyolite. Actually the amount of quartz in the former is much smaller than in an average specimen of the latter, while texturally they are very different. Glass is variable in amount: it may be interstitial to the sanidine microlites in the groundmass, or at the other extreme a wholly glassy trachyte is not impossible. Actually, trachyte-pumice has been copiously erupted, for example from certain Central African Rift Valley volcanoes. Most trachytes are porphyritic, with phenocrysts of feldspar, often of large size. The groundmass is typically microlitic: *i.e.* it consists of closely packed microlites, lath-like in shape, and several times longer than they are broad. On account of flow movements in the magma they are often in parallel alignment. At its best, therefore, the *trachytic texture* involves a close packing of feldspar microlites which often exhibit flow structure (Fig. 103). The amount of coloured silicates in trachytes is small: they are therefore light, both in colour and weight.

In some instances it is possible to subdivide further, on the basis of the dominant alkali, and to classify the rock as a potassic or a sodic trachyte respectively. In the former, the feldspar is sanidine, ideally, and is accompanied by ordinary biotite, clinopyroxene or hornblende. In the latter the feldspar may be albite, but is commonly a sodi-potassic type, while the coloured silicates are the distinctively coloured soda-iron amphiboles and pyroxenes, already noted in the syenites of comparable type: *viz.* aegirine, riebeckite, aenigmatite. It is important, however, to note that in some trachytes, while the feldspar is dominantly potassic, the coloured minerals may be strongly sodic. One such rock has been described, consisting of nearly 90 per cent of sanidine, the rest consisting of aegirine and accessories. It is customary to prefix the rock-name with that of the dominant coloured mineral, and to speak of biotite-trachyte, hornblende-trachyte, augite-trachyte, riebeckite-trachyte, etc.

The Drachenfels trachyte enjoys the reputation, among trachytes, of the Dresden syenite among the syenites; but like the latter it is not an ideal type—it contains a good deal more oligoclase than a typical trachyte should. It is noteworthy for the extremely large

FIG. 103

Trachyte, Solfatara, Naples.

The principal phenocrysts are of sanidine, often forming large crystals. Oligoclase also occurs in less abundance. The groundmass is almost wholly composed of sanidine laths, though sparse interstitial patches of isotropic glass occur. Coloured minerals include pale brown hornblende, and greenish augite, sometimes in poikilitic plates (not shown).

size of sanidine phenocrysts, which measure over an inch in diameter. Much smaller plagioclases also occur, embedded in a light cream-coloured matrix.

A more typical trachyte is illustrated in Fig. 103 which represents a rock of this type from the Solfatara volcano near Naples.

A variety of albite-trachyte bears the name **keratophyre**—unfortunately a name with two meanings. At one time it was customary to use separate names for pre-Tertiary and post-Tertiary igneous rocks, termed palaeovolcanic and neovolcanic respectively, and under this scheme keratophyre was the pre-Tertiary equivalent of trachyte. Now it is customary to apply the name kerotophyre to trachytic lavas and minor intrusives belonging to the spilitic suite (see p. 382). As a rule the coloured minerals in keratophyres are so much altered that it is impossible to identify them. Keratophyres are often somewhat over-saturated, and while in many the feldspar is pure albite, in some it is a strongly sodi-potassic type.

Equally strongly sodic trachytes, probably of Carboniferous age, occur in the Eildon Hills near Melrose, southern Scotland.[1] These, however, are in a much better state of preservation, and contain the sodic amphibole, riebeckite, in the characteristic 'mossy' aggregates. From the Tertiary volcano of Mull,[2] a similar trachyte occurs, but contains approximately equal amounts of aegirine-augite and riebeckite.

Pantellerite is the name of a suite of lavas which straddle across the boundary between Acid and Intermediate volcanic rocks and were described originally in 1881 by Foestner who discovered them in Pantelleria, an island near Sicily. Recently they have come into prominence on account of their occurrence in association with the trachyte—phonolite volcanic assemblage in the African Rift Valley region.[3] They are distinctive chiefly by reason of prominent phenocrysts of anorthoclase containing Ab slightly in excess of Or. An essential part of the original definition was the presence of anorthoclase and the rare amphibole, aenigmatite. The list of coloured silicates in the African pantellerites is long and varied: it includes fayalitic olivine, hedenbergite and sundry NaFe-rich amphiboles (riebeckite, arfvedsonite and katophorite) and pyroxenes as phenocrysts and/or in the groundmass. In the pantelleritic trachytes, with which we are more immediately concerned, a small amount of normative quartz occurs; but in the rhyolitic pantellerites Q rises to nearly 30 and quartz occurs both as phenocrysts and groundmass granules.

[1] Lady M'Robert, *Quart. J. Geol. Soc.*, **70** (1914), 303.

[2] Tertiary and post-Tertiary geology of Mull, *Mem. Geol. Surv., Scotland* (1924), 191.

[3] Sutherland, D. S., The Eburru volcano, Kenya, *J. Geol. Soc.*, **127** (1971), 417.

It may well happen that detailed examination of the African trachytes may bring to light other occurrences of pantellerites: at the moment they are known to occur in Kenya, Ethiopia and French Somaliland near the crystalline Basement forming the Ethiopian escarpment.

(*c*) Feldspathoidal Trachytes and Related Rocks: Phonolites, Leucitophyres, etc.

The old masters of rock classification, Rosenbusch and Zirkel, thought differently concerning the nomenclature of broadly trachytic rocks containing essential feldspathoids. The former used the name phonolite for all associations of alkali-feldspar with *any* kind of feldspathoid;[1] but the latter restricted the name phonolite to nepheline-bearing rocks of this type. Since the earliest editions of this book we have followed the Zirkel tradition in this matter. Nepheline and leucite both belong to the same mineral group; but apart from this, they have nothing in common. The two mineral assemblages, alkali-feldspar, nepheline and mafics on the one hand, and alkali-feldspar, leucite and mafics on the other are worthy of distinctive names—phonolite for the former and leucitophyre for the latter.

For rocks consisting of the trachyte assemblage of minerals *i.e.* alkali-feldspar—sanidine or anorthoclase—plus mafic minerals, with *accessory* feldspathoids, we derive nepheline-trachyte, leucite-trachyte, nosean-trachyte, etc. If the feldspathoid mineral increases in amount and attains to the status of an *essential* component, an independent name is necessary. Thus:

(1) Sanidine + mafics + nepheline = **phonolite;** Phonolite + subordinate leucite = leucite-phonolite;[2]

(2) Sanidine + mafics + leucite = **leucitophyre;** Leucitophyre + subordinate nosean = nosean-leucitophyre: *cf.* nosean-phonolite which contains essential nepheline (by definition) with subordinate nosean in addition.

The complete series (1) above comprises trachyte→nepheline-trachyte→phonolite; and similarly series (2) comprises trachyte→leucite-trachyte→ leucitophyre.

Phonolites are equivalent to nepheline-syenites among the coarse-grained rocks, and like the latter are typically felsic rocks in which feldspars, together with feldspathoids, are dominant over the coloured silicates. The distinctive feature is the occurrence of an alkali-feldspar, most typically sanidine, with a *sodic* feldspathoid,

[1] There would be just as good grounds for arguing that the rock-type we call diorite should really be 'plagioclase-syenite' on the grounds that both contain feldspar, though of different kinds.

[2] Streckeisen, A. L., Classification and nomenclature of igneous rocks, *N. Jb. Miner. Abh.*, **107** (1967), 185.

usually nepheline, and a subordinate amount of amphibole or pyroxene of NaFe-rich type, such as aegirine, riebeckite or the much rarer aenigmatite.

In hand-specimens phonolites are very compact, grey-green rocks which are supposed to give a sonorous ring when struck by a hammer—hence their name, which is a classical rendering of an old term, 'clinkstone', used by A. G. Werner with the same significance. Texturally most phonolites are porphyritic, the phenocrysts including both nepheline and sanidine. A very well-known example which is found in most European teaching collections is figured below (Fig. 104).

The quantity of nepheline in phonolites is very variable, and it is

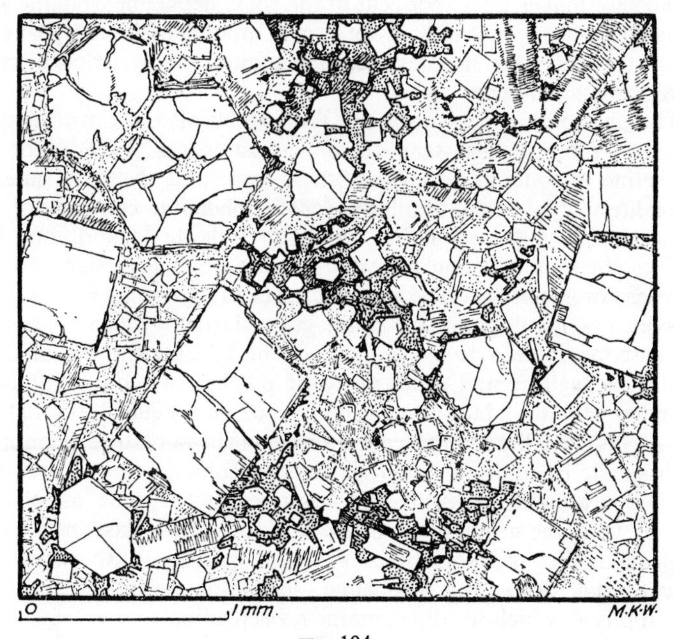

FIG. 104

Phonolite, Brux, Bohemia.

Both sanidine and nepheline occur in crystals of two generations. The former shows alteration parallel to the basal cleavage (*top right*). The latter forms almost square or hexagonal sections, outlined in some cases by dark poikilitic riebeckite aggregates.

clearly necessary to distinguish between those in which the mineral has the status of an essential component, and those in which it is an accessory only. The latter are the connecting links between true phonolites and saturated trachytes. In the sense that they consist of the minerals appropriate to trachyte, with nepheline in addition, the logical name for them is nepheline-trachyte; but as Jonanssen and others unfortunately have already used this term in another sense, the alternative of **phonolitic trachyte** may be used. An example familiar to British petrologists forms the Traprain Law Laccolith in Haddington, Scotland. The greater part of this rock consists of fluxionally arranged laths of alkali-feldspar, and the nepheline, which amounts to only 4 per cent, is elusive in thin sections; but it is supported, as it were, by 20 per cent of analcite and a few grains of sodalite.[1] Strictly the Traprain Law rock is a sodalite-bearing phonolitic analcite-trachyte.

Even closer to trachyte is a similar rock which forms a plug, known as the Bass Rock, in the Firth of Forth, Scotland. Soda-orthoclase makes up 85 per cent of the rock, nepheline, sodalite and analcite all occur, but only in very small amounts, about 1 per cent, while the coloured silicates, aegirine-augite and iron-rich olivine, total some 8 per cent.

The phonolite which forms the Wolf Rock off the Cornish coast is much more typical than these Scottish rocks. In addition to nepheline it contains much greyish nosean, and the term **nosean-phonolite** is applicable. Similarly, **sodalite-phonolite** contains all the minerals of typical phonolite, but with sodalite in addition. The end-product of this line of variation is a rock in which sodalite proxies for nepheline. The volcanic tract in Turkana, Kenya,[2] has provided a good example—a fine-grained rock, which under the microscope is seen to be composed chiefly of laths of sanidine, grains of aegirine and small poikilitic patches of sodic amphibole, together with abundant though small (0·025 mm) euhedral sodalites.

Leucite-phonolite should contain the minerals of true phonolite, with leucite in addition. The rocks of the Roman volcanic province provide examples. Leucite and aegirine-augite occur as phenocrysts, while the groundmass consists of soda-orthoclase, nepheline, aegirine-augite and sometimes sky-blue hauyne. With increasing leucite and decreasing nepheline, leucite-phonolite grades into leucitophyre, which ideally contains no nepheline.

Of rather special interest are the basic, phonolitic trachytes of Kenya type, named **Kenyte** by J. W. Gregory,[3] from the type-locality, Mount Kenya, Kenya, Africa, but known to occur also at Mount Erebus, Antarctica.[4] The distinctive feature is the nature of the feldspar which occurs as large prisms, rhombic in cross-section, of anorthoclase. In shape, if not in composition, these appear to be

[1] Macgregor, A. G., *Geol. Mag.*, **59** (1922), 514.
[2] Campbell Smith. W., *Quart. J. Geol. Soc.*, **94** (1938), 522.
[3] Redescribed by W. Campbell Smith in *Bull. Brit. Mus.* (*Nat. Hist.*), vol. I, no. 1, 3.
[4] Prior, G. T., National Antarctic Expedition, 1901–4, *Nat. Hist.*, i (1907), 101.

identical with the feldspars of rhombic cross-section occurring in the Oslo larvikites and rhomb-porphyries. In this and other respects, both chemical and mineralogical—the occurrence of small quantities

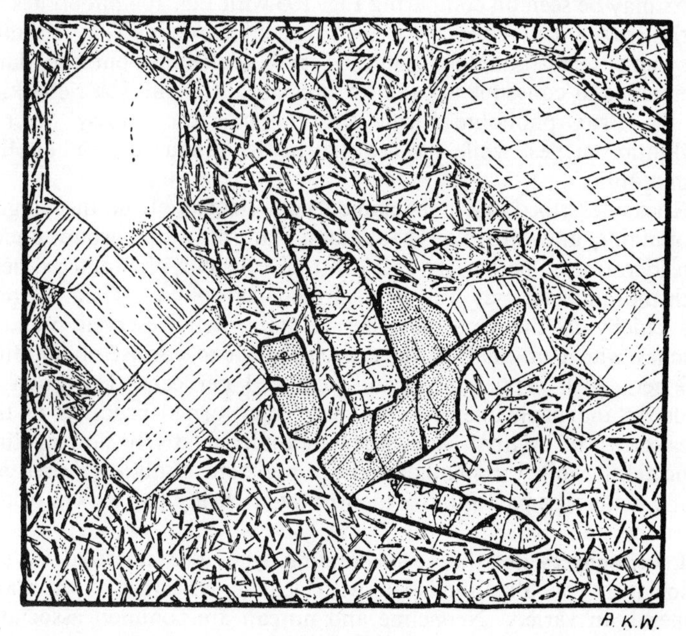

FIG. 105

Porphyritic aegirine-phonolite (Tingua type), Brazil.

Clustered phenocrysts of nepheline (with basal cleavage); sanidine, brilliant green aegirine (*fine stipple*); and sphene (*coarse stipple*); set in a groundmass in which minute aegirine microlites are abundant.

of olivine and nepheline[1]—kenyte closely resembles larvikite and laurdalite, and it is undoubtedly their fine-grained equivalent. Rhomb-porphyry is the medium- to fine-grained dyke-equivalent, and, as might be expected, occurs in association with the kenytes on Mount Kenya.

Falling in the phonolite group as defined above are certain dyke rocks commonly known as **tinguaites**, so named by Rosenbusch (1887) after occurrences in the Tingua Mountains near Rio de Janeiro, Brazil, where they occur in dykes or as marginal facies of larger bodies of nepheline-syenite (Fig. 105). When the name was introduced it was considered necessary to use one name for a particular mineral assemblage occurring in the form of lava, and a different name for the *same* assemblage occurring in a dyke. As

[1] Campbell Smith, W., Classification of some Rhyolites, Trachytes and Phonolites from part of Kenya Colony, *Quart. J. Geol. Soc.*, **87** (1931), 242.

already argued, we consider this duplication of rock-names unwarranted. The principle is outmoded, though many petrologists still call intrusive phonolites, 'tinguaites'.

As may be seen on comparing Fig. 105 with Fig. 104 phonolites of both Brux and Tingua types are distinctive rocks. They contain the same felsic minerals—sanidine and nepheline—but the dark minerals and consequently the textures are different. The riebeckite in the Brux phonolite forms the characteristic 'mossy' micro-poikilitic patches, while the aegirine in the Tingua dyke-phonolite is microlitic.

Rocks described under the name **leucite-tinguaite** from the Tingua Mountains and Bearpaw Mountains fall into the group under consideration. One variety is an intrusive leucite-phonolite; others with more leucite may well be aegirine-leucitophyres. Their interest lies chiefly in the fact that they contain well-preserved pseudo-leucites which on analysis have been shown to consist of sanidine (66 per cent), nepheline 30 and acmite 3 per cent. The potassic feldspar and nepheline are, of course, the normal products of the breakdown of original Na-rich leucite under relatively slow cooling conditions, while the small amount of pyroxene doubtless represents mafic material adventitiously incorporated during the growth of the phenocrysts.[1]

Leucitophyres are the corresponding types characterized by the association of alkali-feldspar with leucite, together with mafic minerals in variety. Nepheline and nosean are common associates with the leucite, and as the feldspar is potassi-sodic and the mafic minerals also contain some Na^{+}, the allocation of alkali atoms has been complicated indeed. The leucite usually forms large phenocrysts, distinctive both in hand-specimens and thin sections. The best-known rock of this type occurs as a dyke near Rieden in the Eifel. In the hand-specimen numbers of small phenocrysts of leucite and nosean are plentifully scattered in a greyish groundmass which also contains small black prisms of pyroxene. Under the microscope the phenocrysts of leucite and corroded noseans with heavy black borders are very striking (Fig. 106). Small nephelines of typical shape occur in the groundmass only, where they are associated with equally small laths of sanidine. Aegirine-augite is abundant, while sphene, apatite, magnetite, and less commonly melanite garnet are usual accessories. The proper designation of this well-known rock is nosean-nepheline-leucitophyre.

Leucitophyres occur together with other related types in the Leucite Hills, Wyoming. One variety consists of equal amounts of leucite, sanidine and coloured minerals. Among the latter, phlogo-

[1] Zeis, E. G., and Chayes, F., Pseudoleucites in Tinguaite from Bearpaw Mountains, *J. Petrol*, **1** (1960), 86–98.

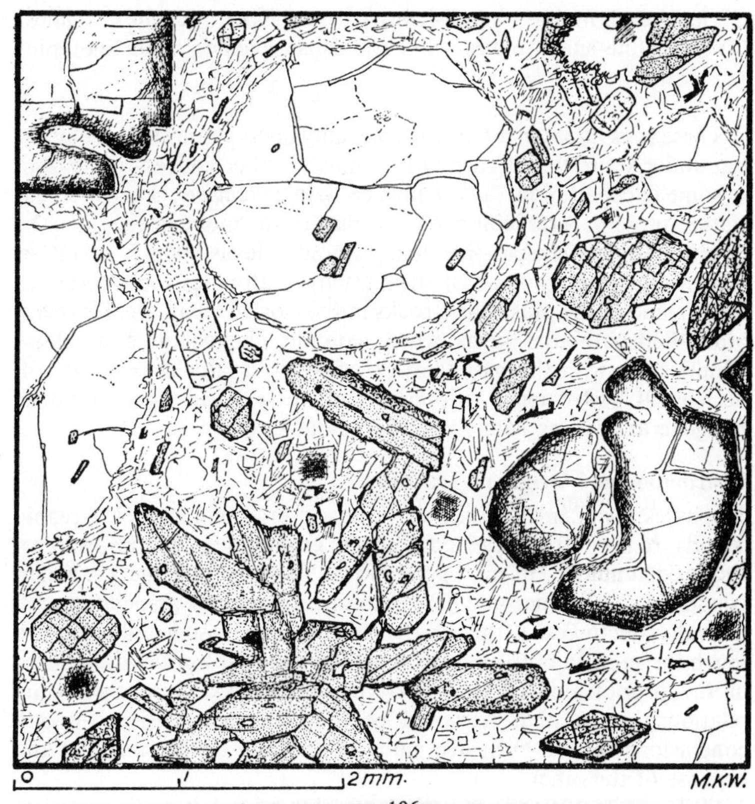

FIG. 106

Nosean-leucitophyre, Rieden, Eifel.

Megaphenocrysts of leucite and nosean are about equally abundant, occurring with green-rimmed aegirine-augite, set in a groundmass of sanidine and minute euhedral nepheline crystals. Apatite and sphene are important accessories.

pite, a rare mineral in normal igneous rocks, is dominant, forming small phenocrysts visible in the hand-specimen. It is accompanied by minute needles of diopside and a katophorite-like amphibole. This is a type of **phlogopite-leucitophyre,** and for its silica percentage is particularly rich in potash.

Summarily the fine-grained equivalents of under-saturated syenites consists fundamentally of two mineral associations:

(1) alkali-feldspar with nepheline (and mafics) = phonolite;
(2) alkali-feldspar with leucite (and mafics) = leucitophyre.
With the addition of another feldspathoid in an accessory capacity we have:

(3) leucite-phonolite—alkali-feldspar, nepheline, (leucite), mafics; (4) nepheline-leucitophyre—alkali-feldspar, leucite, (nepheline), plus mafics.

These rock-types in a sense are partially under-saturated trachytes; but under-saturation is carried very much farther, indeed to the limit in some feldspathoidal lavas which contain *no feldspar*. Their nomenclature presents a difficult problem; but for the moment they may be referred to as feldspar-free feldspathoidal lavas, which is rather clumsy but true. In spite of their trachytic affinities these lavas include some of the most Basic rocks in the world. There are, however, other feldspathoidal lavas which contain plagioclase and whose affinities are with the basalts as described in a later chapter. It will be a great convenience to consider all these lavas together under the general heading Mafic Feldspathoidal Lavas.

Distribution and Origin of Trachytes

Trachytes occur in two different kinds of environment: (1) oceanic islands remote from the continents, exemplified by the Azores, Ascension and St Helena in the North Atlantic, and the Hawaiian Islands in the Pacific; and (2) in continental regions associated with rift valley tectonics. Although the trachytes and associated rocks are much the same in both environments there are substantial differences in the proportions of the rock-types involved and in their time relationships, so that it would be easy to reach entirely wrong conclusions by the exclusive study of the one environment at the expense of the other.

In the oceanic islands as a whole, trachytes (mainly sodic varieties) form only a small fraction of the exposed rocks: basalts, with which they are associated, are overwhelmingly dominant. The visitor to Ascension Island cannot fail to be impressed by the contrast between the two, both in general appearance and relative proportions: the white-weathering trachytes are seen as late-comers on the scene, capping the earlier, black basaltic lava-flows. In the Azores the proportion of basalt to trachyte is higher—the two types may occur in approximately equal amounts; but if so, this must be regarded as an exception to the general rule for oceanic island volcanicity. In the Azores the volcanoes are still actively erupting trachyte, not as lava-flows, but as air-fall ash deposits of trachytic pumice.

Trachytes in both oceanic island and continental settings tend to be undersaturated and grade through phonolitic trachytes into phonolites which are extensively developed in and near the African Rift Valleys. Further, variation involving increase in silica leads, as it were, in the opposite direction, from trachyte into the pantellerite suite, the most acid member of which, quartz-pantellerite—is rich in silica and iron, while maintaining the high degree of alkalinity characteristic of trachytes.

Turning now to the continental occurrences of trachytes, we note that they have been, and locally still are being, erupted on a more impressive scale; but even so, in terms of global distribution they do not bear comparison with the other major eruptive rocks—the basalts, andesites and rhyolites—as regards the volume of lava and 'ash' erupted. Much of what we have to say is based on volcanic activity dating from the Eocene to the present time in eastern Africa, especially in the region of the great rift valleys, in Ethiopia, Kenya, Malawi and Tanzania.

The volcanic rocks can be grouped into three fairly distinct categories. In some parts of East Africa, Ethiopia for example, the earliest phase of activity in Eocene times was marked by extensive outpourings of basalt, mainly as fissure eruptions.[1] Later the nature of the eruptions changed dramatically. In Kenya[2] phonolites and later trachytes were erupted in Middle and Late Tertiary times, forming thick successions of lava-flows which are generally so flat-lying, and extend over such wide areas that they are referred to as plateau- or flood-phonolites and trachytes. The source and mechanism of eruption are not always apparent; but some of the lavas are associated with shield volcanoes with calderas, while others seem to have been erupted from fissures. In view of the extreme viscosity of trachytic magmas one must assume that some fluidizing mechanism has operated similar to that which forms ignimbrites. In the main the trachytes and phonolites are concentrated within areas of down-faulting where the rift structures cross the Kenya 'swell', a gently sloping and very broad doming of the crust caused by heating and expansion of the asthenosphere in the underlying mantle. The last volcanic episode saw a return to basaltic eruptions which are associated with trachytic and phonolitic lavas and pyroclasts in some of the youngest volcanoes which tend to occupy central positions in the Rift Zone. The latest volcanic activity is largely confined to the centre of the Rift Zone and is more varied, trachytes and phonolites being associated with basalts and, in some cases, with Basic to Ultrabasic alkaline lavas including basanites and nephelinites (which are described in due course). The latter rocks occur sporadically throughout much of the Tertiary to Recent volcanic history of the region. The account so far has been factual. It remains now to consider how trachytic magmas originate; and at this point it is necessary to indulge in some speculation, because

[1] Cole, J. W., The Gariboldi Volcanic Complex, *Proc. Geol. Soc.*, **1647** (1968).

[2] King, B. C., Volcanicity and rift tectonics in East Africa, in *African Magmatism and Tectonics*, ed. T. N. Clifford and J. G. Gass (1970), 263–83; Williams, L. A. J., Petrology of volcanic rocks associated with the Rift System in Kenya, *Rep. UMC/UNESCO, Seminar on the E. African Rift System, Nairobi* (1965), 33.

much of the information that would be critical, is beyond the reach of direct observation. This problem is shared by all strongly alkaline rocks because Na and K are, by their nature, influenced more than

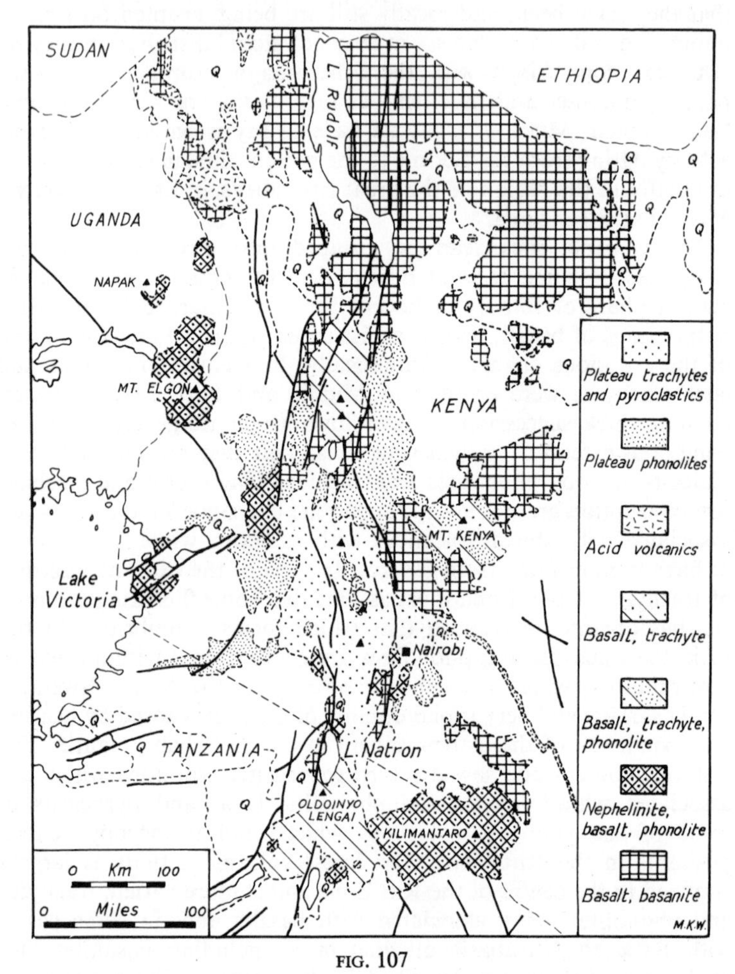

FIG. 107

Volcanic rocks in the Kenya and Tanzania section of the Eastern Rift. (*After L. A. J. Williams* (1965) *and B. C. King* (1970).)

other rock-forming components by factors which are difficult to quantify, such as local variations in load-pressure and different degrees of solubility and mobility in the presence of volatiles.

In the oceanic islands the problem is simplified by the absence of sialic crust and by the overwhelming dominance of basaltic volcanicity. There can be no doubt that some kind of basaltic magma was the ultimate parent from which trachytes were derived. Comprehensive and detailed research has shown that these parent magmas are of alkali olivine-basalt composition, containing some nepheline in the norm. Chemical analysis of the glassy mesostasis of individual basalts of this type shows that in some cases the composition is close to that of trachyte. Of course the proportion of this mesostasis and therefore of the potential trachyte that might theoretically be separated from an initially basaltic magma is very small. During the process of differentiation various intermediate types should be produced—alkali olivine-basalt, hawaiite, benmoreite and trachyte. Ideally these should be erupted in progressively smaller amounts, with differentiates of more extreme composition appearing late in the sequence (see Fig. 75). This predicted pattern is followed sufficiently closely in a number of oceanic islands for one to be reasonably confident that differentiation of alkali olivine-basalt magma is the main process involved in the evolution of trachytic magma. Progressive differentiation based on fractional crystallization cannot, however, be the only process involved because there is a gap between mugearite and trachyte in which there is a marked scarcity of rocks of the appropriate composition.[1] In the final stages of the process the action of water and volatiles is probably critical in assisting by both physical and chemical means the segregation and expulsion of the last fractions of differentiated magma, probably from a magma-chamber situated high in the structure of the volcano. Among the possible mechanisms that have been proposed as operative in this connection are diffusion of alkalies due to their co-ordination with water dissolved in the melt, gaseous transfer of alkalies and so-called gas-streaming.[2]

Some volcanic associations in continental regions are sufficiently similar to those of oceanic islands as to suggest an analogous mode of origin. However, in some cases, notably East Africa, the eruption of large amounts of trachytic rocks associated with closely related phonolites or varieties of rhyolite, has occurred apparently quite independently of basalt. Eruptions of basalts or of trachytes in a particular region are typically separated both in time and composition by a clear-cut hiatus.

The tectonic environment is obviously a significant factor, since trachytes and related alkaline rocks which are so extensively developed in East Africa are characteristic also of other regions of comparable structure notably the Permian Oslo graben, the Rhine graben and the Midland Valley of Scotland.

[1] Chayes, F., Relative abundance of intermediate members of the basalt-trachyte association, *J. Geophys. Res.*, **68** (1963), 1519–34.

[2] The various possibilities are discussed by Turner and Verhoogen in their comprehensive review of the alkali olivine-basalt volcanic association in *Igneous and Metamorphic Petrology* (1960), 184.

The stage is probably set for the eventual eruption of trachytic and related volcanic rocks during the initial doming of the crust prior to the actual rift faulting. This phase, both in Ethiopia and Kenya, was accompanied by extensive eruptions of alkali-rich basalts and feldspathoid-bearing Basic lavas. It is probable that the levels and pressures at which initial melting took place in the mantle during the expansion of the asthenosphere and formation of the 'swells' favoured enrichment of the primitive magmas in alkalies. Even slight differences in the initial composition of such magmas can become greatly accentuated by the time their differentiation products are erupted at the surface.

However, it seems doubtful whether differentiation alone could account for the formation of trachytic and related magmas in such large volumes in areas where basalts are missing. Several lines of evidence suggest, in fact, that the continental crust plays a vital role in the origin of trachytic magmas, either by reaction between crustal rocks and magmas rising from the mantle, or by partial melting at crustal levels. An indication of the latter possibility is provided by unusually high Sr^{87}/Sr^{86} ratios in some members of the pantellerite suite. Fractional melting of crustal rocks to produce trachytic liquids could, in one sense, be regarded as a natural outcome of the composition of the latter, lying, as shown in Fig. 73 in a region of low-temperature liquids in the system nepheline-kalsilite-SiO_2.

The most convincing evidence to the field geologist, however, is provided by the relationship between trachytes and their deep-seated equivalents the syenites, to which we now turn our attention. No major syenitic intrusion comparable in size with the average granitic batholith has been found: normally syenites form only a small part of a larger, usually granitic intrusion. The Dresden syenite, for example, is a relatively silica-poor facies of the Meissen granite. Field relations, coupled with laboratory studies have shown clearly that many syenites were formed by *in situ* modification (desilication) of previously existing siliceous rocks. This process is illustrated by the occurrence of syenites among other products of metasomatic alteration in the outer zones of many intrusive alkali complexes such as that of Fen in Norway, which is considered in detail in the discussion of the origin of the feldspathoidal rocks (Ch. 12).

The converse process involving the acidification of more basic material has also occurred, demonstrably on a small scale, at Kiloran Bay, Colonsay, in the Hebrides. An ultramafic rock consisting largely of hornblende crystals contains numbers of quartzite xenoliths which are surrounded by thin rims of quartz-syenite. The latter has obviously resulted from interaction between hornblendite and the siliceous inclusions. This may be a unique occurrence; but it does at least indicate a possibility. The existence of syenitic rock-bodies beneath at least some of the trachytic volcanoes in East Africa is proved by the occurrence of xenoliths of syenite in some of the latest trachyte flows and in associated pyroclastic rocks. Despite the expected differences in the mineral phases and textures of the trachytes and syenites, their strong chemical similarity leaves no doubt that they both originated in the same way and came from a common source. The connection is confirmed in some cases in East Africa by the field relationships: the eruption of trachytes was accompanied by the formation of calderas which may be interpreted as surface expressions of deeper cauldron subsidences connected with the growth of ring-complexes beneath the volcanoes. Deep dissection in adjacent areas in East Africa has, in fact, uncovered partly syenitic ring-complexes of just the kind that one would expect from this interpretation. We may anticipate fuller discussion of the nature and significance of these complexes by concluding that there is no evidence of the existence of deep-seated reservoirs of syenitic, *i.e.* trachytic magma: on the contrary 'plutonic' rocks analogous to the volcanic trachyte-phonolite association have been formed largely from pre-existing rocks by complicated processes involving ionic diffusion and interchange, intense gaseous activity and ultimately mobilization with the potentiality of movement as magma through the crust. Certainly these syenites did not result from direct fractional crystallization of basaltic magma, nor from simple selective melting of crustal or mantle rocks. We can probably never *know* (though we may infer) how the magma represented by trachytic flows originated; but we do know in some detail how syenites were formed. Trachyte and syenite are co-magmatic; therefore the processes which we see frozen-in, as it were, in the act of producing syenite, must in at least some instances, have operated to produce trachytic magma.

CHAPTER V

DIORITES, MICRODIORITES AND ANDESITES

THE predominance of alkali-feldspars and/or feldspathoids makes it relatively easy to define, classify and name syenitic rocks; when dealing with the plagioclase-dominant suite the matter is complicated by the fact that, by tradition, it is necessary to subdivide into two sub-groups: diorites and gabbros, together with their medium and fine-grained equivalents (microdiorites and andesites) on the one hand and microgabbros (dolerites) and basalts on the other. At different times and by different authorities the division has been made on the basis of the following criteria: (i) silica percentage; (ii) nature of the coloured minerals; (iii) kind of plagioclase present; and (iv) the percentage of coloured minerals—the colour index of the rock. Undoubtedly the easiest way of effecting the separation is to choose arbitrarily a convenient silica percentage, regardless of other considerations. This principle has already been discussed and found wanting; and for reasons already stated, we define rock-types in terms of their constituent mineral assemblages.

The position then is this: there are three *visible* criteria involved—(1) the kind of plagioclase, (2) the kind(s) of mafic constituents and (3) the amount of the latter, *i.e.* the colour index. These features combine to give distinctiveness to the rocks concerned; but it is virtually impossible to reconcile the evidence afforded by all three variables. There will always be the odd specimen which conforms to the definition in two, but fails to do so in regard to the third (see below). The best, indeed the only chance of achieving unanimity in the solution of this problem—the most important of its kind, is to apply one criterion rigidly, but to allow reasonable elasticity in relation to the other two.

The precedent has been established by A. Johanssen, and adopted by many other American petrologists, of using the kinds of plagioclase as affording the most satisfactory means of distinction between the diorite-microdiorite-andesine family on the one hand and the gabbro-microgabbro-basalt family on the other. As the distinction is being made arbitrarily, it does not matter a great deal where the division is made: but in many diorites and andesites the plagioclase is within the oligoclase-andesine range, while typical gabbros and basalts contain labradorite or bytownite. Therefore the logical division to choose is the boundary between andesine and labradorite, *i.e.* An_{50}.

Apart from the kinds of mafic constituents present in these rocks, the amounts are important factors in naming and classification. The difficulty of using colour index for differentiating between the two plagioclase-dominant suites is illustrated by the following facts. Shand in his textbook advocated drawing the division at 30 per cent; but S. E. Ellis[1] after a careful statistical study of the problem concluded that this figure was too low and suggested raising it to 40. The position as we see it is this: most diorites have a colour index below 40, most gabbros above this figure; but surely a rock of dioritic aspect, containing the right kind of plagioclase and mafic minerals, should not be classified as gabbro merely because its colour index is, say, 43.

Colour Index	An_{10} – An_{30}	An_{30} – An_{50}
0		
	Oligoclasite	Andesinite
10		
20	Oligoclase-LEUCODIORITE	Andesine-LEUCODIORITE
30		
40		
	OLIGOCLASE-DIORITE	ANDESINE-DIORITE
50		
60		
70	Oligoclase-MELADIORITE	Andesine-MELADIORITE
80		
90		
	Hornblendite	
100		

Common hornblende is the characteristic mafic mineral in diorites, though biotite, augite and hypersthene are also encountered. Olivine is normally absent. The mafic constituents typical of gabbros are pyroxene and olivine. Let it be imagined for a moment that diorite is defined by the mineral association common hornblende, andesine

[1] *Min. Mag.*, **28** (1948), 447.

feldspar. In a large collection of specimens the proportions of these two minerals would be found to vary within the widest possible limits as suggested in the accompanying table. The same applies to the mineral association pyroxene (augite), labradorite, which defines ordinary gabbro.

Rocks consisting of these pairs of minerals, then, cover the whole range of colour index variation, but not all of them are acceptable as diorites and gabbros: those varieties of extremely melanocratic and leucocratic nature are excluded as shown in the table. They are sensibly monomineralic, consisting essentially of the minerals named, together with a very small amount, limited, it is suggested, to 10 per cent of other minerals. That leaves diorites in the widest sense as having a colour index between 10 and 90—that is very different from Shand's limit of 30, or Ellis' of 40; but it is the lesser of two evils—it does group together all those rocks in the coarse-grain category which consist of the same two minerals, in varying proportions. The diorite family includes members of coarse-, medium- and fine-grain, named diorites, microdiorites and andesites respectively. The grain-size limits are the same as those stated above for the quartz-rich rocks. For general purposes the division between microdiorite and andesite is drawn at the limit of unaided vision; that between diorite and microdiorite is drawn at a limiting grain diameter of 1 mm.

DIORITES

Probably no other rock-name has been used with so many different shades of meaning as 'diorite'. Ideally its meaning should be completely unequivocal: diorite is the Intermediate coarse-grained igneous rock-type consisting essentially of plagioclase ($An_{<50}$) amounting to at least two-thirds (65 per cent) of the total feldspar, associated with unspecified mafic minerals. Thus defined there should be no possibility of ambiguity; but we would emphasize that *all* diorites have compositions represented by points within the diorite field, which is defined by boundaries drawn through P_{65}, P_{100}, Q_0 and Q_{20}.

Although most petrologists concerned with rock classification and nomenclature agree with this definition, the broad diorite field has been 'whittled down' by establishing sub-fields within it until only a small portion remains, occupying the bottom right-hand corner of the QAP triangle. This vestige has most commonly been labelled just 'diorite'; but we cannot have 'diorite *s.l.*' and 'diorite, *s.s.*'. Some petrologists, realizing this, have attempted to get over the difficulty by attaching the qualifiers 'true' or 'normal' to the word 'diorite'; but we regard all diorites, as defined above, as both true and normal.

In quartz-diorites, biotite is often the most abundant coloured silicate, and **quartz-mica-diorite** is a common and easily identified rock-type, even in hand-specimens. However, this mineral assemblage occurs above, as well as below the Q_{20} boundary, and strictly we should name the former 'biotite-granodiorite' and the latter, 'quartz-mica-diorite'. There is no doubt, however, that both, in the field, would bear the same name, based on the visible mineral assemblage. Whether one should stretch the definition of quartz-mica-diorite so as to include those specimens which have strayed, as it were, across the Q_{20} boundary, or rigidly enforce the 'letter of the law' in this matter might be made a matter of principle or of expediency, according to the circumstances.

Tonalite is virtually synonymous with quartz-mica-diorite and therefore different shades of meaning are attached to the former term according to whether one adopts the narrower, or the wider meaning of 'quartz-mica-diorite'. In Britain it is customary to regard

FIG. 108

Biotite-hornblende-diorite, Haute Savoie, France.

Plagioclase, common hornblende (*stippled*), biotite (*fine-ruled*), magnetite and apatite shown. Drawn in plane polarized light, but the twinning of the plagioclase is indicated.

tonalite as an Intermediate, strictly dioritic rock; but in America and elsewhere it is commonly regarded as synonymous with plagioclase-rich granodiorite of our classification, lying above the Q_{20} boundary and between the field boundaries P_{90} and P_{100}.

In the original tonalite from Monte Tonale in the Tyrol biotite accompanies hornblende. In Britain tonalites have been described from the neighbourhood of Loch Awe, south-east of the Ben Cruachan granite.[1] The composition is as follows: plagioclase (An_{30}) 72, microperthite 11, quartz 7, dark minerals and accessories 10 per cent. Similar tonalites occur elsewhere in Scotland as integral parts of the Caledonian plutonic complexes, including those of Galloway.[2]

The diorite illustrated in Fig. 108 represents the ideal type: it is not over-saturated, biotite is subordinate to common green hornblende, the plagioclase, as is commonly the case, is zoned, with the cores more calcic than the outer parts of the crystals. The normal accessory minerals liable to occur in diorites of all kinds include sphene, probably the most prominent, apatite and magnetite. In the more basic diorites the place of hornblende tends to be taken by either clinopyroxene which is colourless in thin section, or by orthopyroxene, usually identified as hypersthene. This mineral is the 'trade-mark' of the Charnockite Series, and as noted above, charnockitic complexes include rocks in which the dominant feldspar is plagioclase. Quartz-poor or quartz-free **enderbites** (Tilley) are hypersthene-diorites in the wider sense, though their charnockitic affinities may be shown by their field relations.

Meladiorites differ from normal diorites only in their higher content of dark minerals. Examples have been described from the Glen Fyne–Garabal Hill Complex in South–West Scotland. They are much more extensively exposed on the coast of South-East Jersey[3] and northern Guernsey, Channel Islands. These rocks are certainly meladiorites in the general sense, but are distinctive among the latter by containing euhedral hornblende prisms, in the Channel Island examples, often with hollow cores, and fluxionally arranged. These **appinites** (E. B. Bailey, 1916) are often pegmatitic, of relatively coarse grain and occur either in 'pockets' in normal diorite or in the upper flux-rich portions of differentiated dioritic sheets. Originally appinite was a dioritic rock-type; but latterly the term has been expanded and now, with special reference to Scottish Caledonian intrusives, the 'appinite suite' includes a wide range of rock-types of

[1] Nockolds, S. R., The contaminated tonalites of Loch Awe, *Quart. J. Geol. Soc.*, **90** (1934), 302.

[2] Gardiner, C. I., and Reynolds, S. H., The Loch Dee complex, *Quart. J. Geol. Soc.*, **88** (1932), 1.

[3] Wells, A. K., and Bishop, A. Clive, An appinitic facies associated with certain granites in Jersey, Channel Islands, *Quart. J. Geol. Soc.*, **111** (1955), 143.

different compositions. The only mineralogical feature common to all members of the suite is the occurrence of hornblende.[1]

In the field meladiorites grade into **hornblendites** by the elimination of plagioclase, and the (nearly) pure hornblende rock is the ultimate mafic differentiation product of the dioritic suite. The nature, origin and crystallization history of the ultramafites in general are considered in detail in connection with the gabbroic suite. They are regarded as 'accumulative' rocks, in the sense that they are believed to have been formed by the accumulation in one place of crystals that were precipitated elsewhere and sorted, so that ultimately a monomineralic mafic rock resulted. It is difficult to believe that this process can apply to monomineralic hornblendite, and one important occurrence suggests a different mode of origin. At Garabal Hill several different kinds of ultramafite occur, chief among them augite-peridotite (dominant olivine associated with augite), which grades into hornblende-pyroxenite consisting of augite, orthopyroxene and hornblende as the third, minor constituent. The pyroxenite in turn grades into a pure, brown hornblendite termed **davainite** by Wyllie and Scott in the original description of the complex. The hornblendite is of particularly coarse grain, the individual crystals being prisms up to 2 inches in length. There can be little doubt that this extraordinary rock represents a facies of the pyroxenite which has been thoroughly reconstituted so that the place of the two pyroxenes has been taken by a single amphibole. Presumably the process was metasomatic, though no agent more chemically active than water-vapour, to provide the necessary (OH), would be required. Finally, the ultramafites noted above form only part of the Garabal Hill Complex,[2] and in so far as energy was needed for the above conversion, it was doubtless provided by the later members of the complex.

Finally, certain very rare rocks may be admitted as having some dioritic affinities though they are far removed from the normal rocks of this series. Their interest is two-fold: they contain nepheline and the rare accessory, corundum; and they grade into marginal facies which are practically pure plagioclase rocks, the feldspar being within the normal dioritic range. The proportions of plagioclase, nepheline and corundum vary considerably in different types: in one, there is much more nepheline than oligoclase, while corundum is of accessory status only. In another, nepheline occurs in small amounts only, while andesine and corundum make up the rest of the rock.

[1] See summary by E. L. P. Mercy on the Appinite suite, in *Geology of Scotland*, ed. G. Y. Craig (1965).

[2] Nockolds, S. R., The Garabal Hill–Glen Fyne igneous complex, *Quart. J. Geol. Soc.*, **96** (1940), 451.

Plumasite, described by Lawson (1901) from Plumas County, California, forms a wide dyke cutting peridotite, and is fundamentally an oligoclase-corundum rock. The corundums are light-bluish in colour, up to an inch in length, and are embedded in a matrix of white oligoclase. Very similar plumasites have been described from Natal[1] and the Transvaal, South Africa, where again they form dykes, and tend to be of pegmatitic facies—the corundums may be several inches in length. The marginal facies of these African rocks may be almost pure oligoclasite, while the corundum becomes an important constituent in the more central parts.

Only slightly different from plumasite is **dungannonite** (Adams and Barlow[2] from Dungannon in the Haliburton–Bancroft area in Ontario, a locality well known for its nepheline-syenites. Dungannonite is, in effect, a plumasite with a rather more calcic plagioclase taking the place of oligoclase. Andesine makes up about three-quarters of the whole rock, and large corundums about half of the rest. White mica, often tending to sheath the corundums, biotite, and perhaps the most important accessory, nepheline, also occur in small quantities. Again, some phases of dungannonite are almost pure andesine-rocks (andesinites). These rocks are of special interest because in the first place they are *under-saturated* diorites; and secondly, except for corundum-syenites (described above), they are the only igneous rocks which contain corundum.

From the petrogenetic point of view these corundum-bearing rocks are of special interest, though the problem of their origin has not yet been solved. One factor involved is demonstrably desilication: in several instances the dyke-rocks carry quartz until they penetrate, or at least come into contact with peridotite (or serpentinite, which is the same thing chemically). At the point where quartz disappears, the wall-rock shows a gain in silica, and corundum appears in the dyke-rock. But the loss of silica by the dyke-rock is not the only factor involved as there is a considerable gain in alumina to be accounted for. Those who have made a special study of certain Russian corundum-bearing rocks believe that the alumina was introduced by volatiles.

Reference back to the table showing the mineralogical relationship between the various members of the dioritic series will remind the reader that at the one extreme of composition occur rocks consisting essentially of plagioclase only. In this sense these rocks may be grouped for the purposes of classification with anorthosites which are more basic. **Oligoclasite** and **andesinite** were both men-

[1] du Toit, A. L., Plumasite . . . from Natal, *Trans. Geol. Soc. S. Africa*, **21** (1918), 53.

[2] Geology of the Haliburton–Bancroft area, Ontario, *Geol. Surv. Canada Mem.*, no. 6 (1910), 315.

tioned in the paragraphs immediately above, and it will be realized that they must be regarded as unusual rocks, of very limited distribution, and probably formed by different processes from those involved in the genesis of labradoritite and bytownitite, the two normal varieties of anorthosite.

MICRODIORITES

The general name microdiorite has been chosen for rocks of dioritic composition, but of medium grain-size, that differ only in texture from the diorites into which they grade. As the grain becomes finer, they pass into andesites. Though aphyric types are known, most microdiorites are porphyritic. To cover these textural variants two types are distinguished:

(i) porphyritic microdiorites; and
(ii) aphyric microdiorites.

The former are frequently named 'diorite-porphyries', but this name has been applied to porphyritic diorites, and is better not used. A synonym which is still widely used is 'porphyrites'; but again there are many objections to its use, and in view of the several quite different usages, it is highly desirable to drop this term in favour of **porphyritic microdiorite.** In textural detail many of these latter rocks closely resemble andesites, and there is no doubt that many so-called 'porphyrites' are as fine grained as andesites, and were named solely on account of their mode of occurrence as minor intrusions. Some of the 'porphyrites' surrounding the Dalbeattie granite, for example, consist of a devitrified glassy matrix in which the abundant and relatively large phenocrysts are embedded. Although hypabyssal in mode of occurrence, such rocks are (intrusive) hornblende-biotite-andesites.

The nature of the dominant coloured mineral may be used for further subdivision, giving mica-microdiorite, hornblende-microdiorite, and less commonly augite- and hypersthene-microdiorites. Those specimens containing quartz (and usually biotite) are closely allied to tonalites and are to be distinguished as **microtonalites.** In such rocks quartz is restricted to the groundmass, where it may be intergrown with orthoclase as micropegmatite. Rocks of much the same general appearance but containing quartz phenocrysts in addition to those of other minerals prove on careful examination to be dacitic, and in general this is a useful means of distinguishing between these two types. A rather special variety of over-saturated porphyritic microdiorite has been named **markfieldite,** from Markfield in the Charnwood Forest area in Leicestershire. A graphic intergrowth of quartz and alkali-feldspar forms the groundmass in which numerous phenocrysts (if they are so to be regarded) of plagioclase and hornblende are closely packed. As the feldspar is red stained, the general aspect of the rock is syenitic. An example of graphic microdiorite of a different type occurs at Penmaenmawr on the North Wales coast. In and around the Harlech Dome, also in North Wales, numerous minor intrusions, presumably of Ordovician age, occur. Many of these are of dioritic composition and some fall in the microdiorite category. They are noteworthy for the prominent phenocrysts of hornblende, plagioclase, and less commonly augite which they contain. Unfortunately they are not particularly attractive subjects for petrographic study for they have experienced a mild regional metamorphism which has, in many instances, completely pseudomorphed the original minerals: hornblende is represented by aggregates of chlorite, epidote and calcite,

ANALYSES OF TONALITES, DIORITES AND MICRODIORITES

	1 Tonalite (Daly's average)	2 Quartz-diorite (Tonale type), Carinthia	3 Diorite (Daly's average)	4 Porphyrite, Glencoe	5 Graphic microdiorite, Penmaenmawr	6 Pyroxene-mica-diorite
SiO_2	62·35	63·09	59·67	65·30	58·45	54·07
Al_2O_3	16·41	18·89	16·68	15·20	17·08	15·36
Fe_2O_3	2·57	3·48	2·93	2·49	0·76	0·98
FeO	3·82	2·02	4·09	2·53	4·61	6·76
MgO	2·83	1·97	3·62	1·80	5·15	6·44
CaO	5·45	6·18	6·22	3·16	7·60	8·09
Na_2O	3·41	3·14	3·50	4·13	4·25	3·54
K_2O	2·13	1·30	2·13	3·37	1·02	1·96
H_2O	—	0·63	—	1·00	1·07	1·15
TiO_2	0·67	—	0·77	0·83	—	1·30
Other constituents	0·36	—	0·39	0·23	—	0·38
	100·00	100·70	100·00	100·04	99·99	100·03

1. Tonalite, Daly's average (1933).
2. Quartz-diorite (Tonalite type), Wistra, Carinthia, Austria (W. Krezmar).
3. Diorite, Daly's average (1933).
4. Porphyritic microdiorite, Glencoe, Scotland (Anal. E. G. Radley), *Ben Nevis Memoir* (1916), p. 183.
5. Graphic microdiorite, Penmaenmawr, North Wales (J. A. Phillips).
6. Pyroxene-mica-diorite, Garabal Hill Complex, Scotland (Nockolds), *Quart. J. Geol. Soc.*, **96** (1941), 451.

while the plagioclase phenocrysts are now composed of white mica zoisite, etc.[1]

Among the many 'porphyrites' of the Scottish Survey officers,

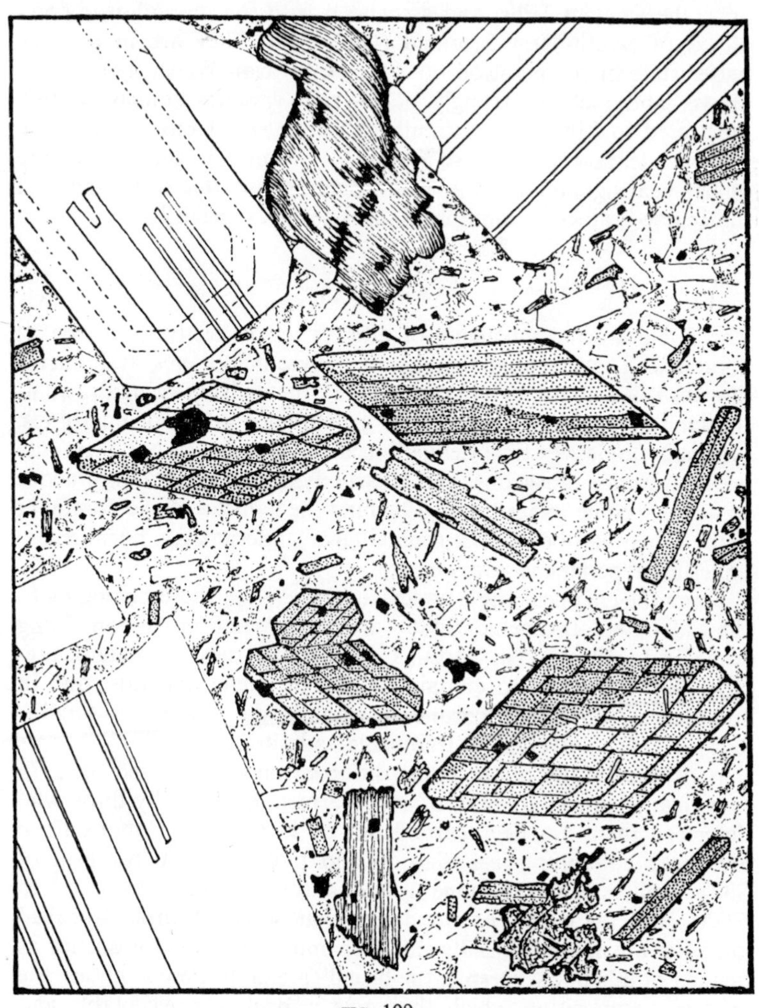

FIG. 109

Porphyritic microdiorite, shore section, Dalbeattie, Scotland.

Euhedral oligoclase-andesine, twinned and zoned; biotite distorted by pressure (top centre); common hornblende mostly twinned on (100), irregular sphene (bottom right centre) and granules, mainly octahedra, of magnetite. *Drawn (A. K. W.) in P.P.L., but with the twinning of the plagioclase indicated.*

[1] Wells, A. K., Geology of Rhobell Fawr, *Quart. J. Geol. Soc.*, **81** (1925), 463; and summary account by the same writer in Matley, C. A., The Harlech Dome . . . , *Quart. J. Geol. Soc.*, **102** (1946), 23.

porphyritic and aphyric microdiorites occur in the dyke swarms related to the late-Caledonian and Old Red Sandstone volcanic centres in southern Scotland, for example in the Glencoe–Ben Nevis area, the Cheviot Hills, and associated with the granodiorite complexes of South-West Scotland (Galloway). They are in a much better state of preservation than the Ordovician Welsh rocks noted above, and include strongly porphyritic varieties similar to that illustrated in Fig. 109, containing phenocrysts of biotite, common hornblende and plagioclase (An_{25-35}). Many of these rocks are quartz-bearing and include microtonalite and the equivalents of the coarse-grained granodiorites, as is to be expected.[1]

ANDESITES

The fine-grained equivalents of diorites are collectively known as andesites. The essential distinction between andesites and microdiorites is the coarser grain of the latter, the division being drawn at the limit of unaided vision for the groundmass grains or microlites. Normally andesites thus defined occur as lava-flows, but they may occur as minor intrusions, particularly dykes. It is common practice to name specific andesites by attaching the name(s) of the dominant mafic mineral(s) before the general term 'andesite', *e.g.* hornblende-andesite, hypersthene-andesite, etc.

From the chemical angle typical andesites—*i.e.* those belonging to the Calc-Alkali (orogenic) Suite—are notably rich in Al, containing some 16 to 18 per cent Al_2O_3. Consequently they are plagioclase-rich; they are liable to contain Al-bearing mafic minerals; and in some cases, normative corundum, though the amount is small—1·5 per cent on average in Cainozoic andesites.

A detailed study of the composition of andesites based on 1775 first-class chemical analyses of rocks termed andesites by those who described them has been carried out by F. Chayes.[2] Among them are rocks containing less than 50 and more than 70 per cent of SiO_2—well outside the range usually allotted to andesites (53·5 to 63·5), and there can be no doubt that some of these so-called andesites are actually basalts and even more of them are dacites as defined in this book. Despite the broad spread of these analyses, the average composition which they give is probably acceptable and significant (see the Table of analyses, p. 321). Although 58·17 per cent SiO_2 is a reasonable figure for an Intermediate rock-type, it is significant that all but a very few andesites are over-saturated with

[1] Phillips, W. J., The minor intrusive suite associated with the Criffell–Dalbeattie Granodiorite Complex, *Proc. Geol. Assoc.*, **67** (1956), 103.

[2] Chayes, F., The chemical composition of Cenozoic andesite, *Proc. Andesite Conf.*, State of Oregon Dept. of Geology and Mineral Industries, Bull., **65** (1969), 1–11. This publication is a valuable source of information on andesites.

silica: they therefore contain normative quartz to a hitherto unexpected degree, the average content being 13·6 per cent, well above the 10 per cent which was formerly used to separate andesites from dacites. This relatively large amount of normative quartz means that the commonest rocks within the andesite field are, in fact, relatively quartz-rich, and it would be logical to speak of them as quartz-andesites. Formerly it was usual to regard andesites, like diorites, as being 'ideally quartz-free'; but in the light of Chayes' results, quartz-free andesite must be a rarity.

Plagioclase is the most significant constituent and, by definition, must amount to two-thirds (65 per cent) of the total feldspar.[1] Theoretically andesites without plagioclase phenocrysts may occur; but normally prominent phenocrysts occur as well as abundant groundmass microlites. The former are typically strongly zoned and are noteworthy for their exceptionally basic cores—often between An_{85} and An_{95}. Significantly, these compositions match those of xenocrysts occurring in andesites from the West Indies. The bulk composition of the plagioclase is difficult to estimate on account of (1) zoning in the phenocrysts, (2) the inevitable difference between phenocrysts and groundmass microlites, and (3) incomplete crystallization due to quenching; but Chayes' average is An_{34}, well inside the limiting value of An_{50} chosen to separate andesites from basalts in our classification.

Yoder has shown experimentally that plagioclase within this range of compositions crystallizes under conditions involving high water-vapour pressure. The characteristic zoning, often oscillatory, could be the result of variation in water-vapour pressure, even at constant temperature.

As a consequence of over-saturation with silica the number of andesites containing normative olivine is very small, only just over 1 per cent of the total, so that the presence of normative enstatite or hypersthene is virtually part of the definition of andesite. Modal olivine in small amount is not uncommon because the excess silica which, under equilibrium conditions would have reacted with early-formed olivines to form orthopyroxene, has been immobilized by quenching. Normally, however, we would view with suspicion any 'andesite' containing olivine—careful examination might well prove the rock to be basalt. It would simplify classification if olivine were to be restricted to Basic igneous rocks.

Orthopyroxene is common in this group of lavas: it may be

[1] We must refer to what we regard as an unjustifiable limitation of the term 'andesite' or 'true andesite' to rocks with a P-value of 90 to 100. *All* rocks within the andesite field, *i.e.* lying between the boundaries defined by Q_0–Q_{20} and P_{65}–P_{100}, are true andesites. Those containing 90P or more are plagioclase-rich andesites: there is no need to try to find a special name for them—this would only cloud the issue.

enstatite or hypersthene, and may occur as well-formed phenocrysts in addition to minute tabular microlites in the groundmass. Clinopyroxene, a pale-coloured diopsidic augite, is the other normal coloured mineral in more basic andesites. Normally the identification of the various pyroxenes should not occasion any difficulty: a thin section of andesite may be relied upon to yield some characteristically shaped basal sections (see Figs. 110, 111); while it is useful to remember the differences in birefringence and the fact that a twinned section must be clino-, not orthopyroxene.

The list of mafic minerals which may occur in andesites is completed by including the OH-bearing silicates mica and hornblende which are more commonly present in andesites than in any other group of lavas. The hornblende is the brown 'basaltic' variety, lamprobolite. Both it and the mica phenocrysts often show in the

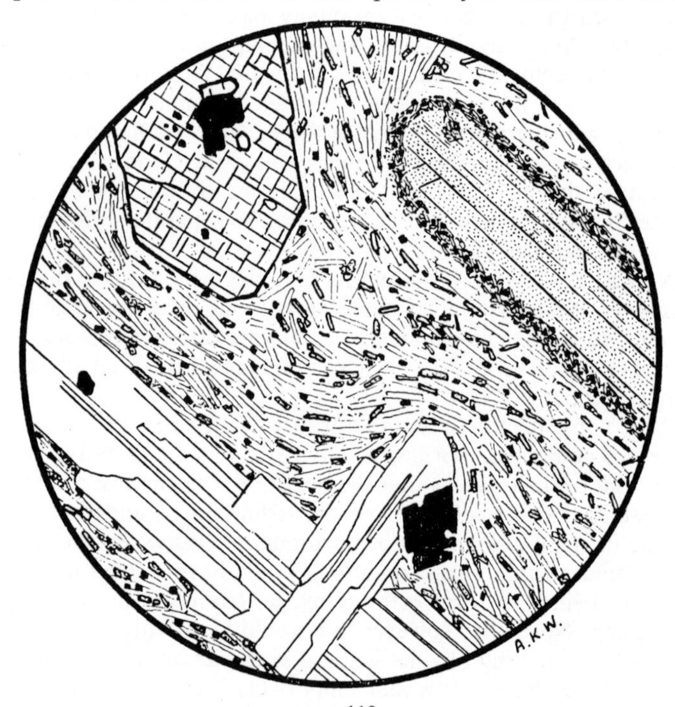

FIG. 110

Hornblende-augite-andesite, Col d'Egremont, Contal, Auvergne.

Phenocrysts of plagioclase show complex Carlsbad-albite-Baveno twinning; a 'basal' section of clinopyroxene encloses a bunch of early magnetite and apatite crystals; a corroded vertical section of lamprobolite (*fine stippled*) with a marginal zone of closely packed recrystallized augite, and octahedra of magnetite; the phenocrysts are set in a fluxional groundmass of second-generation augite, plagioclase and magnetite. *Drawn in P.P.L., but with the twinning of the plagioclase indicated.*

clearest manner that they were not in equilibrium with the magma in which they were carried to the surface. They may exhibit all degrees of magmatic corrosion. At an early stage this may amount to no more than a slight 'peppering' with magnetite granules, but at a later stage of alteration the hornblende is progressively replaced by an aggregate consisting chiefly of granules of nearly colourless clinopyroxene and octahedra of magnetite (Fig. 110). In extreme cases the whole of the mica and hornblende may be so replaced, and only the shapes of the original phenocrysts survive as 'ghost pseudomorphs'. Finally, even the shapes are lost, and indefinite areas rather richer in granular augite and magnetite than the general body of the rock are all that remain. In the same rock, phenocrysts of pale-coloured clinopyroxene may occur, and were apparently quite stable.

Magmatic corrosion and alteration of the phenocrysts in this way is probably due to reactions involving atmospheric oxygen which affect the lava when it reaches the surface, causing a tem-

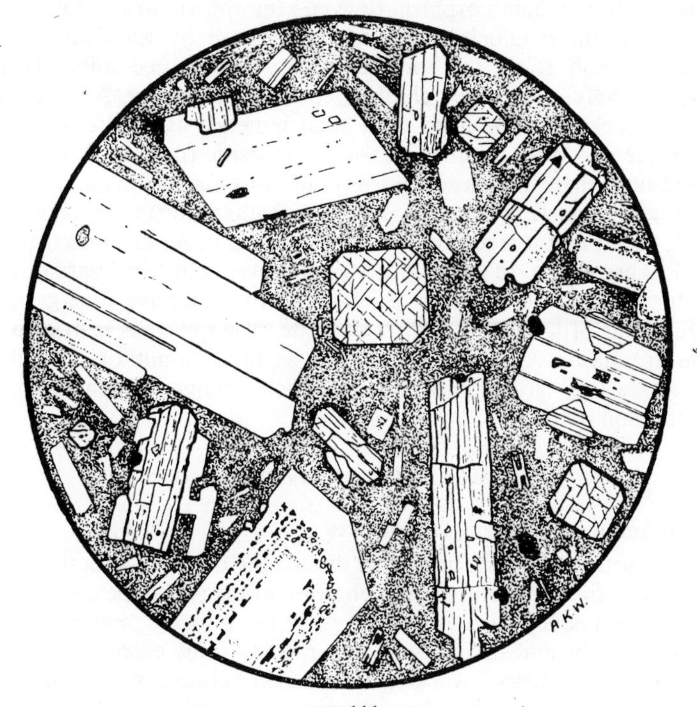

FIG. 111

Enstatite-andesite, Asama-yama, Japan.

Plagioclase (andesine) as plentiful euhedral phenocrysts showing zones of corrosion and growth; some of late formation skeletal with pronged ends. Enstatite also plentiful, showing pinacoidal as well as prismatic cleavage. Magnetite, a few microphenocrysts. Groundmass subvitreous.

porary rise in temperature. Pyroxene replacing amphibole and biotite provides an interesting reversal of the normal sequence of discontinuous reactions of Bowen.

The groundmass in which the phenocrysts are embedded may be wholly crystalline and composed largely of narrow microlites of plagioclase, associated, it may be, with granules of coloured minerals and not much iron ore (Fig. 110). If the microlites are so closely packed as to exclude glass altogether, the texture is said to be **pilotaxitic;** but in some instances wedges of glass lie between the microlites, giving the **hyalopilitic** texture. By increase in the proportion of glass to microlites such andesites grade into true andesitic glass. The latter is brown in thin section and may be perfectly transparent; but with increasing age it tends to lose its transparency through devitrification, just as with rhyolitic glass.

Andesites are somewhat prone to alteration through a variety of causes. Solfataric action in the neighbourhood of an active volcano, slight regional metamorphism or even atmospheric weathering tends to render the plagioclase turbid, to replace it by secondary albite charged with zoisite or epidote, while the coloured minerals are progressively replaced by chlorite, epidote, calcite, etc. Any vesicles in the rocks become filled with such minerals as calcite, chlorite, prehnite and other zeolites. It was to andesitic rocks in this altered condition that the name 'porphyrite' was first applied. A little secondary quartz is liberated during such alteration: there is usually little doubt that it is secondary, on account of its mode of occurrence and its associates. In some andesites, however, a little primary quartz is to be expected on general grounds: the extrusive, fine-grained equivalent of tonalite is quartz-andesite. The quartz is restricted to the groundmass in such rocks; but as the amount of free silica increases they grade into dacites, which commonly contain first-generation quartz.

PETROGENETIC CONSIDERATIONS

Distribution and Origin of Andesites

Before petrologists can profitably discuss the origin of andesites they should be in agreement on the definition of the rock-type concerned; but unfortunately such agreement is, at the moment, conspicuously lacking. Choice of definition is governed to some extent by the experience of geologists working in the field because of the variation that exists between one volcanic region and another in the spread of compositions and therefore the natural grouping of the rocks concerned. To a considerable extent this latitude is acceptable because it helps to expose deficiencies in arbitrary schemes of classification. Obviously adaptation to regional needs must not be carried too far or terminological anarchy would ensue. On the other hand too rigid adherence to a single criterion in classification may also result in serious anomalies. This is strikingly illustrated by the result of restricting the term 'andesite' to lavas in which 90 per cent or more of the total feldspar is plagioclase ($An_{<50}$). Using this

ANALYSES OF ANDESITES

	1	2	3
SiO_2	58·17	58·68	58·05
Al_2O_3	17·26	17·29	17·15
Fe_2O_3	3·07	2·97	3·30
FeO	4·17	3·96	2·54
MgO	3·23	3·14	2·16
CaO	6·93	7·13	5·13
Na_2O	3·21	3·24	4·57
K_2O	1·61	1·27	3·60
H_2O	1·24	1·20	1·64
TiO_2	0·80	0·81	1·10
P_2O_3	0·20	0·17	0·43
Other constituents	—	0·12	0·13
	99·89	99·98	99·80

Molecular norms

	2	3
Ap	0·36	0·91
Il	1·15	1·56
Or	7·67	21·59
Ab	29·72	41·66
An	29·52	15·90
Mt	3·17	3·50
Di	4·39	5·67
Hy	9·85	3·53
Q	14·16	5·67

1. Average of 1775 analyses of Cenozoic rocks defined as andesites, calculated by F. Chayes, *Proc. Andesite Conf.*, (1969), 2.
2. Average of 89 calcic andesites from island arcs, quoted by A. R. McBirney in *The Earth's Crust and Upper Mantle*, ed P. J. Hart (1969) p. 503.
3. Average of 13 analyses of 'alkali calcic' andesites from continental interior regions, A. R. McBirney, (1969).

criterion it has been stated that 'true andesites' are absent from a large part of the Chilean Andes. The rocks which are most abundant in the Andesite Formation of Chile and which we and probably most geologists, would regard as quite normal andesites are designated 'latite-andesites'.[1] This is an instance of the confusion that must arise when the restrictions imposed by an arbitrarily designed classification are in conflict with the dictates of general usage.

It has long been recognized that andesites are characteristically continental rocks and are dominant among the members of the Calc-Alkali Suite in many orogenic regions and in island arcs. Typical andesites, like those of the circum-Pacific belt, are absent from the central parts of the ocean basins and the mid-oceanic rises. However, certain lavas belonging to the Alkali Suite and occurring in oceanic islands have been widely referred to in the past as andesites on the basis of the feldspar they contain: this is andesine in one type and oligoclase in another. In both cases, therefore, the composition of the dominant feldspar falls within the range allotted to the andesites and by strict application of this one criterion they should be classified as andesites. But these rocks are thoroughly Basic: the oligoclase-bearing type commonly has a silica percentage of less than 50—well on the basaltic side of the dividing line drawn at 53·5, and on this showing they are oligoclase-basalts. Further, they contain olivine which underlines their affinity with basalts rather than andesites. Their nomenclature is chaotic. They have been classed as oligoclase-basalts, andesitic basalts, basaltic andesites and even trachyandesites. In view of this confusion we are strongly of the opinion that they should be given independent status and a distinctive name divorced from both basaltic and more particularly andesitic connotation. An obvious choice for the oligoclase-bearing type, largely on the grounds of priority, is **mugearite** defined from Skye in 1904.[2] In our opinion this term might well have been extended to include the andesine-bearing type also; but after reviewing all the relevant factors G. A. Macdonald proposed the name **hawaiite** for the latter and his recommendation has been widely adopted.[3]

With the exclusion of mugearite and hawaiite we are left with andesites which are almost, but not quite, confined as members of the Calc-Alkali Suite to regions of mountain-building and orogenic uplift. An exception is provided by rather subordinate occurrences of Intermediate extrusives the compositions of which lie broadly in the andesite field, though by comparison with andesites of similar silica percentage in the Calc-Alkali Suite, they generally contain distinctly

[1] Pichler, H., and Zeil, W., Andesites of the Chilean Andes, *Proc. Andesite Conf.* (1969), 165–74.

[2] Harker, A., Tertiary igneous rocks of Skye, *Mem. Geol. Surv.* (1904), 264.

[3] Macdonald, G. A., Dissimilarity between Continental and Oceanic rock types, *J. Petrol.*, **1** (1960), 172.

less Al_2O_3 (rarely exceeding 16 per cent), more alkalies and often more iron, so that McBirney for instance, groups them together under the title 'ferro-andesites'. They occur in association with other rock-types characteristic of anorogenic rather than orogenic volcanicity, and may well be of rather diverse origin since some examples provide strong evidence of resulting from hybridism, the mixing of Acid and Basic magmas (p. 332). Although the rocks of this rather ill-defined group are clearly varieties of 'andesite', it is best wherever possible to distinguish them from andesites of the Calc-Alkali Suite by use of special names such as **icelandite**[1]. In the type examples described by Carmichael, the normative quartz-content may be as high as 25 per cent, so that some of the original icelandites fall in the dacite field. The detailed classification of andesites of these different kinds is a matter for specialists. We note in this connection that H. S. Yoder, with andesites in mind said that 'it is not possible to collect an individual specimen of lava and unambiguously classify it . . . without regard to its comagmatic suite'.

An interesting feature concerning the distribution of andesites is their tendency to vary in composition across the structural trend of the regions concerned.[2] In an island arc, close to an ocean trench, the andesites tend on average to be very calcic with an abundance of An-rich plagioclase; while at the other side of the orogenic belt, towards the continental interior they are more alkaline, with a notable increase in potassium as shown by the analyses in the table (p. 321).

Attempts to solve the problem of the origin of andesites provide a fascinating story of petrological detective work. Until the 1960s the only clues available were those derived by direct observation of their petrographic character, their distribution in space and time and their association with other volcanic rocks, more particularly basalts. More recently exciting new indirect evidence has become available as the result of the study of the distribution of trace elements and isotopes, geophysical investigations into the nature of the underlying crust and mantle as outlined in Ch. 2 of Part II, and experiments on melting relationships of rocks and minerals at high pressures.

In the High Andes in South America, the natural 'home' of andesites of course, large numbers of volcanic cones of the so-called strato-type, formed by the accumulation of inter-layered pyroclastic material and lava-flows, project like vertebrae along the spine of the western cordillera; and similar ranges of dominantly andesitic

[1] Carmichael, I. S. E., The petrology of Thingmuli, a Tertiary volcano in Eastern Iceland, *J. Petrol.*, **5** (1964), 435–60.

[2] Kuno, H., Lateral variation of basalt magma-type across continental margins and island arcs, *Bull. Volc.*, **29** (1966), 195–222. Dickinson, W. R., Relations of andesite volcanic chains and granitic batholithic belts to the deep structures of orogenic arcs, *Proc. Geol. Soc., Lond.*, **1662** (1970), 27–30.

volcanoes occur in other regions of orogenic uplift in North and Central America and in many of the island arcs around the Pacific, forming a veritable 'ring of fire'. A glance at the world map (Fig. 55) indicates the vast scale of this type of volcanicity. The dominant volcanic rocks in these orogenic belts are andesites showing only a limited range of variation in chemical composition; and it is therefore tempting to conclude that primary andesitic magmas, however formed, must exist in their own rights.

From the distribution of andesites in the continents and around their margins it was natural to infer that, in some way, continental crust was an essential constituent in the recipe for producing andesites. Taken in conjunction with the common association of basalts with andesites, it was widely believed that andesitic magma was formed by some process involving admixture of continental crust material and basaltic magma. A correct balance of the equation, *crust + basalt = andesite*, is provided by combining equal amounts of basalt and granite (representing continental crust), or alternatively a 60:40 mixture of granodiorite and basalt. This equation holds for the major constituents, but not for the trace elements which have been shown to occur in amounts inconsistent with the admixture hypothesis which is thereby considerably weakened. It is proved to be untenable by the fact that geophysical research has demonstrated that beneath some of the island arcs there is no sialic crust. This is true of the Tonga group, for example, situated thousands of kilometres from the nearest continent: the andesites are there, but one essential ingredient of the above recipe is wanting. Further, the volcanic zone in the Izu–Hakone arc, extending for 800 km, passes from a region in the north underlain by continental crust to one of oceanic crust in the south, without significant change in the composition of the andesites. The same is true of the volcanic regions in Central America.[1] The andesites, erupted from a chain of Quaternary volcanoes running parallel to the coast, cross a clearly defined change in basement rocks from thick continental crust in the north (Guatemala and El Salvador) to oceanic crustal materials, resting directly on mantle-type peridotites of unknown thickness, in the south (Nicaragua and Costa Rica) without commensurate change in composition. Such evidence of the complete independence of andesitic volcanic activity and the nature of the underlying rocks leads inevitably to the conclusion that the generation of andesitic magma does not result primarily from a reaction between crustal rocks and basaltic magma.

This leaves in essence two main possibilities for the origin of

[1] McBirney, A. R., Compositional variations in Cenozoic Calc-Alkaline Suites of Central America, *Proc. Andesite Conf.* (1969), 185–9; and Andesites of North and Central America, *Proc. Geol. Soc., Lond.*, **1662** (1970), 30–5.

andesitic magmas: either they are derived from parental basalt *magmas* by processes of differentiation, or they are produced by fractional melting of *rocks* from the deep crust or upper mantle. We examine each of these in turn.

The reality of differentiation of basaltic magmas by fractional crystallization is not in question: it must play a part, and in some circumstances a major part in the evolution of the Calc-Alkaline Suite.[1] The point at issue is the adequacy of the process to produce basalts, andesites, dacites and rhyolites in the proportions, and showing the space- and time-relationships encountered in major provinces of andesitic volcanicity. Progressive differentiation in a magma reservoir might be expected to result in the eruption of the more Basic rocks first, and in general, the later fractions should follow in sequence in diminishing amounts. In fact this pattern can rarely be recognized. Thus, in the central parts of the Cascade Range in western North America the volcanoes Mts Rainer, Hood and Jefferson have erupted nothing but andesites of monotonously uniform composition; but towards each end of the range andesites—still of the same composition—are associated with subordinate rhyolites and basalts, forming a divergent volcanic series.[2] Both the rhyolites and the basalts appeared late in the sequence and were erupted from satellitic cones on the flanks of the main volcanoes.

In the Andes the lavas are dominantly andesites and rhyolites associated with subordinate basalts. It is very difficult to form an estimate of the relative amounts of andesites and rhyolites because of their different topographical expression: the former were erupted from stratovolcanoes which tower over high-plateau features of vast extent—more than 150,000 km^2—formed mainly by rhyolitic and dacitic ignimbrites. Many other instances might be quoted of volcanic regions where the sequence of eruptions and the proportions of the different lava-types involved appear to be completely wrong for the andesites to have been derived, on a large scale, by differentiation from basaltic magma.

Another major obstacle to accepting the hypothesis of generating andesitic magma by the differentiation of basalt concerns the trend of chemical variation shown by the Calc-Alkali Suite (Fig. 112). This is markedly different from that displayed by the rocks from different levels in the Skaergaard and similar intrusions, where the dominant factor in producing the observed variation is *known* to have been fractional crystallization and differentiation of basaltic

[1] See for example the impressive data relating to differentiation presented by E. S. Larsen: *Amer. Min.*, **21** (1936), 679–701; **22** (1937), 889–905; **23** (1938), 227–57 and 417–29.

[2] McBirney, A. R., Andesitic and rhyolitic volcanism of orogenic belts, in *The Earth's Crust and Upper Mantle*, ed. P. J. Hart, *Am. Geophys. Union* (1969), 501.

magma. A striking feature of especially the earlier stages of the process is enrichment in Fe^{2+} relative to Mg^{2+}, due largely to changes that take place in the composition of olivine and pyroxene crystals in successive fractions.

The conditions which might allow some kinds of basaltic magma

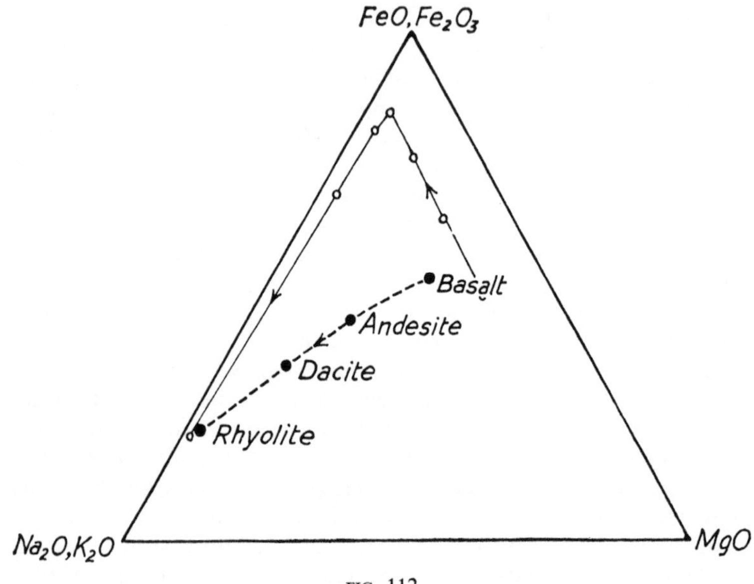

FIG. 112

Trend of differentiation due to fractional crystallization in the Skaergaard complex (open circles) compared with the trend shown by the lavas belonging to the Calc-Alkali Suite. Relevant oxide ratios for the latter are taken from average compositions quoted by R. A. Daly.

to differentiate through fractional crystallization without appreciable iron-enrichment have not yet been fully investigated. It has been suggested, however, that an important factor may be the maintenance of a relatively high concentration of water during crystallization, leading to a higher degree of oxidation of iron. The effects of this are complex and difficult to state in a few words; but one important factor may be summarized by saying that more extensive precipitation of magnetite in the early stages of crystallization removes iron that might otherwise be concentrated in later phases of iron silicates.[1]

This is not the place to go into details concerning the composition of basalts, but it is worth noting that even in andesitic provinces

[1] Osborn, E. F., Experimental aspects of calc-alkaline differentiation, *Proc. Andesite Conf.* (1969), 33–42.

where basalts are present in adequate amounts for them to be regarded as representing a parent magma, their compositions may be so widely and randomly scattered as to deny any such genetic relationship with the rest of the Calc-Alkaline Suite.[1] Several authorities have used the facts in such cases as arguments in favour of the primary magma being andesitic rather than basaltic.

This brings us to a consideration of the third, and probably most widely applicable mode of origin of andesitic magma, *viz.* by selective melting of rocks at depths, which geophysical evidence for Kamchatka, for example, suggests to be of the order of 60–90 km. According to one view, held by many Russian petrologists[2] the rocks available for melting at these levels are mantle peridotites of monotonously uniform composition. However, according to the concepts of plate tectonics, rocks originally of basaltic composition, now converted into amphibolites or eclogites depending on the depth and temperature, are also available. If the basaltic rocks of the original ocean floor are carried down as part of a plate underthrusting the edge of the continent to levels where melting can take place, the mechanism of ocean-floor spreading provides a continuously replenishable source of potential andesitic magma (see Fig. 56).

Green and Ringwood[3] have explored this latter hypothesis by a series of experiments on melts of appropriate basaltic compositions crystallized under various high pressure conditions. When the pressure is above about 20 kbar and in the absence of water, the minerals in equilibrium with the melt are those of quartz-eclogite, *i.e.* a sub-silicic clinopyroxene (omphacite) and garnet rich in the pyrope molecule. The experiments showed that eclogite undergoing partial melting at these high pressures would give rise to liquid fractions of andesitic composition enriched in Si and alkalies compared with the initial rocks. It was found that iron-enrichment is largely suppressed by compensatory changes in the proportions and compositions of the pyroxenes and co-existing garnets in equilibrium with the liquid under the varying conditions of melting. Other experiments at lower pressures (9–10 kbar) and in the presence of water indicate that amphibolites undergoing fractional melting would react in a manner similar to the eclogites at greater depths.

If the source rock at the levels of melting is mantle peridotite rather than the rocks we have just considered, it would consist predominantly of forsteritic olivine, orthopyroxene and clino-

[1] See for example, Tilley, C. E., Some aspects of magmatic evolution, *Quart. J. Geol. Soc.*, **106** (1950), 37–61.

[2] Gorshkov, G. S., Geophysics and petrochemistry of andesitic volcanism of the circum-Pacific belts, *Proc. Andesite Conf.* (1969), 91.

[3] Green, D. H., and Ringwood, A. E., High pressure experimental studies on the origin of andesites, *Proc. Andesite Conf.* (1969), 21–32.

pyroxene. This is the composition of two-pyroxene peridotite, the rock-type lherzolite. Yoder[1] has reviewed the probable course of fractional melting of such a rock under various conditions. At a pressure of 20 kbar, and in the absence of water, the peridotite would begin to melt at 1640° C to a liquid which, if separated, would crystallize as a mixture of sub-calcic augite and olivine.

Under hydrous pressures as low as 7 kbar, however, the same peridotite would begin to melt at 1220° C and from the ensuing liquid fraction, two pyroxenes (ortho- and clino-) together with quartz would crystallize out. This is important in relation to the origin of andesite particularly as the phase relations are not significantly altered by the addition of the components of plagioclase.

The most interesting conclusion to be drawn from these considerations is that, from the same mantle peridotite, magma of basaltic (olivine-normative) composition or of andesitic (quartz-normative) composition may be derived, depending essentially upon the presence or absence of water.

Experimental studies by Yoder of the effects of varying water-vapour pressure on the crystallization of plagioclase has clearly shown that the An-content of plagioclase is controlled by water-vapour pressure. Increasing pressure causes the An-content to increase. It is inferred, therefore, that the extremely calcic composition of plagioclases in xenoliths enclosed in andesites, for example from the West Indies, and of the cores of plagioclase phenocrysts in andesites in general indicates high water-vapour pressures at the site and time of formation.

From a different angle the experiments have shown that in systems involving An (representing plagioclase) and diposidic pyroxene, increasing water-vapour pressures cause a shift in composition of the lowest temperature liquids towards An, showing that high water-vapour pressures may well account for the high plagioclase-content (50 per cent or more) of typical andesites.

Origin and Distribution of Diorites

Diorites among coarse-grained rocks occupy the same position in the scheme of classification as andesites in the fine-grain category. It is tacitly assumed, therefore, that the two types are comagmatic; and that diorites result from deep-seated, slow cooling of andesitic magma. When one is considering the status and mode of origin of diorites in relation to andesites some confusion may well arise from mistaken concepts in the past concerning the definitions of, and relationship between the two types. We have shown that andesites commonly contain unexpectedly large amounts of potential or

[1] Yoder, H. S., Calcalkalic andesites; experimental data bearing on the origin of their assumed characteristics, *Proc. Andesite Conf.* (1969), 77–89.

normative quartz. It follows that 'plutonic' rocks of identical composition will contain at least as much, and possibly in some cases more modal quartz. Such rocks are considerably removed from the conventional 'ideal diorite'—a rock composed essentially of andesine and, say, hornblende. It is important to realize that such rocks ('orthodiorites' in our classification) are comparatively rare, and that the commonest rocks in the diorite field contain significant quantities of quartz.

This helps to explain what may otherwise seem to be an anomaly, namely, that in regions where the dominant lavas are andesites (typical Intermediate rocks) the associated coarse-grained types of major intrusion status tend to be quartz-bearing and often quartz-rich types (granodiorites, tonalites and monzonites). For example, in an area of some 4,000 square miles of the great southern California batholith,[1] over 2,500 square miles of surface area are occupied by tonalites and over 1,100 by granodiorite, these two rock-types together occupying over 90 per cent of the area studied. These are quartz-rich members of an assemblage of undoubtedly dioritic, *i.e.* andesitic affinities. Much of the inequality in the proportion of Acid to Intermediate rocks among the intrusive and extrusive members of a Calc-Alkali igneous province may therefore be more illusory than real, due to accidents of mineralogical classification rather than differences in chemical composition. Even when this is taken into account, however, one is still likely to find that the 'plutonic' rocks may, on average, include more Acid types than their volcanic associates: the plutonic rocks exposed at ground level are likely to be the more siliceous products of differentiation. At greater depth they would probably become less siliceous and more mafic.

In matters concerned with the status of diorites the role of microdiorites, especially the varieties commonly termed porphyrites, is important. Well-developed dyke-suites, covering the whole range of composition of the associated plutonic complexes, provide the essential link between extrusive andesites and deep-seated diorites. Although the fact has not been established on a statistical basis. one gets the impression that the rocks making the dyke-phase are closer to andesite than to the strongly differentiated granodiorite assemblage, in the sense of being less siliceous.

Certain facts of distribution and field relationships show that some diorites are not straightforward products of crystallization in depth of andesitic magma. This may be demonstrated in diorites in the Channel Islands, especially Jersey and Guernsey, where diorites are probably better developed than elsewhere in the British Isles. Jersey is best known as providing fine exposures of a wide variety

[1] Larsen, E. S., Batholithic . . . rocks of Corona . . . California, *Mem. Geol. Soc. Amer.*, **29** (1948).

of rhyolitic rock-types; but lying below these is a considerable thickness of more basic rocks, incompletely described as yet, known collectively as the 'Jersey Andesites'. Some of these contain abundant platy plagioclase phenocrysts showing a well-developed fluxion structure, like that displayed by a well-known Scottish rock—the Carnethy Porphyry—from the Pentland Hills. Granites have broken into the base of this pile of andesites and have incorporated andesitic material forming various dioritic and monzonitic hybrids, many of them quartz-bearing. The granites contain xenoliths of all sizes. Some of them, embedded in alkali-granite, are rimmed with hybrid 'diorite'; but internally the original structure has survived general reconstitution, and it is possible to recognize the fluxional, porphyritic andesite. Further, in this same complex certain diorites have demonstrably been formed by metasomatic acidification of gabbroic rocks: some of the hornblendes contain relics of the original pyroxene, while traces of the original ophitic texture have survived the metasomatism. Finally, in South-East Jersey, where dioritic rocks are extensively exposed along the coast, dioritic members of the complex appear to have been formed by direct crystallization from a magma of their own composition. They are frequently layered and often pegmatitic, containing hornblende prisms up to several inches in length.

This is a convenient place to consider some of the general aspects of hybridization and assimilation which are of such importance in the genesis of andesites and diorites.

Assimilation

Under this heading we shall consider some aspects of the problem of reciprocal reactions which may occur between magma on the one hand and intruded rocks on the other. The word 'assimilation' has been used because it is easily understood and conveys a correct impression of what may happen in certain circumstances: rocks may become incorporated and completely digested or assimilated by magma so that no trace remains. However, assimilation is only one important end-product of the processes of magma/rock reactions to which the comprehensive term **syntexis** may be applied.

The first stage of the process of assimilation involves the fracturing of country-rock during a period of intrusion. Much of the shattering must result from the heating and differential expansion of the rocks in contact with the magma. Wall- or roof-rock which has been mechanically shattered and engulfed in the magma survives, for a time, in the form of angular blocks of all shapes and sizes, known as **xenoliths.** Initially the latter are clearly demarcated from the invading rock, and contrast with it in colour, texture and composition; but as a consequence of heating to magmatic or near-magmatic temperatures and soaking by magmatic fluids, they gradually merge their identity into that of the host rock. Two processes are involved in such assimilation. According to the composition of the xenolith, it may be incorporated by pure melting, or by reaction with the magma.

Bowen[1] has related the reactions between magma and xenolith to those which take place between a magma and its own normally precipitated crystal phases. Thus if a magma is temporarily in equilibrium with a plagioclase of composition, say, An_{50}, which is being precipitated, then the magma will *react* with an inclusion of a more calcic plagioclase (and therefore with a higher melting-point) until it conforms to the equilibrium conditions of the surrounding liquid. The latter contributes sodium and silicon, while the xenocrystic plagioclase yields up calcium and aluminium to the magma, until its composition is changed to that of the stable phase, *i.e.* An_{50}. On the other hand, any included plagioclase richer in sodium cannot exist at the magmatic temperature, which is above its own melting-point. Such a plagioclase must melt, and adds the whole of its substance to the magma. To melt a mineral, however, demands a relatively large amount of latent heat, which has to be provided by the magma. Since the latter probably has no surplus heat, unless it is being continually supplied by convectional transfer from depth, pure melting can only be achieved by a corresponding amount of crystallization—which releases latent heat.

What is true of plagioclase is equally true of all other groups of silicates belonging to continuously variable series, and probably true also of the discontinuous reaction series of Bowen (p. 176). Thus olivine, occurring in a xenolith immersed in magma precipitating biotite, would be progressively converted, in stages, from olivine to pyroxene, then to amphibole, and, if the reaction is not stopped by freezing of the magma, into biotite. Conversely, biotite immersed in olivine-precipitating magma and therefore at a temperature high above its own melting-point, would melt, and in so doing would cause the precipitation of an equivalent amount of olivine, necessary to provide the latent heat of fusion. In short, xenolithic minerals high in the reaction series tend to be 'made over' into those lower in the series, by reaction; while those low in the reaction series may be melted, on incorporation in magma. It follows from these general considerations that the individual component minerals in a xenolith of igneous rock will behave very differently, according to whether their melting-points are above or below the magmatic temperature at the time of incorporation in the magma.

[1] Bowen, N. L., The behaviour of inclusions in igneous rocks, *J. Geol.*, **30** (1922), 513.

All the more obvious effects of reaction between magma and country-rock are confined to the margins and roofs of intrusions. Xenoliths survive only where assimilation has been incomplete because of the exhaustion of the supply of heat energy of the magma. It is reasonable to infer that during the earlier history of an intrusion, and at greater depths, assimilation may not only have been complete, but it may have occurred very much more extensively than the evidence of a limited quantity of surviving xenoliths would suggest. Daly[1] described the process as abyssal stoping and assimilation, which he regarded as being of major importance, particularly as a means by which plutonic intrusions may make room for themselves as they invade higher levels of the crust.

Of course there are strict limitations as to the amount of rock that can be assimilated by any particular volume of magma. This depends primarily upon the heat available in the latter. It is generally accepted that magmas never have a surplus of heat over that needed to maintain them in a molten condition: in other words, they are never super-heated. Any work done by the magma in heating up and reacting with wall-rocks must therefore be at the expense of crystallization of some part of the magma. In some instances the heat available from this source must be very limited, and it is probably unwise to make too sweeping claims for syntexis, particularly in the highest levels of the crust. Here it would be necessary for magmas to use up much of their heat supply simply in raising the temperature of the relatively cold rocks. At greater depths any rocks involved in syntexis would be hotter to start with as a result of the geothermal gradient. Syntexis in depth may also be favoured by the kind of crustal compression and down-buckling which is believed to occur during orogeny.

Hybridism

Rocks of dioritic composition are much less common in non-orogenic than in orogenic volcanic provinces, and like their andesitic equivalents, tend to be chemically rather different from their equivalents in the Calc-Alkali Suite. **Ferrodiorite**[2] is such a rock and is illustrated by an example described under that name from the Marsco Complex in Skye, Scotland, where it forms part of a three-member composite dyke. The mineral composition of the ferrodiorite is complex: the most abundant mineral is andesine, appropriate to diorite, of course, but accompanied by approximately half as much orthoclase, quartz (10 per cent), hornblende, ortho- and

[1] Daly, R. A., *Igneous Rocks and the Depths of the Earth* (1933), especially Part II, Chs. xi and xii.

[2] Wager. L. R., *et. al.* Marscoite and related rocks of the Western Red Hills Complex, Isle of Skye, *Phil. Trans. Roy. Soc., Lond.* (*A*), **257** (1965), 273–307.

clinopyroxene, biotite, and small amounts of olivine, apatite and zircon. The rock is demonstrably a hybrid formed, it is believed, by admixture of Acid magma which contributed the quartz, orthoclase, mica and accessories, and a Basic differentiate which contributed the rest.

Details of the field relationships, composition and petrology suggest in many instances that dioritic rocks of this character have been produced by the mixing of Acid and Basic magmas, the process to which Harker gave the name **hybridism.** The co-existence of two contrasted magmas is most clearly shown by composite minor intrusions, both dykes and sills, in which there is commonly a central, relatively Acid member flanked on either side by Basic rock, often dolerite or basalt. The contact-relationships show that the Basic member was first intruded, followed closely in time by the Acid, central member. Study of the petrography of the rocks may show that, prior to intrusion the magmas suffered some degree of admixture. This is most clearly shown by the presence in the Basic rock of magmatically corroded xenocrysts which are identical with phenocrysts occurring in the Acid rock. These relationships are illustrated by the classic example of the Marsco Complex (and **marscoite** in particular) first described by Harker in the Skye Memoir in 1904.

It is important to note that the final products of hybridization and contamination may be indistinguishable.

CHAPTER VI

THE MONZONITE FAMILY

ALL petrologists recognize the importance and significance of syenites and diorites, but there is not such unanimity concerning the need to distinguish the 'middle series' which shares the characters of both. Theoretically, the members of the monzonite family (monzonite, micro-monzonite and latite = trachyandesite) are defined by the position of the field they occupy within the QAP triangle. Accordingly, the three rock-types must contain alkali-feldspar (A) and plagioclase (P = $An_{<50}$) each accounting for between 35 and 65 per cent of the total feldspar. They may contain quartz, but the amount must not exceed the Q_{20} value. Coloured minerals are not specified, but they commonly include clinopyroxene, hornblende and biotite, *i.e.* the coloured minerals characteristic of the Calc-Alkali Suite to which these rocks belong.

One point concerning the rôle of quartz in monzonites is so important that it will bear repetition. As explained previously in connection with the syenites and diorites, the term **quartz-monzonite** must logically be restricted to rocks lying between the Q_5 and Q_{20} boundaries *within the monzonite field.* Unfortunately this is at variance with the American practice of using the term with the same A/P ratios, but without limitation to the amount of quartz. Thus the quartz-monzonite category in American literature includes the rocks lying above the critical Q_{20} boundary which we classify as adamellites.

MONZONITES

The coarse-grained members of the family are monzonites. Some petrologists prefer the name syenodiorite which indicates the essential compositional features of the rocks so named. The type-rock was described from the Monzoni Complex in the Tyrol. Admittedly monzonite makes up only a part of the whole complex, but this applies to many another valid type. The original monzonite is slightly over-saturated, with 2·5 per cent of free quartz. Nearly two-thirds of the rock consists of feldspar with andesine and orthoclase about equally balanced (32 and 30 per cent respectively). The coloured minerals include augite, biotite and hornblende in order of importance, the colour index amounting to 35.

Olivine-monzonites[1] occur in the Oslo plutonic complex. Oligoclase and alkali-feldspar occur in approximately equal amounts. Although olivine is the most abundant coloured silicate in these rocks it is accompanied by clinopyroxene and biotite: the colour index is 25. The association of olivine with these kinds of feldspar

FIG. 113

Quartz-monzonite ('Banatite') above Pechadoire, Auvergne.

Subhedral plagioclase (twinning formalized), diopsidic augite (*right*), biotite (*close-ruled*), apatite and sphene, all poikilitically enclosed in orthoclase making irregular contact with quartz at edge of section.

is unusual, and olivine-monzonite of this type does not appear to have been recorded elsewhere. Reference back to the description of laurvigite (larvikite) will show that although the list of minerals is the same in both rock-types, the proportions and the relationships between the feldspars are different.

[1] These are abnormal rocks which strictly lie outside the QAP triangle. Their inclusion within the monzonite family is justified by the fact that apart from the occurrence of olivine, they show affinities with monzonites rather than any other family.

In thin section there is usually no difficulty in recognizing the diagnostic features of the monzonite type: the plagioclase tends to form comparatively smaller crystals of rather better shape than the orthoclase, which is poikilitic towards the plagioclase—a single plate enclosing, it may be, a large number of disorientated plagioclase laths as shown in Fig. 113. The pyroxene, when present, is an almost colourless diopsidic clinopyroxene, which in many cases is intimately associated with common green hornblende. The latter is irregularly moulded upon the pyroxene core as a rule, but the two may occur independently. Amphibole of a different kind—a somewhat fibrous 'uralite'—tends to replace the pyroxene in altered monzonites.

MICROMONZONITES

For the sake of completeness we include, as members of the medium grain-size group, rocks which by analogy with the other families should be named micromonzonites. There can be doubt that rocks of this type have frequently been described under other names, and that micromonzonites are actually commoner than existing records would lead us to believe. Thus, under the name 'monzonite-porphyry' (= porphyritic micromonzonites of our terminology) certain rocks occurring as minor intrusions have been described from the well-known locality, Henry Mountains in Utah, where they are associated with and grade into more abundant porphyritic microdiorites.

As noted in the Syenite Chapter some Norwegian larvikites, judged strictly on the basis of their feldspar-content are monzonitic. On the same criterion their medium-grained equivalents should be micromonzonites, though the term 'rhomb-porphyry' is well established and commonly preferred. The fine-grained equivalents of micromonzonites are, of course, latites (trachyandesites.)

LATITES (=TRACHYANDESITES)

In the six-field classification of the fine-grained igneous rocks which we use, the middle member of the Intermediate series, flanked by trachyte and andesite respectively has, over the years, retained the name **trachyandesite** (Michel-Levy, 1894). However, in classifications involving a five-fold division of the base of the QAP triangle, these rocks, although still occupying the middle position, now lie between 'trachyte' on the one hand and, on the other, a group of rocks already bearing a compound name, 'latite-andesite'. A three-part compound rock-name would be unacceptable, so reluctantly we have to drop trachyandesite, and the name **latite** (Ransome, 1898) becomes first choice for the fine-grained equivalents of monzonite.

The essential mineral-composition of latites (especially if we think of them as trachyandesites) is implicit in the position they occupy within the QAP triangle. They should contain roughly equal amounts of alkali-feldspar and plagioclase within the oligoclase-andesine range. Quartz may occur but must not exceed the Q_{20} value. **Quartz-latite** is the correct name to apply to volcanic rocks within the latite field and containing quartz between the limiting values of Q_5 and Q_{20}. In other words quartz-latite is exactly analogous to quartz-monzonite, differing only in grain-size.

It is useful to realize that it is impossible for feldspars of two contrasting compositions occurring in a rock to display the same crystal habit: in the rocks under consideration the plagioclase normally forms phenocrysts while the alkali-feldspar may show either of two relationships to them—it may form rims around the plagioclase, or it may occur as microlites in the groundmass. Latites have been described in which the roles of the two feldspars are reversed: sanidine forms phenocrysts and the plagioclase occurs as microlites. We would view such occurrences with suspicion: the sanidines might be xenocrystic rather than phenocrystic. As noted under the rôle of feldspars in classification, it is inherently very difficult to measure or estimate the proportions of two different feldspars in rocks of these kinds, and ideally one should use a calculated mode or a norm as a guide to the identification of a suspected latite.

Distribution and Origin of Latites and Monzonites

It is impossible at present to form an estimate of the status of trachyandesites (latites) because of lack of uniformity in terminology. Not all petrologists recognize the monzonite-micromonzonite-latite series and consequently it is probable that some latites (as here defined) have been recorded as trachytes and others as andesites. Nevertheless, making allowance for this, it is undoubtedly a fact that by comparison with the major volcanic groups—rhyolites, trachytes, andesites and basalts, latites are poorly represented among volcanic rocks of all ages. This is not entirely unexpected because there is no genetic linkage between trachyte and andesite: they belong to different suites and are developed ideally in different volcanic environments. However, as noted in the 'andesite' chapter, andesites tend to vary in composition across the orogenic belts in which they occur, being most calcic near an oceanic trench, but becoming richer in alkalies—particularly K—towards the interior of the continent. If this trend continues far enough, K-feldspar may increase in amount until it constitutes more than 1/3 of the total feldspar, at which point the K-rich andesite would give place to trachyandesite. A gradual increase in the amount of K may be associated with increasing depths to the Benioff zone, assuming that the composition of the volcanic rocks is determined primarily by the varying depths of origin of the andesitic-trachyandesitic magmas. However, it is difficult to believe that varying thicknesses of continental crust which would have to be penetrated by such magmas would be without effect on the composition of the lavas.

Monzonites are relatively rare rocks compared with, say, granodiorites, tonalities or adamellites with which they are associated in the field as members of the Calc-Alkali Suite, and generally do not form large independent rock-bodies, but occur as marginal facies of, or small intrusions satellitic to, major intrusions of the types named. It will be recalled that some syenites and some diorites (whose substance is combined as it were in monzonite—*i.e.* syenodiorite) originated by *in situ* modification resulting from contamination between magma and wall-rock; and it must be regarded as more likely that monzonites also originated in this way, rather than by crystallization from a magma of their own composition, formed in the course of normal differentiation.

In the absence of uniform terminology and consistent definition it is difficult to gauge the importance of monzonites. We may illustrate the point by referring again to the well-known teaching 'syenite' from Dresden which is an offshoot from the Meissen granite and is itself largely monzonite. Similarly the rock described as 'ferrodiorite' from the Red Hills in Skye has an A/P ratio which puts the rock, strictly speaking, among the monzonites (p. 332).

Monzonites occur among the very variable minor intrusives associated with the Caledonian igneous activity in Scotland which are reported as consisting of all possible proportions of quartz, alkali-feldspars and oligoclase together with pyroxene, hornblende and biotite, often poikilitic. This range must include not only the end-member syenites and diorites, but also monzonites.

CHAPTER VII

THE 'BASIC' IGNEOUS ROCKS: GABBROS, DOLERITES AND BASALTS

Classification and Nomenclature

THE one kind of magma which all petrologists are prepared to accept has the composition of olivine-basalt. This is potentially capable of crystallizing to a coarse-grained aggregate of plagioclase, pyroxene and olivine in a suitable environment. Natural rocks occurring as integral parts of the major layered Basic intrusives contain these minerals in all possible proportions: individual rocks may consist of any one, two or three component minerals together, but the central types are gabbros and norites:

Gabbro consists fundamentally of plagioclase more calcic than An_{50} associated with clinopyroxene;

Norite consists essentially of plagioclase of the same range of composition associated with dominant orthopyroxene.

There is a complete gradation from gabbro devoid of hypersthene to norite devoid of augite. The one-pyroxene rocks we refer to as **orthogabbro** and **orthonorite** respectively. Between these extremes are the two-pyroxene members of the series including those which are fundamentally gabbros (though containing some hypersthene), and others, which are fundamentally norites, though containing some augite as shown in the appended table.

In addition to this line of variation, the rocks under discussion show a complete gradation between extremely leucocratic types on the one hand and extremely melanocratic varieties on the other. In the former, mafic minerals are virtually unrepresented, while in the latter, felsic minerals are practically absent. Thinking in terms of essential constituents, therefore, these extreme types are *mono-mineralic*, consisting in the first case of nothing but plagioclase, and in the second of pyroxene exclusively. They are, therefore, not covered by the definitions of 'gabbro' and 'norite' stated above, and special names must be used for them, as shown in the table. The intervening types, which are far more common, are adequately covered by the terms 'gabbro' and 'norite' coupled with suitable qualifiers. A rational scheme of nomenclature, based on the pro-

Clinopyroxene dominant			Orthopyroxene dominant
ANORTHOSITE			
LEUCOGABBRO			LEUCONORITE
GABBRO	Hypersthene-gabbro	Augite-norite	NORITE
MELAGABBRO			MELANORITE
CLINOPYROXENITE, *e.g.* diallagite			ORTHOPYROXENITE, *e.g.* bronzitite

portions of the two pyroxenes, and on the ratio of felsic to mafic components, is shown in the Table which illustrates the relationships between the types mentioned above.

Any single large Basic complex, for example the Bushveld or Sudbury lopolith, might well provide a suite of specimens to which individually most of the names in the table might be applied: to illustrate this point in teaching, one of us uses a suite of specimens collected from the Bushveld Complex comprising anorthosite, leuconorite, norite, melanorite, pyroxenite, hypersthene-peridotite, dunite, chromitite and bronzite-chromitite, and titaniferous magnetite rock. All these are essential parts of this one rock-body, and were derived by differentiation from a common magmatic source; but an absurdly cacophonous term would result from hyphenating these nine rock-names in attempting to derive *one* name which would at least indicate the essential nature of the complex. The central type, representing the average most closely perhaps, is gabbro, and therefore we rather loosely use the term 'Bushveld gabbro' for the whole rock-body.

Finally, it will be appreciated that all the rocks so far considered in this discussion are silica-saturated and may therefore carry accessory quartz. Others show various degrees of under-saturation by the presence of olivine. When the latter is subordinate in amount the terms **'olivine-gabbro'** and **'olivine-norite'** should be used; but with increase in the proportion of olivine at the expense of pyroxene, a point is ultimately reached at which the latter disappears, and we are left with the two-mineral type, **troctolite,** consisting of olivine and plagioclase only.

Ideally (ortho) troctolite contains no pyroxene; but augite may progressively displace olivine to give continuous gradation from troctolite through varieties containing olivine *and* augite in varying proportions, ultimately into gabbro free from olivine. Similarly, orthopyroxene may displace olivine giving a gradation into olivine-free norite. Any one of these varieties may be represented by leucocratic, mesotype or melanocratic varieties according to the ratio of plagioclase to total mafic minerals.

It will be noted from the Table that the naming of these rocks is

Olivine	Gabbro series	Pyroxene	Norite series	Olivine
	GABBRO (Orthograbbro)		NORITE (Orthonorite)	
10		90		10
	Olivine-gabbro		Olivine-norite	
30		70		30
	Troctolitic gabbro		Troctolitic norite	
50		50		50
	Gabbroic troctolite		Noritic troctolite	
70		30		70
	Augite-troctolite		Bronzite-troctolite	
90		10		90
	TROCTOLITE (Orthotroctolite)			
100		0		100

The figures to the left and right refer to olivine, those in the middle column to pyroxene, indicating the percentages of the total mafics in both cases.

consistent with the principles laid down elsewhere, and with the scheme used for the gabbro–norite series: thus, in the troctolite–gabbro range any rock containing *both* coloured minerals is named gabbro (with suitable modification) if augite is in excess, but troctolite if olivine is the more abundant. Similarly within the troctolite–norite range, so that, for example, a 'troctolitic melanorite' is a coarse-grained aggregate of plagioclase, orthopyroxene and olivine, with the former in slight excess of the latter, while the total mafic silicates amount to 70 per cent or more.

Further, in most troctolitic intrusions, as in gabbroic ones, some form of internal differentiation, shown by layering, is prominent. There is a tendency for olivine to be concentrated towards the bottom of a layered unit, with plagioclase increasing towards the top. Therefore a gradation from monomineralic olivinite through melatroctolite, troctolite and leucotroctolite into anorthosite may be shown in extreme cases.

Finally, mention must be made of a Basic plutonite termed **eucrite** in Survey publications. Properly this is the name of a type of meteorite and should not have been applied to a rock. The only rocks to which the name is applied occur among the Tertiary Basic–Ultrabasic complexes in Scotland. They are of mixed composition in the sense that they are neither clearly defined gabbros nor norites but combine the mineralogical features of both. In addition to olivine, two pyroxenes (hypersthene and augite), occur in approximately equal amounts, while the plagioclase is exceptionally calcic, being bytownite or even anorthite.

Since the name was originally introduced (as *eukrite*) rocks of widely differing composition have been included under this name, and it is quite clear that the rocks in question could be covered by the terms 'hypersthene-gabbro' or 'augite-norite' in specific instances. We thoroughly agree with Le Bas's conclusion that this is one of the rock-names that serves no useful purpose and tends to confuse, rather than clarify the issue.[1]

To indicate the interrelationships between the many rock-types involved it would be necessary to use a tetrahedron, with plagioclase, olivine, augite and hypersthene at the four corners; but this cannot be adequately shown on a two-dimensional diagram.

In the detailed account which follows, the gabbros, norites and troctolites are considered together with their medium and fine-grained equivalents; while the products of extreme differentiation including the several monomineralic types, together with the phenomena of layering, are considered in a later chapter.

I GABBROS, NORITES AND TROCTOLITES

After the above detailed discussion of the classification and nomenclature no difficulty should be experienced in identifying the ordinary gabbroic rocks. There are, however, certain points of detail which arise in the examination of thin sections to which we wish to direct attention.

Mineral Composition of Gabbros and Norites

Plagioclase is normally somewhat in excess of mafic minerals in most gabbros and norites, amounting to about 60 per cent of the whole. Broad albite-twin lamellae are characteristic, combined in many cases with Carlsbad and/or Pericline twinning. The crystals tend to be of platy habit, flattened parallel to the side-pinacoid (010), and consequently in layered intrusions some degree of parallel orientation is commonly observed. The plagioclases are frequently schillerized. In some instances low-power magnification reveals the cause of the phenomenon: minute rods of iron-ore in parallel orientation. In hand-specimens these inclusions may render the crystals quite dark, and in some instances cause the familiar 'play of colour'. The rods often change their direction at twin planes, and may be orientated in more than one direction, causing a fine-grained graticule. When these inclusions are exceptionally small they may cause a general cloudiness which is irresolvable except

[1] Le Bas, M. J., The term eucrite, *Geol. Mag.*, **96** (1959), 497–502.

under the highest magnifications. In certain circumstances—low-grade regional metamorphism or late-stage (deuteric) alteration—the plagioclase breaks down to an aggregate of Ca-rich minerals including typically zoisite or epidote, and calcite, set in a matrix of secondary albite.

Clinopyroxene is commonly augite (the variety diallage); but in some cases is the lilac-tinted, slightly pleochroic titanaugite. This also may be schillerized in much the same fashion as the plagioclase; but a special point to look for is the presence of regularly orientated sheet- or film-like inclusions of orthopyroxene. Similarly the **orthopyroxene,** usually hypersthene or bronzite, may show inclusions of clinopyroxene, easily identified (when the lamellae are thick enough to produce an optical reaction) by the stronger birefringence of the augite as compared with the host mineral. These are features of exsolution as described previously under 'Crystallization of Pyroxenes'.

Under conditions involving incipient metamorphism, pyroxene alters in one of two ways: it may change to amphibole or chlorite. The former type of change is often termed uralitization, though the amphibole commonly present is a light-coloured cummingtonite of fibrous habit, which develops first along the cleavages and round the periphery of the crystal grains.

Olivine is a safe mineral to identify even though only a grain or two may occur in a complete thin section.[1] It displays no special features; but it may be perfectly fresh or replaced to any degree by serpentine, iddingsite or bowlingite. Although on account of the high temperature at which it crystallizes olivine is an early silicate to separate from a basaltic magma, euhedral crystals are seldom seen in these coarse-grained rocks on account of crystal$\leftrightharpoons$liquid reactions during slow cooling.

Accessories are few in number and generally limited to iron-ore, often identified (erroneously) as magnetite. In most instances it is a combination of ilmenomagnetite and ilmenite forming complex aggregates.

Textural Range

Many gabbros and norites show the ordinary xenomorphic granular texture, typical of coarse-grained rocks in general; but some develop a very distinctive relationship between plagioclase and pyroxene, termed *ophitic texture*. When well developed, ophitic texture involves the enclosure of euhedral, disorientated plagioclase 'laths' in extensive 'plates' of augite (as shown in Fig. 114). This texture is by

[1] These may easily be overlooked unless the observer takes the precaution of examining the *whole slide* with the naked eye or a low-power lens *before* placing it on the stage of the microscope.

FIG. 114

Olivine-gabbro with ophitic texture, Kofayi, Nigeria.

Three small olivines partly altered to serpentine embedded in strongly schillerized clinopyroxene in ophitic relationship to labradorite. Iron-ore rare, in rounded patches and minute octahedra in the serpentine. (Size of section 6 × 9 mm.)

no means restricted to gabbros: if anything, it is more characteristic of dolerites and is therefore discussed more fully under that heading.

Orbicular norites and gabbros have been described from several localities including Romsaas in South Norway, the Kenora District in Ontario, San Diego County, California[1] and Corsica. Under the name 'corsite' (Zirkel) or 'napoleonite' the rock from the last-named locality is probably the best known of all orbicular rocks, though it is often referred to as orbicular diorite, on the grounds that it consists of amphibole and plagioclase. Actually it is thoroughly Basic,[2] with 46 per cent of silica; the plagioclase is bytownite (An_{75}), while the amphibole is, in part, paramorphic after pyroxene. Approximately three-quarters of the rock consists of bytownite and one-quarter of amphibole. The orbs average an inch in diameter,

FIG. 115

'Olivine-gabbro' (bojitic), Wolf Cave, Jersey. Olivine, titanaugite, 'barkevikite' and plagioclase.

Note the irregular fretted outlines of the barkevikite against the titanaugite: evidently the former is paramorphic after the latter. Actually in this field barkevikite is more abundant than pyroxene, and the rock is bojitic rather than gabbroic. (*Reproduced by courtesy of the Council of the Geologists' Association.*)

[1] Schaller, W. T., Mineralogical Notes, series 1, *Bull. U.S. Geol. Surv.*, no. 490 (1911), 58.

[2] Tröger, E., Quantitative Daten einiger magmatischer Gesteine. *Tsch. Min. Petr. Mitt.*, **46** (1934), 167.

though they may reach three inches, and a central section shows a core of normal texture, surrounded by alternate shells of plagioclase and amphibole. The matrix is of normal texture.

In olivine-gabbros the reaction relationship between the various coloured components is often particularly well displayed. Olivine is rimmed with pyroxene, amphibole envelops the latter, and in turn is surrounded by biotite (Fig. 115). The order of the successive rims is that of the minerals comprising Bowen's discontinuous reaction series. Reaction rims of a quite different nature provide one of the most fascinating textural characters of the igneous rocks. They are termed 'coronas', and take the form of narrow mantles often of singularly uniform width, sometimes single, but in other cases double, round olivine crystals embedded in plagioclase. These

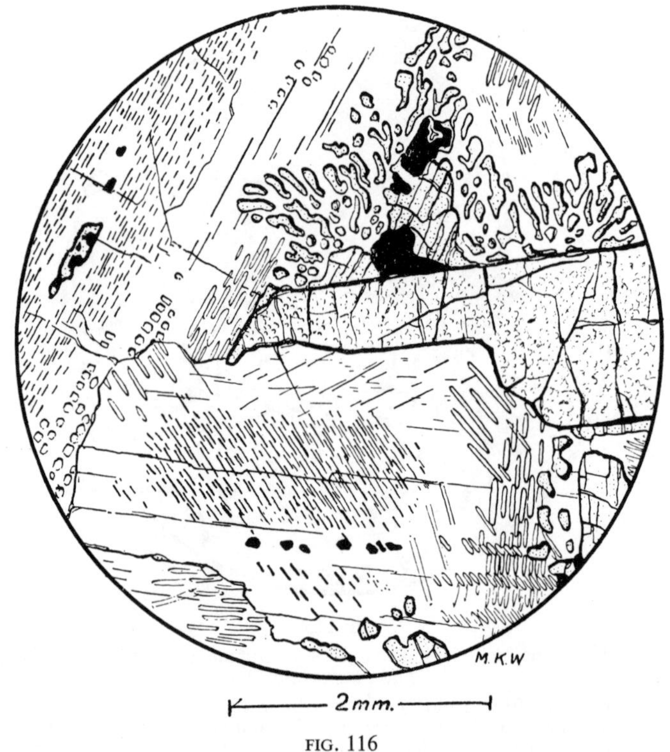

FIG. 116

'Norite', Hitteroe, Norway.

A large apatite is in contact with light green clinopyroxene associated with iron-ore. The pyroxene is fringed with a myrmekite-like intergrowth of the same pyroxene and plagioclase, lobed into schillerized labradorite. The latter contains regularly orientated short rods of iron-ore, and also colourless tubular inclusions whose orientation is controlled by the twinning (*bottom right*).

textures therefore tend to be more uniformly developed in rocks of troctolitic type, and are more fully described under that heading.

A characteristic feature of certain noritic rocks is the development of **symplectic intergrowths** along intercrystal boundaries. They consist of a myrmekite-like bulbous outgrowth of plagioclase, occasionally, though rarely, twinned in a normal manner, riddled with vermicular inclusions of orthopyroxene. In the example illustrated in Fig. 135 the vermicules have grown away from vein-like masses of titaniferous magnetite rimmed against labradorite with grains of fayalitic olivine associated at one point with poikilitic biotite. The olivine itself is locally intergrown with vermicular magnetite. There is no doubt that this phenomenon dates from a late stage in the crystallization of the magma. Similar intergrowths of the same two minerals may be seen in the well-known European norite from Hitteroe, Norway (Fig. 116).

The foregoing description of mineralogical and textural features relates principally to gabbros and to a lesser extent to norites. The

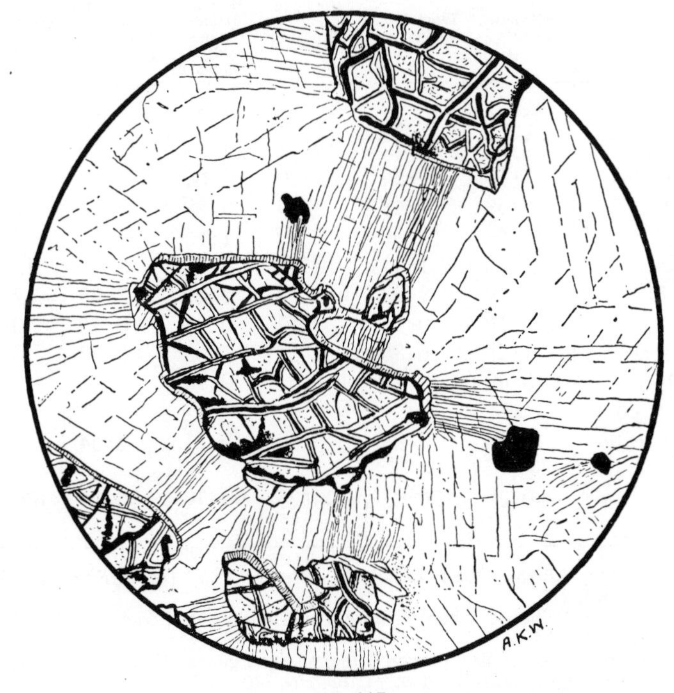

FIG. 117

Troctolite, Belhelvie, Aberdeen, Scotland. Olivines, veined by serpentine with separation of magnetite, in plagioclase traversed by systems of closely spaced fractures. Occasional primary magnetite crystals. A thin veneer of orthopyroxene surrounds the olivines which measure 2–3 mm.

latter are generally of less importance than the former. Troctolites are sufficiently distinctive from the gabbroic rocks in several respects as to merit separate description.

Troctolite.—A typical specimen of troctolite, especially if somewhat weathered, is a striking-looking rock, the grey plagioclase-aggregate being studded with black, brown or reddish olivines or pseudomorphs after olivine. This accounts for the popular name, 'troutstone', often applied to this rock. In many cases the olivines are seen to be insulated from the plagioclase by reaction rims of orthopyroxene, usually very thin, but in special types they may be thicker and double. If the olivine has been serpentinized, the expansion resulting from the change in composition causes intense shattering of the surrounding plagioclase. The fractures radiate out from the olivine nuclei (Fig. 117).

ANALYSES OF GABBROS, NORITES, ETC

	1 Allivalite Rhum	2 Troctolite, Belhelvie	3 Marginal gabbro, Skaergaard	4 Hypersthene-olivine-gabbro, Skaergaard	5 Ferrohortonolite-ferrogabbro, Skaergaard	6 Norite, Huntly
SiO_2	45·56	40·21	48·01	45·48	44·61	49·18
Al_2O_3	21·17	24·22	19·11	16·41	11·70	16·00
Fe_2O_3	1·10	0·54	1·20	2·09	2·05	0·02
FeO	5·59	4·48	8·44	9·29	22·68	8·22
MgO	11·48	9·30	7·72	11·65	1·71	9·47
CaO	11·42	11·41	10·33	10·46	8·71	12·54
Na_2O	1·99	1·92	2·34	2·06	2·95	2·04
K_2O	0·16	0·16	0·17	0·27	0·35	0·26
H_2O	1·28	6·86	0·60	1·03	0·42	0·48
TiO_2	0·40	0·10	1·51	0·94	2·43	1·24
Other constituents	0·14	1·03	0·82	0·11	2·54	0·85
	100·29	100·23	100·25	99·79	100·15	100·30

1. Allivalite, Allival, Rhum (Anal. M. Brown) *Phil. Trans. Roy. Soc.* **240** (1956), 47.
2. Troctolite, Belhelvie, Aberdeenshire (Anal. F. Stewart), *Quart. J. Geol. Soc.*, **102** (1947), 474.
3. Olivine-gabbro, chilled marginal facies, Skaergaard, E. Greenland (Anal. W. A. Deer), Skaergaard Mem., *op. cit.*, 140.
4. Olivine-hypersthene-gabbro, base of layered series, Skaergaard (Anal. W. A. Deer), *op. cit.*, p. 92.
5. Ferrohortonolite-ferrogabbro, Skaergaard (Anal. W. A. Deer), *op. cit.*, 106.
6. Norite, Huntly, Aberdeenshire (Anal. E. G. Radley), Huntly, *Mem. Geol. Surv.* (1923), 115.

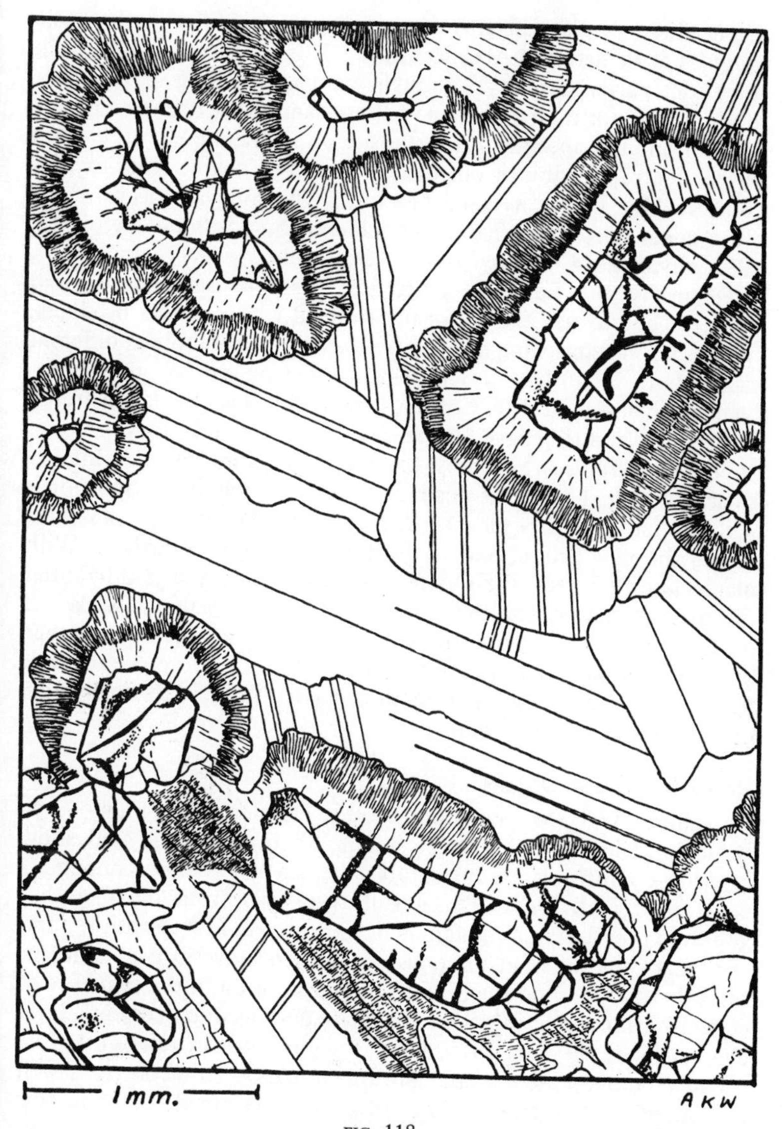

FIG. 118

Augite-cummingtonite-olivine-norite, Risor, Norway.

Corona structure is finely developed around cores of olivines. The coronas consist of orthopyroxene with a sharp boundary against an amphibole which in places is cummingtonite, but elsewhere the place of the latter is taken by a very light green symplectite, crowded with irregular vermicules. Cummingtonite occurs in association with bright green spinel around iron-ore, while the former occurs as a very thin but regular zone around schillerized clinopyroxene (lower part of the drawing.)

The finest display of troctolites in Britain is found in the basic complex at Belhelvie in Aberdeenshire.[1] This is a layered concordant intrusion, partly ultrabasic, though mainly consisting of troctolite grading into norite, hypersthene-gabbro, and even, if one considers a small enough specimen, anorthosite and olivinite. In troctolites proper the proportion of plagioclase (An_{70}) to olivine (Fo_{76}) varies considerably. In an average specimen, the analysis of which is given under no. 2 in the Table, the mineral composition is bytownite 70·5, olivine 28·4, and pyroxene 0·6 per cent.

Texturally the Belhelvie troctolites are interesting on several counts. The olivines are uniformly rimmed with reaction products, the inner zone consisting of orthopyroxene or amphibole in different cases—both colourless varieties—while the outer zone is a hornblende-spinel symplectite. Corona structure of this type is illustrated in Fig. 118, and is believed to be due to the action of liquid residua at high temperatures. Secondly, the degree of idiomorphism exhibited by the two chief minerals depends upon their relative proportions: with olivine in excess of a certain amount, this mineral separates before the plagioclase, and is therefore euhedral towards it. With plagioclase in excess, the olivine crystallizes late, and is interstitial towards the earlier, and therefore better formed, plagioclase. When present in eutectic proportions both crystallize together, neither has the advantage over the other, and neither therefore is euhedral. Finally, in places troctolites develop a pegmatitic facies, with crystals of olivine up to 2 cm in diameter.

Relationships and Origins

As indicated at the beginning of this chapter, almost any large Basic intrusion, whether it be a lopolith or thick sill, will show layering comprising a considerable degree of variation among its constituent rock types. This results primarily from the fact that gabbroic (*i.e.* 'basaltic') magmas have relatively low viscosities, and crystals of the different minerals are able to move in it at different rates depending upon their densities and position in the sequence of crystallization. Layering phenomena are described in detail in a later chapter since they are of particular significance in relation to the origin of monomineralic and ultramafic rocks. However, it is well to remember that most of the rock types described so far in this chapter are likely to have been affected to some extent by processes of crystal accumulation. In a relatively small intrusion there may be little evidence of internal differentiation and the composition of the whole rock body may be that of an average olivine-gabbro (containing about 60 per cent plagioclase, 30 per cent pyroxenes and

[1] Stewart, F. H., The gabbroic complex of Belhelvie, Aberdeenshire, *Quart. J. Geol. Soc.*, **102** (1947), 465.

10 per cent olivine), differing little perhaps from the composition of the magma from which it crystallized. But, of course, once any degree of sorting of the crystal fractions has occurred, each of the constituent rocks of an intrusion will differ in composition from that of the initial magma. It is no use, for instance, chemically analysing a troctolite from a layered complex and hoping to deduce from this something of the composition of the magma from which the olivine and plagioclase crystals were precipitated. It is interesting in this respect to read the description of the Ultrabasic complex of the island of Rhum, one of the Tertiary volcanic centres in West Scotland (p. 529). A repeated succession of peridotite, troctolite and anorthosite layers is thought to have been developed by crystal accumulation in a high-level magma chamber beneath a volcano: the chamber acted as a settling tank for the highest temperature crystals precipitated from successive pulses of magma, most of which are believed to have been erupted at the surface to form olivine-basalt flows of perhaps quite average composition.

The closest equivalent to a typical basalt among the coarse-grained rocks is a gabbro, with clinopyroxene generally appreciably more abundant than orthopyroxene. When orthopyroxenes become very abundant, therefore, it is usual to look for some special explanation. Similar considerations apply to hornblende, which under certain conditions may take the place of pyroxenes. We will deal briefly with each of these cases in turn.

The essential chemical requirement for the crystallization of orthopyroxene in excess of augite is that only a limited amount of Ca should be available for the pyroxenes compared with the amount of Mg and Fe. This can be achieved if a sufficient proportion of the total Ca in the magma can be diverted, so to speak, to enter into the composition of calcic plagioclase. As explained in connection with the calculation of a CIPW norm (p. 189), the amount of the anorthite molecule that can crystallize from a magma depends on the proportion of Al available. If the amount of Al is increased, the amount of anorthite in the plagioclase is increased and less of the total Ca is available to enter into the composition of the clinopyroxene.

H. H. Read made a special study of the problem of the origin of norites and concluded that this extra Al required to satisfy the conditions outlined above was provided by argillaceous country-rocks reacting with, and assimilated by, gabbroic magma. His special knowledge of these rocks resulted from studies of Basic (gabbroic and noritic) intrusives occurring in eastern central Scotland. Among them are certain contaminated norites which yield clear evidence of magmatic assimilation of *argillaceous* materials. They contain very striking crystals of cordierite (illustrated in Fig. 53) associated with swarms of dark green octahedral spinels and red-brown mica additional to the basic plagioclase and pleochroic hypersthene proper to norite. Cordierite and spinel of this kind are metamorphic minerals, typical of thermally metamorphosed argillaceous sediments. They represent local and temporary excesses of Al resulting from the assimilation of xenoliths of nearby rocks of this composition. Therefore for the production of these particular norites the recipe was: local gabbroic (basaltic) magma + argillaceous xenoliths = cordierite-norite. It is significant in this connection that the greatest noritic complex in the world—the Bushveld lopolith—contains in its basal 'envelope' rocks closely similar to the Scottish cordierite-norites in juxtaposition with high-grade sillimanite-hornfels representing aluminous sedimentary rocks below the base of the complex. The two occurrences (on very different scales) strengthen the validity of Read's conclusions.

Magma of basaltic composition normally crystallizes at very high temperatures, of the order of 1000° C during the formation of a typical gabbro. Most gabbros, therefore, crystallize at temperatures above the limit of stability of hornblende and well within the stability field of pyroxenes. If the temperature of crystallization is lowered and adequate H_2O is available—conditions sometimes met in the deeper levels of the crust—hornblende may take the place of some or all of the pyroxene. In the latter case a rock is produced which requires a new name.

Bojite (E. Weinschenk, 1899), like gabbro, norite and troctolite is essentially a two-mineral rock. It consists of hornblende and plagioclase (labradorite-bytownite), the latter amounting to more than 65 per cent of the total feldspar. Therefore, on this criterion alone, it falls in the gabbro field, though it is important to note that this does not make it a hornblende-gabbro. A rock properly so named would consist of the gabbroic assemblage—plagioclase ($An_{>50}$) and clinopyroxene—with *subordinate* hornblende. The proportions of the two essential minerals in bojite may vary in exactly the same way as noted above for gabbros and norites: some facies of bojitic rock-bodies are leucobojites, others, relatively rich in hornblende, are melabojites. In mineral composition bojite is close to hornblende-diorite, the two only differing in the An-content of the plagioclases. Actually, re-examination of the type bojite has shown that the plagioclase contains An_{40}, so the rock is really a diorite. That does not alter the fact that bojite, as defined above, is an authentic Basic coarse-grained igneous rock deserving of recognition in its own right; but there is a distinct possibility that the hornblende may be paramorphic after original clinopyroxene. Nevertheless, the rock must be judged by its present mineral composition, not what it may have been during a previous incarnation (cf. Fig. 115).

Bojites appear to be far less common than gabbro, norite and troctolite; but they are known to occur among the Cadomian intrusions so finely exposed round the coasts of Guernsey and Jersey. One of these is the layered 'gabbroic' complex near St Peter Port, Guernsey.[1]

II MICROGABBROS (DOLERITES)

The representatives of the gabbroic rocks which fall within the limits of the medium grain-size group should logically be termed microgabbros, as they are exactly comparable with microgranite, microsyenite and microdiorite. Unfortunately the logic of this was not realized sufficiently early, and two other names are in vogue at the present time. In Britain many petrologists use the name **dolerite** instead of microgabbro, though with different shades of meaning. To some the name implies *all* medium grained rocks of gabbroic composition, regardless of age, texture and mode of occurrence. Others restrict it to rocks of the appropriate composition exhibiting ophitic texture only; while yet others use it to cover such rocks provided they are of hypabyssal mode of occurrence. In our view, if 'dolerite' is to be used instead of 'microgabbro', the name should cover *all* rocks of the appropriate composition irrespective of textural features, which can in any case be adequately covered by qualifiers. In America 'diabase' replaces 'dolerite'; but in this country, in spite of a veto by the Committee on Petrographic Nomenclature, some petrologists use the former term in the same sense as Rosenbusch did, for pre-Tertiary dolerites. That is, a **diabase** in this country is a rock of doleritic composition, which is altered to such an extent that few, if any, of the original minerals have survived. In view of these inconsistencies, there is a strong case for the general adoption of 'microgabbro'.

Mineral Composition

So far as mineral contents are concerned, microgabbros (dolerites) closely resemble gabbros, norites, etc., and there is consequently no need to describe them in detail. The central type consists of plagioclase, near to labradorite in composition, clinopyroxene (usually common augite, though titanaugite often takes its place), and iron-ore, which in different specimens may be magnetite, titano-magnetite, or ilmenite. The addition of olivine gives olivine-micro-gabbro (olivine-dolerite), while the incoming of quartz gives quartz-microgabbro (quartz-dolerite). The latter rocks are, of course, over-saturated, and in them orthopyroxene may replace, wholly or in part, the augite of typical microgabbro (dolerite). It is customary

[1] Roach, R. A., The layered structure of the St Peter Port gabbro, Guernsey, Channel Isles, *J. Geol. Soc.*, **127** (1971), 295.

to lump together all dolerites containing hypersthene as 'hypersthene-dolerites', regardless of the amount present. By comparison with their coarse-grained equivalents 'hypersthene-dolerite' should indicate a rock of doleritic composition containing *accessory* hypersthene; but if the latter is of *essential* status, micronorite is the correct term to use. According to the degree of silica-saturation we may recognize quartz-micronorite, micronorite and olivine-micronorite. Ortho- and clinopyroxenes may be present in all proportions, and as for the coarse-grained equivalents, the division is most conveniently drawn quite arbitrarily, at 50 per cent. If therefore orthopyroxene is dominant, the rock is micronorite; but if clinopyroxene predominates, it is microgabbro. The qualifiers 'mela-' and 'leuco-' are used to signify respectively richness or deficiency in coloured minerals.

The occurrence of two or more pyroxenes in these rocks is a common feature. The clinopyroxene is frequently strongly schillerized parallel to (001) and when the crystal is in addition twinned on (100) it exhibits a distinctive herring-bone structure. The orthopyroxene is commonly hypersthene, appreciably, though faintly pleochroic in some sections, devoid of twinning unless it has inverted from pigeonite, and as a rule easily distinguished by its optical characters. Some crystals contain irregular cores of normal hypersthene which are surrounded by an irregular mantle of 'hypersthene-perthite'. This consists of hypersthene riddled with plates or irregular vermicules of clinopyroxene thrown out as a consequence of ex-solution. All three types of pyroxene may be studied in sections of the Palisades sill, New York, which may be described as a quartz-hypersthene-microgabbro. The quartz in rocks of this type is particularly distinctive: it is intergrown graphically with orthoclase and occurs in angular interspaces between the lath-shaped labradorite crystals. In Britain this over-saturated type of micronorite is well represented by the famous Whin sill of northern England, and by the late-Carboniferous 'quartz-dolerite' dykes in the Midland Valley of Scotland.

Usually hornblende does not occur in microgabbros; though a brown amphibole in different cases identified as barkevikite, lamprobolite or kaersutite, does occasionally occur, the last-named in **minverite**, described from the parish of St Minver in Cornwall.

Textural Variation

As regards texture, microgabbros and micronorites are very variable. Without question the most characteristic texture is the **ophitic,** and some petrologists will not apply the name dolerite unless the rock is ophitic.

This texture, developed in a rock of the same composition but of coarse texture, is illustrated in Fig. 114. It will be noted that the plagioclases are euhedral, they are randomly arranged, and thus contrast with the clinopyroxenes which are relatively large and of irregular shape. Even in rocks of medium to fine grain individual pyroxenes may reach a diameter of as much as one inch and are therefore conspicuous in hand-specimens largely owing to the reflection of light from the cleavage surfaces (lustre-mottling).

If the plagioclase crystals penetrate into, but are not enclosed in, the pyroxenes, the texture is described as *subophitic*. Feldspars enclosed in the pyroxenes are frequently smaller than those outside, indicating that the periods of crystallization of the two minerals overlapped. This is probably the general case. Simultaneous crystallization produces this particular texture because of the inherent tendency for plagioclase to nucleate more readily than pyroxene and thus to set up many centres of crystallization, as compared with few, much more widely spaced centres of pyroxene-crystallization, the difference being of the order of twenty to one, as suggested by Hess, and indicated in Fig. 114. Less commonly the euhedral plagioclases may show a parallel orientation which must have been impressed at an early stage of crystallization as a result of flow movements in the melt. The orientation and size of the plagioclase crystals tend to be uniform both inside and outside the pyroxenes, suggesting that, in such cases, the latter developed at a somewhat later stage in the crystallization sequence.

The mineral composition affects the extent to which the ophitic texture is developed: whether plagioclase precedes or follows pyroxene is determined by the relative concentration of these two components in the magma. Plagioclase and pyroxene in a magma display a cotectic relationship as explained on p. 166, therefore whichever is in excess of the cotectic proportions will begin to crystallize first. In olivine-rich dolerites the crystallization of pyroxenes may be delayed relative to plagioclase with the result that the pyroxene is partly post-feldspar.[1]

Ophitic texture is not restricted to dolerites: some gabbros display this texture to perfection (Fig. 114), while in the fine-grain category certain basalts are 'micro-ophitic'.

Similarly, not all dolerites are ophitic. The term *intergranular* is used when the pyroxene forms grains interstitial to the plagioclase. This is frequently the case with dolerites of finer grain which grade into basalts and, like the latter, may be porphyritic. The term 'labradorite-porphyry' is a legacy from the past and is best forgotten: 'porphyritic dolerite' (or microgabbro) conveys all the necessary information in such cases.

[1] Walker, W. F., Ophitic texture and basaltic crystallization, *J. Geol.* **65** (1957), 1–14.

Although some microgabbros (dolerites) of all ages are ideally fresh, many others are highly altered as a consequence of their subsequent treatment. Participation in earth movements, a mild degree of metamorphism and, naturally, weathering may lead to the replacement of any or all the original components, in the manner described above for the gabbroic rocks. Saussuritization, albitization, chloritization and epidotization may all contribute to the conversion of the original pyroxene, labradorite, iron-ore, etc., into albite, chlorite, epidote, calcite, leucoxene and quartz; but some of these minerals, though of late formation, are not secondary in the sense of having replaced pre-existing minerals. Thus although chlorite is widespread as an alteration product of pyroxene (with epidote, calcite, etc., as by-products), it occurs also in sharply defined interstitial areas between plagioclase and clinopyroxene which show no trace of alteration, and in these cases is primary. Rather special interest attaches to the quartz in these rocks. As noted above, it is a widespread primary constituent, normally intergrown with alkali-feldspar; but in addition it occurs also in irregular grains closely associated with other obviously secondary minerals, and is evidently an alteration product. Thirdly, quartz may be xenocrystic (see below, under 'basalt'). Clearly it would be a mistake to refer to such rocks, with secondary or xenocrystic quartz, as quartz-microgabbros, or quartz-dolerites. Such terms should apply only to rocks containing the *primary* mineral.

Although amphiboles are rare primary constituents of the rocks under consideration, they are widespread as alteration products. Thus late-stage alteration tends to convert the original pyroxene into actinolite, an early stage showing perhaps merely a fringe of acicular crystals, but the alteration is progressive until all trace of pyroxene is lost, and a pseudomorph of closely packed fibres of pale-green amphibole ('uralite') is produced. In the process of dynamothernal metamorphism a compact common hornblende is produced from the pyroxene in the conversion of microgabbro into hornblende-schist or amphibolite. It follows that as a result of these changes the original labradorite-pyroxene combination gives place to common hornblende, a less basic plagioclase and various minor constituents to strike a balance. Thus the rock, although chemically gabbroic, in mineral contents is, in a broad sense, dioritic. To such rocks the term **epidiorite**[1] has been applied. Typical specimens occur among the sills in the South-West Highlands of Scotland.

It would be unprofitable at this stage to attempt a summary

[1] 'Epidiorite' (von Gumbel, 1874) is sometimes used in a wider sense, for example by Wiseman, J. D. H., who applies the term to all non-schistose rocks produced by dynamothermal metamorphism from basic igneous rocks, whether intrusive or extrusive. See The Central and South-West Highland Epidiorites: A study in Progressive Metamorphism, *Quart. J. Geol. Soc.*, **90** (1934), 354.

statement concerning the distribution of dolerites. Just as basalts are predominant in the extrusive, fine-grain category of igneous rocks, so dolerites are overwhelmingly dominant in the medium-grained rocks which normally are hypabyssal in mode of occurrence. This is emphasized by the immensely abundant dyke swarms and sill complexes which riddle the crust over vast areas.

III BASALTS

The term 'basalt' is one of the few rock-names familiar to the man in the street, and is one of the oldest in petrology. It is applied collectively to the fine-grained equivalents of gabbros and norites. Basalts are the most widely distributed of all volcanic rocks and occur not only as lava-flows but also as cone-sheets, dykes and other minor intrusions. The general position of basalts in relation to other kinds of volcanic rocks has already been indicated; but one important aspect of classification has to be clarified at this point. On the basis of the kind of feldspar they contain, with plagioclase predominant over alkali-feldspar, $P > 65$ and $A < 35$, some petrologists place basalts in the QAP triangle (p. 195) in the field already occupied by andesites. As explained in the general discussion of classification we consider that the ratio of alkali-feldspar to plagioclase is a minor consideration for rocks of high colour index—for basalts the range is commonly 40 to 50. Further, relatively few basalts contain normative quartz, so the majority have compositions which plot on the base of the QAP triangle, or outside it because they contain normative olivine and in some cases nepheline. As already explained (p. 198), these objections are overcome by promoting the An-content of the plagioclase to the status of a primary factor in classification and recognizing An_{50} as marking the boundary between andesites which lie within the QAP triangle and basalts which are excluded from it.

Mineral Composition

For the purposes of definition basalts may be regarded as composed essentially of plagioclase (usually within the range labradorite-bytownite) and pyroxene.[1] In addition, opaque ore-minerals—magnetite and ilmenite—are always present in considerable abundance, so that although they are traditionally referred to as 'accessory minerals' they really rank as essential components in most basalts.

[1] Brown, G. M., Mineralogy of basaltic rocks in *Basalts. The Poldervaart Treatise on rocks of Basaltic Composition*, ed. H. H. Hess and A. Poldervaart, vol. 1 (1967), 103–62. The two volumes of this work provide a valuable source of information on all aspects of the petrology and petrogenesis of basalts.

Special mention needs to be made of the rôle of olivine in basalts. Olivine is present in most, but not all, basalts, often in considerable abundance. Since olivine is virtually absent from andesites, its common occurrence in basalts may be regarded as one of the important criteria by which the two groups of volcanic rocks can be distinguished. The relative abundance of olivine in basalts stems from two causes: firstly and most important, basaltic magma is commonly under-saturated with silica so that precipitation of a certain amount of olivine is inevitable; but secondly, due to the relationship of olivine to MgFe-pyroxenes resulting from incongruent melting (see p. 43), some olivine is liable to crystallize at high temperatures even from somewhat oversaturated magmas, and cooling is generally too rapid for this excess olivine to be 'made over' into pyroxene at lower temperatures. In other words, the disequilibrium conditions obtaining during cooling of basaltic magma lead the rocks, in many cases, to contain a larger amount of olivine than that to which they are chemically entitled. This anomalous behaviour of olivine often tends to confuse those encountering the problems of basalt mineralogy and nomenclature for the first time; but it is a matter to which we shall return particularly in connection with the chemical classification of basalts and the recognition of basaltic magma-types. Meanwhile it should be noted that a basalt of any kind which contains an appreciable amount of olivine is termed olivine-basalt.

As noted above, the plagioclase in basalts typically lies within the labradorite-bytownite range. In many basalts it occurs in two generations: as phenocrysts, often of relatively large size and commonly zoned; and as microlites of somewhat more sodic composition in the groundmass.

Most basalts contain two kinds of pyroxene, the one kind being relatively calcic, and the other calcium-poor: the former is augite, and the latter may be pigeonite or an orthopyroxene depending upon the temperature of crystallization (see p. 41). Augite frequently occurs as phenocrysts which may show zoning and hour-glass structure; while second-generation crystals of minute size are commonly abundant in the groundmass. Pigeonite and orthopyroxenes are generally less conspicuous since they occur as phenocrysts only in basalts lying within quite a restricted range of composition. Pigeonite quite commonly occurs in the groundmass, though it is then extremely difficult to distinguish from augite.

One other factor tends to suppress pigeonite and orthopyroxenes: due to rapid chilling, olivine stands a better chance of survival than in a slowly cooled rock, in which the olivine may be converted by reaction into orthopyroxene. The survival of olivine in basalts must obviously affect the composition of the clinopyroxene. If magnesium silicate is locked up in the olivine, it is not available for the formation of pigeonite: in olivine-rich basalts, therefore, common augite tends to fill the role of pigeonite. Other things being equal, however, in an olivine-free basalt, the pyroxenes must be richer in magnesium, and pigeonite or orthopyroxenes become more important.

Some varieties of olivine-basalt, described in more detail below, contain a single variety of clinopyroxene which is generally titaniferous, grading into titanaugite, and may be rimmed with an outer zone of greenish aegirine-augite, indicating some degree of alkali-enrichment in the last-crystallizing phases, or, it may be, in residual glass.

Hornblende is rare in basaltic rocks, but biotite in small quantities is not uncommon. Among the minor constituents iron ore is conspicuous, and is usually titano-magnetite in small octahedra; but dendritic iron ore often separates from the glassy base present in some basalts during the final stages of consolidation (Fig. 43C). Apatite is plentiful, though usually the crystals are minutely acicular. Secondary minerals are very varied. Olivine may show all stages of alteration to serpentine, talc, iddingsite, chlorophaeite, limonite or rhombohedral carbonate; pyroxenes are replaced progressively by chlorite with or without calcite and epidote; while the plagioclases undergo decomposition as described above for the gabbroic and microgabbroic rocks. Further, on account of the conditions under which they are erupted, basalts (even some dyke-basalts), tend to be vesicular, and although in recent specimens the vesicles are gas-filled, in the course of time they become filled with such minerals as chalcedony, agate, chlorite, calcite and especially zeolites including natrolite, phillipsite, heulandite and analcite (Fig. 62).

Occasionally isolated and much-corroded quartz grains occur in basalts which otherwise appear quite normal. These crystals are xenocrysts, caught up during the uprise of the magma. They are often surrounded by a reaction rim of sorts, consisting usually of closely packed granules of pyroxene. It is incorrect to call such rocks quartz-basalts: they should be termed **quartz-xenocryst-basalts.** Examples occur among the Permian lavas in Ayrshire, and in England among the Exeter lavas of the same age,[1] but such xenocrysts are liable to occur in any basalt, of any age.

Xenoliths (incorporated *rock-fragments* as distinct from crystal grains) are not uncommon and are of significance in that they provide valuable evidence of the nature of the rocks through which the magma passed in transit from its place of origin. Of special significance in this connection are the rock-fragments mistakenly recorded in the past as 'olivine-nodules', but which are proved by

[1] Cf. Tidmarsh, W. G., The Permian lavas of Devon, *Quart. J. Geol. Soc.*, **88** (1932), 741.

examination of thin sections to be magmatically corroded pieces of the coarse-grained, ultrabasic rock, peridotite, consisting of the four minerals olivine, light greenish enstatite (practically indistinguishable from the olivine in hand-specimens), vivid green chrome-diopside and the accessory, chromite. The special significance of these peridotitic xenoliths is discussed below when dealing with magma types.

Textural Range

In texture basalts are very variable: every gradation is represented between the vitreous—basalt-glass or **tachylyte**—and the holocrystalline. A large number of basalts contain glass forming minute angular patches between the crystals of the groundmass. Considerable interest attaches to the composition of this interstitial glass, and it is a feature that repays careful study with high magnification. Glass which may appear black and 'dusty' may then be seen to contain innumerable minute octahedra of titano-magnetite, sometimes in parallel growth or arranged in a dendritic pattern. Equally minute granules of pyroxene are also found in some cases. These minerals indicate the strongly ferruginous character of some basalt glasses. Generally speaking iron-rich glass of this type constitutes only a very small proportion of the total rock, possibly because Fe tends to decrease the viscosity of silicate melts, and thus an abundance of Fe reduces the chance of a residual melt solidifying as glass. If there is an abundance of glass—and particularly if it has a low refractive index—there is a strong likelihood that the glass will be siliceous rather than ferruginous, because SiO_2 increases the viscosity of melts as witnessed by the glassy nature of so many Acid volcanics. Naturally a siliceous and generally alkali-rich residuum is specially characteristic of the most siliceous basalts, the over-saturated tholeiitic types. The character of the glassy mesostasis is of great significance in relation to differentiation of basaltic magmas and is discussed under that heading below.

Apart from forming these interstitial patches in otherwise crystalline basalts, tachylyte also forms the marginal parts of thin dykes and sills, and in the case of very thin sheets injected into cold rock, the whole may consist of tachylyte. Lava-flows of basalt glass are rare, though not unknown, for example on Hawaii. The extraordinary material known as **Pelée's Hair** also comes from Hawaii. It consists of hair-like fibres of basalt glass, of an attractive golden brown colour, with occasional black swellings enclosing minute olivine crystals. It represents basaltic magma erupted as lava spray.

Variolitic texture is limited to quenched basalts and is equivalent to spherulitic texture in the rhyolites. The essential feature is the occurrence of delicate brush- or fan-like 'sprays' of radially disposed

FIG. 119

Variolitic basalt, near Rhobell Fawr, Merionethshire. Skeletal microphenocrysts of plagioclase set in variolitic groundmass of feldspar and augite. Small vesicles are present in the lower part of the field.

fibres of feldspar or less commonly of pyroxene (Fig. 119). Variolitic texture is especially characteristic of the crusts of pillow-lava. In some cases the varioles tend to stand out as small knobs on the weathered surface of flows of this kind.

Many basalts are porphyritic, with phenocrysts of any or all of the constituent minerals, generally identifiable in hand-specimens. They vary considerably in size and relative abundance, and different named types of basalt are distinguished by their phenocrysts, in local successions, for example the Scottish Carboniferous basalts (see p. 508).

Classification of Basalts

As the result of differentiation, differences in composition which may be very small in a parent magma, can give rise to major differences in the volcanic derivatives. Therefore, for basalts more than any other group of rocks, attempts at classification have been dominated by petrogenetic considerations, and the results, based largely on chemical characteristics, are in many respects quite distinct from those of a standard classification based on mineral proportions.

The simplest and most obvious mineralogical criterion for subdividing basalts is the presence or absence of olivine. This depends on the degree of silica-saturation in relation to the available Mg and Fe in the magma. Two main categories may be distinguished:

(1) over-saturated basalts, and (2) under-saturated basalts containing essential olivine.

The over-saturated basalts should theoretically contain no olivine, for there is sufficient silica to convert all of it into orthopyroxene. Quenching may prevent the reaction, however, and some olivine may survive. In this case surplus silica will be locked up in a residuum which, in fresh specimens, is glass. Separation and analysis of the latter have shown it to be of granitic composition, containing up to 70 per cent of silica. These basalts are therefore the fine-grained equivalents of quartz-dolerites which, as we have seen, are characterized by the occurrence of interstitial mesostasis, consisting of alkali-feldspar and quartz, usually intergrown graphically. Because of the non-occurrence of olivine, the pyroxenes in over-saturated basalts are rich in MgFe—either pigeonite, or orthopyroxene or both, occurring as phenocrysts and small greenish prisms in the groundmass.

The name **tholeiite** (Steininger, 1840), was given originally to a basalt containing an abundant glassy mesostasis. Although the type-rock is over-saturated with silica it actually contains a little modal olivine. In the period 1920–30 petrologists working on the volcanic rocks of the North of England and Scotland extended the use of the term to include all basalts (and their doleritic equivalents) that conformed broadly with the description of over-saturated basalts given above. The name was given special prominence by Kennedy who adopted it for one of his two basaltic magma types (see p. 370) and emphasized that tholeiites, by reason of their generally over-saturated character and tendency to develop a residuum enriched in silica, are particularly characteristic of continental regions of flood basalts such as Snake River, Oregon and the Deccan in India. In more recent years emphasis has changed in the definition of tholeiite and the term now covers a whole series of basaltic rocks ranging from siliceous types to those containing abundant olivine, but *all* containing significant amounts of Ca-poor pyroxenes, principally pigeonite or hypersthene. These are the pyroxenes which may be said to have a reaction relationship with olivine.[1] The petrochemical considerations which have led to

[1] Tilley, C. E., and I. D. Muir, Tholeiite and tholeiitic series, *Geol. Mag.*, **104** (1967), 337–43.

this change of emphasis are examined below. At the moment we are concerned only with the fact that the tholeiitic series as at present defined, cuts right across the boundary based on the presence or absence of olivine. The distinction between basalts containing no olivine and those containing *essential* olivine must still be made and is still valid; but in addition it is necessary to decide whether a given olivine-basalt belongs to the tholeiitic suite, or to the Alkali Suite. In the former case the rock is termed 'olivine-tholeiite', but the latter is widely known as 'alkali olivine-basalt'.

Alkali olivine-basalts are distinctive among olivine-basalts in general by reason of the presence of sufficient alkalies, especially Na, in relation to the Al and Si available to ensure the appearance of some nepheline in the norm. The rock (and the magma it represents) is therefore undersaturated in two senses, both as regards the ferromagnesian components (shown by the presence of olivine) and the alumino-silicates (shown by nepheline in the norm). In an

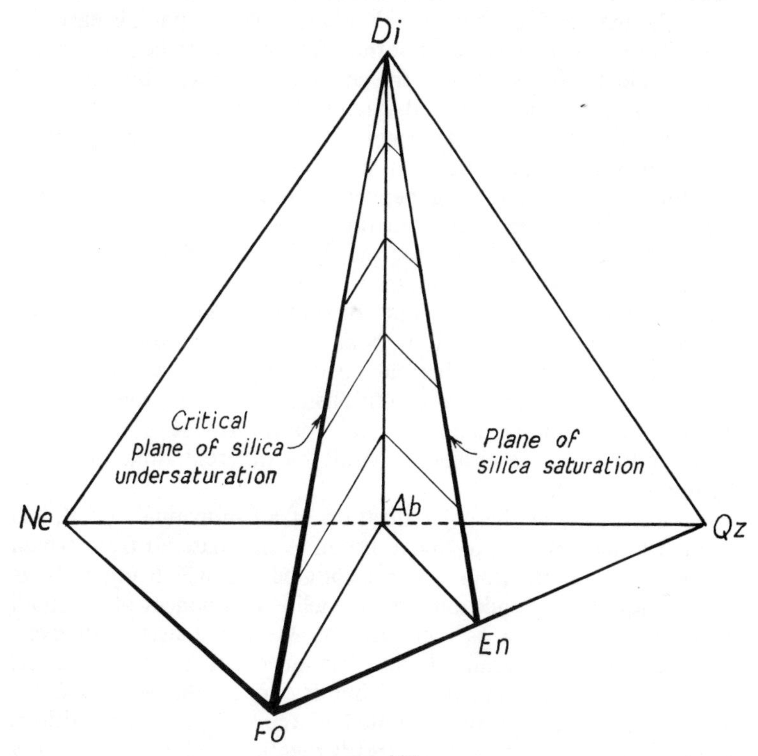

FIG. 120
Compositional boundaries within the system Di–Fo–Ne–Qz. For explanation, see text. (*After Yoder and Tilley* (1962).)

outstanding analysis of the problems involved, Yoder and Tilley[1] have shown the relationship very clearly by use of a tetrahedral diagram (Fig. 120) in which the basalt compositions are plotted in terms of selected normative mineral components. The latter are simplified equivalents of the actual minerals present in basalts, thus revealing the reaction relationships more clearly: Fo + Qz = En, and Qz + Ne = Ab. Diopside substitutes for a clinopyroxene such as augite which, it will be noted, is present in all basalts.

The 'critical plane of silica saturation' is effectively the boundary between alkali olivine-basalts on the left and tholeiitic basalts on the right, and its importance lies in the fact that the nepheline component of the one group is incompatible with the enstatite of the other group.

Experiments carried out on various synthetic combinations of basaltic minerals show that under relatively low-pressure conditions the liquidus surfaces of the components on the nepheline side of the clinopyroxene-olivine-plagioclase (Di-Fo-Ab) plane are separated by a 'thermal divide' from the liquidus surfaces on the enstatite side. The details are complex, but the effect of this can be appreciated in principle by following up the analogy suggested by the word 'divide'. When a basalt liquid having a composition just on the nepheline side of the divide begins to crystallize, the composition of the remaining liquid moves in a direction away from the divide into the fields of increasing nepheline. On the other side of the divide movement of the liquid fractions is, albeit tortuously, towards quartz. It is difficult to overstate the significance of this in terms of differentiation: alkali olivine-basalt magmas can differentiate towards alkali-enrichment; but tholeiitic magmas differentiate towards silica-enrichment. It is believed, however, to be difficult if not impossible for the thermal divide to be crossed during differentiation under pressure conditions obtaining in the crust. A parent magma on the nepheline side of the divide, for example, cannot give rise to a tholeiitic basalt by differentiation involving the removal of olivine.

In the context of the crystallization of an individual basalt, the residual liquid referred to above becomes the material from which the later crystal fractions are precipitated, or which congeals as glass. Under these conditions the nepheline component of an alkali olivine-basalt is often as inconspicuous as the quartz is in over-saturated tholeiitic basalts. It may remain occult in the liquid, or, if the amount of normative nepheline is very small, say 1 or 2 per cent, all of it may be incorporated in the complex compositions of the clinopyroxene, which, as already mentioned above, is generally

[1] Yoder, H. S., and Tilley, C. E., Origin of basaltic magmas: an experimental study of natural and synthetic rock systems. *J. Petrol.*, **3** (1962), 342–532.

titaniferous and also contains significant amounts of Na and Al. Since the great majority of alkali olivine-basalts contain only small amounts of normative nepheline, the most satisfactory guide to the identification of these rocks (in the absence of chemical analyses) is often provided by the nature of the pyroxenes. Because of the incompatibility between nepheline and enstatite, it is generally impossible for any non-calcic pyroxene to crystallize, and normally only a single main phase of calcic pyroxene accompanies the olivine. To the right of the plane of silica saturation lie the over-saturated basalts comprising the main part of the (continental) tholeiites as originally defined. In the centre field occur the olivine-basalts covered by the extended definition of tholeiites as explained above. They are particularly abundant in oceanic volcanoes. We now know that the plane of silica-saturation, loosely regarded as the boundary between 'tholeiites' and 'olivine-basalts' of Kennedy, is petrogenetically not nearly so significant as the critical plane of undersaturation. It should be realized, however, that this latter boundary is often not easily or clearly defined because of the gradation that exists among olivine-basalts. However, an alkali olivine-basalt contains within itself the potentiality of alkali differentiation; and this fact, together with the need to use some distinguishing term is the only justification for using the prefix 'alkali'.

As the proportion of feldspathoid components increases the alkali olivine-basalts pass firstly into feldspathoidal basalts, and ultimately into very Basic and also strongly alkaline rocks which do not fall within the general definition of basalts and are considered separately in Chapter 11.

The essential minerals in the basaltic assemblage—plagioclase, pyroxene and olivine may occur as phenocrysts individually or in pairs, or all three may occur together. Evidence strongly suggests that all the main mineral phases were precipitated more or less simultaneously over a limited range of temperatures (100 to 200° C). This suggests, as emphasized by Yoder and Tilley (1962), that basaltic magmas normally have compositions corresponding broadly to mixtures of the several components which have minimal melting temperatures. There are fields of relatively low-melting components in basalts just as we have seen to be the case with certain mixtures of quartz and alkali-feldspar in granites; but the comparison is valid only in part. Basaltic minerals are all high-temperature members of their respective reaction series; and temperatures of basaltic lavas in eruption are correspondingly high—commonly between 1100 and 1200° C—compared with those of granitic magmas in 'petrogeny's residua system'.

The basalt systems are too complex to be treated adequately here; but the principle involved can be understood by reference to the system diopside-anorthite, described on p. 166. The initial crystallization of whichever of the components of the melt is in excess of the eutectic compositions illustrates the way in which phenocrysts of

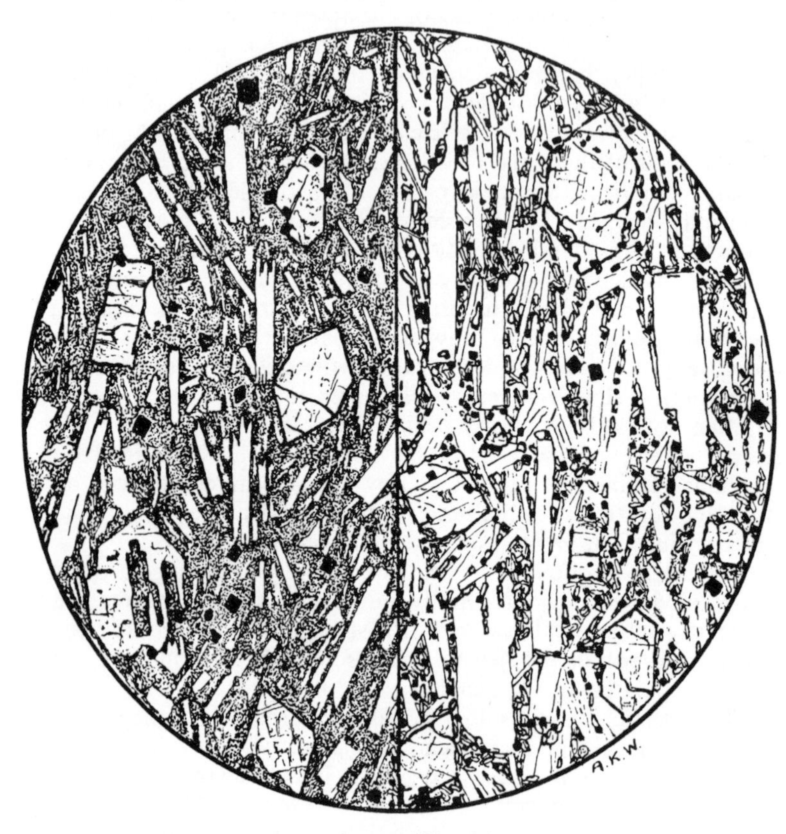

FIG. 121

Olvine-basalts from the Auvergne, France.

Left: Small phenocrysts of fresh olivine and plagioclase, many of the latter having forked ends. The groundmass is partially devitrified glass.

Right: Phenocrysts of olivine and plagioclase embedded in holocrystalline groundmass of lath-shaped plagioclase microlites, granules of augite and minute octahedra of magnetite. The texture is porphyritic, intergranular.

just one mineral may be formed ahead of the others in a basalt. The presence of quite a high proportion of one constituent, say olivine, may be accounted for in this way; but if the proportion is very high, it is safe to infer that crystals of olivine have been added to, and concentrated in, the magma. This is the case, for instance with the so-called **picrite-basalts** or **oceanites** which are very melanocratic rocks, typically belonging to the tholeiite series and in which up to about 50 per cent of the total volume may consist of euhedral olivine crystals, packed so closely that they may be touching one another like detrital grains in a sediment.[1] These rocks are therefore olivine-enriched melabasalts. The name picrite-basalts which has been applied to them does not mean that, in composition they are half way between picrite and basalt, but that they are of basaltic parentage, though enriched in mafic constituents and correspondingly depleted in plagioclase, so that in bulk composition they correspond with the Basic coarse-grained ultramafite, picrite. The synonym 'oceanite' is not a great improvement on 'picrite-basalt' because basaltic derivatives of this kind are not restricted to oceanic islands as originally thought and as the name implies: they occur also in continental settings, for example, in the flood-basalt region of the Deccan in India.

Ankaramites also are melabasalts, but enriched in pyroxene crystals. The character of the pyroxenes and of the groundmass shows that ankaramites have affinity with alkali olivine-basalts rather than tholeiites. Both oceanites and ankaramites are relatively rare rock types.

Finally we have to consider the possibility of plagioclase enrichment. Porphyritic basalts containing an abundance of plagioclase phenocrysts (hence 'plagiophyric') are common in the vicinity of the central intrusive complexes in Mull and elsewhere in the Scottish Tertiary volcanic province, described by the authors of the Mull Memoir as belonging to the 'Central Porphyritic Basalt' magma-type (p. 369). There is little doubt, however, that these rocks resulted from the addition of plagioclase phenocrysts to basalt magma of more average composition. Moreover, in other parts of the world, notably in island arcs like Japan and often associated with calcic andesites, *aphyric* basalts occur which have compositions very close to those of the plagiophyric basalts of Mull. These rocks have been named **'High-alumina Basalts'** by Kuno,[2] who regarded them as products of an independent magma-type comparable in status with the alkali-olivine-basalt and tholeiite types. This view has been questioned on two main grounds: firstly, concerning the primary status of a kind of magma which may have been modified by the addition and resorption of crystals of plagioclase; and secondly, because of inadequate chemical and mineralogical distinctiveness of the high-alumina basalts. Although these rocks are richer in Al_2O_3 and plagioclase than other kinds of basalt with comparable silica percentage, in other respects they appear to be transitional,

[1] For simplicity it is assumed that only one mineral is added, but actually subordinate amounts of other minerals would also normally be involved.

[2] Kuno, H., High-alumina basalts, *J. Petrol.*, **1** (1960), 121–45.

CHAPTER VIII

DISTRIBUTION AND ORIGIN OF BASALTS

BASALTS hold a dominant place in igneous petrology not only from the point of view of their abundance, but also their significance in relation to the origin of other types of igneous rocks. Turner and Verhoogen (1960, p. 165) summarize the situation by saying that 'all volcanic associations have this in common: one of the essential members—and usually the dominant member—is basalt'. It follows that, in the course of petrological research a vast amount of data should have become available concerning basalts, and that ideas regarding their significance and origin should have undergone quite radical changes. In fact, no other aspect of igneous petrology has had to be so completely revised during the writing of successive editions of this book. The extent and pace of development may well confuse a newcomer to the subject. For this reason we think that it will prove useful to recapitulate some of the more important concepts, particularly those relating to the significance and origin of basaltic magmas and magma-types, before discussing the present state of knowledge.

Many eminent geologists, among them Daly[1] and Bowen[2] have deduced that virtually the whole range of igneous rock-types could be derived, either by differentiation or syntexis, from one kind of magma, essentially of basaltic composition. Subsequently this concept of the parental status of basaltic magmas has been elaborated, mainly towards the recognition of a number of chemically distinctive basaltic **magma-types.**

The first use of this term was made by the authors of the Mull Memoir, who distinguished three basaltic magma-types which they regarded as parental to the lavas erupted in Tertiary times from the Mull volcano.[3]

The three types were named: (1) Plateau-Basalt Magma Type; (2) Central Non-Porphyritic Magma Type; (3) Central Porphyritic Magma Type. The choice of names was unfortunate in several

[1] *Igneous Rocks and the Depths of the Earth* (1933), pp. 189, 304, 395.
[2] *The Evolution of the Igneous Rocks* (1928), pp. 5, 21, 320.
[3] Bailey, E. B., and others, The Tertiary and Post-Tertiary geology of Mull, *Mem. Geol. Surv. Scotland* (1924).

respects,[1] notably so in the case of the 'Plateau Basalt' type. Although it is convenient to apply the term plateau basalt collectively to basalts which build the great lava plateaux of the world, it can strictly be used only in a topographic, not petrographic, sense: for this reason 'flood basalt' is a better general term. So far as is known *any* variety of basalt may occur as plateau basalt, although probably on average over-saturated tholeiitic basalts predominate (*e.g.* in

	Olivine-basalt magma-type	Tholeiitic basalt magma-type
SiO_2	45	50
Al_2O_3	15	13
$FeO + Fe_2O_3$	13	13
MgO	8	5
CaO	9	10
Na_2O	2·5	2·8
K_2O	0·5	1·2

Oregon). It can only lead to confusion, therefore, to use the term plateau basalt petrographically. In the case of Mull the anomaly is increased by the fact that the plateau basalts there are predominantly under-saturated olivine-basalts and therefore atypical.

The possibility of linking the various dominant suites of associated lavas with parental magmas from which they were derived was explored by Kennedy who concluded from the facts of distribution that two such magma-types were involved.[2] These were identified with two of the Mull magma-types as follows: Olivine-Basalt Magma type (equivalent to the so-called Plateau type of Mull), and Tholeiitic Magma type (equivalent to the Central Non-Porphyritic type).

The essential chemical characteristics of the two basaltic magma-types recognized by Kennedy are shown in the accompanying Table.

As explained in the previous chapter, basalts belonging to the tholeiitic magma-types were regarded as being dominantly over-saturated with silica. Tholeiites of this character, often containing residual glass of 'granitic' composition[3] or segregations of quartzo-

[1] See Wells, M. K., and Wells, A. K., Magma-types and their Nomenclature, *Geol. Mag.*, **85** (1948) 349–57.

[2] Kennedy, W. Q., Trends of differentiation in basaltic Magmas, *Amer. J. Sci.*, **25** (1933), 239–56; Kennedy, W. Q., and Anderson, E. M., Crustal layers and the origin of magmas, *Bull. Voc., Séries ii, vol. ii* (1938), 24.

[3] Vincent, E. A., The chemical composition and physical properties of the Kap Daussy tholeiite dyke, East Greenland, *Min. Mag.*, **29** (1950), 46–52.

feldspathic intergrowths, constitute the dominant lavas of the two largest plateau basalt volcanic regions, those of the Deccan and the Columbia River.[1] In addition, dolerites of identical composition occur extensively in dyke swarms and in sills typified by those of the Karroo in South Africa.[2]

Kennedy pointed out that dominantly tholeiitic flood basalts were accompanied by occasional flows of rhyolite, and he contrasted this with the ocean basins where the basalts are dominantly under-saturated and Acid volcanic rocks are virtually unknown. From this it was natural to conclude that the tholeiitic magma-type was confined to continental regions and that it could give rise to rhyolitic differentiates. Kennedy's conclusions concerning the relationship between the two magma types and the main suites of volcanic rocks is best summarized in his own words:

'Tholeiitic magma type—andesite—rhyolite;
Olivine-basalt magma type—trachyandesite—trachyte—phonolite.'

These proposals have proved of great value in stimulating and giving direction to petrological research, though as we have already seen in our discussion of the classification of basalts and the chapters on andesites, trachytes and rhyolites, virtually every item has had to be modified in the light of new data, much of which has come from studies of Hawaii and other oceanic islands, and more recently from material obtained from the ocean floor and by deep-sea drilling. A good insight into the evolution of knowledge concerning Hawaii may be gained by reading successive reviews of petrology and petrochemistry written by G. A. Macdonald in 1949 and 1968.[3] In the former paper the olivine-basalts are dealt with more or less collectively: but in the latter, a distinction is made between olivine-tholeiites and alkali olivine-basalts. This distinction is based on the different proportions of total alkalies to silica[4] in the two groups of lavas (Fig. 122B). The boundary line is drawn in an arbitrary position to suit the Hawaiian data; but as Yoder has pointed out in a statement quoted by Macdonald and Katsura, the line corresponds fairly closely to compositions on the plane of critical under-saturation proposed as the basis of separation by Yoder and Tilley.

[1] Washington, H. S., Deccan traps and other plateau basalts, *Geol. Soc. Amer. Bull.*, **33** (1922), 765.

[2] Holmes, A. and Harwood, H. F., The tholeiite dykes of Northern England, *Min. Mag.* **22** (1929), 1–52; Walker, W. F. and Poldervaart, A., Karroo dolerites of the Union of South Africa, *Geol. Soc. Amer. Bull.*, **60** (1949), 591–706.

[3] Hawaiian petrographic province, *Geol. Soc. Amer. Bull.*, **60** (1949), 1541–96. Composition and origin of Hawaiian lavas, in *Studies in Volcanology, Geol. Soc. Amer. Mem.*, **116** (1968), 477–522.

[4] A helpful statement on the criteria used for distinguishing between the different categories is given by G. A. Macdonald and T. Katsura, Chemical composition of Hawaiian lavas, *J. Petrol.*, **5** (1964), 82–133.

The Hawaiian volcanoes are built up chiefly by the extrusion of large numbers of flows of very fluid lava, most of which are olivine-tholeiites of relatively uniform composition. This eruptive phase closed with the development of calderas which were infilled with thick and massive flows of tholeiitic composition, alternating with the earliest alkali olivine-basalts. The latter mark the transition to a post-caldera phase during which alkali olivine-basalts and subordinate differentiates of the Alkali Suite were erupted, at times explosively, to form a relatively thin capping to the caldera and the summit of the massive shield volcano. Finally, after a long erosion interval, lavas of a distinctly more Basic and alkaline character were erupted, relatively rich in nepheline, sometimes containing

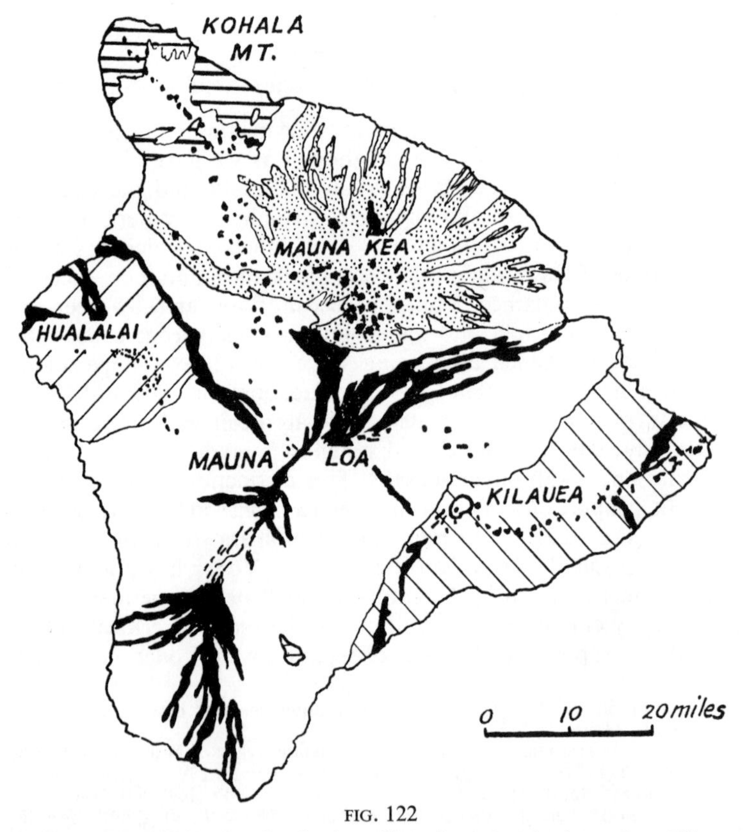

FIG. 122

A. Map of Hawaii showing distribution of lava flows, ranging in age from Pliocene (*horizontal ruling*) to Recent. Lavas belonging to the five main centres are shown by different ornament, while Recent flows are shown in black. Note the caldera of Kilauea within which is the Halemaumau Pit—a lava lake near which is the Volcano Observatory.

melilite, and belonging to what Macdonald calls the 'nephelinic series'.

It should be noted that the proportion of rocks belonging to the Alkali Suite on Hawaii is small compared with the tholeiitic basalts; while members of the nephelinic series are absent from the youngest and still active volcano, Kilauea, and present in amounts varying

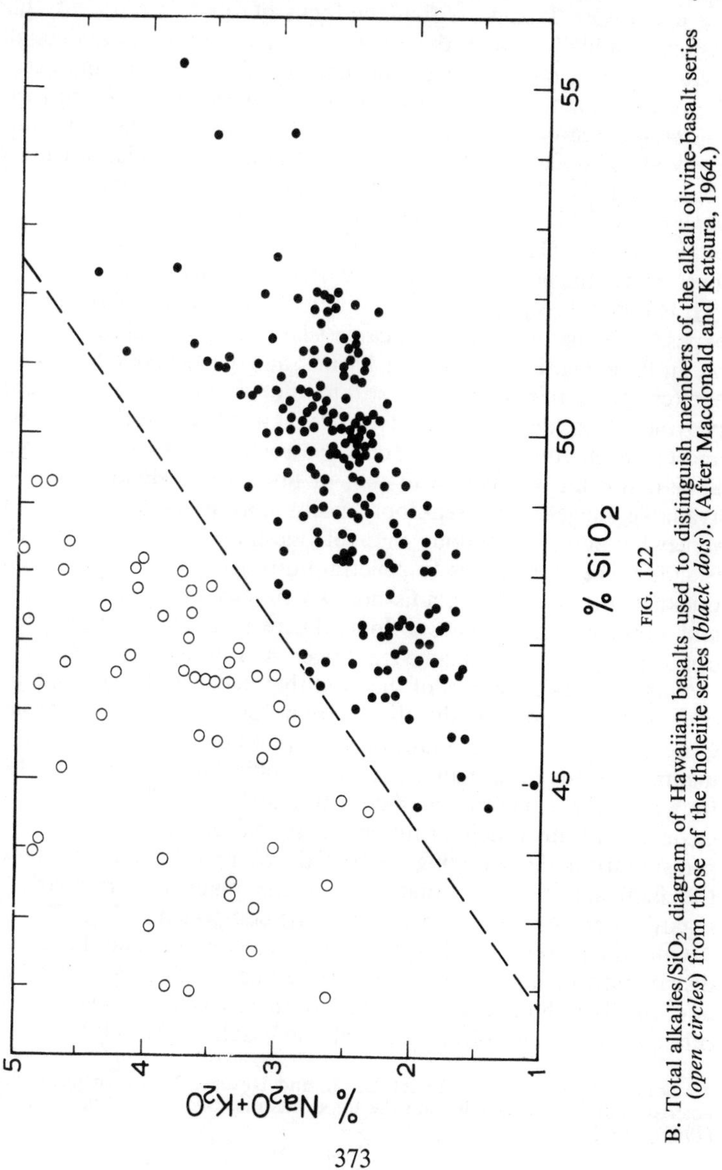

FIG. 122

B. Total alkalies/SiO_2 diagram of Hawaiian basalts used to distinguish members of the alkali olivine-basalt series (*open circles*) from those of the tholeiite series (*black dots*). (After Macdonald and Katsura, 1964.)

from a trace to only 1 or 2 per cent in the older volcanoes. Examination of other islands in the Hawaiian chain shows that the proportion of alkaline rocks tends to increase away from Hawaii; and this can be correlated with the ages of the volcanic islands which range from Recent (Hawaii itself) at the south-eastern end of the chain, to about 5 million years for the island of Kauai, 600 km towards the north-west. It appears that the focus of volcanic eruptivity has shifted gradually along the length of the chain—approximately 2000 km—as a result of plate movements of this part of the ocean floor. The oldest volcanoes have subsided gradually, either to form submerged sea-mounts, or to leave just the tops of the volcanoes exposed. This fact illustrates how misleading the evidence can be concerning the distribution and proportion of different types of volcanic rocks in the ocean basins. One has to remember that even in the case of Hawaii itself, rising to over 13,000 feet above sea level, by far the greater part of the volcano is submerged.

It is difficult to judge from the limited data available how far the exposed olivine-tholeiites of oceanic islands like Hawaii differ from ocean-floor basalts. The ocean basins, and particularly the globe-encircling oceanic ridges, constitute the world's largest volcanic province. Considered in these terms our knowledge of the character and chemical composition of fresh bed-rock basalts from the ocean floor is not far in advance of that of lunar rocks. Evidence so far available suggests that sea-floor basalts may be poor in K_2O and somewhat more aluminous than olivine-tholeiites of the volcanic islands.[1] The magma has to penetrate through perhaps a minimum thickness of ocean crust via fissures which must allow relatively easy and rapid access to the sea-floor. There is probably little or no opportunity for the magma to linger and hence differentiate in shallow magma chambers of the kind that occur in the volcanic pile of Hawaii. Presumably, therefore, ocean-floor basalts may approach quite closely to the composition of one kind of primitive basaltic magma. On the other hand, data are insufficient to judge the extent to which the chemical characteristics may be influenced by the environment in which eruption takes place: contact with, and pressure from, the overlying water of the ocean is bound to influence the final stages of consolidation of the magma, particularly in regard to processes involving its volatile components.

Opinion is very divided as to the true nature and hence the distribution of high-alumina basalts. Some basalt flows which are abnormally rich in Al_2O_3 are found in most volcanic provinces; but often, as in the case of the plagiophyric basalts in Mull (the Central

[1] See *e.g.* Engel, A. E. J., Engel, C. G., and Havens, R. G., Chemical characteristics of oceanic basalts and the upper mantle, *Geol. Soc. Amer. Bull.*, **76** (1965), 719–34.

CHEMICAL ANALYSES OF TYPICAL BASALTS

	1 Tholeiite	2 High-alumina basalt	3 Picrite-basalt	4 Alkali olivine-basalt
SiO_2	49·78	50·19	46·62	45·40
Al_2O_3	15·69	17·58	8·68	14·70
Fe_2O_3	2·73	2·84	2·04	4·10
FeO	9·20	7·19	10·52	9·20
MgO	7·79	7·39	20·86	7·80
CaO	11·93	10·50	7·15	10·50
Na_2O	1·21	2·75	1·41	3·00
K_2O	0·29	0·40	0·28	1·00
H_2O	—	—	0·23	—
TiO_2	0·68	0·75	1·71	3·00
P_2O_5	0·07	0·14	0·14	0·40
MnO	0·35	0·25	0·14	0·20
Other constituents	0·29	—	0·22	—
	100·01	99·98	100·00	99·30

1. Average of 3 tholeiites, Japan and Korea, quoted from H. Kuno, *Journ. Petr.*, **1** (1960), p. 141.
2. Average of 11 high-alumina basalts, Japan and Korea, quoted from H. Kuno, *loc. cit.*
3. Picrite-basalt, average of three analyses, Hawaii, quoted from R. A. Daly (1933), p. 397.
4. Alkali olivine-basalt, average of 35 analyses. Hawaii, quoted from G. A. Macdonald, (1968).

Porphyritic Basalt magma-type referred to above) this can be ascribed to enrichment of an original magma in plagioclase crystals. However, it is only in certain orogenic and island arc environments of dominantly andesitic volcanicity that high-alumina basalts (many of them aphyric) are erupted in significant amounts. They have been studied in the greatest detail in Japan (Kuno, 1960) where they are found to be concentrated in a fairly narrow zone which overlies the boundary between regions on the ocean trench side of the islands, characterized by tholeiitic basalts, and those on the continental side where alkali olvine-basalts occur. Thus the distribution, like the composition and petrology, tends to be intermediate between tholeiites and alkali olivine-basalts, and in fact overlaps with them. Possibly the high-alumina basalts represent a primitive type of magma generated under different conditions from the other magma types. The differences, for example, might be in depths of origin in the mantle. We are inclined to think, however, that the distinction is not so fundamental as this and that some factor such as local availability of water in depth may be critical in causing magmas whose compositions would otherwise be close to the tholleiitic/alkali olivine-basalt boundary to deviate towards enrichment in the plagioclase components. This would be consistent with views concerning the origin of andesites (p. 328) with which the high-alumina basalts are typically associated.

Origin of Basaltic Magmas

This is perhaps the most speculative aspect of theoretical petrology and in a book such as this, with its strong petrographic emphasis, it is impossible to do more than introduce the subject in the briefest outline.

It is now established beyond reasonable doubt that basaltic magmas originate in the upper mantle by partial melting. The composition will depend mainly upon chemical composition of the mantle rocks, and also on the depth at which melting occurs, even if the chemical composition is constant. The rocks of the mantle vary in their mineralogical composition with depth, as the stabilities and relationships of the mineral phases respond to increases in pressure and temperature. This naturally affects the conditions of partial melting and the composition of the liquids formed at different depths. Other variable factors include the amount of water in the mantle (see *e.g.* the discussion on andesitic magmas) and the degree of melting which is achieved before the new magma, which we may refer to as a *primitive* type of magma, is able to separate from the remaining solid components of the mantle. Estimates of the minimum proportion of liquid required to allow independent mobility of the magma fraction vary: a realistic figure may be between 10 and 20 per cent.

The compositions of the original mantle rocks from which the magmas are derived can never by known precisely. Evidence from ultramafic xenoliths in basalts and other kinds of Basic or Ultrabasic volcanic rocks e.g. kimberlite, and from mantle-derived ultramafic intrusions of the 'alpine peridotite' type, can be combined with data from high-pressure experiments to show that the rocks in the upper mantle must in the main be peridotites composed dominantly of varying proportions of olivine, clinopyroxenes, orthopyroxenes and garnet with spinel as a subordinate component. In some regions, particularly beneath continental crust, the upper mantle may also comprise a significant proportion of eclogite, essentially an omphacite, garnet rock.

It may be assumed that most of the samples of mantle rocks which appear as xenoliths are of a residual character, normally left in the uppermost part of the mantle as solids after the magma fraction has been separated from the primary mantle components. Green and Ringwood have coined the useful term, **pyrolite,** to refer to mantle components in their primary state. Estimates of the composition of pyrolite must be based on what can crudely be expressed as an 'equation': pyrolite = residual mantle rock + derived magma. The solution of this equation, which these two authors have chosen as a basis for their experimental work and discussion, is a mixture of three parts alpine peridotite (with 20 per cent orthopyroxene) and one part Hawaiian basalt.[1]

Green and Ringwood[2] have conveniently summarised the results of their own research and that of many other authors in a form which traces the changes in the phase relationships of mineral assemblages of pyrolite composition that occur with increasing depth. They pay particular attention to the minerals most directly involved in liquid/crystal reactions under the different conditions of initial melting, or as they call them, near-solidus conditions.

At shallow depths, less than 15 km, the mineralogy of pyrolite consists of the assemblage: olvine, orthopyroxene, clinopyroxene, plagioclase and chromite. Because of incongruent melting of orthopyroxene to produce residual olivine and a quartz-rich liquid, a small degree of melting would allow the formation of an oversaturated tholeiitic magma, leaving either residual dunite or clinopyroxene-peridotite. In fact it is unlikely that at such shallow depths the geothermal temperature would reach levels necessary for partial melting to occur except under rather abnormal and localized conditions.

At depths greater than about 15 km, a significant change occurs in the phase relationships and orthopyroxenes melt congruently, so that the mechanism described above (whereby partial melting of an undersaturated rock can give rise to an oversaturated magma), no longer operates. The most significant change to be noted is in the aluminium content of both clino- and orthopyroxenes: this increases steadily with pressure and has an important bearing on the compositions of magma fractions produced at depths greater than about 15 km, until at around 100 km depths garnet becomes increasingly important and gradually displaces pyroxene as the chief Al-bearing phase available for melting.

Between about 15–35 km the mineralogy of pyrolite at near-solidus temperatures consists of olivine, moderately aluminous pyroxenes and minor plagioclase. Partial melting to yield 20–25 per

[1] See *e.g.* Ringwood A. E., Composition and evolution of the upper mantle, in *The Earth's Crust and Upper Mantle*, ed. P. G. Hart, Amer. Geophys. Union, Monogr., **13** (1969), 1–17.

[2] Green, D. H., and Ringwood, A. E., The origin of basalt magmas, *Amer. Geophys. Union Monogr.*, **13** (1969), 489–94.

cent basalt magma leaves residual olivine, low-alumina enstatite and possibly a clinopyroxene relatively low in calcium. The magma fractions have the compositions of olivine-tholeiite with fairly high

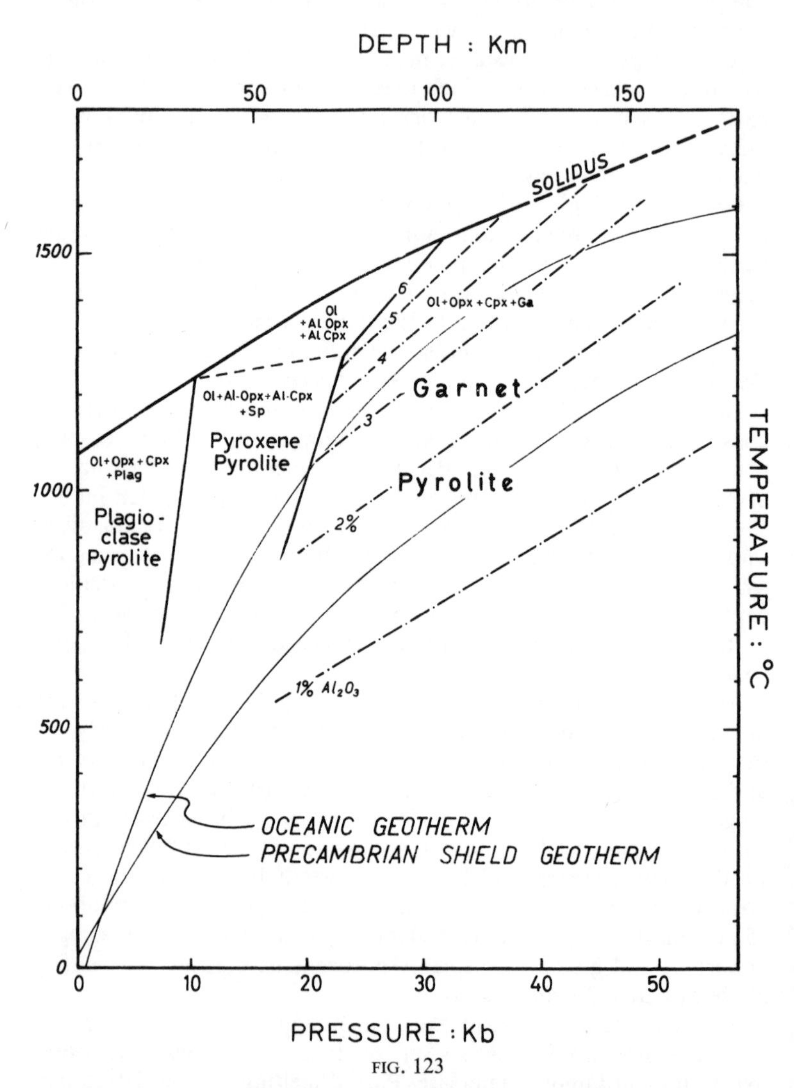

FIG. 123

Diagram to show the effects of temperature and pressure (depth) on the mineral composition of pyrolite. The Al_2O_3-content in orthopyroxene in equilibrium with garnet in garnet-pyrolite shown by broken lines. The geothermal gradients must be regarded as conjectural. Al-Opx = aluminous orthopyroxene; Cpx = clinopyroxene; Sp = spinel. (*After Green and Ringwood*, (1967).)

alumina. It is inferred that this is essentially the mode of origin of basaltic magmas of the mid-oceanic ridges, where the seismic evidence and high heat flow suggest relatively shallow depths of melting of pyrolite which is continuously replenished by convective or other kinds of movements in the mantle.

Below about 35 km the relatively low-density structure of plagioclase is no longer stable and so disappears from the list of near-solidus minerals of pyrolite: this becomes aluminous enstatite, aluminous Ca-poor augite and olivine. The composition of the liquid that can be produced depends on the extent of the partial melting, since this naturally governs the proportions of these three components which enter the melt phase. With only a small degree of melting (less than 20 per cent), the liquid compositions are relatively far removed from the composition of the total original rock, and in this case would be in the range of alkali olivine-basalts, leaving residues containing fairly high proportions of the aluminous pyroxenes, and of course, olivine. More extensive melting dissolves a greater proportion of the pyroxenes, so that with an increase in the degree of melting the liquid compositions shift gradually from alkali olivine-basalt to an olivine-rich tholeiite. Here we are obviously dealing with conditions which are very relevant to the origin of Hawaiian basalts.

The final item in Green and Ringwoods' review concerns the effects of melting of pyrolite at depths greater than 100 km. The near-solidus mineral phases include garnet (mentioned above), somewhat aluminous pyroxenes and olivine. Garnet dissolves at a very early stage of melting leaving liquids in equilibrium with pyroxenes and olivine. These liquids will be very Basic, rich in MgO and enriched in virtually all the alkalis present in the original mantle rocks. The proportions of alkalis in the mantle at these depths will be very minute; but as long as all of this enters only a small volume of melt, the concentration can become relatively high. Magmas rich in olivine, nepheline and melilite are probably formed in this way.

From this review it is apparent that much of the main kinds of Basic magmas, quartz-tholeiite, tholeiite and high-alumina basalt, alkali olivine-basalt and nepheline-bearing types, may be derived by variations in the depths and degree of melting of mantle rocks of pyrolite compositions. This model fits the data of volcanic activity so well that it can probably be accepted as close to the truth, particularly for oceanic environments. However, probable differences in the character and chemical composition of the upper mantle, particularly beneath areas of ancient continental crust, make it virtually certain that the processes involved in the origins of basaltic magmas can be very varied.

Yoder and Tilley, for example, have discussed the effects of fractional melting of eclogite under varying pressure conditions. Eclogite consists predominantly of omphacite, and a pyrope-rich garnet, the former including within its composition the normative components of diopside, albite and nepheline, and the latter the components of olivine, orthopyroxene and anorthite. They suggest that with increasing pressure the proportion of omphacite which enters into the composition of initial melts increases relative to the garnet. They conclude from this that 'liquids which would produce alkali olivine-basalt at the surface would come from greater depths than those liquids which would produce tholeiite at the surface' (1962, p. 507).

One of the factors which is most difficult to evaluate is the role of water. In general terms free water (*i.e.*, not combined in a hydroxyl-bearing mineral like hornblende or phlogopite) has the effect of lowering the temperatures of melting perhaps by as much as 200° C, and this, of course, must affect the compositions of the melts produced. At appreciable depths in the mantle, however, the amount of water available must be minute and figures of the order of 0·1 per cent are quoted. Even these small amounts will lower the temperatures of melting; but only a very small fraction of initial liquid will be produced, no more than would be represented by a pore liquid in an essentially solid rock. Most of the water will enter this liquid phase. Pore liquids of this kind will tend to form in advance of more general melting which may not take place until the temperature is raised a further hundred degrees or so. From a geochemical point of view, the chief interest in these incipient melts is that they must be enriched not only in water, but also in elements like potassium which, because of their ionic radii or some other reason, are said to be incompatible, and have no convenient 'home' in any of the stable mineral structures in the mantle.

Perhaps the behaviour of potassium may be analogous to, say, boron or some other element that is 'incompatible' in normal granite minerals, but becomes concentrated in water-enriched pegmatitic and hydrothermal fluids. If the alkalis and potassium in particular, display in the mantle anything like the mobility shown by boron in the crust, this could be an important factor in the formation of some kinds of Basic and abnormally potassic magmas.

In hypotheses relating exclusively to partial fusion in the mantle, no account is taken of possible differences in composition between primitive magmas at their source and the lavas which are erupted at the surface. As emphasized by O'Hara,[1] these differences may be substantial, due to shifting in the phase-equilibria between liquid

[1] O'Hara, M. J., Primary magmas and the origin of basalt, *Scot. J. Geol.*, **1** (1965), 19–40.

and crystals as the magma loses heat and is subjected to lower pressures. Successive crystal fractions may be precipitated, leading to differentiation of rising magma. Furthermore, if the initial magma is charged with crystals derived from the mantle, it may quite simply abandon some of its load during ascent.

This situation may obtain in Hawaii where, as Macdonald has shown, (1968 p. 509), evidence exists from both geophysical data and from a convergence of the variation trends of the alkali olivine-basalts and tholeiites, to indicate that in depth the primitive magmas for both these series may be very olivine-rich and of picrite-basalt compositions. There is no difficulty in accounting for such compositions by partial melting: all that is required is a greater degree of melting at an appropriate level in the mantle, so that more of the most refractory component of the original pyrolite—olivine—is incorporated in the melt. But if the possibility of picrite-basalt magmas is admitted, it means that much of the difference between the two basalt series at the surface may be the result of differences in differentiation, perhaps of a *common* type of primitive magma in depth as Macdonald suggests, rather than derivation of two primary magma-types originating under different conditions of partial melting.

Recent researches have provided an extremely valuable framework for hypotheses concerning the origin of basalt magmas; but as the Hawaiian example shows, there is still a great deal to learn.

CHAPTER IX

SPILITES AND ALLIED ROCK-TYPES

THE status and significance of spilites are matters of dispute among petrologists, while their position in a scheme of classification is problematical.[1]

Spilites (Brogniart, 1827) are Basic, fine-grained rocks, with SiO_2 percentage averaging 40, occurring as pillow-lavas of various ages in many parts of the world. Petrographically they are distinctive as the feldspar they contain is albite, in spite of their basicity. Secondly, the place of the usual ferromagnesian silicates is taken by chlorite, often of an iron-rich type, though augite does occasionally occur.

Intermediate and Acid lavas and minor intrusions are associated with spilites in the way that basalts are accompanied by subordinate trachytes and rhyolites. These are known as keratophyres and quartz-keratophyres respectively.

Further, in areas where spilites, keratophyres and quartz-keratophyres occur as lavas, certain sills and dykes of spilite occur, passing into albite-dolerites of coarser grain, while some are of sufficiently coarse grain as to fall into the albite-gabbro category. A variant of albite-dolerite is minverite described above. Other members of the suite, of ultramafic composition, include augite-picrite at Menheniot, Cornwall, and hornblende-augite-picrite at Molenick in the Plymouth district. Thus there is a group of rocks proved by their distribution to be comagmatic, linked to one another by distinctive chemical and mineralogical features. These rocks constitute the Spilitic Suite of H. Dewey and Sir John Flett who emphasized the strongly sodic character of the several members of the suite.[2]

With regard to classification, the whole range is alkaline with

[1] Wells, A. K., The problem of the spilites, *Geol. Mag.*, **60** (1923), 63; and Sundius, N., On the spilitic rocks, *Geol. Mag.*, **67** (1930), 1; Vuagnat, M., Sur quelques diabases suisses: contribution a l'étude du problème des spilites et des pillow lavas, *Bull. Suisse Min. Pétr.*, **26** (1946), 155. Bailey, E. B., and McCallien, W. J., Some aspects of the Steinmann Trinity, mainly chemical, *Quart. J. Geol. Soc.*, **116** (1960), 365–95. The reader is recommended to study Turner and Verhoogen's chapter in *Igneous and Metamorphic Petrology* (1960), 258.

[2] Some British pillow-lavas . . .' *Geol. Mag.*, **8** (1911), 202–9 and 241–8.

special emphasis on high soda-content and notable deficiency in potash. A point we must emphasize is one that is liable to be overlooked: the terms keratophyre and quartz-keratophyre are not synonymous with sodic Intermediate lavas (soda-trachytes) and sodic Acid lavas (soda-rhyolites) respectively. Among Na-trachytes, for example, one type is appropriately termed keratophyre; but there are other types which are definitely *not* keratophyres. To be entitled, so to speak, to the name 'keratophyre', an albite-trachyte must be associated with other members of the Spilitic Suite in the right kind of environment.

There are considerable difficulties in the way of stating the true facts regarding distribution, since many petrologists refuse to recognize spilite as a valid type, and claim that these rocks are merely 'altered basalts'. It may well be, therefore, that spilites and other members of the suite are more widely distributed than available records would suggest. The distinctive chemical features, especially the high sodium content, are claimed to have been impressed upon magma developing in a geosynclinial environment and subsequently erupted, perhaps from fissures opening in the sea-floor. Certainly spilites are interbedded with sediments of geosynclinial types, while the frequency with which they have developed pillow structures proves them to have been of submarine origin. The material filling the interstices between the pillows is dark blue-black mudstone in some occurrences, but chert in others. In some instances the chert is a banded silica-haematite rock of jaspilite type. Occasionally jaspilite occurs *within* the lava pillows proving that it was caught up and incorporated at the time of eruption. The most extensive occurrences of spilitic rocks are found in New South Wales and New Zealand and have been thoroughly studied by W. N. Benson and others. In Britain spilites have been described from among the Pre-Cambrian, Ordovician and Devonian-Carboniferous volcanic assemblages. Spilites are well developed on several horizons in the Mona Complex of Pre-Cambrian age in Anglesey.[1] A division of the Dalradian at Tayvallich in Argyllshire includes spilitic lava-flows and intrusive basic sills of epidiorite. Notable occurrences are those of Ordovician age to be found on the Girvan coast in south-western Scotland; those of the same period in parts of North Wales, notably on Cader Idris in the Dolgellau area, and of Carboniferous age in parts of Devon (Chipley) and Cornwall. In the latter county a spectacular display occurs at Pentire Point near Padstow.

With regard to origin it is now more generally agreed than formerly that spilites and the Spilitic Suite do really exist, though opinion is

[1] Striking photographs as well as a detailed account of the field relations and petrography will be found in Greenly, E., *Memoir on Anglesey*, vol. 1 (1919), 54 and 71.

divided as to their mode of origin. The main question at issue is whether their distinctive compositional character is a magmatic quality or one impressed later as a consequence of metasomatism involving albitization.

Spilites *are* restricted to geosynclinal regions; they were erupted under submarine conditions. They are markedly vesicular, and although the vesicle-infilling materials include various forms of silica (chalcedonite and quartz), chlorite and calcite, the last-named is the most abundant and accounts largely for the relatively high CO_2 (10 per cent) shown on analysis. It cannot be ascertained whether these substances were derived from internal or external sources; but certain spilites in Anglesey have been extensively carbonated, in some cases to such an extent that only a shell of spilite remains. The rest is chiefly carbonate. Similarly some spilites in the Dolgellau district are completely pseudomorphed by carbonates though the characteristic texture has survived. Therefore in the latter cases the replacement of the original minerals by calcite was a secondary, post-crystallization process.

Some of the calcite, epidote, etc., in spilitic rocks may have resulted from *internal rearrangement* of components. The most significant possibility concerns the albite and calcite present in the rocks. If the Ca were available for building feldspar instead of being immobilized in the calcite, etc., it would allow crystallization of a more normal basaltic type of plagioclase.

Many of the petrographic features of spilites are comparable with those seen in basalts that have been subjected to low grade regional metamorphism. Although relics of primary augite and calcic plagioclase may be present they are often entirely absent. In the latter case there may be no evidence to show whether the albite is of secondary origin or has crystallized directly from the magma.

Much of the albite, calcite and epidote may have resulted from a purely internal rearrangement of the components of calcic plagioclase, so that the chemical composition of a spilite may not be greatly different from that of basalt. Spilitic magma, by inference, may be regarded as essentially basaltic, but with an abnormally high H_2O—and probably CO_2—content. Crystallization of such a magma would follow a course different from that appropriate to normal basalt, particularly if the escape of volatiles were prevented either by the rapid formation of an impermeable envelope of glass in pillow-lava, or by some other factor associated with a submarine environment.

It may be noted that P. Eskola and others[1] successfully brought about albitization of calcic plagioclase by the action of Na_2CO_3

[1] Eskola, P., Vuoristo, U., and Rankama, K., An experimental illustration of the spilite reaction, *Comptes rendus Soc. Géol., Finlande*, no. 9 (1935), 1.

solutions, which proved that albitization such as that seen in basalts may be effected at temperatures as low as 250° C. This is a suggestive experiment, though it does not prove that all, or indeed that any, spilites were actually formed in this way.

It is very significant that shales marginal to spilitic minor intrusions (of albite-dolerite, etc.) may be converted into **adinole.** This is a metasomatic, compact, white-weathering rock consisting of up to 90 per cent or more of albite. Adinoles provide the field evidence for the existence of sodic solutions emanating from the spilite magma and capable of metasomatizing the country-rock. Similar mineralogical changes brought about by the solutions trapped within the spilites may be regarded as auto-metasomatic.

It appears, therefore, that, as a consequence of down-warping in the early stages of the development of a geosyncline, basaltic magma is locally generated by selective fusion of a peridotitic earth-shell. During uprise through unindurated sediments containing juvenile water this magma developed its hydrous and essentially spilitic quality; and crystallization under the special environmental conditions obtaining during the submarine eruption of pillow-lavas accounts for the distinctive mineralogical features of spilites. It has been suggested that under hydrous conditions metastable augite would crystallize together with albite, followed by primary Fe-rich chlorite and quartz. Retention of volatiles is essential, and neither introduction of Na from some extraneous source nor the existence of a special kind of magma appears necessary.[1]

[1] Amstutz, G. C., Spilites and spilitic rocks, in *Basalts*, vol. 2, ed. H. H. Hess, and A. Poldervaart (1968), 737–53, with references.

CHAPTER X

SYENOGABBROS AND TRACHYBASALTS

SYENOGABBROS

AMONG both the basic plutonites and lavas of the same composition rocks occur in which the characters of syenites and gabbros on the one hand, and trachytes and basalts on the other, are combined. Within the general basic framework, affinity with the latter should be shown by occurrence of plagioclase within the gabbroic range, *i.e.* $An_{>50}$. The fundamental facts, then, are two in number: firstly these rocks are thoroughly Basic as shown chiefly by their high colour index. Secondly, alkali- and calc-alkali feldspar occur in association, and ideally in nearly equal amounts. These facts suggest that the best group name available is syenogabbro for the coarse-grained, and trachybasalt for the fine-grained, members.

The alkali-rich portion of any member of this family may, in theory, consist of an alkali feldspar, or its unsaturated equivalent, leucite, nepheline or analcite. The gabbroic (or basaltic) portion consists of plagioclase ($An_{>50}$) plus mafics, of which pyroxene is the commonest.

Saturated varieties are by no means common rocks in the coarse-grain category; but the Scottish type kentallenite is typical syenogabbro and a very distinctive rock.

Kentallenite[1] was discovered at Kentallen near Ballachulish in western Scotland. It is a heavy, dark-coloured rock, with prominent patches of bronzy mica and an abundance of black crystals embedded in a rather meagre feldspathic base. In thin section olivine is very prominent, chiefly as it is densely charged with a separation of magnetite, as a fine dust that imparts a general grey colour to the crystals, and also as irregular dendritic patches. The most abundant mineral is a light greenish augite, in euhedral crystals. The feldspar is not easy to deal with; but careful examination will usually show that both plagioclase and orthoclase are present, involved in the same sort of relationship as in monzonite, but rather less obviously. All of these features are illustrated in Fig. 124. The

[1] Hill, J. B., and Kynaston, H., On kentallenite and its relation to other rocks, *Quart. J. Geol. Soc.*, **56** (1900), 531.

relationships between the mafic minerals are interesting and unusual. Olivine crystals are not notably euhedral: they are sometimes bounded by crystal-faces, but in the main give the impression of

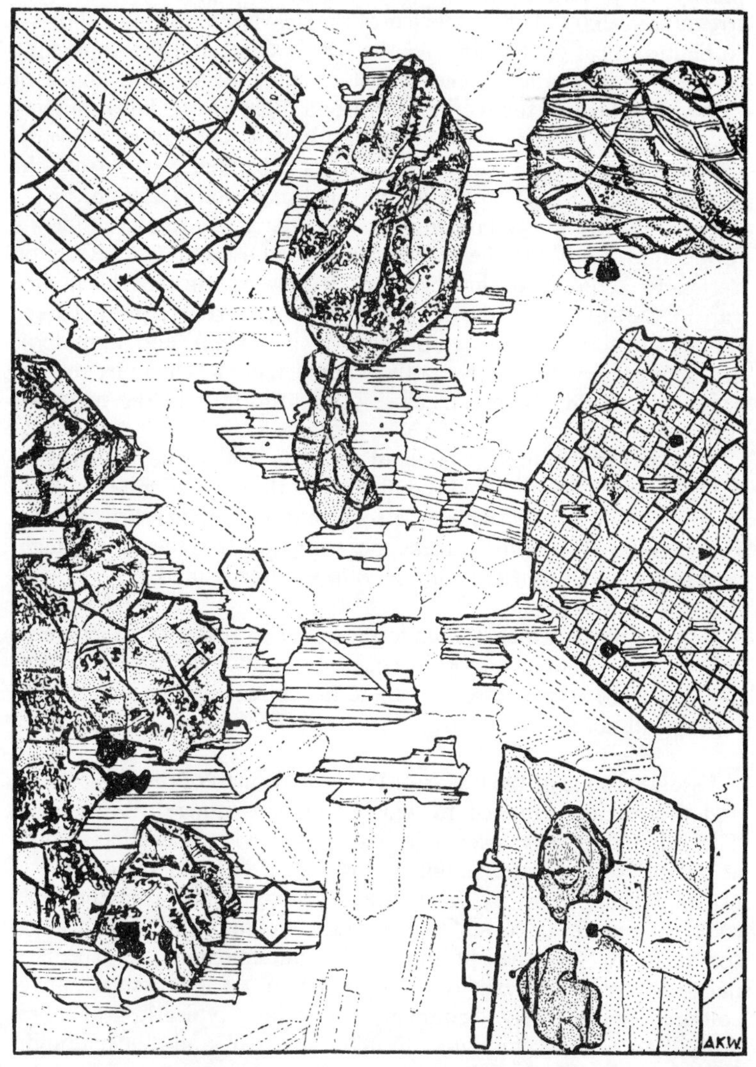

FIG. 124. Kentallenite, the type-rock from Kentallen, Argyllshire, Scotland.

Olivine (*close stipple*) enclosing dendritic magnetite; augite (*light stipple*); biotite (*lined*); plagioclase shown in outline with traces of the twinning; orthoclase left blank. Accessories are magnetite, in octahedra, some in parallel growths; also large euhedral apatites shown in basal and vertical section (the latter, *bottom right*.) See text.

being somewhat rounded and 'corroded'. By contrast the greenish augites are singularly well-formed. The abundant mica is obviously related to the olivines which in many cases are partially or completely enveloped in the former, which is poikilitic and highly irregular in shape. The parallelism of the cleavage of the mica shows it to be in optical continuity over the whole slide; and the apparently isolated 'shreds and patches' must have connected up, above or below the plane of the section, to form a highly irregular, ramifying crystal unit. The mica grew concurrently with the feldspar towards which it is poikilitic. The mica drew its substance partly from the olivines and partly from the K- and hydroxyl-rich residuum.

Normally in a crystallizing Basic magma the mafic minerals follow one another in the order of Bowen's Discontinuous Reaction Series. In kentallenite, however, the sequence is broken by a wide gap in the middle of the sequence: orthopyroxene, clinopyroxene (in its normal reaction relationship with olivine), and amphibole are unrepresented: on the other hand the last member of the sequence (biotite) is seen in close juxtaposition with the first—the olivine.

It is extremely unlikely that all rocks to which the term syenogabbro might be applied conform to the kentallenite pattern. Kentallenite is a very distinctive rock, but all syenogabbros are not kentallenites. The rarity of these rocks suggests that they are brought into being by an unusual combination of circumstances.

Syenogabbros which are undersaturated in respect of their felsic constituents include gabbroic rocks containing nepheline and/or analcite. The types essexite, theralite, teschenite, crinanite and lugarite all fall in this category.

Essexite is one of the least satisfactorily defined of the Basic rocks, and the name is applied to widely different mineral assemblages. The original essexite occurs in association with nepheline-syenites at Salem Neck,[1] Essex County, Massachusetts, and was defined in rather general terms by J. H. Sears. On re-examination the type-rock proved to have suffered contact metamorphism and metasomatism by an adjacent intrusion. Fundamentally essexite is a dark-coloured gabbroic rock in which a varied assortment of coloured silicates may accompany plagioclase, typically near labradorite, and small amounts of alkali-feldspar and feldspathoids. With regard to the last named, opinion seems to have been far from unanimous as to the amount of nepheline and/or analcite which should be present. Certain so-called essexites from Norway, des-

[1] Washington, H. S., The petrographical province of Essex County, Mass., *J. Geol.*, **7** (1899), 53.

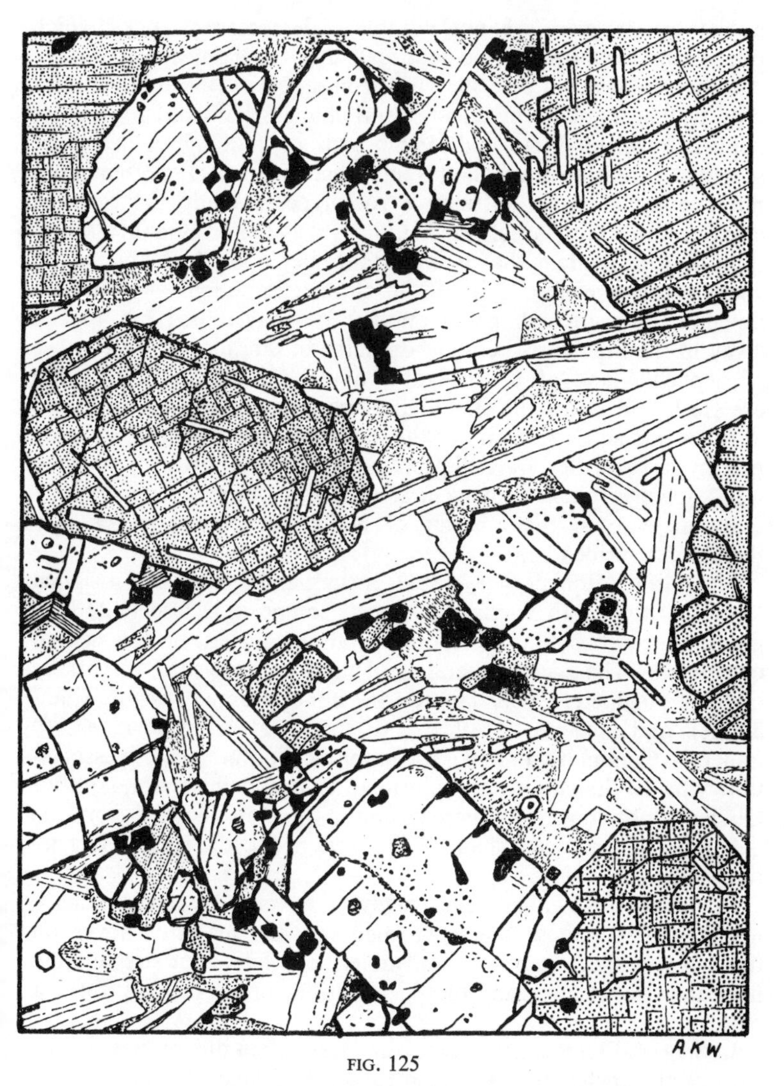

FIG. 125

Essexite, Crawfordjohn, Scotland.

Euhedral olivine, clear and fresh with a slight peppering of iron-ore granules; titanaugite (*stippled*) containing zonally arranged small plagioclase 'laths' and commonly twinned on (100); labradorite, pinacoidal, flattened parallel to (010) largely post-titanaugite; all set in a residuum of analcite, mostly fresh and clear, containing small euhedral nephelines, turbid through incipient alteration, (two basal sections slightly above centre left); magnetite microphenocrysts clustered chiefly near olivines; apatite long, prismatic and occasionally cored.

cribed by Brögger, are not notably different from rocks which others would term olivine-gabbros: they contain no nepheline. Another well-known essexite from Rongstock in Bohemia contains only trifling amounts of that mineral; but in the Scottish essexites nepheline can be discovered in thin sections, although it has to be searched for. But although the mere list of minerals present does not suggest a highly distinctive rock-type, actually the Scottish essexites are well characterized, though it is textural quality rather than mineral content that is important (Fig. 125). In so far as the rocks consist chiefly of labradorite, titanaugite, olivine, apatite and ilmenite, they are like many another olivine-gabbro; but tucked away in the interstices between the laths of labradorite are patches of nepheline and of unmistakable analcite. Further, the olivines are fresh and clear, the titanaugites particularly well-formed, and at once attract attention in the hand-specimen. These strongly porphyritic essexites are well represented by the Crawfordjohn rock well-known to many Scotsmen, as it is a favourite 'curling-stone'.[1]

By increase in the amount of nepheline essexite passes into theralite, described below.

Teschenite (Hohenegger, 1861) is so named after the original locality, Teschen, in Silesia. In mineral composition teschenite is in general like gabbro, but differs in containing analcite as an essential constituent. The type is therefore defined as consisting essentially of basic plagioclase, near labradorite, clinopyroxene (usually titan-augite), and analcite, together with the usual gabbroic accessory minerals. Commonly barkevikitic amphibole occurs in close association with the titanaugite, and biotite often occurs in small quantity.[2] Olivine is not constantly present, and is therefore omitted from the definition of the type; when olivine does occur, the rock is **olivine-teschenite.**

In texture teschenites vary widely: they may be ultra-coarse—teschenite-pegmatites, coarse to medium—the majority hover on the brink between gabbroid and doleritic, while some are notably finer than the main body of the intrusion, of which they form a part.

In Scotland teschenitic rocks form large, often differentiated sills, such as the Inchcolm sill[3] which forms the island of that name in the Firth of Forth (see map, Fig. 148), and the well-known sheet exposed at Salisbury Crags, Edinburgh (Fig. 60). Certain of the Scottish teschenites contain some alkali-feldspar in addition to

[1] Scott, A., The Crawfordjohn essexite and associated rocks, *Geol. Mag.*, **12** (1915), 458.

[2] Tyrrell, G. W., Classification and age of the analcite-bearing rocks of Scotland, *Geol. Mag.*, **60** (1923), 249–60; Walker, F., Notes on the Scottish and Moravian teschenites, *Geol. Mag.*, **60** (1923), 242–9.

[3] Campbell, R., and Stenhouse, A. G., *Trans. Geol. Soc. Edin.*, **9** (1907) 121.

labradorite. Usually it does not amount to more than 10 per cent of the whole rock, but it is a significant constituent, and Johanssen considers that a distinctive name, **glenmuirite,** from the locality in central Ayrshire where the exposures of 'teschenite' in the Lugar Sill occur, is justified.

Crinanite,[1] named after a Scottish locality by J. S. Flett (1911) is linked with teschenite: it is so close, in fact, that there is little doubt that crinanites have been recorded as teschenites and vice versa. Both names are widely used in official Survey publications. The original crinanites were collected from (presumably) Tertiary dykes in the neighbourhood of Loch Crinan, Argyllshire. Stress was laid, in the description, on their relative fineness of grain and perfect ophitic textures. Teschenite and crinanite both consist of the same mineral assemblage, although the proportions are different: in crinanite the very small amount (2 per cent) of analcite is interstitial to the feldspars; but in teschenite zeolites (chiefly analcite) are more abundant and the plagioclases themselves are analcitized. Crinanite strictly cannot be classed as syenogabbro or microsyenogabbro: the 'syeno-' part of its composition is insignificantly small, and the rock should not be distinguished by an independent type-name. It is merely an analcite-bearing olivine-dolerite. The fact that most specimens are ophitic is of no genetic significance.

An allied type in which nepheline displaces analcite is **theralite,** consisting of plagioclase within the usual gabbroic range, clinopyroxene and nepheline, together with sundry accessories. Rosenbusch proposed the name (1887) for the plutonic representative of nepheline-tephrite—a fine-grained volcanic rock consisting of this mineral assemblage. The rock originally described under this name by J. E. Wolff from the Crazy Mountains, Montana, does *not* correspond with nepheline-tephrite as it is rich in orthoclase, and must be classed as malignite.

Typical theralites are not common rocks, but specimens closely agreeing with Rosenbusch's conception of what a theralite should be have been described from the Cordilleras of Costa Rica, the Bohemian alkali province and from Scotland. Perhaps the best known theralite, which figures in many teaching collections, comes from Duppau in Sudetenland. Equally typical are certain late-Carboniferous examples occurring in Ayrshire, notably in the Lugar composite sill, in which theralite is accompanied by teschenite. In Scottish theralites titanaugite is the most abundant component (35 per cent), while labradorite and nepheline occur in approximately equal amounts (16 per cent each in the Lugar theralite). An amphibole, 'barkevikite' in the Scottish rocks, biotite and olivine may all be present in varying amounts, but are not covered by the

[1] Cf. Walker, F., The term 'crinanite', *Geol. Mag.*, **71** (1934), 122.

definition. If olivine is present in significant amounts, there are many precedents for calling the rock **olivine-theralite.**

Kylite (G. W. Tyrrell, 1912), is an olivine-rich melatheralite which is correspondingly poor in nepheline.

Lugarite (G. W. Tyrrell, 1912)[1] is one of the very few unique British rock-types. It takes its name from the stratified differentiated sill at Lugar, Ayrshire, where it occurs as veins cutting the other members of the complex—teschenite and theralite. Quantitatively it is of minor importance; but from the mineralogical point of view it is of outstanding interest. Long prisms of 'barkevikite' and titan-augite are conspicuous, embedded in a greenish grey matrix, originally thought to be altered analcite, together with relics of much-altered labradorite and traces of nepheline. Subsequently fresher material became available for study, through boring operations, and the chief felsic component was proved to be nepheline. The assemblage: nepheline, clinopyroxene, labradorite is theralitic; but this rock is so distinctive, chiefly on account of the 'barkevikite' it contains, that retention of the term lugarite seems justified: it is, in effect, barkevikite-theralite. In lugarite two accessories are conspicuous: abundant prisms of apatite and ilmenite showing stages of conversion into brownish translucent leucoxene (Fig. 43).

Rocks which technically fall in the corresponding medium grain-size group and which have actually been recorded are few in number. Relatively finer-grained parts of a kentallenite intrusion would be distinguished conveniently as **microkentallenite,** and similarly in the case of the other types described above. Tyrrell applied the name teschenite-basalt to the fine-grained marginal facies of the teschenite forming part of the Lugar Sill; but this term seems to imply a rock which is between basalt and teschenite in composition, and a better term, in our opinion, is **microteschenite.**

TRACHYBASALTS

Trachybasalts are the fine-grained equivalents of syenogabbros. They are essentially rocks of basaltic aspect combining the distinctive compositional features of both trachytes and basalts. This should be a quite unequivocal definition, but the reader is warned that the term trachybasalt is often applied to rocks other than those conforming strictly with this definition. We have to note serious discrepancies between the theoretical concept of what trachybasalts *should* contain, and their actual mineral composition. In so doing we emphasize that the principle of classifying rocks

[1] A boring through the Lugar Sill, *Trans. Geol. Soc. Glasgow,* **21** (1948), 157–202 with references.

solely on their A/P ratio is unsatisfactory when applied to Basic igneous rocks. The rock-type **ciminite** (pronounced 'chiminite') is instructive in this respect. It was described by Washington from a volcano in the Roman Volcanic Province, and is rich in potassic alkali-feldspar (51 per cent) accompanied by subordinate labradorite together with clinopyroxene (23 per cent), olivine (11 per cent) and accessories. We may regard the alkali-feldspar, making up half the rock, as the trachytic contribution to the bulk composition of ciminite. The rest, comprising the labradorite, clinopyroxene, olivine and magnetite constitutes a typical basaltic assemblage. Therefore, taking into account the *whole* mineral composition of the rock, we may regard it as a 50 : 50 'mixture' of trachyte and basalt, that is, as an ideal trachybasalt. If the rock is classified on the basis of its A/P ratio (83 : 17), it would be grouped with the trachytes which would be quite irrational.

It is useful to remember that analcite and nepheline represent two different degrees of undersaturation of albite; and that in teschenite, had a little more silica been available, the analcite would have given place to albite. However, the latter would not have appeared as discrete crystals, but its substance would have been incorporated in the labradorite, so that a less calcic plagioclase would have crystallized. Had this been andesine in a typical basaltic assemblage of mafic constituents, the rock-type would have been **hawaiite**. The same argument may be applied to **mugearite** (with oligoclase), and benmoreite. From this angle, therefore, these three rock-types are trachybasalts, although the 'trachytic contribution' is not immediately apparent. In discussing the origin of trachytes we referred to the hiatus between the parental basalt and its ultimate differentiation product, trachyte. Hawaiite, mugearite and benmoreite are in the direct line of descent between the two extremes and in the genetic sense must be regarded as trachybasalts which in a frequency curve would lie in the trough between the two maxima representing basalt and trachyte respectively.

As regards distribution trachybasalts in variety occur in 'continental' suites, for example in the mountains of North America, notably in the Yellowstone Park area, where several varieties were described by Iddings and given Indian names. Washington added ciminite and vulsinite to the list of named types in his studies of certain Italian volcanic districts. Trachybasalts are important in Jan Mayen among the later Tertiary lavas; and they also occur and compare closely with Yellowstone Park types in certain of the central African volcanic regions.

The most interesting (and the most elusive) thing about trachybasalts is the significance of their mineral composition. The name trachybasalt and an appreciation of its systematic position in the scheme of classification naturally leads one to think in terms of admixture of a trachytic residuum with basaltic magma. It is generally believed that, as a consequence of fractional crystallization, a trachytic residuum may be produced from olivine-basaltic magma. Such a residuum might be incorporated in a later influx of basaltic magma, forming a hybrid magma from which trachybasalt might crystallize directly. Alternatively in slightly different circumstances the trachyte might be represented by aggregates or clots of alkali feldspar embedded in a basaltic matrix, and obviously of mixed origin. Rocks of this type occur among the trachybasalts of Jan Mayen. If this is a true indication of what happens, these rocks should display a wide variation in the proportion of basaltic to trachytic material in different types. One of Washington's types—vulsinite—contains over 70 per cent of Na-orthoclase (the trachytic contribution) as against anorthite phenocrysts 6 per cent, and 12 per cent of labradorite in the groundmass; so that this type is three-quarters trachyte and only one-quarter basalt.

Undersaturated trachybasalts include a number of rock-types bearing a superficial resemblance to basalts: they are dark-coloured, fine-grained rocks occurring either as lava flows or in dykes. As in certain basalts, phenocrysts of augite and olivine may be visible even in hand specimens; but there the resemblance to basalts ends, and the differences are much more significant than the resemblances. The feldspathoid minerals nepheline (or the chemically similar analcite) and leucite occur, in place of the alkali-feldspar of the saturated trachybasalts.

Classification is based upon the kind of feldspathoid mineral present, and the presence or absence of olivine.

Olivine-free types are termed **tephrites** (Cordier, 1816) of which there are three varieties, distinguished by prefixing nepheline-, leucite- or analcite- as circumstances demand. The addition of olivine, by analogy with many other rock-types, should give olivine-tephrite; but long-established precedent demands that they shall be called **basanites** (Brongniart, 1813), again with the appropriate mineral prefix giving nepheline-, leucite- and analcite-basanite. It would be not inappropriate if varieties containing only *small* (accessory) amounts of feldspathoids were termed nepheline-basalt, leucite-basalt or analcite-basalt respectively. These are the logical names to use; but unfortunately they have been wrongly applied in the past to rocks which are not basalts as they contain no plagioclase.

With regard to the details of mineral composition and texture little need be added to what has already been written. Thin sections convey the impression that one is looking at basalt, with nepheline analcite, and/or leucite added. The rocks are generally porphyritic. Probably the most widely distributed variety is leucite-basanite, typically represented by well-known lavas from Monte Somma and Vesuvius, containing phenocrystic leucite, olivine, augite and plagioclase, embedded in groundmass which may include all of these minerals as second generation microlites, with magnetite the commonest accessory. In the nephelinic varieties nepheline may occur as easily recognized phenocrysts, but when it occurs only in the groundmass it is less readily identified. Of the analcite-bearing varieties only analcite-basanite is at all common, and is, in effect, olivine-basalt with a variable amount of analcite interstitial to the groundmass microlites.

As regards origin, all that was said above about trachybasalts in general applies also to these rocks. The trachytes, which are regarded as ultimate differentiates from olivine-basalt magma, are often somewhat undersaturated and range from orthotrachytes (saturated) through nepheline-trachytes to phonolites. By analogy the trachytic portion of a trachybasalt may be represented by alkali-feldspar (sanidine or anorthoclase) or its place may be taken by nepheline, analcite, or leucite, according to the balance between K^+ and Na^+.

The close mineralogical analogy between tephrites and especially basanites on the one hand and teschenites, theralites and essexites on the other will be apparent to the reader; and again it will be noted that there are no precise analogues of leucitic varieties among the rocks in the coarse-grain category because leucite cannot survive the deep-seated conditions of crystallization.

Finally, it is worth noting that all the rocks described in this Chapter are related to alkali olivine-basalts. It will be remembered that an essential feature of the latter is the presence of some nepheline in the norm. The proportion of feldspathoids and/or alkali feldspars tends to increase as a result of differentiation and other processes affecting magmas originally of alkali olivine-basalt composition.

CHAPTER XI

THE MAFIC FELDSPATHOIDAL LAVAS AND RELATED TYPES

UNDER this heading are included several rock-types which superficially resemble basalts: they are dark in colour, fine in grain, Basic to Ultrabasic in composition and occur either as lava-flows or as petrographically identical dyke-rocks. Phenocrysts, often visible in hand specimens, include olivine and augite, heightening the resemblance to basalts; but in other types megascopic leucite may be a distinctive feature, while in thin sections other feldspathoids are visible.

Classication is effected on the basis of the kinds of feldspathoids present, the presence or absence of plagioclase, and of olivine. In the first instance distinction is made between plagioclase-bearing and feldspar-free varieties. This distinction is fundamental and cannot be over-emphasized. The plagioclase-bearing varieties—tephrites and basanites—were discussed above as under-saturated trachybasalts. Examination of thin sections of these rocks conveys a strong impression that they are fundamentally basaltic, with a variable amount of feldspathoid minerals added: tephrites and basanites *without* their nepheline, analcite or leucite would be just basalts or olivine-basalts respectively.

No such impression is conveyed by the feldspar-free types: compositionally they have nothing in common with basalts as is clearly indicated in the varieties illustrated. The absence of plagioclase automatically excludes these lavas from the basalt family. By definition a basalt must contain plagioclase as one of its essential minerals, in combination with pyroxene. With more silica available in the magma leucite, kalsilite and nepheline would have been represented by alkali-feldspar, therefore the affinities of these rocks are with the syenite-microsyenite-trachyte series. They are compositionally allied to phonolites and leucitophyres, with which they are sometimes closely associated in the field.

It may be useful to remember that we are dealing with quenched rocks containing unstable mineral associations. This is particularly true of the kalsilite- and leucite-bearing varieties, the coarse-grained equivalents of which would be found among the shonkinites and certain syenoids. The corresponding nephelinic lavas are analogous with the ijolite—melteigite series.

Difficulties of nomenclature arise when the feldspar-free lavas are considered. Among the rock-names in common use some will probably shock the reader as being illogical and conveying entirely false impressions of the mineral composition. This is particularly true of the terms 'nephelinite', 'leucitite' and 'melilitite'. If it were possible to start again there would be no difficulty in finding suitable names; but a hundred and fifty years ago when these names were first introduced, petrographic nomenclature was in its infancy; the urtite-ijolite-melteigite series was unknown, so too were the various monomineralic rocks upon the logical naming of which a good deal depends. Now, however, the (nearly) pure nepheline rock is known, as the end-member of the urtite-ijolite-melteigite series, and there is no logical alternative to the name nephelinite. The facts speak strongly for themselves. It is inconsistent with the rules to which we are working to continue using these names in an inadmissible sense, so that in the previous Edition we ventured

Mineral composition	Suggested names	Names in current use
Kalsilite plus mafic minerals	Kalmafite	Mafurite, katungite
Leucite plus mafic minerals	Leumafite	Leucitite, ugandite
Nepheline plus mafic minerals	Nemafite	Nephelinite
Melilite plus mafic minerals	Melmafite	Melilitite, melilite-basalt

to introduce a new set of terms which roll easily off the tongue, and which adequately suggest the mineral composition of the rocks concerned. They will prove acceptable, it is believed, to the general geological reader. These things can be said of very few rock-names. To avoid confusion we propose to bracket the old with the new terms.

The rocks under consideration are mafic, and the feldspathoid minerals are diagnostic of the different varieties. Both facts may be combined in the rock names as shown in the Table.

Of the many rock-types involved, the leucitic varieties are the most familiar and the most easily identified on account of the distinctiveness of leucite in thin section as well as in some of the hand specimens. In the following account we attempt to cover the necessary petrographic detail and the facts of distribution together.

Mafic Leucitic Lavas

The most familiar leucitic lavas occur in the Roman Volcanic Province which includes Vesuvius and Monte Somma. At one time they were thought to be restricted to this region; but they are now known to occur in all the continents, notably in the aptly named Leucite Hills, Wyoming, U.S.A., the Kimberley District in Western Australia, Celebes in Indonesia, Antarctica and several parts of Africa including the neighbourhood of the Western Rift, the former Belgian Congo, Uganda and adjacent parts of Tanzania.

The central and eastern African region is by far the most important of these. The different occurrences are being actively explored and studied at the present time and knowledge concerning them is growing fast. The leucitic lavas include leucite-tephrites, leucite-basanites, the less basic leucitophyres and phonolites, as well as plagioclase-free leumafites ('leucitites'). Some of the latter are associated with and grade into kalsilite-bearing lavas; while nepheline occurs, too, in an accessory capacity. An augite-rich olivine-leumafite (termed 'ugandite' by Holmes, 1945) is an uncommon variety which grades into leucite-melabasanite with the incoming of plagioclase.

With regard to the Roman Volcanic Province it may truthfully be claimed that the first feldspathoidal lavas to be recognized as such came from the neighbourhood of Vesuvius, and they remain the main source of supply of teaching material. Euhedral leucites occurring in these rocks are particularly distinctive and well known. Large first-generation leucites display their characteristic polysynthetic twinning in thin sections, but the smaller crystals of the second generation are often singly refracting. The small leucites illustrated in Fig. 126 are of more than usual interest on account of the minute glass inclusions which they contain. In their arrangement they are identical with those described by Bowen and Morey from leucites obtained by quenching during their classical experiments on the melting of orthoclase. As will be seen from the illustration, the inclusions are often significantly grouped in threes or fours about the points marking the emergence of triad or tetrad axes of symmetry.

A unique leucitic rock from the Roman Volcanic Province was named **italite** by Washington (1920),[1] though leucitite would have been accurate and informative, as leucite is the only identifiable mineral in the rock. It occurs as blocks in an agglomerate in the Alban Hills and consists of large euhedral leucites cemented to-

[1] Washington, H. S., Italite, a new leucite rock, *Amer. J. Sci.*, **1** (1920), 33.

gether by a small amount of whitish material which may be altered glass. If it is, the rock must be regarded as a phenocryst-rich extrusive. Evidently italite is an accumulative rock, and there can

FIG. 126

Leumafite ('leucitite') Capo da Bove, near Rome.

Small euhedral leucites, clinopyroxene, a little mica, accessory apatite and magnetite. The groundmass is colourless glass.

be little doubt that it represents a roof- (as distinct from a floor-) accumulation of sorted phenocrysts. As we have argued in the case of other monomineralic rocks, anorthosite for example, such an accumulation could neither be intruded nor extruded as lava, but would almost inevitably be disrupted by volcanic vent explosions, as happened in this instance.

'Leumafites' in variety occur in the West Kimberley area in Western Australia.[1] Four types connected by intermediate varieties have been described, emphasis being placed on the mafic minerals

[1] Wade, A., and Prider, R. T., The leucite-bearing rocks of the West Kimberley Area, Western Australia, *Quart. J. Geol. Soc.*, **96** (1940), 39.

which they contain. Phlogopite is perhaps the most significant as being equivalent to leucite and olivine combined. This Mg-rich mica makes up to a quarter of the whole rock in one type; it is accompanied by a manganese-amphibole of rare type in another, while a third contains prisms of clinopyroxene and olivine pseudomorphs. It is therefore an olivine-leumafite ('olivine-leucitite'). The most interesting fact about these Australian rocks is that, although they are described as the most richly leucitic rocks in the world, chemically speaking they should contain no leucite as there is plenty of silica available in the rocks to have converted it all into sanidine. The silica needed for this conversion is present in a copious glassy base, but was made inoperative by quenching. One chemical feature of the magma from which these lavas crystallized is a high content of titanium: much of it is present in the mica, but some occurs as the uncommon accessory, rutile, which amounts to as much as 6 per cent in one type.

Several varieties of leucitic lavas occur in the Leucite Hills, Wyoming.[1] Most of them are leucitophyres; but one of them, on account if its mafic character, may be considered here, especially in view of what has been stated in the immediately preceding paragraph. The rock was termed **madupite** by Cross (1898). It consists of diopsidic pyroxene making up almost half the rock, phlogopite comes second in importance, while the distinctive accessory, perovskite, also occurs. It will be noted that leucite does not occur in the list of minerals; but up to one-third of the rock consists of glass having the composition of leucite, so that *potentially* leucite is present. The rock is named and classified on the basis of a mineral composition that it would have had, if crystallization had not been arrested by quenching. Whether or not this is a sound principle is a debatable point. The rock would be described as a phlogopitic leumafite ('leucitite') containing leucite-glass. A short phrase is needed to convey the right impression of the nature of this rock: the single word 'madupite' certainly does not.

Mafic Kalsilite-bearing Lavas

These are the rarest of lavas, at present known to occur in only two areas in Africa. The mineralogical characteristics of kalsilite and its relationship to other feldspathoids have been already discussed. Although these rocks are feldspar-free and Ultrabasic in composition they are particularly noteworthy on account of the variety of unsaturated minerals which they contain—felsic, mafic and accessory. The several varieties have been named after type-localities.

[1] Cross, W., The igneous rocks of the Leucite Hills, etc., *Amer. J. Sci.*, **4** (1898), 126, 134 and 139.

An olivine-kalmafite termed **mafurite** by Holmes[1] contains small phenocrysts of olivine and augite set in a matrix of green augite and accessories including abundant perovskite, with interstitial kalsilite. Had there been a little more silica available in the magma kalsilite would have given place to leucite and the mineral assemblage would have been olivine, augite, leucite and accessories, already encountered as olivine leumafite ('olivine-leucitite').

A very distinctive variety named **katungite** after the extinct volcano Katunga, in Uganda, contains kalsilite, melilite, olivine, perovskite, titanomagnetite and interstitial glass. With slight increase in silica katungite grades into varieties less completely undersaturated and containing leucite instead of kalsilite and augite instead of melilite. This variety, distinguished as **ugandite** by Holmes, is particularly rich in mafic minerals, otherwise the mineral assemblage is that of a melanocratic olivine-leucitite.

Kalsilite-bearing lavas of exceptional interest have been discovered in the Belgian Congo.[2] In general they are close to katungite —the same minerals occur in both, but with nepheline most intimately associated with kalsilite in the Congo varieties. These two minerals form glomerophyric aggregates of phenocrysts which show varying degrees of unmixing of the two components. The distribution of alkalis in these lavas makes an interesting study. Thus there is a reciprocal relationship between the amount of $KAlSiO_4$ in the nephelines, and the quantity of leucite in the rock: in specimens relatively poor in leucite, the associated nephelines are rich in K-ions; but if the rock is leucite-rich the accompanying nephelines are deficient in K-ions.

In the permanent magma lake of Nyiragongo in the Congo the accumulation of aggregates of phenocrysts swept up by turbulent currents in the magma may be observed. This is a feature of special significance in connection with the formation of 'accumulative' rocks including phenocryst-rich lavas which display a condition of unstable equilibrium between phenocrysts and magma now represented by groundmass.

Mafic Nepheline-bearing Lavas

Nepheline-bearing varieties comprise the commoner nepheline-tephrites and basanites and the much rarer nemafites ('nephelinites') and olivine-nemafites ('olivine-nephelinites'). The nemafites at their best are very distinctive rocks. Nepheline may occur in two generations, both euhedral, or the second generation, instead of

[1] Holmes, A., A suite of volcanic rocks from S.W. Uganda, *Min. Mag.*, **26** (1942), 197; also Holmes, A., and Combe, A. D., The kalsilite-bearing lavas of . . . S.W. Uganda, *Trans. Roy. Soc. Edin.*, **61** (1945), 359–79.

[2] Sahama, T. G., Kalsilite in the lavas of Mt Nyiragongo, Belgian Congo, *J. Petrol.*, **1** (1960), 146.

FIG. 127

Nemafite (nephelinite), summit of Lodwar Hill, northern Turkana, Kenya.

For description see text. Pyroxene, stippled; magnetite, black; sphene, heavy stipple. Phlogopite with cleavage indicated. There are two generations of pyroxene.

forming the characteristic small square and hexagonal sections as in Fig. 127 may be interstitial to the other constituents and is then less easy to identify. Indeed a staining test or even X-ray analysis may be necessary before the mineral is positively indentified. The nemafite drawn in Fig. 127 is a striking example of its kind, with the nephelines abundant and euhedral. Mafic minerals include large titanaugites with pale green outer zones and extensive poikilitic phlogopitic mica. Of the accessories sphene is abundant, and its relationship to the other minerals is unusual: some of the crystals are euhedral, but in the main it forms irregular crystal grains poikilitically enclosing the groundmass microlites, especially nephelines. This sphene is evidently late in the crystallization sequence. Two textural features call for comment: the zoning of the pyroxenes and the late growth of the outer zones so that nephelines are enclosed in them. This is true also of the phlogopite crystals.

All the nephelinic lavas referred to above contain identifiable nepheline; but there are in addition certain rocks of the same chemical composition which contain 'occult' nepheline: *i.e.* potential nepheline represented by glass of the same composition. In the case of a rock having the chemical composition of nepheline-basanite, for example, if the nepheline were occult in the glass, it might be termed a basanitoid. Such a rock would be, in effect, a nepheline-glass basanite.

Melilite-bearing Lavas and Dyke-Rocks

Melilite, like the feldspathoid minerals, may 'proxy' for feldspar in undersaturated rocks, particularly those which have been quenched. It differs fundamentally of course, as it contains Ca^{2+} (and often Mg^{2+}) instead of K^+ and Na^+ and therefore takes the place of plagioclase instead of alkali-feldspar. Until the discovery of the alkali-rich lavas in parts of Africa, notably South-West Uganda and the Belgian Congo, melilite-bearing lavas were represented by one type only which was miscalled 'melilite-basalt', and later 'melilitite'.

The original **melmafite** ('melilite-basalt', Hochbohl, 1883) occurs in the Swabian Alps. It is thoroughly ultrabasic, with a silica percentage of only 34. Nearly half the rock is olivine, approximately 40 per cent is melilite, while small quantities of nepheline, perovskite, apatite and calcite also occur. 'Calcitic nepheline-melmafite' concisely describes the type rock. Similar lavas, with or without olivine, occur as rarities in several localities including Tasmania, but more commonly in Africa including Namaqualand and the central African volcanic region in Kenya and south-west Uganda. In the latter area melilite is accompanied by kalsilite in the variety termed **katungite** by A. Holmes, described above.

Although the rock called **alnöite,** from Alnö, an island off the coast of Sweden, was originally classed as melilite-basalt, the name was changed by Rosenbusch to alnöite, *on the grounds that it is intrusive.* Although we strongly disagree with Rosenbusch's reason for changing the name, alnöite is sufficiently distinctive in itself to justify its name. Fundamentally the type is one-third each of melilite and biotite, the remainder of the rock being made up of pyroxene, calcite and olivine in order of importance, with various minor accessories. At Isle Cadieux[1] between Montreal and Ottawa, a small intrusive complex of alnöitic rocks occurs. Two types in particular are outstanding: one contains the mineral monticellite in addition to ordinary olivine (chrysolite); while the other, occurring as mere streaks in the alnöite, consists almost wholly of biotite and melilite.

Alnöite has been recorded also from certain of the African ring-complexes associated with carbonatites, notably at Chilwa, Malawi. Actually in this rock melilite is represented by carbonate pseudomorphs.

As a matter of convenience certain melilitic rocks of coarse grain may be included here. Three varieties have been described under the names turjaite, okaite and uncompahgrite.

Turjaite (Ramsay, 1921) forms a small intrusion at Turja in the Kola Peninsula. In hand-specimens large crystals of titaniferous biotite are conspicuous, embedded in a matrix of melilite which makes up nearly half the rock. Nepheline, too, is an essential component, while apatite (very abundant in some facies of the rock), magnetite, and, as might be expected, perovskite, are important accessories. Although olivine and melanite garnet are usually rare, they also become important in some specimens. The frequent occurrence of calcite, as in certain nepheline-syenites, and the association of melilite-bearing rocks with carbonatites are significant facts bearing on the genesis of this rock.

Of still more extraordinary composition is a rock, occurring in a volcanic plug near Quebec, but varying in grain-size from coarse to fine, termed **Okaite** by Stansfield. In its mineral composition okaite is very close to turjaite, but melilite is slightly more abundant, biotite less so, while the place of nepheline is taken by hauyne. Okaite is a great rarity, and is noteworthy as having a particularly low silica percentage, only 29, while it contains 25 per cent of CaO.

A very rare coarse-grained melilitic rock has been recorded under the name **uncompahgrite** from a locality in Colorado. The most conspicuous feature is titaniferous biotite occurring as large reddish brown crystals, together with aggregates of titanomagnetite, perov-

[1] Bowen, N. L., Genetic features of alnöitic rocks at Isle Cadieux, Quebec, *Amer. J. Sci.*, **3** (1922), 1.

skite and apatite, all embedded poikilitically in melilite amounting to two-thirds of the total. Thus essentially the mineral composition is biotite plus melilite—the same as that of the rock occurring in the Isle Cadieux complex, noted above. It has the composition of a biotite-melilitite, carrying accessories which seem to be characteristic of melilitic rocks.

Finally reference may be made to a British occurrence described by C. E. Tilley, from Scawt Hill near Belfast, where a plug of Tertiary dolerite penetrates chalk. Both nepheline and melilite occur in a hybrid zone formed by interaction between the magma and the carbonate rock.

With regard to the origin of the melilitic rocks the following facts appear to be significant:

(1) the close association of many, but not all, the rocks concerned with carbonatites (Alnö and Chilwa) and the occurrence of accessory primary calcite in some of them;
(2) the exceptionally low silica percentage and the extremely high lime content, which is the highest in all igneous rocks excluding carbonatites;
(3) experimental studies have proved that at high temperatures nepheline and augite combine to form olivine and melilite with monticellite as an intermediate product of the reaction;
(4) when calcium carbonate is added to fused basalt, melilite crystallizes from the melt; and
(5) melilite occurs in the hybrid reaction zone between chalk ($CaCO_3$-rock) and basalt at Scawt Hill, Co Antrim.

Of these facts we would emphasize (3), (4) and (5) as being specially significant. The association of nepheline and augite is one of the commonest in the nephelinic feldspathoidal lavas. The experimental results suggest that this association is essentially the same as melilite with olivine, encountered far less frequently. This indicates that physical conditions, rather than exceptional composition, determine whether the association nepheline-augite, or melilite-olivine will separate from the magma in any given case. The latter association is characteristic of quenched rocks, while the former is the more stable combination, and that which is much more frequently encountered. However, chemical factors must also play a part. The evidence suggests that desilication of basic magma by reaction with carbonates—whether as limestones or carbonatites is immaterial—probably plays a critical part. Thus the origin of melilite may be closely analogous to that of nepheline, as discussed in the following chapter.

Finally reference must be made to those mafic lavas which contain analcite. The only variety which calls for recognition is **analcite-**

basanite, comparable with nepheline-basanite, but with analcite substituted for nepheline. Thus the mineral assemblage comprises olivine, pyroxene, basic plagioclase, analcite and accessories. This is the mineral assemblage of the Scottish essexites and teschenites which are the coarse-grained equivalents of these analcite-bearing lavas. It will be noted that it is also the mineral assemblage of normal olivine-basalt, with analcite in addition; and the reader is warned that small quantities of analcite are easily overlooked in a thin section, especially when it is interstitial. A British example occurs at Calton Hill, Derbyshire.[1] The rock is very similar in general appearance to olivine-basalt, but the analcite is rather more obvious than usual as it forms small spherical ocelli.

[1] Tomkeieff, S. I., The volcanic complex of Calton Hill, Derbyshire, *Quart. J. Geol. Soc.*, **84** (1928), 715.

CHAPTER XII

ORIGIN OF THE FELDSPATHOIDAL ROCKS AND CARBONATITES

HAVING in the preceding chapter concluded our review of the petrography of the many varieties of feldspathoidal rocks, it remains to consider the evidence relating to the origin of some of them. The diversity of rock-types is so great that obviously no single mode of origin can cover them all, and the discussion which follows is necessarily selective. It is mainly the strongly alkaline rocks rich in feldspathoids with which we are concerned and which provide the major problems. Two aspects of the problems of origin are examined here:

(1) the origin of nepheline-syenites and related feldspar-free types (the urtite, ijolite, melteigite series) and their relationship to the equivalent lavas;

(2) the nature and origin of the closely associated carbonatites.

(1) The Origin of Nepheline-syenites and Related Feldspar-free Types

The majority of nepheline-syenites are closely associated with, and apparently intrusive into, granitic rocks. At one time this association was interpreted as demonstrating that the former had been produced from granitic magma by a process of desilication. The desilicating agent was thought to be limestone which in many occurrences was closely involved with the granite and nepheline-bearing rocks. R. A. Daly formulated a hypothesis according to which reaction between limestone and granitic magma would produce Ca- and CaMg-rich silicates which, on account of their high specific gravity, would tend to sink, leaving a desilicated magma-fraction from which nepheline could crystallize. Although Daly proposed the idea and gave it wide publicity, S. J. Shand had a major share in developing the hypothesis largely as a result of his detailed studies of alkali complexes including that at Loch Borolan in Assynt and several in Africa, especially Spitzkop and Palabora. These and other examples are discussed in considerable detail in his *Eruptive Rocks* which all students interested in feldspathoidal rocks should study.

In the following discussion three alkali complexes are considered: all three involve 'granites' and carbonate rocks and have been used to illustrate the Daly–Shand hypothesis. The first two are

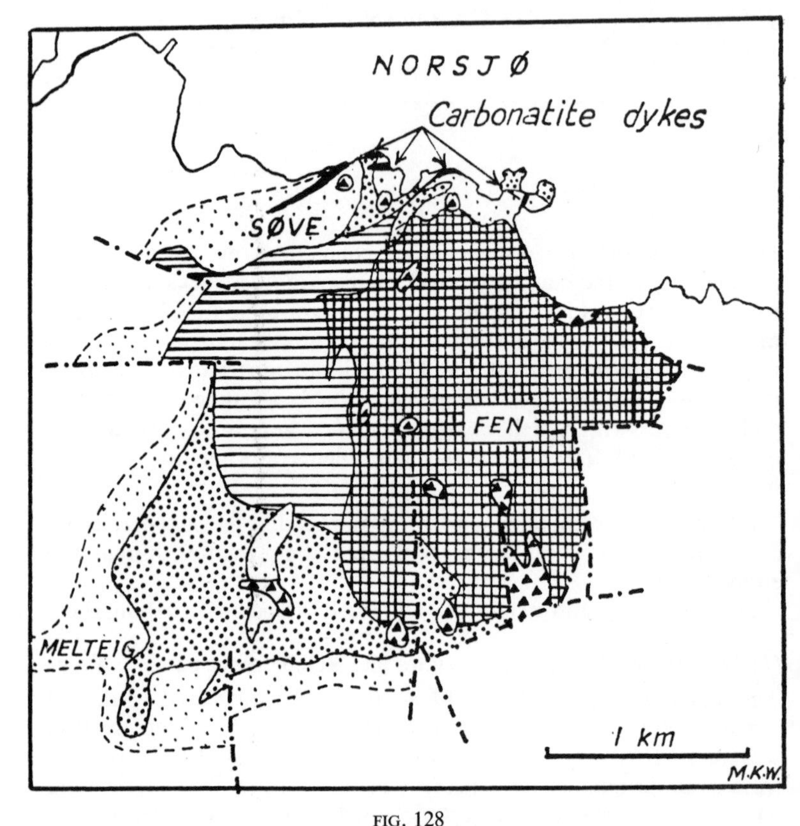

FIG. 128

The Fen alkaline igneous complex.

Precambrian left blank; fenites (metasomatic syenites and nepheline-syenites), light stipple; rocks of the ijolite–melteigite series, heavy stipple; mixed carbonate–silicate rocks, horizontal ruling; calcite, dolomite, ankerite and haematite rocks (in part metasomatic), cross-ruling; carbonatite dykes, black; kimberlite breccia-vents, triangular ornament. (*After E. Saether.*)

situated in the Baltic Shield; the Alnö Complex[1] forms part of an island off the Swedish coast, while the Fen Complex[2] lies just west of the Oslo Graben in southern Norway. Both are ring-complexes,

[1] Von Eckermann, H., The alkaline district of Alnö Island, *Sver. geol. undersok.*, no. 36 (1948), 176.

[2] Saether, E., The alkaline rock province of the Fen Area in Southern Norway, *Kong. Norske Vidensk. Skr.*, no. 1 (1957). See also summary account in *Geology of Norway* (Oslo, 1960), ed. Olaf Holtedahl.

the alkali rocks enclosing central masses of carbonate rocks originally thought to be sedimentary and termed limestone, but now known to be **carbonatite.** Very poor exposure is a characteristic of many alkali-complexes, including these two, and must be held responsible for the early misinterpretation of the field relationships. The two complexes are so similar in their essential features that they may be described together. Both are set in 'granitic' environments, the country-rock being granite-gneiss forming part of the Basement Complex. The central carbonatite forms a plug about which the other rock-bodies are arranged concentrically. They comprise (1) an inner zone of feldspathoidal rocks in variety; (2) a zone of nepheline-syenites formed by metasomatism of the gneiss *in situ*; (3) an outer zone of saturated syenites also resulting from metasomatic alteration of the gneiss; and (4) a zone of fractured gneiss.

The first stage in the formation of the complex involved the explosive fragmentation of the rocks above an advancing plug of magma. Progressive metasomatism of the brecciated rocks produced the outer zone of syenites by elimination of free quartz, or rather its incorporation in newly crystalline silicates which include a distinctively coloured amphibole, though aegirine is the characteristic mafic mineral. No feldspar was actually introduced into the gneiss, though new alkali-feldspars of both sodic and potassic types were developed. To what extent they owe their origin to redistribution of material already in the gneiss it is impossible to say because the original gneiss is not available for analysis. Some ions must have been introduced. The new potassic feldspar contains Ba^{2+}, so that in a manner of speaking it contains celsian.

Desilication beyond the point where free quartz was eliminated gave rise to nepheline-syenites which were formed from solid gneiss by metasomatism in place. This is proved by the fact that structural features may be traced from the gneiss into the nepheline-syenites. The latter, therefore, are nepheline-syenites in composition though they are definitely not magmatic in the strict sense. Brögger[1] introduced the term **fenite** for rocks of this nature, while the term fenitization is widely used for the metasomatizing processes involved: the latter, as we have seen, resulted in the progressive elimination of silica (desilication) and the accession of alkalies. Lying inside the encircling fenites occur feldspathoidal rocks including nepheline-syenites and feldspar-free types belonging to the urtite-ijolite-

[1] Brögger, C. W., Die Eruptivgesteine des Kristianiagebietes—IV. Das Fengebiet in Telemark, Norwegen, *Vidensk. Selsk. Skr.* 1, *Math. Nat. Kl.*, no. 9 (1920), 1921. This is a classic monograph containing the original descriptions of a large number of feldspathoidal and carbonate rocks to which, unfortunately, he gave an equally large number of names of the type-locality variety, among them melteigite, fenite and sövite.

melteigite series. These rocks form a major part of both the Alnö and Fen Complexes. No relict structures inherited from pre-existing rocks are visible, and therefore there is no evidence suggesting derivation from the latter. On the contrary they exhibit sharp and cross-cutting contacts with adjacent rocks and are sometimes clearly intrusive. Further, some of these rock-types occur also in radially disposed dykes which are regarded as magmatic. One view concerning the origin of these apparently magmatic nephelinic rocks is that, after metasomatism, the material was rendered mobile (mobilization is the term commonly used for the process) and thus achieved a magmatic condition. Thus the metasomatic stage is a prelude to the formation of magmatic rocks.

The sequence of events outlined above provides the general pattern for a large number of similar complexes, especially in Africa. Naturally they differ in detail. The country-rock is Basement Gneiss in many, but not in all cases: it may be sandstone or rhyolite tuff; but in most cases desilication like that observable at Alnö and Fen has led to the formation of syenites and/or nepheline-syenites. Fenites are frequently, though not invariably visible; while in one case of outstanding significance, denudation has bitten into the rocks less deeply than usual with the result that the remains of an extrusive phase including lavas and tuffs have survived, as well as the more usual coarse-grained rocks of the intrusive phase.[1] The importance of these relationships will not be lost on the reader, of course.

In the Haliburton-Bancroft area in Ontario a very different type of occurrence of nepheline-syenites is found. The rocks form an integral part of the metamorphic complex of the Canadian Shield. It is a classic region; but once again the field relations are very differently interpreted. As compared with Alnö and Fen, the fundamental difference is the association of granites with nepheline-syenites and limestones of sedimentary origin—the very extensive Grenville Limestone. Formerly the nepheline-syenites were regarded as having originated by desilication of granitic magma in the manner postulated by Daly and Shand; but Tilley has made a detailed study of some of the nepheline-bearing rocks, and has shown that the granites were intruded after the formation of the former, so that the limestone-syntexis hypothesis may not apply.[2] The nephelinic rocks are again interpreted as being metasomatic; but the process involved silication of the limestone which has been converted into

[1] King, B. C., The Napak area of Southern Karamoja, Uganda: a study of a dissected late Tertiary volcano, *Geol. Surv. Uganda*, Mem. 5 (1949).

[2] Tilley, C. E., Problems of alkali rock genesis, *Quart. J. Geol. Soc.*, **108** (1958), 323. This valuable review consists of two parts: the first deals with general problems; the second is a petrological study of an occurrence of nephelinic rocks east of Bancroft.

varied assemblages of plagioclase and nepheline associated with hornblende, biotite, microcline and other minerals. The process whereby the nephelinic rocks were generated (aptly termed nephelinization[1]) was clearly metasomatism of the reverse kind to that which was operative at Alnö: there desilication of 'acid' rock took place; but at Bancroft silication of very basic material occurred. In both areas nephelinic rocks were the products of metasomatism *in situ*, but nepheline-syenite magma was available, too, though whether it should be regarded as cause or effect is a baffling problem.

It is noteworthy that, of the materials needed to build-up nepheline, all were originally present in the granite-gneiss at Alnö—they only needed redistributing; but at Bancroft none of the necessary cations were present—they all needed to be introduced from outside. The source of these is problematical: Tilley considers it probable that they were derived from a basic magma in depth; but granitic magma was on the spot, so too was another magma represented by the magmatic nepheline-syenites: their interrelationship is still quite unknown.

(2) **Carbonatites**

In the foregoing account passing reference has of necessity been made to the more or less pure carbonate rocks now widely known as carbonatites.[2] Up to 1968 about 200 occurrences of carbonatites had been recorded from 90 different localities spread over all the continents except Australia; but half of these occur in Central or Southern Africa. Notable among them are carbonatites associated with nephelinite-tuff shield volcanoes, certain craters in the Western Rift where carbonatite lavas are preserved and, by contrast, deeply eroded plugs in S.W. Africa and the Transvaal.

In addition to the central plugs, carbonatites form dykes radiating outwards from the volcanic centres, and also cone-sheets concentric with the latter. In all three modes of occurrence carbonatites are comparable with, say, dolerites, and a magmatic origin is inevitably suggested by the demonstrable field relations. The carbonatite cone-sheets at Alnö, for example, are exactly comparable with the doleritic cone-sheet complexes in the British Tertiary volcanic province.

In spite of these structural analogies there has been widespread

[1] Gummer, W. K., and Burr, S. V., Nephelinized paragneisses in the Bancroft Area, Ontario, *J. Geol.*, **54** (1946), 137–68.

[2] Campbell Smith, W., A review of some problems of African carbonatites, *Quart. J. Geol. Soc.*, **112** (1956), 189–220; King, B. C., and Sutherland, D. S., Alkaline rocks of eastern and southern Africa, *Sci. Prog.*, **48** (1960), Pt. I, 300–21, and Pt. II, 504–24; *Carbonatites*, ed. Tuttle, O. S., and Gittings J., (1966); Deans, T., World distribution of carbonatites in relation to volcanism, *Proc. Geol. Soc.*, **1647** (1968), 59.

reluctance to accept carbonatites as magmatic for two chief reasons: (1) the high melting point of calcite (1,339° C); and (2) the very high vapour pressure of CO_2 which, it is thought, must have been involved. With regard to the latter, it is significant that the intrusion of carbonatites has been accompanied by intense brecciation of the country-rock, as for example at Alnö, proving that gas pressures were indeed very high. It may be mentioned in passing that the suggestion has been made, by Bowen among others, that carbonatites might be replacive after silicate rocks; but this is generally not supported by the features displayed in thin sections. Critics of the carbonatite-magma hypothesis emphasize the point that observed intrusive relationships, which they do not query, do not prove a magmatic origin. That is agreed. In Iran in particular rock-salt has punched its way through thousands of feet of sedimentary rocks, developing a strong cross-cutting relationship towards them in so doing. It has been extruded like lava from cones aping those of genuinely volcanic origin; and underground has formed intrusions to which such terms as laccoliths might be applied. In spite of these facts nobody finds it necessary to invoke a rock-salt magma. On account of the ease with which plastic deformation and flow may be induced in calcite, there is no doubt that in a suitable environment calcite could behave like rock-salt. Actually the two cases are not analogous. Where the rock-salt originates and how it reaches the surface are reasonably well understood; but where and how carbonatite originates is still a complete mystery, and its almost invariable association with nepheline-syenites and related rocks is not understood.

Viewed simply as rock-types, carbonatites are very variable. Some are pure carbonate rocks consisting of coarsely crystalline calcite, dolomite or ankerite in different instances. At Alnö, for example, one set of cone-sheets is calcitic while a second set, converging on a different focus, is dolomitic. Apatite and magnetite are frequent accessories, and there is a gradation into silicate rocks. At the one extreme are such rocks as that which has been given a distinctive type-name (tuvinite, Yashina, 1957) and consists of nepheline 75 to 95 per cent and calcite 5 to 25 per cent. Nepheline-syenites of several types contain small amounts of accessory calcite, undoubtedly primary and often 'armoured' with cancrinite where calcite abuts against nepheline.[1] The silicate minerals in carbonatites are usually patchily distributed and often the rock shows a streakiness simulating flow-structure parallel to the walls of the intrusion. Further, in some instances carbonatite plugs are composite, showing a textural variation between one member and another, in the manner of granitic

[1] This is useful indication of the presence, in a thin section, of nepheline which otherwise might pass unnoticed.

complexes.[1] Some carbonatites are demonstrably eruptive, though one hesitates to call them igneous for obvious reasons. Whether they may legitimately be regarded as magmatic depends upon how magma is defined; but it is impossible to escape from the conclusion that carbonatites which form parts of the eruptive complexes described above are magmatic in the broadest sense. By accepting the term carbonatite magma we do not imply that the latter consisted of nothing but fused carbonates: on the contrary it contained not only the components of the latter, but also of the associated silicates, as well as the volatiles responsible for the metasomatic changes. In view of these facts it is evident that one of the chief objections to the carbonatite magma hypothesis—the high melting point of the carbonate minerals—has nothing to do with the problem. It may be noted, further, that Wyllie and Tuttle[2] have shown by experiment that the presence of an adequate amount of water-vapour lowers the temperature at which calcite-rich liquids can occur very greatly—to the region of 650° C even under moderate pressures. Thus the supposed physico-chemical objections to a carbonatite magma do not apply, and a reconciliation has been achieved with the field evidence.

Chemical analysis often reveals the magmatic nature of carbonatites by demonstrating relatively large amounts of such elements as barium, strontium, niobium, cerium, etc., which give rise to minerals of economic importance, such as pyrochlore, the chief source of niobium, and, on a much greater scale, of apatite. In addition to these relatively rare elements it is evident that the magma was rich in H_2O, CO_2, Ca^{2+} and alkalies, especially, but it was notably deficient in silica.

From the foregoing account certain important generalizations emerge. Some nepheline-syenites and rocks of ijolite type have resulted from metasomatism involving in some cases granitic rocks. Although the metasomatic aureoles frequently surround plugs of carbonatite, this is not always so, though it may be argued that, although not exposed at present ground-level, it may well be hidden below. When the nephelinic rocks are associated with visible carbonatites, the latter are often seen to intrude into the former. It has been argued that, as the carbonatite demonstrably

[1] Almost inevitably, but regrettably, the wide diversity of carbonatites in regard to their essential carbonate minerals, the proportions of silicates present, and even their mode of occurrence, has led to the introduction of a number of new rock-names, all based on type-localities. All that is required, in fact, is the name of the dominant carbonate mineral used as a prefix to 'carbonatite'. The problem of classification of carbonatites is currently under investigation by the international commission already referred to.

[2] Wyllie, P. J., and Tuttle, O. F., The system $CaO–CO_2–H_2O$ and the origin of carbonatites, *J. Petrol.*, **1** (1960), 1–46.

post-dates the fenitization at the present level of erosion, it cannot have been the cause of the metasomatism; but evidence of this kind may all too easily be misinterpreted. At any stage before the ultimate immobilization of the magma, its uprise would have been preceded by a wave of metasomatizing emanations which fenitized the country-rock surrounding the conduit. The more slowly uprising magma might then develop intrusive relationships towards the fenites; or alternatively, mobilized nephelinic rock-material might intrude into the material occupying the conduit. The problem is three-dimensional and the time-factor is important.

With regard to the nature of the magma responsible for these manifestations, it was evidently very different from that connected with normal volcanic activity. On general grounds it appears to have been much more Basic than normal basaltic magmas: it very effectively desilicated the surrounding rocks as described above. Further, apart from the rarer elements already noted, it was rich in volatiles of which CO_2 was the most significant, and which controlled the whole sequence of events and determined the course of crystallization. The high CO_2 content resulted in the elimination of Ca^{2+} from the silicate rocks, with the suppression of the anorthite and diopside constituents. Hence any potential plagioclase in the original magma is represented by desilicated albite—nepheline—while the associated pyroxene is generally aegirine instead of augite. The elimination of Ca^{2+}-bearing components 'boosted-up' the alkalinity of the rocks, though there is the possibility of further increase in alkalis by gaseous transfer from deeper sources.

The real nature of the magma must remain uncertain: no samples of the complete magma are available for study, and we can judge only from the effects it produced upon the contiguous country-rocks. In complexes like those at Alnö and Fen the site of the volcanic conduit is now occupied by carbonatites, as though a carbonatite magma-fraction had flushed out its forerunners, or had replaced them in some other way. There is no possible way of deciding to what extent the composition of the main stream of magma was modified by fenitizing reactions. The problem may be likened to that of deciding on the composition of liquids which had passed through the waste-pipe of a chemical laboratory, on the evidence of the effects of corrosion of the pipe itself.

One of the major problems is to decide from the evidence of the rocks just what volatiles were available in the original magmas. A fascinating glimpse of the possibilities is provided by the data relating to the 1966 eruption of the active carbonatite volcano, Oldoinyo L'Engai in Tanzania[1] (Fig. 107) when sodium-, not

[1] Dawson, J. B., Sodium carbonate lavas from Oldoinyo L'Engai, Tanganyika, *Nature, Lond.*, **195** (1962), 1075–6, and Volcanic activity of Oldoinyo L'Engai, Tanzania 1966, *Proc. Geol. Soc.*, **1644** (1968), 268.

calcium-carbonatite lava was erupted, followed by violently explosive eruptions of Vulcanian and Plinian types, giving rise to extensive ash-falls. The chemical composition of the lava is remarkable in several respects. Of silica it contains only a trace; alumina (0·08), total iron (0·26), and magnesia (0·49) are exceptionally low. By contrast, Na_2O (29·53), CaO (12·74), K_2O (7·58), SrO (1·24), BaO (0·95) and H_2O (8·59) are exceptionally abundant; so too, are the volatiles, especially CO_2 (31·75), Cl (3·86), F (2·69) and SO_3 (2·00). But for the accident of an arid climate the sodium carbonate would rapidly have dissolved and evidence of considerable petrogenetic significance would have been lost and probably never even been suspected.

CHAPTER XIII

THE LAMPROPHYRES

THE lamprophyres were so named by von Gümbel (1887) when describing the 'mica-traps' of the Fichtelgebirge, and had reference to the lustrous character of some types due to the presence of abundant phenocrysts of biotite. The term lamprophyre is now applied to a group of melanocratic dyke rocks irrespective of the nature of the ferromagnesian minerals of which they largely consist.

In any scheme of classification it is difficult to fit in the curious rocks grouped under this heading, for some are associated in the field, and are therefore genetically linked, with each of the major rock-groups dealt with in the previous pages. Specific types might therefore have been dealt with under the headings granitic, syenitic, dioritic or gabbroic rocks; but something is to be gained by considering them together as one group. As rock specimens, most are unconvincing. They may closely resemble normal igneous rocks in the medium and fine-grain groups, and from the examination of hand-specimens alone would almost inevitably be misidentified. Identification is possible only with knowledge of the field relations and microscopic characters.

Speaking generally, the lamprophyres are strongly porphyritic with abundant phenocrysts of any of the following: dark mica, pyroxene, amphibole and olivine. These are set in a groundmass of alkali-feldspar normally, though in one group it is plagioclase near andesine in composition, while another group comprises the feldspar-free lamprophyres.

Undoubtedly the most striking phenocrysts occur in the mica-bearing lamprophyres: the abundance and large size of the biotite crystals causes a very distinctive appearance in the hand-specimen, while under the microscope the perfect idiomorphism of the mica is unique. In common with all the other mafic constituents, the biotite may be corroded to almost any degree; while internal bleaching with the development of a complementary dark margin is a constant feature. Less commonly they may display a striking colour zoning (Fig. 129).

The amphibole in lamprophyres is of two very different types: it

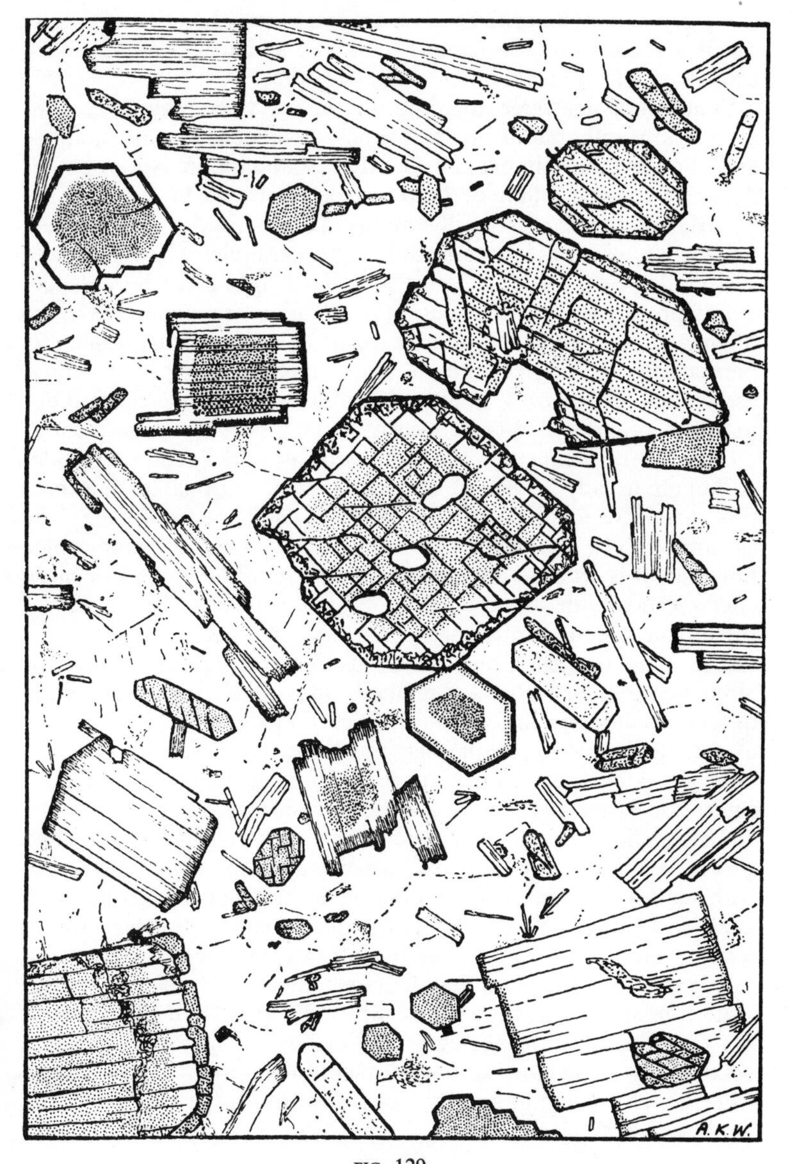

FIG. 129

Augite-minette, South Hill, St Helier, Jersey.

Phenocrysts of diopsidic pyroxene subordinate to biotite, some crystals of which are beautifully zoned. Hornblende rare. The chief accessory is apatite in two generations: large euhedral phenocrysts (*e.g. bottom, centre*) and minute needles. Orthoclase forms the matrix in which all the other components are embedded; it is ideally fresh.

is a green common hornblende in lamprophyres of dioritic affinities; but is a sodic variety, originally identified as barkevikite,[1] in those associated with sodic syenites such as foyaite.

Similarly a light-greenish diopsidic augite is commonly seen in association with biotite in lamprophyres associated with granites and diorites; but titanaugite is often abundant in lamprophyre-dykes of foyaitic affinities. In the rock illustrated in Fig. 129 the pyroxene is zoned, a greenish core being surrounded by a colourless outer zone, while the margins are slightly granulated and in places surrounded by a reaction rim of granular bright green hornblende.

Lamprophyres of all types may contain pseudomorphs after olivine, as isolated euhedral crystals, or as synneusis aggregates. A particularly striking feature of many lamprophyres is the close association of perfectly unaltered biotite with completely pseudomorphed olivines. Although serpentine may occur in these pseudomorphs, very frequently the latter consist chiefly of carbonate. Although usually identified as calcite, it is reasonable to expect this carbonate to be magnesite, ankerite or at least dolomite. The carbonate is by no means restricted to the olivine pseudomorphs, however; it is often so abundant in the body of the rock as to cause effervescence with dilute acid. In even the freshest lamprophyres the olivine is consistently pseudomorphed; but actually these rocks are, in many instances, highly altered. In extreme cases they consist almost entirely of secondary minerals such as chlorite, carbonates, quartz, chalcedony and limonite. It is difficult to decide to what extent this is due to weathering; but there is little doubt that it is largely a late-stage, autometamorphic effect, due to a residual concentration of magmatic water, carbon-dioxide, etc.

The mafic minerals described above may occur as a second generation in the groundmass, associated in the commoner types with abundant alkali-feldspar and, in the feldspar-free group, with analcite. Occasionally the feldspar is quite fresh, water-clear in thin section, and poikilitic towards the other components. More commonly it is microlitic and may be much altered. Rarely the groundmass consists of light-brown glass which devitrification converts into fibrous subspherulitic aggregates of orthoclase microlites. Minute octahedrons of magnetite may be plentifully scattered throughout, while a high content of apatite, varying from stout prisms to delicate needles, is characteristic.

Classification and Nomenclature

Since in so many lamprophyres it is difficult or impossible to identify

[1] As explained in the account of the amphiboles, specific identification of brown amphiboles is difficult, and many varieties originally called barkevikite are proving to be Ti-rich, comparable with kaersutite.

the feldspar accurately, while on the other hand the most striking and most easily identified components are the mafic minerals, the most useful general classification is based on the latter, thus: mica-lamprophyre, hornblende-lamprophyre, augite-lamprophyre. But when the rock is sufficiently well-preserved to allow identification of the groundmass minerals, each of these main categories may be subdivided in the manner suggested by Rosenbusch, as shown in the Table.

ESSENTIAL MINERAL COMPOSITION OF SOME TYPES OF LAMPROPHYRES

	With orthoclase	With plagioclase	No feldspar
Biotite	MINETTE if with augite, augite-minette)	KERSANTITE (if with augite, augite-kersantite)	ALNÖITE (with melilite)
Common Hornblende	VOGESITE	SPESSARTITE MALCHITE (= aphyric spessartite)	
'Barkevikite' and/or Augite		CAMPTONITE ('barkevikite' in type-rock from Campton Falls, New Hampshire)	MONCHIQUITE (with analcite)

The advantage of Rosenbusch's classification lies in the fact that the relationship between a lamprophyre and the parental magma from which it was derived is indicated. Thus the types with dominant orthoclase—minette and vogesite (Fig. 130)—are associated with granites; the plagioclase-bearing types, kersantite and spessartite, are allied to diorites, and are often associated in the field with microdiorites. Finally, camptonite and monchiquite are associated with highly alkaline deep-seated rocks such as foyaite and other sodic syenites.

Petrographic Characteristics and Petrogenesis

The chemical characteristics of lamprophyres are tabulated below. It will be noted that, although they contain less alkali than the aplites, in proportion to their silica percentage they are rich in alkalies, while the CaO is very high. In the mica-lamprophyres a proportion of the K_2O goes to form biotite, and in consequence of the high content of this basic silicate, such rocks with a silica percentage of under 50 may contain a good deal of free quartz.

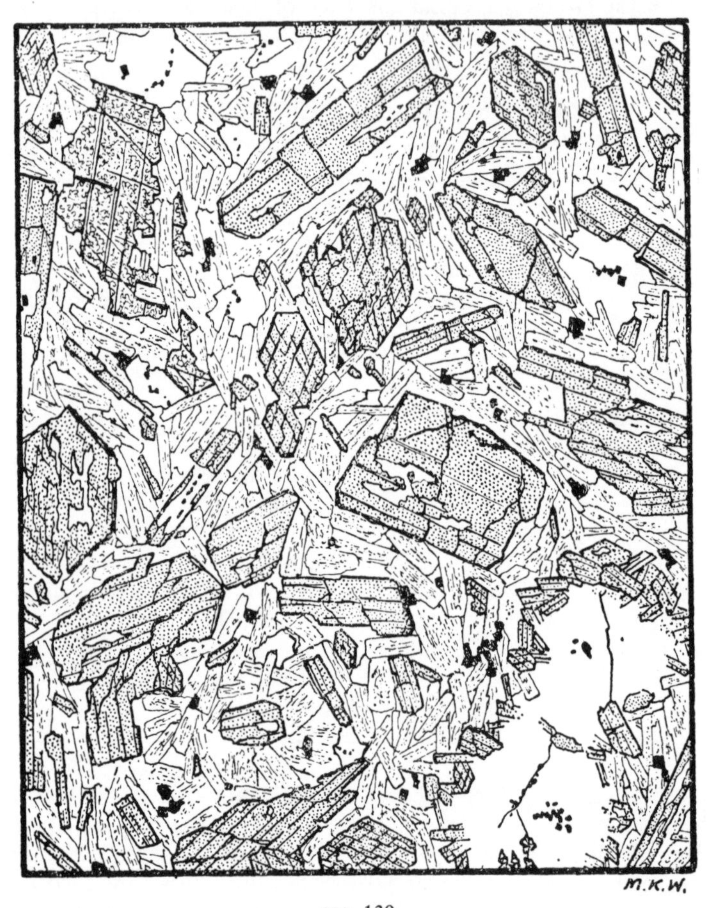

FIG. 130

Vogesite, Ards Peninsula, Ireland.

Common hornblende, stippled; magnetite in small octahedra; small areas of quartz, the largest with a rim of small hornblendes. Orthoclase chiefly in small lath-shaped crystals. Rhombohedral carbonate, *top left*.

The presence of much CO_2 and water is characteristic, and in this and other respects lamprophyres stand in contrast to aplites, with which they may be closely associated in the field, and to which they are in a sense complementary.

The strongly corroded and highly altered state of the phenocrysts in many lamprophyres proves the former to have been 'out of sympathy with their environment', and this fact is interpreted as implying different origins for phenocrysts and groundmass. Bowen regards the phenocrysts as crystal accumulations, and stresses the alkali-rich nature of the matrix in which they are embedded. Tidmarsh, in describing some of the Exeter lavas, which show many of the characters of lamprophyres, invoked mixing and interaction between two residua: the one, a 'depth residuum', highly charged with crystals of mafic minerals; the other, an alkaline residuum, such as might, in the pure state, have crystallized as aplite.

Thus lamprophyres are abnormal rocks: they represent small highly-specialized magma-fractions, and must be carefully distinguished from dyke-rocks which have originated in other, more normal ways. There is a modern tendency, much to be deplored, to use the term 'lamprophyric' as an adjective to describe any relatively melanocratic facies of syenitic or dioritic dyke-rocks. Thus hornblende in association with an 'intermediate' plagioclase constitutes the normal assemblage to be expected in dyke-rocks derived from a dioritic source. Just as the parental rock-types vary widely in the proportion of light to dark constituents, so their dyke equivalents must be expected to show similar variation. Some are melanocratic, but lack the other attributes of true lamprophyres, the corroded phenocrysts and evidence of interaction between accumulated crystals and an alkali residuum. They are, in fact, merely melanocratic microdiorites.

The lamprophyres which agree most closely with the postulated conditions of formations are the mica-lamprophyres.

In the (so-called) feldspar-free lamprophyres the **monchiquites** are outstanding. The name has reference to the Caldas de Monchique in Portugal, whence the original monchiquites were described by Rosenbusch (1890). These rocks are compact and black in hand-specimens, and superficially resemble basalts, but often contain very large phenocrysts (or xenocrysts) of bronzy biotite. Ideally they contain no feldspar, but otherwise resemble other types of lamprophyre, though probably they are even richer in dark minerals. The latter may be olivine, augite often titaniferous, amphibole or biotite. These minerals are embedded in an isotropic base, which in some cases has been identified as analcite, though in others it appears to be glass and may have the composition of melilite. In the original monchiquite the base has lately been shown to have the composition of a mixture of basic plagioclase and nepheline. Obviously the mere presence of an isotropic base cannot be considered the diagnostic feature of a rock-type, and re-examination of this group seems necessary. If the suppressed phase had been allowed to crystallize, the true affinities of the rock would have been clear. If, for example, it was plagioclase, then the appropriate name to apply would be camptonite: the term monchiquite should be restricted to basic lamprophyres with an analcitic base. These rocks frequently contain analcite in a more obvious form—as small circular areas resembling vesicles, but termed ocelli.

Monchiquites are not common rocks in Britain. Interesting examples have been described from the Orkney Islands,[1] and include a biotite-monchiquite containing many phenocrysts of olivine associated with large biotites. Augite occurs as smaller crystals making up a large part of the groundmass, while the base has been

ANALYSES OF SOME TYPES OF LAMPROPHYRE

	1 Minette (Osann, Clarke)	2 Kersantite (Osann, Rosenbusch)	3 Vogesite (Osann)	4 Camptonite	5 Monchiquite (Osann)	6 Alnöite, Alnö, Sweden
No. of analyses	10	20	4		16	
SiO_2	49·45	50·79	52·62	46·46	45·17	27·30
Al_2O_3	14·41	15·26	14·86	14·45	14·78	8·95
Fe_2O_3	3·39	3·29	3·60	3·79	5·10	8·87
FeO	5·01	5·54	4·18	9·87	5·05	7·01
MgO	8·26	6·33	8·55	4·59	6·26	12·34
CaO	6·73	5·73	5·86	7·48	11·06	17·18
Na_2O	2·54	3·12	3·21	4·23	3·69	0·38
K_2O	4·69	2·79	2·83	1·96	2·73	2·99
H_2O	2·43	3·10	2·70	2·05	3·40	5·27
TiO_2	1·23	1·02	0·54	4·01	1·90	3·68
Co_2	0·61	2·61	—	—	—	2·17
Other constituents	1·25	0·42	1·05	0·97	0·86	4·06
	100·00	100·00	100·00	99·86	100·00	100·20

4. Camptonite, Skaergaard, E. Greenland. E. A. Vincent, *Quart. J. Geol. Soc.*, **109** (1953), 38.
6. Alnöite, Alnö, Sweden (Anal. R. Blix). H. Von Eckermann, *Sver. Geol. Undersök.*, **36** (1948), 103.

[1] Flett, J. S., *Trans. Roy. Soc. Edin.*, **29** (1900), 387.

decomposed to a mixture of calcite and fibrous zeolites: its original nature remains unknown. A monchiquite dyke is intrusive into the Old Red Sandstone of Monmouthshire.[1] As is often the case, this lamprophyre contains many corroded xenocrysts, including augite crystals up to 6 inches in length, plates of biotite 2 inches across and grains of quartz. The groundmass is analcitic, so the rock is a true monchiquite.

The **alnöites** are sometimes regarded as a type of feldspar-free, biotite-rich lamprophyre, allied to monchiquite. The type has already been referred to with other melilite-bearing rocks.

Examples of the feldspar-bearing lamprophyres are widespread in Britain, and a comprehensive list would occupy much space. A few typical occurrences are noted, to serve as illustrations only. The North Country lamprophyres occur as narrow dykes over an area extending from Teesdale to Furness, and from Bassenthwaite to Ingleton—a circular area with a diameter of 50 miles, centred about Shap Fell. On account of their radial disposition about the Shap granite, the lamprophyres are considered to be genetically related to it. The types represented are minettes and kersantites, though the distinction is not easily made on account of alteration of the groundmass. Mica-lamprophyres are well represented among the latest minor intrusions in the island of Jersey,[2] and are noteworthy on account of the ideally fresh condition of some of them. The example illustrated in Fig. 129 is one of the most attractive. Most of the Jersey lamprophyres are narrow vertical dykes of minette, but one or two have been identified as camptonite and monchiquite, though, in the latter case, the diagnostic analcite does not seem to have been observed.

As might be expected the dyke phases connected with the Caledonian complexes of southern Scotland include hornblendic types—vogesites and spessartites together with less common kersantites.[3] A typical example of the former is illustrated in Fig. 130, from the Ards dyke-swarm, in North-East Ireland. This rock shows a tendency for both chief minerals to be euhedral—the 'pan-idomorphic texture'. The corroded quartz xenocryst surrounded by small prisms of hornblende (bottom right-hand corner) is characteristic.

[1] Boulton, W. S., On a monchiquite intrusion in the Old Red Sandstone of Monmouthshire, *Quart. J. Geol. Soc.*, **67** (1911), 460.

[2] Smith, H. G., New lamprophyres and monchiquites from Jersey, *Quart. J. Geol. Soc.*, **92** (1936), 365.

[3] Phillips, W. J., The minor intrusive suite associated with the Criffell-Dalbeattie Granodiorite Complex, *Proc. Geol. Assoc.*, **67** (1956), 103.

CHAPTER XIV

THE ULTRAMAFITES, ANORTHOSITES AND THE PHENOMENA OF LAYERING

INCLUDED in this chapter are the descriptions of a very varied group of rock-types, most of which are represented in layered intrusions, and some of which are monomineralic. They are confined to plutonic environments and have no extrusive equivalents, so that they share a number of common problems in regard to their origin. This provides the justification for including anorthosite—a notably leucocratic rock-type—in with the various ultramafic rocks.

The account of layering phenomena may be read either in conjunction with the petrographic description of gabbroic and noritic rocks; or with reference to the bearing of layering on the origin of monomineralic rocks, so many of which are products of extreme differentiation of 'gabbroic' (*i.e.* basaltic) magma.

THE ULTRAMAFITES

In the account of gabbros, norites and troctolites we were concerned chiefly with rocks of average composition. Those of extreme composition referred to briefly in the introductory paragraphs in that chapter remain for consideration. They may be divided into the following groups:

(1) peridotites, in which olivine is dominant and is associated with other mafic minerals, *feldspar being excluded*;
(2) pyroxenites and hornblendites;
(3) picrites which contain the mineral assemblages of peridotites, but with *accessory plagioclase* in addition. They are thus the connecting links between olivine-gabbros and olivine-norites on the one hand and peridotites on the other.

A large majority of these rocks are cumulates derived generally from olivine-tholeiite magma, or less commonly from alkali olivine-basalt magma, and characteristically form parts of layered Basic to Ultrabasic complexes. In addition, however, other intrusive bodies of comparable mineral composition occur, mostly peridotites or their serpentinized equivalents, quite independently of other rock bodies which might be regarded as comagmatic differentiates of less extreme composition. They invariably occur in tectonic belts of intense thrusting, and on account of their modes of occurrence, are termed Alpine type peridotites. These are of special interest because, in at least some cases, they may represent material from the mantle, mechanically emplaced, by thrusting, at high levels in the crust. In the following account, which is essentially petrographic, we are concerned mainly with ultramafites belonging to the first category —those forming parts of layered complexes.

PYROXENITES

The general relationship between gabbros, norites and pyroxenites is discussed on p. 339. Monomineralic varieties may consist, theoretically, of any of those clino- and orthopyroxenes which are relatively Mg-rich. Actually described varieties of reasonably wide distribution are bronzitites and hypersthenites.

Naturally the reader will expect 'augitite' to occur in this list; but this name is not available as it was long ago applied to a type of feldspar-free lava containing prominent augite phenocrysts. **Diallagite** is fully described by its name. Of the pyroxenites the recently described monomineralic **aegirinite** from the Kola Peninsula has the distinction of being the only rock of its kind of distinctly alkaline (sodic) type.

Pyroxenites are quantitatively important rocks in some of the great stratified lopoliths, particularly the Bushveld and Stillwater Complexes and the Great Dyke of Rhodesia. In the first named, hypersthene- and bronzite-pyroxenites, some of them virtually monomineralic **bronzitites,** form sheets of great extent: even when only a few feet thick, a bronzitite scarp-feature in the Bushveld may be visible extending along the strike for many miles until it ultimately passes over the horizon. One such pyroxenite contains the famous 'Merensky Reef' which yields platinum in exploitable amounts. In the Stillwater Complex a thickness of 2,500 feet consists of bronzite-pyroxenite almost identical with the Bushveld occurrences, associated at the base of the lopolith with bronzite-peridotite ('harzburgite'). In both occurrences chromite is a constant accessory, and layers of chromitite (described below) are exploitable commercially.

The most striking accessory in some varieties of bronzite-pyroxenite is the bright emerald-green clinopyroxene, usually termed chrome-diopside, though chrome-augite is, in most cases, a truer name. These rocks are two-pyroxenites and the most satisfactory way of naming them is to combine both mineral names, using the dominant one for the rock-type, and the minor one as the qualifier. Thus in **augite-hypersthenite,** augite is second in importance to dominant hypersthene. Another type contains both diallage and hypersthene in unspecified proportions; and according to the relative amounts it is feasible to distinguish between **diallage-hypersthenite** and **hypersthene-diallagite**—both terms are self-explanatory, and in our opinion superior to 'websterite' (G. H. Williams 1890) the name chosen for this particular pyroxenite from a locality in North Carolina, and, to the confusion of students, still in use.

Mica-pyroxenites are chemically not very different from some kinds of mica-peridotites: the latter are slightly more under-saturated, but both types are relatively rich in K^{+} and Al^{3+}.

That these rocks lie beneath some of the existing lava fields is proved by their occurrence as xenoliths in the lavas themselves which include such distinctive types as the leucite- and kalsilite-bearing types in the (former) Belgian Congo and in Uganda. Augite-peridotite and even olivinite also occur in this association, so that the alkaline affinities of the latter are established in these particular cases. Biotite-pyroxenite is represented among the ejected blocks of deep-seated rocks for which Monte Somma is well known. In Northern Ireland a member of the Newry Complex has the composition: biotite 45, augite 27, hornblende 20, actinolite 2, iron-ores 3 and apatite 3. A rock of this mineral composition is difficult to name. In the description of the complex[1] the name chosen was 'biotite-pyroxenite', but as there is nearly twice as much mica as pyroxene, perhaps 'augite-hornblende-biotitite' gives a truer indication of the mineral composition of the rock.

PERIDOTITES

The essential features of peridotites as noted above, are their exclusively ultramafic character—no plagioclase is allowable—and the dominant role of olivine. This is the sense in which the term peridotite has been used since redefinition by Rosenbusch in 1877. One type is monomineralic and consists of olivine as the only essential constituent. It is normally not olivine-rock in the sense of consisting of the pure mineral; a constant accessory is chromite, and it is only reasonable to permit a very small quantity (say up to 5 per cent) of other accessories, without the necessity of changing the name. The generic name for these rocks is **olivinite,**[2] while such

[1] Reynolds, D. L., The eastern end of the Newry Complex, *Quart. J. Geol. Soc.*, **90** (1934), 585.

[2] Unfortunately olivinite is very differently defined by Johannsen (vol. IV, 402). He uses this name for 'dunites carrying a considerable amount of pyroxene, or amphibole, or even biotite'. In our classification such rocks are typical *peridotites*.

specific names as forsteritite, hortonolitite etc. may be used as appropriate. Instead of olivinite, the less satisfactory name **dunite** (Hochstetter, 1859) is still often used. The type takes its name from a New Zealand locality, Mount Dun, and consists of olivine with chromite as an accessory. Olivinites of this type have been recorded as one of the minor rock-types occurring in layered Basic complexes, in which the occurrences vary from sheets a few feet in thickness to films only a few crystals thick.

All olivinites do not occur in the form of sheets in layered complexes, however. Some occur as pipes and dykes rising from the upper surface of larger olivinite sheets, while in the Bushveld Complex carrot-shaped bodies cross-cutting the layered members of the Complex present an interesting problem. The detailed forms of these cone-olivinites have been very thoroughly studied in the course of mining operations for platinum. Obviously cone-olivinites must have originated under conditions significantly different from those obtaining during the formation of sheet-olivinites. The latter, by general consent, represent accumulations of olivine crystals which settled out of a body of magma, under gravity control; but

FIG. 131

Olivinite ('Hortonolitite'), Mooihoek, Bushveld Complex, South Africa. Anhedral grains of hortonolite showing cleavages and dendritic plates of exsolved iron-ore regularly orientated in two directions.

the material forming cone-olivinites must either have displaced or replaced its own volume of the layered rocks which it cuts. From the textural angle there is nothing in these rocks to indicate that they were formed otherwise than by direct crystallization *in situ* from a 'melt' of olivinitic composition (see Figs. 131 and 132). On

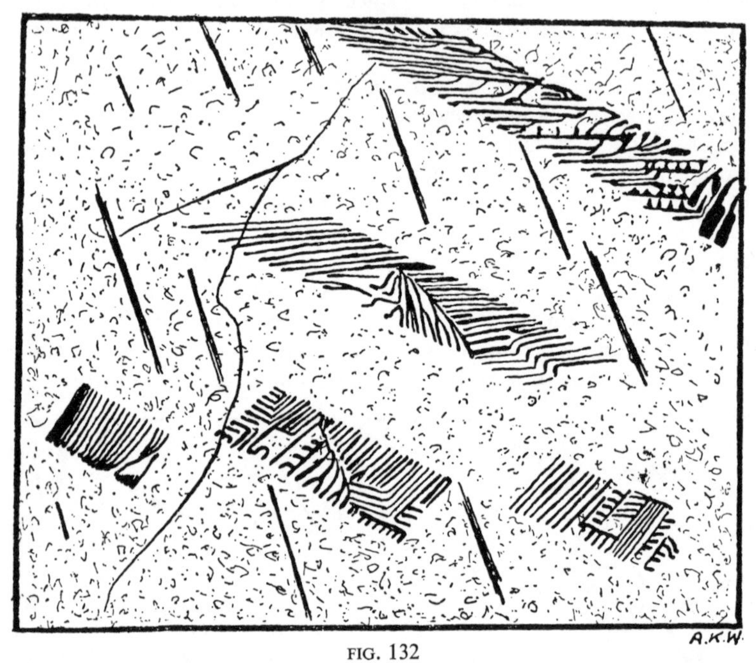

FIG. 132

Regularly orientated plates of dendritic exsolved iron-ore in hortonolite. Same rock as that illustrated in Fig. 131, drawn under ¼-inch objective.

the other hand Hess describes as **'secondary dunites'** certain irregular bodies occurring in the Stillwater Complex, embedded in, and formed from, serpentinite, by a process of reconstruction *in situ.*[1] Bowen and Tuttle have shown experimentally that such reconstruction can be effected at temperatures as low as 500° C, under pneumatolytic conditions.[2] The Stillwater 'secondary dunites' are somewhat richer in iron and coarser in grain than the sheet-dunites, and are stated to be almost identical with the hortonolitite-cones of the Bushveld. There we leave the matter for the moment: it is certainly a thought-provoking problem.

In chemical composition olivinites show analyses like those of

[1] Hess, H. H., The Stillwater igneous complex, Montana, *Geol. Soc. Amer.*, Mem. 80 (1960), 62.

[2] Bowen, N. L., and Tuttle, O. F., The system $MgO–SiO_2–H_2O$, *Amer. J. Sci.*, **29** (1935), 157–217.

the individual minerals. With the incoming of a little alumina, this combines with MgO or FeO to give the almost constant accessory, spinel. With rather more silica, orthopyroxene appears.

Orthopyroxene-peridotites. In view of the chemical relationship between olivines and the corresponding orthopyroxenes, it will be realized that, with a slightly less degree of under-saturation, mono-mineralic olivinite will give place to enstatite-peridotite, bronzite-peridotite or hypersthene-peridotite, according to the Mg:Fe ratio of the rock concerned. In most occurrences (especially the Bushveld and Stillwater Complexes)[1] the orthopyroxene falls within the bronzite range. Although no actual figures are available, it is probable that **bronzite-peridotite** is the most abundant type of peridotite, not only in the two complexes named above, but in all parts of the world. Unfortunately all peridotites are prone to alteration into 'serpentine', and in many occurrences little trace of the original minerals is left: this is the case, for example, in the Lizard Complex in Cornwall; but little difficulty is experienced in inferring from the structure of the serpentine that the 'bastite-serpentine' was originally bronzite-peridotite. Ideally fresh rock of this type occurs in a very different setting elsewhere. In the Kimberley district in South Africa deeply eroded volcanic pipes filled with kimberlite (described below) yield an extraordinary number and variety of blocks of ultramafites, some of very large size. Among them are bronzite-peridotites, quite unaltered, consisting of dominant olivine in closely packed crystal grains showing peripheral granulation due to movement during accumulation of the crystal mush. Embedded in the olivine are scattered 'phenocrysts' of orthopyroxene, somewhat rounded, while translucent spinel is a constant accessory.

Not uncommonly a second pyroxene, in an accessory role, may accompany the hypersthene or bronzite. It may be augite or the much rarer chrome-augite which plays the same part in the corresponding pyroxenites, as noted above. Examples are to be found among the blocks occurring in the kimberlite pipes; but of greater significance are so-called olivine-nodules which occur in basaltic lavas in many parts of the world.[2] Specimens are to be found in most teaching collections, often as mineral specimens of olivine; but even with a hand-lens, and more clearly under the microscope it is seen that these are rock-fragments, *i.e.* xenoliths of peridotite of a distinctive kind. They consist of a coarse-grained aggregate of olivine, enstatite, chrome-augite and spinel—a typical peridotitic

[1] These bronzite-peridotites are the 'harzburgites' in H. H. Hess's account.

[2] Obviously the term olivine-nodules as applied to these bodies is a misnomer. A nodule is a concretionary body formed *in situ* in various kinds of rocks. They have nothing in common with the peridotitic xenoliths occurring in basaltic lavas and dykes.

mineral assemblage. They often show clear evidence of selective magmatic corrosion.

Augite-peridotites do not appear to be widely distributed, but they are represented in the Garabal Hill–Glen Fyne Complex. Augite is the most abundant mineral, occurring as large irregular crystal grains poikilitically enclosing numbers of olivines and associated with small amounts of both hornblende and dark mica. The addition of plagioclase to this assemblage gives augite-picrite, a much more widely distributed rock-type.

Hornblende-peridotite might well figure in a list of possible types of peridotite, though it is indeed difficult to see how a rock consisting exclusively of olivine and hornblende could be formed, in view of the wide separation of these two minerals in the discontinuous reaction series. However, hornblende does occur, with other mafic minerals in some rare types of peridotite, including parts of the Cortland Complex in New York State.[1] Shand has pointed out that there is continuous variation in several directions, and in a rock-body of such variable composition, it is unprofitable to attempt restricting the definition of **cortlandtite:** it is a hornblendic hypersthene-peridotite, containing in addition to the three chief components some malacolite (a clinopyroxene colourless in thin section) and mica. It is reminiscent of scyelite, described below.

Mica-peridotites.—Of the peridotites those which carry abundant mica are evidently as strongly alkaline as any peridotites can be. After olivine, mica (typically a strongly phlogopitic variety), is the most abundant mineral and accounts for the relatively high content of K^+ and Al^{3+}. The most widely distributed and the most important of the mica-peridotites is, beyond doubt, **kimberlite,**[2] which takes its name from the diamond-mining area in the Transvaal, where it occurs in a series of deeply eroded volcanic pipes. Unfortunately the kimberlite itself is very thoroughly brecciated and 'altered' and as the 'blue ground' of the miners, it does not lend itself to detailed petrographic study. Actually the kimberlite pipes are of far greater interest on account of an extraordinarily varied assortment of xenoliths of ultramafic rock-types, including garnet-peridotites and eclogites, etc., which represent material brought up from great depths. For this reason, N. V. Sobolev, leading expert on Siberian kimberlites has stated that kimberlite pipes provide 'a window through which we can observe conditions in the mantle'. The pipe-filling material, the dykes and country-

[1] Shand, S. J., Phase petrology of the Cortlandtite Complex, *Bull. Geol. Soc. Amer.*, **53** (1942), 409.

[2] Dawson, J. B., A review of the geology of kimberlites in *Ultramafic and Related Rocks*, ed. P. J. Wyllie (1967), 241–78.

rock are heavily carbonated. The kimberlite is commonly altered to a mixture of chlorite, talc and carbonates; but occasionally it is solid enough to analyse and some of the olivine is fresh. Outstanding chemical features are the extraordinarily low SiO_2 percentage (25 to 30), high magnesia (30 to 35), TiO_2 (3 to 4), and CO_2 (up to 10 per cent). Diamond-bearing kimberlites occur also in Equatorial

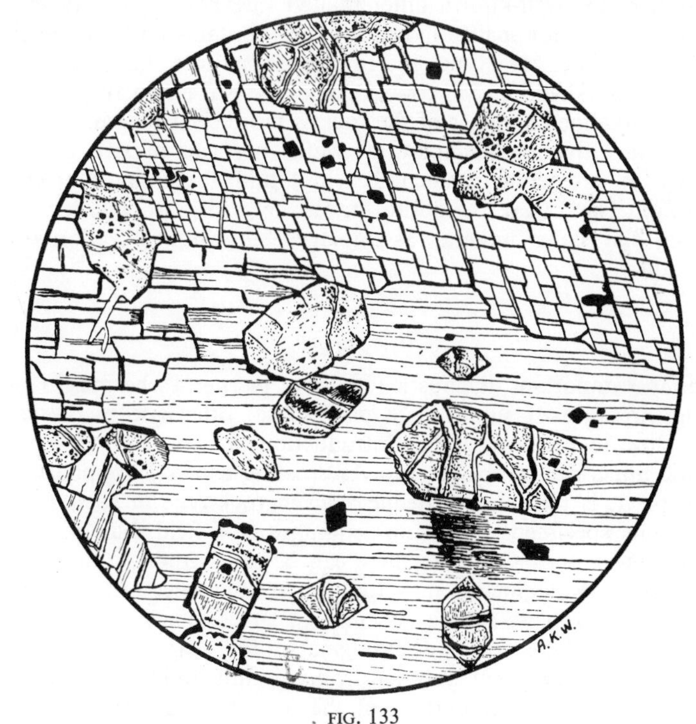

FIG. 133

Scyelite, Loch Scye, Sutherland.

Upper part of the field occupied by basal and vertical sections of nearly colourless amphibole; the lower part by phlogopitic mica. Serpentine pseudomorphs after olivine enclosed poikilitically in both minerals. Magnetite in octahedra and peppering the serpentine. Hexagonal plates, seen end-on, causing schiller structure in the mica (*bottom right*.)

and West Africa and in eastern Siberia (Yakutsk), where again they are noteworthy for the ultramafic xenoliths which they contain, and rarely in the eastern Canadian Shield, where the pipes penetrate through the Keewatin lavas. Diamonds occur rarely in the nearby Drift. Carbonate-bearing kimberlite dykes have been discovered around some of the carbonatite ring-complexes, for example, at Alnö.

Among peridotites of more mixed composition **scyelite** (Judd, 1885) named from Loch Scye in Sutherlandshire is a phlogopite-hornblende-peridotite. In hand-specimens scyelite is a very distinctive, heavy, dark-coloured rock, with large (one inch) mica and hornblende crystals containing black olivines. The texture is poikilitic and the consequent lustre-mottling is most striking. The features seen in thin section are illustrated in Fig. 133.

A characteristic peridotite with alkaline affinities forms the basal portion of the well-known differentiated Lugar Sill from Ayrshire. In addition to dominant olivine this rock contains both titanaugite and barkevikite, sometimes the former, sometimes the latter being in excess. Barkevikite is, of course, one of the NaFe-rich amphiboles; but the alkaline character of the rock is more clearly demonstrated by the fact that the complex, of which it forms a part, consists largely of teschenite (analcite-bearing) and theralite (nepheline-bearing). The Lugar peridotite grades upwards into picrite, containing the same mineral assemblage, but with labradorite in addition.

PICRITES

The most important diagnostic feature of picrite (Tschermak, 1866) is the presence of subordinate plagioclase; otherwise picrites are very close to peridotites. The introduction of feldspar into a peridotitic mineral assemblage involves crossing a phase boundary. In principle it is unsound to define a given rock-type in such a way that some specimens fall in one compositional field, but others in an adjoining one. Therefore there can be no latitude so far as the occurrence of plagioclase in peridotites is concerned. At the point where the bronzite-olivine assemblage gives place to bronzite, olivine *and plagioclase*, we pass out of the peridotite, into the picrite, field. The point may be profitably illustrated by an actual example. In the basal portion of the Stillwater Complex, among the dominant bronzite-peridotites are variants carrying olivine and bronzite (two thirds of the whole rock) with chrome-augite and plagioclase accounting for the rest. The amount of plagioclase is not stated; but that is really beside the point. *Some* plagioclase occurs and therefore the assemblage is picritic, not peridotitic according to definition—our definition admittedly. The rock-name applicable is 'chrome-bearing bronzite-picrite'.

Variants of this type are distinguished as **enstatite-picrite** or **hypersthene-picrite** according to the composition of the pyroxene.

A well-known picrite, mineralogically much like the Lugar picrite referred to above, forms part of the small island, Inchcolm, in the Firth of Forth. This rock contains a large assortment of minerals and is therefore much used for teaching purposes. Olivine, partly altered into a vivid yellow serpentine, is abundant, and is accompanied by titanaugite, in reaction relationship with subordinate barkevikite, and abundant phlogopitic mica, now largely represented by bluish green chlorite. The felsic minerals in the Inchcolm picrite are difficult to identify. The amount of basic plagioclase is, in any case, small, and it is further obscured by analcitization. Even the analcite is turbid and usually has to be 'taken on trust'.

Calc-alkaline **hornblende-picrites,** containing common hornblende as the chief mafic component additional to olivine, occur, for example, in the Plymouth district in Cornwall, Anglesey in North Wales and Colonsay and Glen Orchy in Scotland.

Although there are excellent grounds for precisely defining the boundary between peridotites and picrites, there are no similar reasons for limiting the amount of plagioclase acceptable in a picrite before it grades into melagabbro. The dividing line is arbitrary. This point may be illustrated by reference to the 500-ft layered sill which forms the Shiant Isles off the Scottish coast.[1] In the lowest visible portion a so-called picrite occurs consisting of olivine (59), augite (10), plagioclase (26), iron-ores (2) and zeolites (3 per cent). With over a quarter of the whole rock consisting of plagioclase and with a colour index of only 71, the rock is on the borderline between melagabbro and picrite.

Before leaving the picrites it will be well to remember that they are the only ultramafic coarse-grained rocks which have their equivalents among the lavas, namely the picrite-basalts including both olivine-rich (oceanites) and augite-rich varieties (ankaramites).

MICAITES (GLIMMERITES)

Under these synonymous terms we group certain rocks which are no less monomineralic than olivinite, or pyroxenite, though they are unrelated to the suites considered above. The name for a rock consisting essentially of dark mica is difficult to choose: micaite is the obvious choice and is in use, but some people prefer the Germanic form, **glimmerite** which is certainly more euphonious. The specific terms biotitite and phlogopitite are self-explanatory but must be used only if the mica is of the appropriate composition: a 'biotitite consisting of phlogopite' has been recorded as quoted, but is certainly not accurately named!

How these mica-rocks originate is problematical. They are certainly of deep-seated origin, and as they do not occur in igneous layered complexes it is possible that they should be regarded as metamorphic rather than igneous rocks. They are represented among the ultramafic xenoliths in the kimberlite pipes in the

[1] Walker, F., Geology of the Shiant Isles, *Quart. J. Geol. Soc.*, **86** (1930), 355.

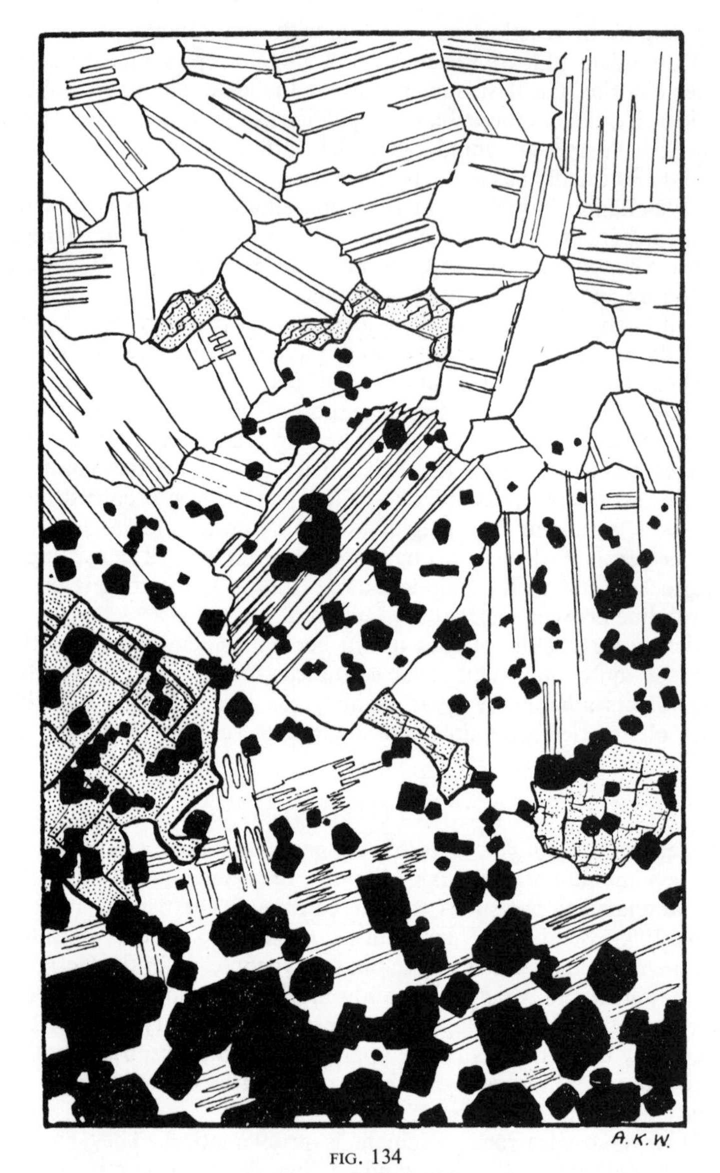

FIG. 134

Contact-zone between leuconorite and chromitite, Dwars River, Bushveld Complex, Transvaal, South Africa.

The outlines and nature of the twinning in the plagioclase are indicated. Bronzite *stippled*; chromite in single crystals and groups of octahedra increasing in size and concentration downwards.

Kimberley district, and occur also among the granulites and gneisses forming the Basement Complex in southern Nyasaland. There is no rock in the fine-grain category which closely matches these coarse-grained rocks, though some minettes (mica-lamprophyres) are exceptionally rich in biotite, to the extent of up to 80 per cent.

CHROMITITES

The amount of chromium calculated as Cr_2O_3 in undifferentiated noritic or gabbroic magma is only a fraction of 1 per cent; but in a small specimen from a chromitite layer the figure may rise to a high value. Apparently the magma quickly becomes effectively supersaturated with Cr_2O_3 and a swarm of cumulus chromites is precipitated. These crystals filter down through the magma and accumulate on a 'floor' where further movement is impeded by crystallization of silicates. The intercumulus magma at this early stage must be capable of precipitating pyroxene and plagioclase, and, in the case illustrated, both these minerals are seen acting as hosts to the chromite octahedra (Fig. 134). Had the magma remained fluid for a longer time segregation would have been more complete, the chromites would have been more closely packed, while the intercumulus material would have been proportionally less. It will be noted that where the concentration of chromite is highest, the crystals are largest and so are those representing the interprecipitate material: a single plagioclase crystal grain spans the field at the bottom of the diagram.

Another Bushveld type contains the same high concentration of chromites, associated in this case with dull greenish bronzite crystals in a whitish plagioclase matrix. This rock has been enriched in two components, firstly in chromite, and secondly in bronzite—both presumably by gravity differentiation. The rock is chromite-enriched melanorite, complementary in composition to the first type. A third type known to the writers is virtually a two-mineral rock, and consists of a coarsely crystalline three-dimensional 'mosaic' of large irregular bronzites, of pegmatitic dimensions—up to half an inch or more in diameter—in which the closely packed chromites are embedded poikilitically, so that hand-specimens display a distinctive lustre-mottling. The rock is in effect a chromite-enriched bronzitite, whose formation may be explained as follows. A few small bronzites were enmeshed in the chromite crystal concentrate. As no other silicate minerals were present at this level, these seed crystals gave bronzite an advantage over the other silicates; a diffusion-gradient was set up and as a consequence, Mg- and Fe-ions migrated towards the rapidly enlarging bronzites, while ions which could not be incorporated migrated in the opposite direction—

chiefly upwards. From the large size of the bronzite crystals a high concentration of volatiles, favourable to pegmatitic crystallization, may safely be inferred. The rock illustrated in Fig. 134 presents some features of interest especially in the field as seen at Dwars River in the Bushveld. The chromitite layers are more sharply defined than is suggested in the figure; they are parallel to the general stratification of the Complex and are interlayered with leuconorite. Further, individual sheets change their horizons suddenly and bifurcate in a manner closely simulating a series of intrusive sills. The field relations are extremely well displayed, and it is difficult to avoid the conclusion that here a chromite concentration was sufficiently lubricated by the interprecipitate material as to develop an intrusive relationship towards 'rafts' of plagioclase crystals with which it became associated before finally freezing-in.

MAGMATIC IRON-ORE ROCKS

Magnetite and ilmenite display many features similar to those described above for chromite in their occurrence in layered intrusions. Thus magnetite and ilmenite both occur in small quantities in gabbroic and noritic rocks and their derivatives, and, like chromite, may become concentrated into layers which consist almost exclusively of ore-minerals, either ilmenite or magnetite separately, or more commonly of the two in close association. There is, however, one significant difference between natural concentrates of chromite on the one hand and of iron-ores on the other. The former is probably always an early precipitate, while magnetite commonly forms late in the crystallization sequence. Further, iron-enrichment is proved by the changes in composition of successive crops of olivine and pyroxene crystals precipitated at successive stages in the formation of a complex such as that of Skaergaard (in which this phenomenon has been particularly studied). It is underlined, so to speak, by the evident iron-enrichment in the interstitial glassy residuum in certain olivine-basalts as already described. The potentiality of the formation of a magma-fraction notably enriched in iron should be borne in mind; this may well be important in connection with the mode of origin of some of the layered iron-ore rocks.

But whatever the composition of the melt, some of the iron must become oxidized from the ferrous to the ferric state, before magnetite can be precipitated. The state of oxidation depends largely on the concentration of water-vapour in the melt.[1] If the latter is 'dry' a large proportion of the silicates must crystallize from it before the water content of the residuum reaches a value sufficiently high

[1] Kennedy, G. C., Equilibrium between volatiles and iron-oxides in igneous rocks, *Amer. J. Sci.*, **246** (1948), 529.

for magnetite to crystallize. This naturally affects the textural relationships between the iron-ores and the silicates. Frequently in gabbroic rocks ilmenomagnetite forms poikilitic or interstitial crystal growths which are evidently of late origin. It follows, further, that the iron ores will be more closely associated with the later, rather than the earlier-formed silicates. This means that magnetite-ilmenite layers in the Bushveld and Freetown Complexes (to quote two examples known to the writers) are closely associated with either anorthositic or pegmatitic facies of the rocks concerned. Some of the largest ilmenite deposits occur in anorthosites, as for example, in the Adirondacks. It will be remembered that chromite concentrations are normally associated with early-formed members of a complex, consisting largely of olivine and pyroxenes.

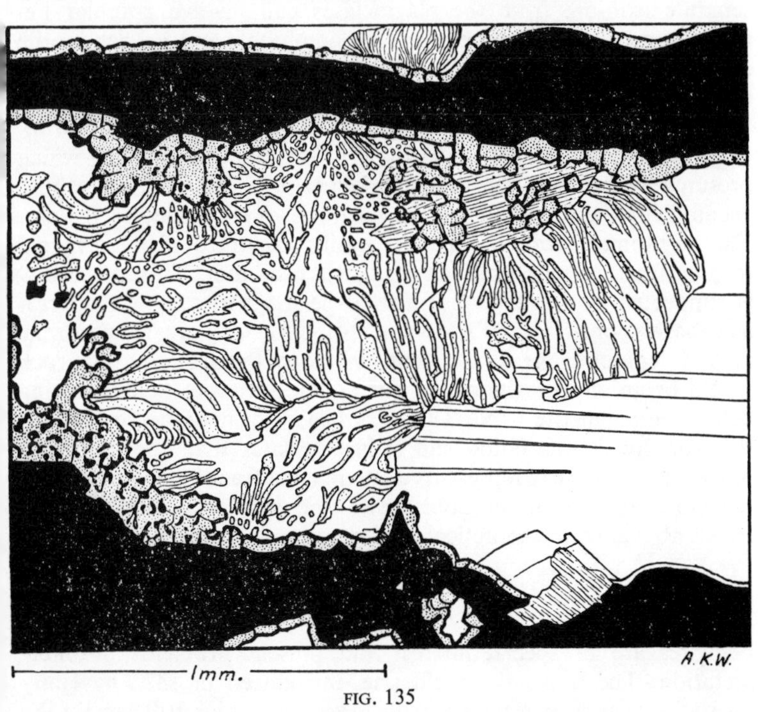

FIG. 135

Reaction Phenomena and Symplectic Structure in Maghemite-Anorthosite, Magnet Heights, Sekukuniland, South Africa.

In this section titano-maghemite runs between plagioclase crystals, while a distinctive feature is the reaction rim of fayalitic olivine which insulates the iron-ore from the feldspar. Myrmekite-like symplectites of orthopyroxene and plagioclase are lobed into the labradorite. Mg-rich mica is intimately associated with the olivine at one point, while the latter contains dendritic magnetite. Orthopyroxene, *light stipple*; fayalitic olivine, *close stipple*; mica, *lined*; iron-ore, *opaque*. Twinning indicated in the plagioclase.

Probably the most fascinating examples of titano-magnetite rocks occur at the aptly-named locality, Magnet Heights, in Sekukuniland in the Transvaal. The iron-ore rocks are interstratified with anorthosite and norite at a low level in the Upper Zone of the Bushveld Complex. The contact between an underlying white anorthosite and the overlying black iron-ore rock is sharp and striking; but although the lowest parts of the latter are almost pure, strongly magnetic iron-ore, plagioclase crystals appear in the ore, at first widely scattered, but increasing in numbers upwards. They are strongly fluxioned, lying with their pinacoidal faces parallel to the top and bottom of the layer. The ore-mineral is titanomaghemite containing Fe_2O_3 in large excess over FeO (60 : 10) with up to 18 per cent TiO_2 and some V_2O_3. Mention should be made of 'reaction rims' which separate iron-ore from the plagioclases and involve granular Fe-rich olivine, shells of cummingtonitic amphibole and Fe-rich biotite as well as symplectic pyroxene-plagioclase intergrowths (growing like myrmekite) shown in Fig. 135. The writers cannot believe that these maghemite rocks represent gravity-sorted concentrates of iron-ore crystals which accumulated *in situ*: the movements which presumably brought the different layers into juxtaposition would account for the fluxioning of the plagioclases embedded in the ore. The phenomena would be most easily explained if the original magma-fraction was itself an iron-rich residuum. This possibility is supported by the fact that above the magnetite occurs a **ferro-anorthosite** consisting of bytownite (An_{63}) 87, iron-ore 11, pyroxene 1, and lepidomelane 1 per cent. Without the iron-ore the rock would be an anorthosite of the same composition as that occurring below the magnetite band. This rock has a normal igneous texture; the iron-ore is subophitic and is largely later than the dominant plagioclase. It clearly represents a late magma fraction and illustrates the close association of anorthosite and iron-ores which we inferred above, from theoretical considerations.

ECLOGITES

Eclogites are of special interest and present problems of interpretation. The name is an old one, introduced in 1822 by Haüy for rocks occurring in the Fichtelgebirge, and which still remain the most typical eclogites. Eclogite is of notably high specific gravity; and consists of bright red garnets set in a bright green crystalline matrix. Two components are essential in eclogites—red garnet of the pyrope-almandine type, and the bright green pyroxene, omphacite. Accessories sometimes present include brilliant emerald-green chrome-diopside, quartz and kyanite. It has been shown by analysis that the eclogite has much the same chemical composition as olivine-gabbro, but contains none of the normal gabbroic minerals: plagioclase has been completely eliminated and the other components have been broken down and redistributed. Labradorite has been split, as it were, into albite and anorthite 'minals'; the former has been desilicated into jadeite, which is incorporated in the pyroxene, omphacite. The silica released may appear as free quartz in spite of the basicity of the rock. The anorthite molecule goes into the garnet which, though fundamentally a type rich in magnesium derived from olivine and original pyroxene, does contain a certain proportion of the 'lime-garnet' molecule ($Ca_3Al_2Si_3O_{12}$). In Britain eclogites occur in northern Scotland in the so-called Lewisian inliers near Glenelg—the first record in this country (J. J. Teall, 1891)—in central Sutherland and central Ross-shire. They there form small outcrops over a wide area, and are closely associated with injection gneisses which were originally Basic igneous rocks impregnated by pegmatite. True eclogite forms only a small part of the eclogitic bands which contain cores of the typical red-garnet, omphacite rock.

Eclogites are the high-temperature and very high-pressure metamorphic equivalent of igneous rock of basaltic composition. One aspect of this relationship is illustrated in Fig. 56, in which the rocks of the oceanic crust (primarily basalts, dolerites, and gabbros) are shown being converted into eclogite as the crust sinks down along a Benioff zone. This particular relationship would normally have little interest for the igneous petrologist were it not for the possibility of the eclogite formed in this way being a likely source of andesitic magmas, derived by partial melting (p. 327). Evidence from xenoliths erupted with volcanic rocks, and found in particular abundance in volcanic pipes of kimberlite, shows that ecologite must be widely distributed as a component of the mantle beneath the continental crust. Whether these eclogites were derived from original Basic igneous rock or crystallized directly from a magma of similar composition under great pressure is rather an academic point: once again the chief interest to the igneous petrologist is in the possibility of eclogite melting as a source of Basic magmas (p. 380).

SERPENTINITES

Although as a general rule special names are not considered necessary for 'altered rocks', serpentinites are the exception to the rule. These consist essentially of mixtures of the different serpentine minerals, chrysotile (fibrous), antigorite (platy), lizardite (massive) and bastite in varying proportions. Popularly these rocks are often named 'serpentines'; but it is awkward and in a sense incorrect to use the same name for a mineral and for the rocks composed largely of that mineral: therefore we use 'serpentinite' for the rocks and 'serpentine' for the mineral.

Serpentinites are compact, variously coloured rocks, often banded, blotched and streaked with bright green and red, the general body of the rock being any colour between light greyish-green and greenish-black. On account of their attractive colouring and because they can be shaped and polished easily, they are used for ornamental purposes.

In some specimens relics of original minerals may remain and give an indication of the nature of the unaltered rock. From such evidence it is clear that some serpentinites correspond in composition with, and were presumably produced from, ultramafites of various types. In some, nothing but olivine (and chromite) can be inferred from the existing texture, and the term 'dunite-serpentinite' has been used. In others, bastite pseudomorphs after bronzite or other orthopyroxenes are prominent and are embedded in other forms of serpentine which clearly indicate that they are pseudomorphous after olivine, and the rock was evidently a bronzite-peridotite. It may safely be inferred, therefore, that in general serpentinites represent ultramafites, dominantly peridotites of various kinds, also picrites and pyroxenites which have been profoundly altered, the changes being of a simple type, involving only hydration of the original silicates. Sometimes the alteration has been so complete that there is no indication that the minerals of which the rocks are composed are other than in their original condition: in other words some serpentinites *look* as if they were intruded as serpentinite magma from which the constituent serpentine minerals separated as primary constituents.

With regard to distribution, serpentinites occur in two very different associations: (1) in the basal parts of layered complexes as, for example, in the Ultrabasic Zone of the Stillwater Complex; and (2) as massive 'Alpine-type' ultramafic bodies in orogenic zones, and frequently bounded by thrust-faults. Outstanding examples of the latter are afforded by the Great Serpentinite Belt in New South Wales, and the comparable belt in southern New Zealand, where individual masses may extend for 400 miles and may be on average 4 miles thick. They are almost vertical and deeply dissected, and the wholesale serpentinization is a most impressive feature. These sheets are in part peridotites but largely serpentinites. Similarly serpentinites forming 'belts', 'lenses' or 'masses' extend throughout the central tectonic zone in Japan. They are of special interest in that they are associated with dykes of two complementary groups: (1) leucocratic—comprising albitite, quartz-albitite, anorthosite, and trondjemite; and (2) melanocratic—including hornblendite, titanaugitite and some gabbroic types.

British examples of serpentinites are typical of the 'Alpine' category in that they occur in tectonic settings of intensive shearing and overthrusting. The Lizard Complex in Cornwall has been studied in the greatest detail and we have chosen it as a good example to illustrate some of the problems involved (see p. 469).

With regard to origin the problem is similar to that of the spilites: serpentinites are altered rocks, but where, when and how the alteration was effected are matters of uncertainty.[1] Without the water they contain, and allowed to crystallize normally, serpentinites in the main would be converted into peridotites: the 'secondary dunites' of the Stillwater Complex are believed to have originated in this way. In dealing with orogenic, as distinct from layered-complex serpentinites, we are concerned with peridotites *not* associated (in the same rock-bodies) with other rock-types from which they might have been derived by gravity-controlled crystal accumulation. We are led to infer, therefore, that the orogenic serpentinites were derived from primarily intrusive peridotite. The physical state of the latter at the time of its intrusion provides one of the major problems of petrogenesis on which a considerable amount of light has been thrown by an investigation of the system $MgO–SiO_2–H_2O$.[2] This was the first study of its kind involving high water-vapour pressures, equivalent to a depth of 6 miles below the surface and temperatures up to 1000° C. Even under these conditions the silicates forsterite and enstatite remained completely crystalline. Therefore water-vapour alone does not lower the melting points sufficiently to make a peridotite magma of this composition a reasonable proposition, and makes it almost certain that orogenic serpentinites were moved and intruded under compressional conditions as lubricated crystal mush. This is consistent with the almost negligible metamorphic effects observable around these bodies. Further, the experiments showed that crystalline olivine remains unaltered in the presence of any quantity of water down to temperatures of 500° C. This is, therefore, the highest temperature at which serpentinization can take place.

Alpine-type peridotites and serpentinites are often closely associated in the field with spilitic volcanic rocks, forming the so-called 'ophiolite' assemblage characteristic of a geosynclinal environment. It requires only a little imagination to see how this association *might* arise in the general context of plate tectonics and orogenesis. In fact, interpretation is very difficult because tectonic transport and recrystallization generally combine to obliterate the evidence needed

[1] Hess, H. H., Serpentinites, orogeny and epirogeny in *Crust of the Earth* Symposium, *Geol. Soc. Amer. Special Paper*, **62** (1955).
[2] Bowen, N. L., and Tuttle, O. F., The system $MgO–SiO_2–H_2O$, *Bull. Geol. Soc. Amer.*, **60** (1949), 439–60.

to decide the derivation of any particular ultramafic body; whether directly from the upper mantle as crystalline rock or peridotite magma; or from basal crystal cumulates of some differentiated basaltic magma. It is generally safe to assume that any isolated body of ultramafic rock that may now be found closely associated with rocks characteristic of the upper part of the crust is far removed—possibly in time as well as space—from its original source with rocks which accumulated in a geosynclinal environment. Up to a point they share certain chemical characteristics, *i.e.* high water and CO_2 content, but they are widely separated in time: spilites are generated in the manner described above during an early phase in the development of a geosyncline; but serpentinites are associated with the orogenic phase following the collapse of the geosyncline; for example in this country the well-known Cambro-Silurian geosyncline dates from the opening of the Cambrian period; the earliest spilitic eruptions occur in the Arenig Stage of the succeeding Ordovician period, and the main period of folding coinciding with emplacement of serpentinites followed at the end of the Silurian.

ANORTHOSITES

As already noted anorthosites are pure or nearly pure plagioclase rocks: in a sense they are plagioclasites, consisting usually of labradorite or bytownite, but the composition of the feldspar depends to some extent upon the nature and environment of the anorthosite. Plagioclase from sheets low down in a layered sequence tends to be more calcic than that from higher levels, while andesine, indicating dioritic rather than gabbroic affinities, is widespread in some of the large plutonic anorthosite bodies. Although some anorthosites are very light coloured rocks, in keeping with their low colour index, others are quite dark as the feldspar is schillerized, as in the well-known rock from Newfoundland which is so much used for demonstrating this property, while some of the Norwegian anorthosites are purplish brown. According to the definition adopted here, up to 10 per cent of minerals other than plagioclase may occur: with more than this amount anorthosite grades into leuconorite, leucogabbro or leucotroctolite according to the dominant mafic mineral present. Actually these mafic minerals are often so widely scattered that it is difficult to estimate proportions.[1] The nature of the coloured silicate tends to vary with the composition of the plagioclase: thus olivine accompanies bytownite, augite and/or bronzite tend to accompany labradorite, while hornblende may occur in the dioritic

[1] Some authors apply the term anorthosite to rocks containing more than this amount of coloured and accessory minerals: thus the Stillwater anorthosites contain 85 to 100 per cent plagioclase.

anorthosites containing andesine. Magnetite and ilmenite, usually intergrown, are constant accessories, and are sometimes concentrated into thin layers and segregations in exploitable amounts. Many anorthosites display nodular weathered surfaces due to widely spaced masses of pyroxene in ophitic relationship to the feldspars. The 'mottled anorthosites' of A. L. Hall from the Bushveld are of this type.

Anorthosites are found in two strikingly different environments: firstly they are widespread in layered basic complexes, occurring in sheets varying in thickness from mere streaks to layers many yards thick; and secondly and on a much more important scale they form major rock-bodies of batholithic dimensions in Pre-Cambrian shield areas, where they may outcrop over vast areas. Thus in the Canadian Shield the Saguenay anorthosite occupies approximately 5800 square miles of surface area. These rock-bodies ensure that anorthosite is by far the most important quantitatively of all the monomineralic and near monomineralic rocks.

With regard to origin, much that has been said concerning the ultramafic rocks including olivinites and pyroxenites applies also to anorthosites, especially those occurring in layered complexes. Sheet-anorthosites to a large extent must represent layers of sorted plagioclase crystals: this is clear from their field relations with other gabbroic or noritic derivatives. This does not imply, of course, that the crystals sank through the magma from which they were precipitated: the specific gravity of these basic plagioclases is not very different from that of ordinary Basic magma, so that conceivably the crystals might be buoyed up forming a 'plagioclase raft' or if they came under the influence of rising convection currents they might form a roof accumulation. The general picture involves a layer of plagioclase crystals out of which any mafic silicates had been separated by gravity, while interstitial magma might rise or be squeezed out.

The stage of crystallization with which we are concerned is important: much MgFe would have been extracted from the magma by the crystallization of olivine and much of the pyroxene. Thus what is left must have become enriched in plagioclase-building components and also in volatiles. H. S. Yoder[1] has shown experimentally that the addition of H_2O to melts of diopside and anorthite has a profound effect upon the eutectic relationship between the two minerals, lowering the freezing temperature as might be expected, and changing the composition of the lowest freezing point mixture (the eutectic) far towards enrichment in anorthite. Experimental data for the melting temperatures of anorthite and diopside in a dry

[1] Synthetic basalt, *Ann. Rep. Dir. Geophys. Lab., Carnegie Inst., Washington* (1953–4), 106–7.

system and under conditions of high water vapour pressure are shown in Fig. 69. Under natural conditions involving plagioclase, augite and other components, melting temperatures will be somewhat lower than in the synthetic systems. In general terms, a differentiating gabbroic magma containing the ingredients of plagioclase and pyroxene has its composition changed by increasing concentration of water, becoming progressively more feldspathic. Now the actual magma with which we are concerned is interstitial to a plagioclase crystal concentrate; its composition will almost certainly lie well within the plagioclase field and therefore the first phase precipitated from it must be plagioclase, and with free diffusion a pure anorthosite will be formed. Even without postulating free diffusion and thus conveniently dispersing the unwanted components, anorthosite within the terms of the definition used in this book could be formed from a 60 per cent plagioclase crystal concentrate provided that the interstitial liquid contained 70 per cent of plagioclase substance: the anorthosite would be 90 per cent pure, the remainder would be pyroxene and iron ore. Provided that the pyroxene crystallized from the interstitial liquid and not from suspended nuclei, it would form widely dispersed, large poikilitic masses of the kind already noted: there would be no suspended seed crystals which would act as nuclei for outgrowth, and therefore the resulting mode of crystallization is in complete contrast with that of the plagioclase. These large, irregular ophitic pyroxenes may thus enclose large numbers of plagioclase crystals. By contrast the plagioclase precipitated from the interstitial liquid forms either outgrowths from the original crystals or a second generation, easily recognized in thin sections. These textural features are very distinctive.

The explanation summarized above has been applied to the Freetown Complex, Sierra Leone,[1] where the conditions for the formation of a residual feldspathic melt appear to have been particularly favourable. The anorthositic layers are underlain by troctolites, troctolitic gabbros and other olivine-rich rocks. The downward filtration of early precipitated olivines is clearly demonstrated by the field-relations, as indicated in Fig. 136, and it can be seen that the anorthosite layers represent complementary accumulations of precipitated plagioclases. The textural features seen in thin sections demonstrate the two stages of crystallization noted above. The question now arises as to whether the same processes may reasonably be expected to have operated in the case of the great batholithic anorthosites of the Pre-Cambrian shield areas. It is tempting to argue that as these rocks are petrographically identical

[1] Wells, M. K., and Baker, C. O., The anorthosites in the Colony Complex near Freetown, Sierra Leone, *Col. Geol. Min. Res.*, **6**, no. 2 (1956), 137–58.

with sheet-anorthosites in layered complexes, they must have been formed in the same way; but this is not necessarily so. Naturally it is necessary to examine the possibility of these rocks having been formed by direct crystallization from a melt of their own composition. This has already been ruled out as impossible for other monomineralic rocks on account of the abnormally high temperatures involved. The same objection applies to anorthosites as far as dry

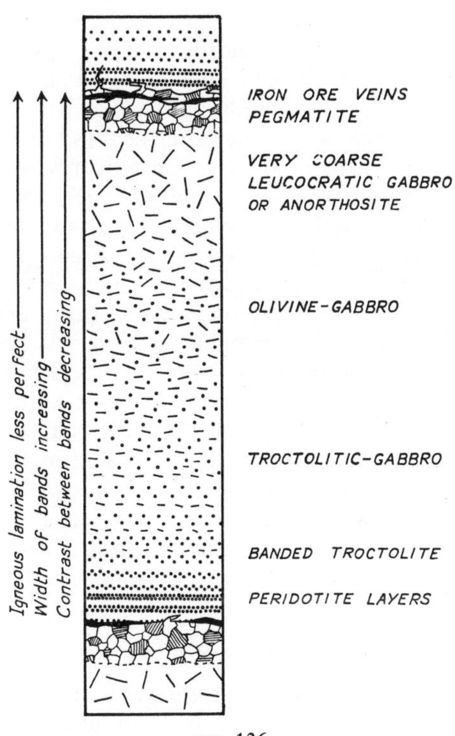

FIG. 136

Idealized layered unit of the Freetown intrusion. Individual units range from 200 to 500 feet in thickness.

melts are concerned. However, Yoder's experiments have shown that with adequate water, magmas of leucogabbro composition (*i.e.* approaching that of anorthosite) may form. Since such magmas would involve high water-vapour pressures they could only be formed under plutonic conditions. It is significant that no anorthositic lavas occur. Although magma approaching an anorthositic composition is likely to be formed under certain conditions, this does not constitute an entirely satisfactory explanation of the batholithic anorthosites of the Pre-Cambrian, because in many instances they are associated with only very limited metamorphic effects. It still seems necessary to postulate some mechanism of crystal-accumulation as in the case of the layered anorthosites. Gravity sorting of the components is a possibility, though an argument against this is the apparent absence of the enormous amounts of basic and complementary ultramafic rocks which the hypothesis demands: they are not associated in anything like the requisite amounts, so far as can be determined. True, the argument is not wholly conclusive because practically nothing is known as to the depths to which anorthosite extends underground, nor what underlies it. Nevertheless, it is generally agreed that by some means the latter has become separated from its former associates, presumably by movement in the form of a lubricated crystal mush during compressional earth movements. During such movement some granulation of a cleavable mineral seems inevitable: it is significant therefore that peripheral granulation is often seen in thin sections, indeed in many instances a pseudoporphyritic texture has resulted, with large, somewhat rounded plagioclases being embedded in a 'groundmass' consisting of granulated material. The texture is cataclastic. The necessary deformation and squeezing of largely crystalline material is possible under orogenic and plutonic conditions though, of course, this has obviously not been a factor in the formation of the layered anorthosites described above.

To prove in any specific case whether or not such movement of a crystal mush has taken place involves detailed study of the structural relationships within the rock-body and its surroundings. Such studies have been carried out in the Adirondacks, for example.[1]

LAYERING PHENOMENA

The most striking feature of the great gabbroic and noritic complexes is their composite character: they are built up of a number, often a large number, of rock-types. In some cases it has proved possible to establish the fact that differentiation by fractional crystallization has affected magma of 'average' basaltic composition after injection of the latter. In other cases it appears likely that successive injection of magma has been followed by differentiation in place. As a consequence the major units become split into smaller ones forming layers, seen as bands of different composition, texture or colour on a rock-face. Consequently such terms as 'banded gabbros' have become commonplace since first used by Alfred

[1] Balk, R., Structural survey of the Adirondack anorthosite, *J. Geol.*, **38** (1930), 289–302; and Buddington, A. F., Adirondack igneous rocks and their Metamorphism, *Geol. Soc. Amer. Mem.*, **7** (1939).

Harker when describing the igneous rocks of Skye.[1] 'Banding' is not a wholly appropriate term to apply to a three-dimensional structure—'layering' is preferable.

On the grandest scale layered units are recognized in the great Basic complexes enabling the whole to be divided into 'zones' such as the Critical Zone of the Bushveld Complex[2] and the basal Ultrabasic Zone of the Stillwater Complex.[3] These are of the order of several thousand feet in thickness. Each such zone is capable of subdivision into thinner lithologically distinctive units which to a varying degree are repetitive. Thus in the Freetown Complex, Sierra Leone, troctolitic gabbros alternate with anorthosites (Fig. 136); while in the Stillwater Complex the latter alternate with gabbros, the individual rock-types forming layers some hundreds of feet thick. Not all major layering of this kind is repetitive: individual distinctive layers of extraordinarily uniform character and thickness may be of very wide extent and must underlie large areas, measurable in hundreds of square miles. The Merensky Reef in the Bushveld and the Chromitite Layer in the Stillwater are notable examples, the former being traceable for 200 miles.

So far we have been concerned with major units down to layers of recognized well-known rock-types; but the layering to which the term is commonly applied is on a smaller scale and involves varying proportions of the gabbroic minerals, olivine, pyroxene, plagioclase and accessories, which may occur singly in monomineralic layers or in close association. The thicknesses of individual layers vary from a fraction of an inch to perhaps about a foot. There may be mere streaks of chromitite in thin layers of anorthosite or olivinite; or the layering may take the form of manifold repetitions of half-inch layers of light-coloured plagioclase-rich gabbroic rock (leucogabbro) alternating with correspondingly pyroxene-rich melagabbro. These thin layers are complementary in respect of their mineral composition and hundreds of such units may be visible in a single exposure.

In small-scale layering when it is only the proportions of the component minerals which vary it is appropriate to speak of the variation as involving **rhythmic mineral layering** when it is of the kind 1-2-3: 1-2-3, etc. A gradation may be seen from a clearly defined dark base enriched in, say, olivine or pyroxene, passing up gradually into a leucocratic feldspathic top, as in the thinner differ-

[1] Tertiary igneous rocks of Skye, *Mem. Geol. Surv.* (1904). Apart from a full description of the banded gabbros, the illustrations are most useful and convincing.

[2] Hall, A. L., The Bushveld igneous complex of the Central Transvaal, *Geol. Surv. S. Africa.*, Mem. 28 (1932), 560.

[3] Hess, H. H., The Stillwater igneous complex, Montana, *Geol. Soc. Amer.*, Mem. 80 (1960), 230.

entiated layers in Fig. 137. This is comparable with grading in sedimentary rocks and suggests by analogy that these layers were built up from the base by crystal accumulation.[1]

FIG. 137

Gravity differentiated layers separated by layers of average rock in the Skaergaard Complex, in eastern Greenland. (*After Wager and Deer*, (1939), *plate* 8.)

The problem of accounting for layering phenomena is not a simple one. The scale, details and kinds of variation are widely different and no one process can be applicable to all cases. Successive injections of magma-fractions of contrasted composition has frequently been suggested as the probable cause of some kinds of layering. The acceptability of the process depends upon the compositions of the magma-fractions involved: in some cases it is manifestly impossible. Some of the layers are monomineralic, and it is generally agreed that melts having the composition of pure plagioclase, olivine, pyroxene or chromite do not exist: the temperatures involved are the melting points of these minerals and are several hundreds of degrees higher than known magmatic temperatures. A process which can be seen to have operated in many cases is *crystal accumulation*. Much of the evidence favouring the process stems from the interpretation of the textures of the rocks which are termed **cumulates.**[2]

There can be no doubt that, in general, in sheet-like intrusions

[1] Wager, L. R., Layered Intrusions, *Medd. Dansk. Geol. Foren.*, **12** (1953), 335–49.

[2] Wager, L. R., *et al.*, Types of igneous cumulates, *J. Petrol.*, **1** (1960), 73–85.

including lopoliths, crystallization commences in the coolest part, close under the roof, but accumulation of the precipitated crystals takes place on the then-existing floor, where they are spread out to form a layer. The effectiveness of the process depends upon the difference in specific gravity of the precipitated crystals and of the magma from which they have separated: it is most marked in the case of the accessories, including chromite, and among the silicates, olivine, which separates first and at highest temperatures. Further, there may be a definite interval between the separation of olivine and the commencement of pyroxene- or plagioclase-crystallization. This gives olivine a good start over the other minerals named, and other things being equal, should give a relatively sharp, clear-cut surface of separation in the case of olivinite (dunite) layers. This process makes no great demand on the imagination to be visualized in operation. But actually it is not quite so simple as that. The crystals sink through magma into liquid which is displaced by the growing accumulation, but never completely: some interstitial magma remains and acts as a lubricant, so that before it solidifies the crystals may be subject to disturbances like those which affect sedimentary materials before their induration on the sea-floor. Consequently structures akin to those covered by the term sedimentation structures may be developed, and include slump structures due to changes in the slope of the floor perhaps. The layer may be uparched or down-warped; it may show false-bedding, washouts and channels. In fact the aping of sedimentary structures is so close that if it were not for independent evidence proving a magmatic origin, it might well be thought that the whole complex is a sort of pseudomorph after a bedded sequence of sedimentary rocks.

A given monomineralic layer therefore represents an accumulation at a particular level of a concentration of crystals, all of one kind. It resulted from **crystal sorting,** the all-important process operative in such cases. It is not merely a matter of a slight difference in specific gravity between a particular kind of crystal and the magma from which it was precipitated: several minerals may well be involved over the same period of time. They will certainly sink at different rates due to differences of two kinds—in specific gravity, and, less obviously, in crystal habit, which will affect the buoyancy of the crystals, as will the viscosity of the magma. Convection also plays its part, which is believed in certain circumstances to be a major one. 'Convective overturn' is the term used for the process whereby a roof-accumulation of crystals, after being buoyed up for a time, is perhaps quite suddenly carried down to the floor. There it may still be under the influence of convection currents which may spread and winnow the crop of crystals, for in a major magma chamber convection currents move down, along and up, so that in some places these currents will co-operate with gravity, but elsewhere the two agents of crystal-sorting will work in opposite directions, so that certain crystals may be prevented from sinking or may even be carried upwards to form a roof accumulation.

Wager and Deer[1] have sought to explain the magnificant display of rhythmic layering in the Skaergaard intrusion, East Greenland, by sudden changes in the velocity of convection currents: rapid flow causes turbulence and maintains crystals of all the minerals uniformly in suspension, while slower movements allow the denser minerals to settle. Evidence from the Freetown intrusion suggests a process of mechanical sorting by laminar flow which passed into shearing movements as the magma crystallized. In most layered intrusions effects of flow movements are indicated by the preferred orientation of crystals, particularly of tabular plagioclases which lie with their most prominent (010) faces parallel to the layer surfaces. Fluxion is the general name given to this structure; but if the evidence points to an origin by the settling of crystals uninfluenced by lateral flow movements (as in normal sedimentation), then the structure may be termed **'igneous lamination'** (Wager and Deer, 1939).

Nearly all the accepted explanations of rhythmic layering invoke mechanical principles of crystal sorting; but we should not leave the subject without stating our belief that chemical processes, such as some kind of pulsatory crystallization, may ultimately be found to provide more convincing explanations of some of the puzzling phenomena involved,[2] particularly when the layers are thin and two minerals only are involved.

One other point must be made: it is relatively easy to suggest a reasonable explanation of the mineral variation observed in *one* layered unit; but the elusive feature is the rhythmic *repetition* of the particular set of conditions involved (whether purely mechanical, physico-chemical or both), on perhaps a hundred separate occasions. Whether the repetition resulted from periodically applied external stimuli, or the repeated build-up, during protracted crystallization, of conditions favouring the precipitation of different minerals in a definite sequence, remains problematical.

Cryptic layering, aptly so termed as it is not visible to the naked eye, involves systematic changes in the compositions of the individual minerals throughout a layered series. The minerals are precipitated from successively lower-temperature fractions of the

[1] Wager, L. R., and Deer, W. A., Geological investigations in East Greenland. III. The petrology of the Skaergaard intrusion, Kangerdlugssuaq, East Greenland, *Medd. om Grønland*, **105**, no. 4 (1939), 209.

[2] Cf. Wager, L. R., Differing powers of crystal nucleation as a factor producing diversity in layered igneous intrusions, *Geol. Mag.*, **96** (1959), 75–80.

magma as the layered series is built up, so that there is a tendency for lower-temperature members of each of the isomorphous mineral groups—plagioclases, olivines, orthopyroxenes and clinopyroxenes—to appear in the higher layers of the series. In other words, one can observe in a layered intrusion a wide range of products of differentiation by fractional crystallization, all laid out as a continuous series in vertical sequence upwards. The most notable example is provided by the Skaergaard Layered Series, in which the changes of composition within each mineral group follow almost perfectly the trends indicated by the experimental researches of Bowen and his colleagues. As the plagioclase becomes progressively more sodic in the ascending sequence, so the ferromagnesian minerals show iron enrichment. The changes shown by the more important mineral groups involved in the Skaergaard Complex are summarized in the accompanying Table.

Height above base of layered series (metres)	Plagioclase	Olivine	Orthopyroxene	Clinopyroxene
2,400	An_{30}	Fa_{96}		$Wo_{30}En_{2}Fs_{68}$
1,800	An_{40}	Fa_{60}	En_{26}	$Wo_{34}En_{32}Fs_{34}$
1,100	An_{45}		En_{39}	$Wo_{32}En_{32}Fs_{36}$
600	An_{56}	Fa_{37}		$Wo_{42}En_{40}Fs_{18}$
Early (marginal) phases	An_{65}	Fa_{19}		$Wo_{41}En_{42}Fs_{17}$

Changes of the same kind, though not necessarily of the same range, characterize most major basic complexes; but in some no cryptic layering can be discerned, while in others it is repeated. In the latter case it appears likely that the complex was built up of a number of successive influxes of magma, each separate injection becoming in effect a unit of cryptic layering.

A special feature of cryptic layering has been termed **phase layering** by Hess, and involves the sudden appearance or disappearance of a particular mineral in the rock sequence, reflecting the fact, of course, that the composition of the magma had at that time passed the appropriate phase boundary. Thus in the Ultrabasic Zone of the Stillwater Complex the pyroxene is almost exclusively bronzite: precipitated augite first appears at a level of 5600 feet above the base and represents the important stage in the evolution of the complex when, for the first time, the CaAl concentration was such that augite instead of bronzite could crystallize.

ORIGIN OF MONOMINERALIC LAYERS

As already noted, among the rocks forming layered complexes monomineralic types are encountered, and the first important step in the process of formation has already been described: it involves the accumulation in the magma of a layered concentration of crystals of a single type of mineral.

The second stage in the formation of a monomineralic rock involves the crystallization of the intercumulus (or interprecipitate) magma. It is impossible to pack crystals so tightly as to exclude completely the liquid magma into which they sank. The composition of the latter must vary with the passage of time according to the minerals which have been separated from it; but at an early stage it could not be very different from the original magma, and therefore potentially any of the minerals appropriate to it could crystallize from the interstitial liquid trapped in the interstices in the crystal concentrate. To deal with the problem more explicitly we will consider the formation of olivinite. Olivine is precipitated so early and at so high a temperature that a layer of self-supporting olivine crystals near the floor is a foregone conclusion. Further, the interstitial magma at this early stage and at this level would not be appreciably different from the original magma before intrusion. Let it be supposed that this magma is noritic. It is therefore potentially capable of crystallizing as a 'mixture' of olivine, pyroxene (say bronzite), and plagioclase.

Hence selective crystallization must operate, allowing the precipation of further olivine, while preventing the formation of bronzite and plagioclase. In the case of a pure olivine concentrate, the crystals would present a relatively enormous surface area against the enveloping liquid, and this would give olivine an advantage over pyroxene and plagioclase as these would have to set up independent nuclei as a prelude to crystallization. Therefore the settled olivines grow by enlargement, the outgrowths being in optical continuity with the original crystals. The impetus given to the crystallization of olivine might be such as to set up a diffusion gradient which would ensure the flow of the necessary MgFe-ions from the adjacent overlying magma to the growing crystals, and in the ideal case this might continue until the interstices were completely filled, and olivinite would result. This is adcumulus growth and the resulting rock is an adcumulate.

In some cases careful examination of thin sections of monomineralic rocks shows slight, but perceptible, differences between the original crystals of the concentrate and the secondary outgrowths; but this is not invariably so, and there may be no visual evidence of the two stages of crystal growth.

It is fitting to end this account with reference to a book[1] which incorporates the results of much of the very fruitful research on layered igneous rocks which was initiated by Wager and Deer's work on the Skaergaard Complex and has been extended with notable success to the Tertiary gabbros and Ultrabasic intrusions in Skye and Rhum. Apart from providing very full and copiously illustrated accounts of the Skaergaard and Hebridean intrusions, the book includes also descriptions of the Stillwater and Bushveld Complexes (based in part on the authors' own research) and thus constitutes essential reading for any student of layered gabbros.

[2] Wager, L. R., and Brown, G. M., *Layered Igneous Rocks* (1958), 588.

CHAPTER XV

THE PYROCLASTIC ROCKS

THE term pyroclastic implies fragmentary material resulting from volcanic activity of explosive type. The material forming pyroclastic accumulations is of the most varied character: it may include, or consist of, volcanic bombs or lapilli of lava (plastic when erupted), or of any type of sedimentary, metamorphic or igneous rock lying beneath the volcano, in fragments varying in size from the smallest dust particles to large blocks some feet in diameter.

Classification of pyroclasts is difficult as there are several variables including: (1) the composition of the component fragments; (2) their size; (3) their lithic character; and (4) their mode of origin. For the purposes of mere record (1) and (2) are readily combined and such terms as basaltic (or andesitic or rhyolitic) agglomerate (or tuff) are self-explanatory. Actually most pyroclasts are Acid to Intermediate in composition as a consequence of the relatively higher viscosity of magmas of these types as compared with Basic (basaltic) ones.

With regard to variation in grain-size it would simplify matters to adopt the scheme of classification used in sedimentary petrology, distinguishing as far as possible between deposits consisting of blocks (boulders), cobbles, pebbles, coarse, medium and fine sands, silt grains and clay particles in order of decreasing grain-size. In the coarse-grained category **volcanic breccias** are analogous with, and may closely resemble such sedimentary breccias as those of Permian age well exposed along parts of the eastern Devon coast. Cobbles and the larger pebbles compare in size (and often in shape) with volcanic bombs and lapilli. The general name **tuff** is applied to medium or fine-grained pyroclastic *rocks*, while 'ash' is the term loosely applied to such deposits in their unindurated condition. If precision in nomenclature is desired, then a particular tuff may be described as being of coarse, medium or fine-sand grade. The silt and clay grades are also represented by volcanic dust particles which are prone to wind-transportation over long distances before deposition.

The variation in grain-size may be correlated with distance from the eruptive centre, the coarsest and heaviest material being deposited first and close to the volcano, while at the other extreme volcanic dust may be carried high into the air and may ultimately settle thousands of miles away. Fragments of pumice may be floated by currents for long distances until they ultimately become waterlogged and, on sinking, contribute to the material forming the deep-sea deposits.

To cover variation in lithology Pirsson[1] introduced three useful terms: (*a*) vitric tuffs (or ashes), composed of glass fragments; (*b*) lithic tuffs or ashes, composed of rock fragments; and (*c*) crystal tuffs, composed of crystals, whole or fragmentary. It is only rarely that a particular tuff consists wholly of one of these classes of material: usually it is an admixture and may be very heterogeneous.

Pyroclasts composed dominantly of angular fragments of large size are volcanic breccias. The term **'agglomerate'** is practically synonymous and is widely used without restriction as to grain size. Normally the angular rock-fragments of which they are composed are widely variable in size—they are completely unsorted—and in the rock-types represented. Agglomerates occur in many volcanic successions, particularly at or near the base of the pile of interbedded lavas and pyroclasts. These are distinguished as bedded aggomerates. Lithologically identical rocks occur within circumscribed, roughly circular outcrops, and in three dimensions must have an approximately cylindrical form. These are distinguished as vent-agglomerates, and are believed to represent infillings of volcanic vents. **Explosion breccias** are special types, of greater interest than their name suggests.[2] They consist of blocks of country rock (sedimentary, metamorphic or igneous) varying in size from an inch to several feet in diameter, embedded in a matrix consisting of finely comminuted rock- or mineral-fragments. In the marginal parts of the rock-body the blocks are sharply angular and scarcely displaced; but away from the contacts they become more widely separated, the amount of matrix increases and they lose their angularity, the edges and corners becoming rounded.[3] It is suggested that, after the initial fragmentation, gas streamed through for a long period, carrying with it, in suspension, mineral fragments which in these circumstances could actively abrade the larger blocks. This mechanism was first suggested by D. L. Reynolds[4] who termed

[1] *Amer. J. Sci.*, **40** (1915), 193.

[2] Hughes, C. J., The Southern Mountains igneous complex, Rhum, *Quart. J. Geol. Soc.*, **116** (1960), 111, with discussion.

[3] By definition breccias consist of *angular* fragments; but the most distinctive feature of these explosion-breccias is *rounding* of the included blocks. In spite of this anomaly there is no other term available, and it is appropriate when the whole rock-body is included under the term.

[4] Reynolds, D. L., Fluidization as a geological process . . ., *Amer. J. Sci.*, **252** (1954), 577.

it 'fluidization', by analogy with an industrial process involving a suspension of small solid particles in a gas stream. It may well be that close examination of the agglomerates filling certain volcanic vents, for example those forming such a distinctive feature of the geology of the Fifeshire coast in Scotland, may reveal rounding of the included blocks and other signs of a similar mode of origin.

Intrusive tuffs have recently been described from Tertiary[1] ring-complex fault zones within which they can be seen to intrude, and even 'finger' into country-rock. Although the material consists largely of rounded fragments of rock-forming minerals including quartz, feldspars, augite, etc., of about 0·1 mm diameter, larger rock fragments also contribute, together with blocks comparable in size with those occurring in explosion breccias, but fewer in number. The conditions of formation are believed to have been essentially the same as for the explosion breccias; but the explosive fragmentation, followed by gas-streaming, occurred at greater depths and under higher containing pressures. The gas operating in the fluidization process is believed to be supercritical water-vapour.

The phenomena described above are of particular interest and importance as providing examples of intrusion of non-magmatic (non-liquid) material. Prior to the demonstration of the effects of gas-streaming and fluidization, intrusive contacts were regarded as providing infallible evidence of the magmatic nature of the intrusive material. To this extent the problem has been complicated, particularly in the case of fine-grained rocks of granitic composition, and it is essential to examine the evidence of each case critically. It is particularly unfortunate that late-stage recrystallization affecting rocks of this composition—whatever their mode of origin—may give rise to an uninformative felsitic texture. A valuable clue may be provided by the virtual absence of metamorphic effects adjacent to an Acid rock which originated as an intrusive tuff, because of the dissipation of thermal energy in doing the work of fluidization.

Crystal tuffs are relatively uncommon and demand rather special conditions of formation. They represent phenocrysts brought up from depth by magma which must have been notably rich in crystals of the kinds which occur in the tuffs. An explosion in the vent would disrupt the column of magma and the crystals would be separated from the liquid through friction with the air. In this way accumulations of loose crystals, often of perfect form, may be built up on the flanks of the volcano, providing a profitable hunting ground for mineral collectors. Leucite and augite crystals around Vesuvius provide a striking example of the process. Not all crystal accumulations have originated in this way, of course; the olivine

[1] Hughes, C. J., (1960), 120.

sands occurring locally on the Hawaiian beaches consist of perfectly euhedral crystals of olivine resulting from the rapid disintegration of olivine-rich basalts which have been subjected to active marine erosion. In view of the prevalence of feldspar phenocrysts in several widely distributed types of lava, it might be anticipated that feldspar-crystal tuffs would, among rocks formed in this way, be relatively common. They have been recorded, for example, among the Ordovician volcanics of the Lake District and North Wales, and these serve to illustrate a difficulty encountered when volcanic ashes of any kind are deposited under water in a marine environment. In such circumstances the 'ashes' show normal sedimentary structures—stratification, current bedding, ripple marking and grading, together with possible admixture with normal sediment appropriate to the environment. Thus feldspar-crystal tuffs may closely resemble, grade into, and therefore be difficult to distinguish from, feldspar sands.

Reverting to the coarse-grained pyroclasts, J. F. N. Green, when studying the Borrowdale Volcanic Series in the Lake District, drew attention to the fact that rocks resembling volcanic breccias may result from autobrecciation (flow-brecciation), involving the breaking up of the crust of a lava-flow followed by the incorporation, within the still-fluid portion, of the solid rock fragments. The latter, on exposure to weathering, become etched out (or in, as the case may be, depending upon whether the blocks prove to be more or less durable than the matrix). Superficial examination of such rocks during a field survey might lead to their misidentification as agglomerates; but the study of thin sections should show that, in the case of an autobrecciated lava, both included blocks and matrix consist of lava, of the same type, though differing in minor points of detail.

On account of their mode of origin, their occurrence close to active volcanoes, and their fragmental, often finely divided condition, pyroclasts are particularly vulnerable to alteration by circulating volatiles and solutions of volcanic origin. Minerals characteristic of late-stage (deuteric) and hydrothermal processes develop, including low-temperature quartz, chalcedony and opal, also chlorite and, in special circumstances, garnet and late sphene. Of more than general interest are certain agglomerates in the Edinburgh district, Scotland, which contain silicified plant remains. These were thoroughly soaked in colloidal silica so soon after burial that the details of botanical structure are perfectly preserved. Silicification of pyroclastic material in an extremely finely divided condition has resulted in the formation of halleflintas—exceedingly tough, compact and durable rocks, well represented in this country among the Pre-Cambrian volcanics of the Charnwood Forest area.

Little reference has been made in this account to pyroclasts consisting of basaltic glass, for reasons which should be self-evident. Palagonite-tuff falls in this category. The basaltic glass fragments of which it was originally composed have suffered devitrification and conversion into a dull dark green substance of doubtful composition named palagonite (Penck, 1879).

It would be pointless at this stage to attempt to summarize the distribution of pyroclasts in Britain: but they will be referred to as opportunity offers in the last part of this book.

As we have noted previously, the great majority of pyroclastic rocks are of relatively siliceous composition (p. 247), and in fact almost all Acid volcanic rocks are pyroclastic in character. It was in recognition of this fact that we have transferred the account of ignimbrites and ash-flow tuffs to the chapter on the Acid igneous rocks: we may say that the *normal* extrusive equivalent of a granite is some form of air-fall ash, ash-flow or welded tuff of rhyolitic composition and *not* a lava.

The Study of Pyroclastic Rocks

Except for ignimbrites, which have been studied intensively in recent years, it is generally true to say that pyroclastic rocks have constituted the Cinderella of igneous petrology: they have not received the attention they deserve. It is quite easy to see why they have been neglected. In the first place they occupy an ambiguous position in the sense that they are part igneous and part sedimentary in character. A second, and frequently more important, reason lies in the fact that they are relatively unattractive rocks for petrographic study. Their initially porous character renders them liable to decomposition by weathering, and even if the minerals or glassy fragments are not altered in this way they are likely to have been attacked by volcanic gases. The optical characteristics and crystal shapes which enable minerals to be identified in, say, an intrusive rock, are often missing, and much of the rock may be of an indeterminable character. However, what individual pyroclastic rocks lack in petrographic attraction is compensated by the value of complete pyroclastic formations and successions studied in a stratigraphical context. It will readily be appreciated that a fall of fine ash, blown by the wind, may blanket an area several hundred square miles in extent, and if it has a sufficiently distinctive lithology will provide almost the ideal stratigraphical marker horizon.

Careful analysis of the lithological succession of pyroclastic deposits, and of course any lavas that may occur, can provide a most useful record of the sequence of events affecting a volcano. Estimates can be made of the volumes of different kinds of lava and pyroclastic material erupted, and how these have changed in amount and composition through time. From this data it may be possible to build up a picture of processes operating in a magma chamber, for instance concerning the progress of crystallization and of differentiation. In particular, the pyroclastic rocks provide the best evidence available for ascertaining the role of volcanic gases in the volcano's history.

Some of the advantages of studying pyroclastic formations quantitatively are shown by an account of tuffs composed largely of trachytic pumice occurring in the Azores.[1] Measurement of variation in grain-size of the tuffs has enabled a clear picture to be obtained of the explosive nature of the eruptions.

A second illustration is provided by Smith and Bailey's[2] description of the Bandelier Tuff, associated with the Valles Caldera in New Mexico (Fig. 96). As the title of their paper suggests, from a detailed study of the stratigraphical succession in the ash-flow, it has been possible to reconstruct events taking place in the magma chamber.

[1] Walker, G. P. L., and Croasdale, R., Two Plinian-type eruptions in the Azores, *J. Geol. Soc.*, **127** (1971), 17–55.

[2] Smith, R. L., and Bailey, R. A., The Bandelier Tuff: a study of ash-flow eruption cycles from zoned magma chambers, *Bull. Volc.*, **29** (1966), 83–104.

PART IV

IGNEOUS ACTIVITY IN THE BRITISH ISLES

CHAPTER I

PRE-CAMBRIAN TO DEVONIAN IGNEOUS ACTIVITY

ALTHOUGH it is neither advisable, nor even possible, to divorce stratigraphy entirely from an account of the igneous history of Britain, the details of the former may be largely suppressed except in so far as they bear directly on the sequence of events and the quality of the magma intruded or extruded, as the case may be during any igneous episode. In this brief account emphasis will be laid only upon those occurrences which are of special interest, in that they have some bearing upon the problems of petrology and petrogenesis.

The **igneous cycle** when complete comprises:

(i) a volcanic phase (outpouring of lava),
(ii) a phase of plutonic (or major) intrusion,
(iii) a phase of minor intrusion ('dyke-phase').

The products of every cycle are not necessarily completely exposed at the surface at the present time: the active period of the volcanoes may have been of short duration; the amount of lava actually extruded may have been small; subsequent erosion may have been profound. These circumstances have combined in some cases to destroy all record of the volcanic phase. In like manner the plutonic intrusions may have failed to penetrate sufficiently near to the surface to be exposed by subsequent denudation, as happened with the Carboniferous igneous cycle in the Midland Valley of Scotland, the cycle in such cases consisting of lava-flows and minor intrusions only.

The most constant phase is that of the minor intrusions. There are, indeed, seemingly unattached dyke-phases that can only with difficulty be assigned to a particular cycle; while individual sills and dykes extend to such great distances from their volcanic centre that there is considerable difficulty in correlating them.

In Britain, volcanic activity played an important role in the Pre-Cambrian, Ordovician, Devonian, Carboniferous and early Permian periods, and in the Tertiary epoch.

One of the most significant developments since the last edition has been the great increase in radiometric age determinations.[1] It is impossible in the brief account which follows to discuss in any detail the effects that new age data have on the placing of igneous rocks in their stratigraphical context. For extrusive rocks, except for those of Pre-Cambrian age, the problems are generally not great because the relative geological age of most of these rocks in known with reasonable certainty from their position in the stratigraphical column. However, the problem can become acute for intrusive rocks, particularly granites. It is seldom possible to date an intrusion accurately on stratigraphical evidence and reliance has to be placed upon radiometric age determinations. 'Dates' obtained in this way have to be treated with caution because of the many factors which may influence the isotopic ratio from which the ages are determined. For reasons of space, we have stated determined 'ages' of a number of intrusions without giving any indication of either the experimental error or possible errors in geological interpretation of the results. This is admittedly a reprehensible practice; but we think it is justified in the limited space available. In several instances, isotopic ages are widely at variance with previous estimates based on geological evidence. Thus, a number of granitic intrusions in the Lake District which are intruded into Ordovician strata were thought at one time to have been associated with the eruption of Ordovican volcanic rocks which form the Borrowdale Volcanic Series (see p. 481). They are now know to be much younger than this and to have been emplaced during the Caledonian Orogeny. It would have been logical in one sense, therefore, to have transposed the account of these intrusions into the section dealing with the Caledonian igneous activity. In this case, and in some others which will become apparent in due course, we have given priority to regional consideration and described these intrusions in the section of the account dealing with the Ordovician, merely adding the isotopic 'ages' as a corrective.

PRE-CAMBRIAN

Of the several areas in Britain where Pre-Cambrian rocks are visible, by far the most important are the Scottish Highlands and particularly the unique natural region of the **North-West Highlands** of Scotland where British geological history begins. The beginning of the Phanerozoic is clear-cut and coincides with the outcrop of a distinctive white-weathering quartzite, which is the local basal member of the fossiliferous Cambrian. This rests with striking unconformity

[1] Data for British rocks are conveniently summarized in a series of compilations of abstracts, edited by P. E. Sabine and J. V. Watson under the general title Isotopic age-determinations of rocks and minerals from the British Isles . . ., 1955–64, 1966, 1967–8, in *Quart. J. Geol. Soc.*, **121** (1965), **123** (1968) and **126** (1971).

on the Torridonian, a 20,000 foot-thick, essentially arkosic, formation consisting very largely of the wastage of a vast amount of granitic rock. The Torridonian is not yet accurately dated, but suggestions range around 800 to 1000 m.y. A profound unconformity, representing an immense span of time separates the Torridonian from the underlying Lewisian Complex. Further, the former has been unaffected by any subsequent metamorphism; but in dramatic contrast, the Lewisian rocks are chiefly high-grade metamorphic gneisses of widely variable composition.

The **Lewisian Complex**[1] is conveniently separable into two main divisions, to some extent geographically, but spectacularly so in the time sense. One of the most striking features shown on geological maps of the North-West Highlands is a post-Lewisian, pre-Torridonian dyke-swarm of North-West to South-East trend. This dyke phase is critical in the interpretation of the geological history of Scotland as it divides Lewisian time into two dated orogenic episodes: the older, carrying us back immensely far in time is termed **Scourian** extending from 2600 to 2200 m.y. ago. The Scourian consists of banded gneisses covering the widest possible range of chemical composition, from ultra-acid granitic pegmatites to ultrabasic, some of peridotitic, others of pyroxenitic composition. The Basic-Ultrabasic rocks occur as concordant sheets or lenses and it may well be that originally they formed parts of layered intrusives comparable with those referred to in earlier chapters. A charnockitic 'flavour' is given to the series by the important role of hypersthene among the mafic constituents. What is described as the largest mass of anorthosite in Britain occurs in Lewis in the Outer Hebrides.[2]

The reader is invited to use his imagination in visualizing the changes in environment, stress conditions in the crust, and the time element involved between the Scourian metamorphism and the injection of the Basic dyke-phase. The dykes themselves have a radiometric age of 2200 m.y.; they are (or were originally) subophitic tholeiitic dolerites chiefly, but they are accompanied by W.N.W.-E.S.E. ultramafic dykes, rich in olivine and augite. In those parts of the area which have been unaffected by later orogenies the dykes have maintained their original characters suprisingly well in view of their great age; but elsewhere they were involved in a much later orogeny, termed the **Laxfordian** and dated as 1600 to 1200 m.y. old. This profoundly affected the Scourian rocks as well as the dykes, of course. The latter were folded, faulted, reorientated and completely reconstituted into amphibolites or green schists.

[1] See summary account by Janet Watson in *Geology of Scotland*, ed. G. Y. Craig, 1965, 49–77, with references.

[2] Dearnley, R., The Lewisian Complex of South Harris, *Quart. J. Geol. Soc.*, **119** (1963), 243–307.

During the Laxfordian orogeny the older gneisses were completely reconstituted, they were migmatized with the injection of much granitic material—apparently trondjhemitic as oligoclase and quartz are much in evidence, and subjected to a K-metasomatism evidenced chiefly by the formation of microcline porphyroblasts. Small bodies of eclogite occur in the Laxfordian in Glenelg. Again, as in the Scourian, granite-pegmatite is much in evidence.

All these Pre-Cambrian rocks are pre-Caledonian: they were involved in orogenies which took place far back in Pre-Cambrian time. Except in so far as they provide glimpses of the foundations on which Caledonian structures were erected they belong to much earlier chapters of geological history. The sequence of events can be 'read' only in the 'unmoved foreland', *i.e.* the coastal strip of the North-West Highlands lying west of the overthrust zone. To the east of the latter the rocks were involved in the several phases of folding and metamorphism which constitute the Caledonian orogeny. The main conclusion, that the latter occurred broadly during the time-interval late-Silurian, early-Devonian, was reached long ago from stratigraphical evidence, and in general has been confirmed by geochronological data.

In the rest of the area now occupied by the British Isles igneous activity must have been widespread since most of the areas where Pre-Cambrian rocks occur as inliers contain volcanic, and often contemporary intrusive rocks. Outstanding features are the prevalence of rhyolitic lavas, the development at different levels of spilitic pillow-lavas, the almost invariable presence of albite instead of more calcic plagioclase in the Basic lavas. We include in this necessarily brief account some reference to the Pre-Cambrian igneous rocks of Anglesey in North Wales, Snowdonia, coastal Argyllshire, Shropshire, the Charnwood Forest in Leicestershire, the Channel Islands and the Lizard area in Cornwall.

In **Anglesey**[1] volcanic rocks occur on three horizons in the Mona Complex. The lavas are predominantly spilitic lavas, those in the Gwna (pronounced Goona) Group being particularly well exposed (Fig. 61.) The spilites include devitrified glassy and variolitic varieties and are in some areas accompanied by rhyolites. In some specimens the coloured silicates are rather better preserved than usually, and pseudomorphs after olivine also occur. When traced into regions of intense regional metamorphism the spilites grade into a rare type of schist consisting essentially of glaucophane and epidote. These glaucophane-schists are restricted to one small area in Anglesey. An outlier of the Mona Complex occurs in the western extremity of the Lleyn Peninsula in Carnarvonshire, and here also spilites are

[1] Greenley, E., Geology of Anglesey, vol. 1, *Mem. Geol. Surv.*, (1919) 54, and 71.

well developed. Plutonic rocks of Pre-Cambrian age forming part of the Mona Complex range from alkali-granites to dunite-serpentinites, and include pyroxenite, gabbro, dolerite and subordinate masses of diorite. The latter are regarded as marginal modifications of the granites. The Sarn granite in Lleyn, for long regarded as an Ordovician intrusion, is also Pre-Cambrian, and formed part of the ancient land surface on which the basal Ordovician sediments were deposited.

In **Shropshire** lavas and tuffs, penetrated by intrusions of leucogranite, form the Uriconian rock groups, occurring both east and west of the plateau of the Longmynd. The eastern belt includes the hills of the Wrekin and Caer Caradoc; while the western culminates in Pontesford Hill, long famous, together with the Lea Rock near Wellington, for their devitrified perlitic and spherulitic rhyolites. Andesitic tuffs and flows of mugearite also occur.[1] Although the centres of eruption have not been exactly located there is no doubt that these lavas were erupted from volcanoes of the central type, some 638 m.y. ago.

Lavas probably of nearly the same age constitute the Warren House Volcanic Series in the **Malvern Hills.** Although poorly exposed they show much the same range of composition as the Uriconian lavas, and include sodic rhyolites, sodic trachytes and variolitic spilites, together with tuffs.

Pebbles of lava closely resembling the Uriconian types have been recorded from the Torridonian sedimentary rocks of the North-West Highlands. The northerly provenance of the bulk of the Torridonian sediment, however, renders it highly improbable that these fragments were carried by coastal drift or otherwise from Shropshire to Scotland. They appear to be the only traces remaining of sheets of rhyolite, the original location of which is unknown. Probably they were contemporaneous with the Uriconian lavas of the more southerly localities.

The Channel Islands consist in part of sedimentary rocks, *e.g.* the Jersey Shales, which are correlated with the Brioverian of mainland France, and presumably rest on a gneissic floor locally exposed in Sark. The Jersey Shales are overlain by a thick volcanic series, divisible into an older Andesite Series and a younger Rhyolite Series. Both are well exposed in fine coastal sections, but no modern accounts are available. The andesites have been extensively quarried especially around St Helier where the sea-wall provides a museum-like display of several different types including aphyric and porphyritic varieties.

The rhyolites are more widely distributed, and are well known on account of their beautiful spherulitic and lithophysal textures.

[1] Boulton, W. S., *Quart. J. Geol. Soc.*, **60** (1904), 470.

Micro-phenocrysts of quartz and feldspar are sometimes present; flow banding is strongly emphasized by streams of small spherulites, generally the size of a pin's head, but occasionally much larger (Boulay Bay). Some of the thicker flows have a perfect columnar jointing, as near Ann Port. Several of the rhyolites, *e.g.* near Mount Orgeuil, formerly regarded as flows, prove to be sills and dykes penetrating the granite. These rhyolites bear a superficial resemblance to the Uriconian lavas in Shropshire. Similar lavas are well exposed in a belt of country in North Wales lying immediately north of the 'slate belt' and crossed by the Llanberis Pass. This is the Padarn Ridge where the rhyolites are seen to be overlain unconformably by the basal Cambrian conglomerates which largely consist of boulders and pebbles of these lavas: they may be late Pre-Cambrian in age.

Plutonic intrusions, mainly of granitic rocks, are important in the Channel Islands, especially Jersey. Stratigraphical evidence of the ages of these plutonics is virtually non-existent. From their geological setting and geographical location they were generally regarded as Armorican in age and as a matter of convenience, their description under the heading of 'Armorican Intrusions' (p. 518), has been retained. Preliminary results of radiometric dating suggest, however, that they are much older than this. Some may be of Caledonian age, while a dioritic complex of South-East Jersey yielding an apparent age of about 580 m.y. may have been intruded in late Pre-Cambrian or possibly early Cambrian times.[1]

We take up the story again to include the volcanic and associated intrusive rocks in the Tayvallich Peninsula in South-West Scotland. In this area Basic magma was erupted on two occasions during the Upper Dalradian, on the first occasion tentatively, but on the second on a more important scale to form the Tayvallich Lavas exposed in the neighbourhood of Loch Awe. These lavas were submarine and display well-developed pillow structure. They are older than the regional metamorphism; but their unmistakeable structural features have survived. The lavas are spilitic. As regards age, the Upper Dalradian on a somewhat higher horizon contain a Middle Cambrian fauna and the only natural base for the Cambrian would be considerably lower than the Tayvallich volcanic horizon, so an early Cambrian age is not impossible.

In **Leicestershire** the Charnian succession[3] consists largely of volcanic material, particularly in the lower part of the Middle Group, the Maplewell Series. This includes two spectacular agglomerates: a lower, 'Felsitic Agglomerate' and an upper, 'Slaty

[1] Adams, G. J. D., *Earth & Planetary Science Letters*, **2** (1967), 52.

[2] Bennett, F. W., Lowe, E. E., Gregory, H. H., and Jones, F., Geology of Charnwood Forest, *Proc. Geol. Assoc.*, **39** (1928), 241.

Agglomerate'. The latter includes blocks up to five feet in maximum diameter, which indicate proximity to nearby violently explosive volcanoes. These Charnian rocks are associated with a varied suite of intrusions, some few of which are represented by fragments in the agglomerate mentioned above, and their age is therefore definitely fixed. These intrusions include dacite and porphyritic microdiorite (Peldar Tor). The most extensive intrusions in the area are essentially sub-acid, and include the well-known Markfield and Groby rocks. The former gives its name to 'markfieldite', a quartz-diorite with micrographic groundmass. The Pre-Cambrian age of these intrusions was proved by quarrying operations in the Nuneaton district, where the basal Cambrian conglomerate is seen to rest unconformably upon an irregular surface of markfieldite, and includes large rounded boulders of the latter.[1]

The third group of intrusives in Leicestershire comprises the much-quarried Mountsorrel granite and its associated more Basic derivatives. K/Ar age determinations show that the granite is Caledonian—368 m.y. old.

The complex of metamorphic and intrusive rocks which outcrop on the **Lizard Peninsula**[2] forming the southernmost projection of the Cornish coast, presents problems of dating at least as acute as those of the Channel Islands rocks. The difficulty arises from the fact that all the varied crystalline rocks of the Lizard Complex are separated by overthrusts from the sparsely fossiliferous Lower Palaeozioc rocks included in the Meneage Crush Zone and the Gramscatho and Mylor Beds (Fig. 138). Isotopic dating is of little help in this case, since the ages obtained by this means almost certainly bear little relationship to the time of the original emplacement and crystallization of the intrusive Lizard rocks: rather, they are likely to record later metamorphic events, possibly during Caledonian and again to some extent during the Armorican earth-movements. It is quite possible that the intrusive rocks *may* be Pre-Cambrian, but there is obviously no certainty that this is so. Interest in the Lizard Complex centres on the mass of serpentinized peridotites which are dominant in the Complex, covering an area of approximately twenty square miles. It is the largest such body in the British Isles, and serves as an excellent example of Alpine-type peridotite.

The plutonic rocks of the Lizard range in composition from ultrabasic peridotites (the earliest intrusions) to alkali-granites (the latest intrusions). Three varieties of serpentinite were recognized by the Geological Survey: bastite-serpentinite, tremolite-serpentinite

[1] Wills, L. J., and Shotton, F. W., The Junction of the Pre-Cambrian and Cambrian at Nuneaton, *Geol. Mag*. (1934), 13.

[2] Flett, J. S., Geology of the Lizard, *Proc. Geol. Assoc*., **24** (1913), 118 and Geology of the Lizard and Meneage, *Mem. Geol. Surv*., 2nd edn., (1946).

and dunite-serpentinite. The last two form discontinuous zones round the central core of coarse-grained bastite-serpentinite. By a careful examination of the pre-serpentinization assemblage of minerals, it has been possible[1] to demonstrate that the peridotites from which the serpentinites were derived crystallized initially at very high pressures and temperatures. Critical evidence of their original state is provided by the compositions of the pyroxenes: even the

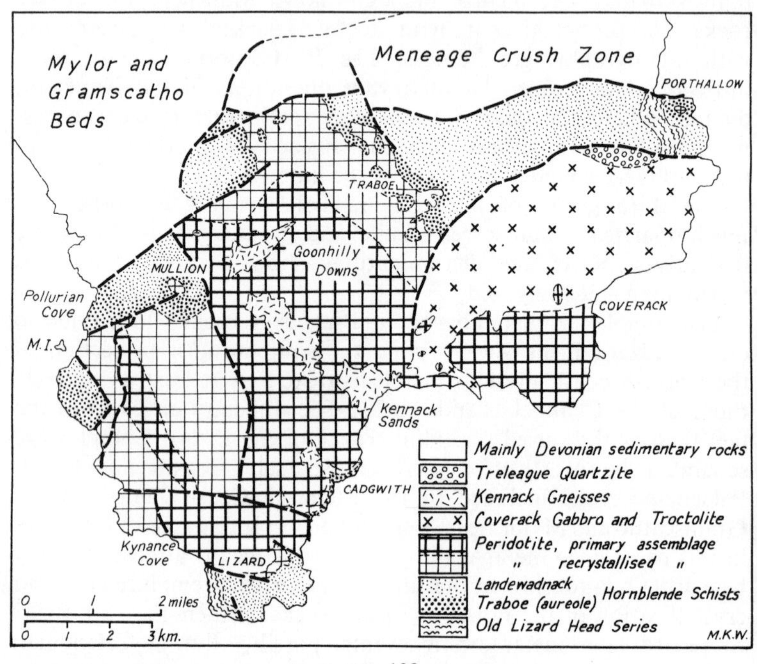

FIG. 138

Map of the Lizard Complex, Cornwall, showing the distribution of the primary and recrystallized facies of the peridotite in their pre-serpentinized state, and the metamorphic aureole (*heavy stipple*) in the hornblende-schists. M.I. = Mullion Island. (*After D. H. Green*, (1964).)

orthopyroxene (from which the bastite pseudomorphs were derived) is very rich in Al_2O_3. Since orthopyroxenes in normal igneous rocks are virtually Al-free, the high Al_2O_3-content is regarded as strong evidence of initial crystallization at great depth, at least approaching that of the mantle. Traces of a high-grade metamorphic aureole (shown by the heavy stipple on Fig. 138 confirm that the temperature at the time of emplacement was high. Subsequent to its emplacement

[1] Green, D. H., The petrogenesis of the high-temperature peridotite intrusion in the Lizard area, Cornwall, *J. Petrol.*, **5** (1964), 134–88.

the peridotite mass has been drastically modified in two stages. In the first, occurring at a high temperature, 'tremolite' developed especially in the marginal parts of the mass. The second stage of alteration was serpentinization which affected the whole peridotite body. As a result of the tectonic forces which were active during the emplacement and cooling of the peridotite mass, marginal cataclasis produced a fine-grained mylonitic rock equivalent to the 'dunite-serpentinite' of earlier descriptions.

The serpentinite is penetrated at Coverack by a small intrusion of troctolite, which in turn is cut by veins of gabbro. These are offshoots from a massive intrusion of gabbro which broke through the north-eastern part of the serpentinite mass, and penetrated it in a maze of veins and dykes. The coarse-grain of quite thin veins of gabbro proves this later injection to have taken place while the serpentinite was still hot.

After the cooling of the gabbro, a large number of Basic dykes (the so-called 'black dykes') were injected. Many of these are normal olivine-dolerites, surprisingly fresh for their age, but all stages occur in the conversion of these into hornblende-schists. Others contain significant streaks and blotches of red granitic material which increases in amount in the Kennack gneisses, these having been forced into position under great pressure, and consisting of bands and lenticles of alternating granitic and gabbroic composition.

Finally, red granite-gneiss, similar to the red streaks in the Kennack gneisses, was intruded into the central serpentinite as dykes and small bosses.

Igneous activity in Britain during the early part of the Phanerozoic has to be dealt with in two different contexts. The first concerns volcanic episodes which interrupted the normal stratigraphical record of the areas concerned. These areas lie outside the region of maximum structural complexity, deformation and intense metamorphism consequential upon the Caledonian orogeny. This main theatre of intense tectonic activity embraces Scotland (= 'Caledonia') north of the Highland Boundary Fault; and here ordinary methods of stratigraphical dating are inapplicable and age relationships are difficult to establish. South of the Highland Border the Lower Palaeozoic rocks, although often quite strongly folded, are not metamorphosed and the history of igneous activity may be interpreted using the ordinary tools of stratigraphy. We propose to deal with the latter areas first; and here we are concerned mainly with Ordovician volcanicity chiefly in Wales, the English Lake District and the non-metamorphosed parts of Scotland.

ORDOVICIAN

The Cambrian Period was one of quiescence so far as volcanic activity is concerned. With the opening of the succeeding period, however, occurred the first great outburst that ushered in the widespread and long-continued activity of the Ordovician. Some of these early volcanoes were grouped about the northern and southern shores of the geosyncline, which at this time occupied the British area, but others, in the centre of this tract, were localized by movements of elevation—the prelude to the Caledonian orogeny.

The eruptions that gave rise the the pillow-lavas (spilites) of Mullion Island off the Cornish coast were possibly contemporaneous with the spilites and dolerites in the Highland Border Series occupying an analogous position on the north of the geosyncline; but the exact age of both series is still a little doubtful.

In early Arenig times eruptions in Pembrokeshire, Carmarthenshire and southern Merionethshire gave rise to the Trefgarn, Llangynog and Rhobell Fawr Volcanic Groups respectively.

While the Trefgarn group consists essentially of andesites, the calc-alkaline rocks of Rhobell Fawr include dacites and basalts, in addition to dominant horblende-and augite-andesites. Emanations connected with the waning phase of activity effected a complete pseudomorphism of most of the original minerals, and have riddled the rocks with epidote and pyrite. Numerous sills crowded together west of (originally beneath) the volcanic pile indicate a well-developed 'dyke-phase' to which the lithologically similar sills, dykes and small laccoliths occurring in the Harlech Dome and intimately associated with the gold-bearing quartz veins in the Dolgellau district, undoubtedly belong. The rocks of the minor intrusions are those that should be associated with andesitic lavas, being hornblende-microdiorites of porphyritic and aphyric types.[1]

In later Arenig times volcanic action became more general. To this episode belong the Coomb Volcanic Series of Llangynog in Carmarthenshire, the Skomer Island Volcanic Series, forming the island of that name off the coast of Pembrokeshire, the Lower Acid Series of Merionethshire, the Ballantrae and Sanquhar Volcanic Series in the Girvan district. Of these several series, those in Merioneth and Carmarthen consist essentially of Acid lavas and tuffs, those of Girvan, Scotland, are spilites, with well-developed pillow-structure, magnificently exposed on the coast near Ballantrae and associated with coarse agglomerates.

Following a short period of quiescence during which normal sedimentation took place, the Llanvirn period witnessed the maximum of Ordovician volcanic activity; and simultaneous eruptions

[1] Wells, A. K., *Quart. J. Geol. Soc.* (1925), 463.

occurred at several centres in North and South Wales,[1] South Shropshire and the English Lake District. In each of these localities the Llanvirn volcanic rocks rest upon blue-black shales containing the *D. bifidus* fauna, which accurately dates the eruptions and proves them to have been submarine. In South Wales the volcanoes soon became extinct; in Shropshire activity ceased in *bifidus* times but was renewed in the Caradocian period; in North Wales they were continuously active in one locality or another until late Caradocian times. Towards the close of the period represented by the zone of *Nemagraptus gracilis*, the volcanoes of southern Merionethshire became finally extinct, but new eruptive centres were established northwards in Snowdonia and southwards in the Wells country of East-Central Wales (Llanwrtyd Wells, Llandrindod and Builth).[2]

North Wales. (*a*) **The Volcanic Rocks.**—The Ordovician volcanics in North Wales are of three different kinds: rhyolites (potassic and sodic), andesites and spilites. A noteworthy feature is the great development of pyroclasts of all degrees of coarseness. These occur on all the volcanic horizons, and usually bulk far larger than the actual flows. The Basic lavas are spilites, characterized by pillow structure which attains its most perfect development in the Lower Basic Group near Arthog (Merionethshire) and in the Upper Basic Group on Cader Idris. The former has a maximum thickness of 1,500 feet and the latter of about 500 feet. In addition to typical spilitic pillow lavas these Basic Groups include bomb-tuffs, vitric tuffs, feldspar-crystal tuffs and rocks in which pyroclastic and sedimentary materials are mixed in all proportions.

To the north, the spilites give place to hypersthene-andesites, which were erupted from the Arenig Mountain centre. The fact that the local volcano had built itself up above sea-level accounts for the absence of pillow structure: but in addition the lavas are of andesitic, not spilitic, type.

In the Cader Idris district Acid lavas and tuffs were erupted in post-*hirundo*–pre-*bifidus* times, and again towards the close of the volcanic episode, forming in the latter case the Upper Acid Group, with a maximum thickness of 1,500 feet[3] which includes lavas of variable composition: some are potassic rhyolites, others sodic (quartz-keratophyres) while some are less rich in SiO_2 and are trachytes and keratophyres (albite-trachytes). To the north this group persists into the Arenig Mountain district.

The great majority of the rhyolitic lavas of Carnarvonshire are of

[1] Thomas, G. E., and Thomas, T. M., Volcanic rocks of the area between Strumble Head and Fishguard, *Quart. J. Geol. Soc.*, **112** (1956), 291–314.

[2] Jones, O. T., and Pugh, W. J., *Quart. J. Geol. Soc.*, **104** (1948), 71.

[3] Davies, R. G., The Cader Idris granophyre and its associated rocks, *Quart. J. Geol. Soc.*, **115** (1959), 189.

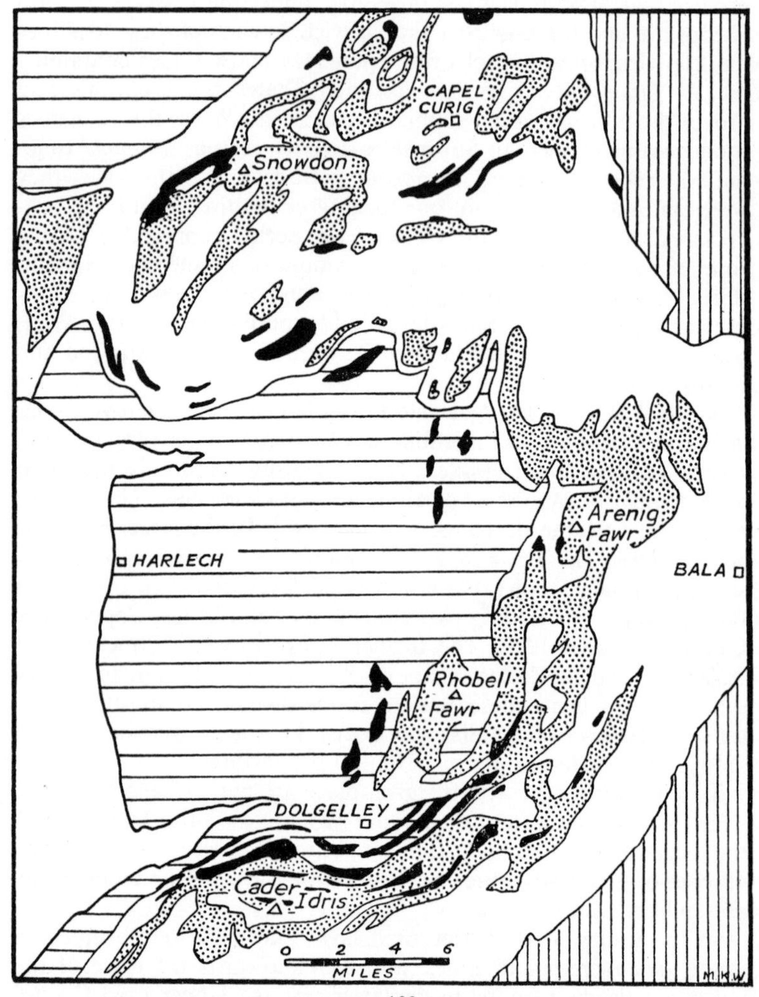

FIG. 139

Sketch-map of part of North Wales showing the distribution of the igneous rocks: volcanic, *stippled*; intrusive, *black*; Cambrian, *horizontal ruling*; Ordovician, *plain*; Silurian, *vertical ruling*. (*After Geol. Surv., A. H. Cox, A. K. Wells and others.*)

Bala age. Howel and David Williams[1] have shown that no lavas occur on Snowdon below the zone of *Nemagraptus gracilis* (the zone to which the highest lavas of Cader Idris and Arenig belong). Further,

[1] Williams, H., Geology of Snowdon, *Quart. J. Geol., Soc.*, **83** (1927), 346; Williams, D., The geology . . . of Nant Peris . . ., *Quart. J. Geol. Soc.*, **86** (1930), 191.

a thickness of some 1,600 feet of unfossiliferous, and therefore undated, strata intervene between the fossiliferous horizon and the actual base of the Snowdonian volcanics. In the Dolwyddelan Syncline east of Snowdon the highest volcanics are immediately overlain by black graptolitic slates, in the zone of *D. clingani*. It follows that the volcanic episode was of relatively short duration—perhaps within the period of one zone.

Within recent years much resurveying of parts of N. Wales,[1] combined with re-examination of the igneous rocks themselves, has modified many of the older views concerning the nature and structure of the volcanic rocks, notably in Snowdonia. In particular the Snowdon(ian) Volcanic Series has proved to consist largely of rhyolitic ignimbrites. This does not alter the basic fact that a large volume of rhyolitic magma was available in this part of Wales in Upper Ordovician (Caradocian) times. The only change has been in interpretation of the style of eruption.

The Capel Curig Volcanic Series lies below the main Snowdonian Volcanics. A feature of great interest has been the discovery of extensive sill-like bodies of intrusive tuff, which disrupt the adjacent wall-rocks, produce contact-metamorphic effects, and display chilled margins. Their significance and mode of emplacement pose interesting problems for the future to solve. The Snowdon massif[2] is shown to be a Caradocian volcano of Central type. The rocks erupted from it are dominantly rhyolitic subaerial tuffs associated in the middle part of the eruptive sequence with subordinate Basic lavas and tuffs. The new account reads almost like that of one of the Tertiary complexes. The initial uprise of volatile-rich rhyolitic magma was accompanied by the formation of a dome dating from early in the Ordovician Period. The magma first saw daylight with the eruption of the well-known Pitt's Head ignimbritic rhyolites and this was followed by caldera collapse. The caldera was partly buried beneath the Lower Rhyolitic Tuffs, which are ignimbrites near the area of eruption but grade into bedded subaqueous tuffs. Further subsidence ensued, followed by the formation of the Middle Basic Series, largely submarine and spilitic but including rhyolitic material. The close association of rhyolitic and spilitic volcanics is noteworthy. Finally, uplift of the area was accompanied by a reversion to rhyolitic magma and the ignimbritic mode of eruption, forming the Upper Rhyolite Series.

[1] Rast, N., Ordovician structure and volcanicity in Wales, in *The Pre-Cambrian and Lower Palaeozoic Rocks of Wales*, (1969); Bromley, A. V., Acid plutonic igneous activity in the Ordovician of North Wales, *op. cit.*, 1969; Fitch, F. J. Ignimbrite volcanism in North Wales, *Bull. Volc.*, **30** (1967) 199; and The recognition of ignimbrites in North Wales, *Proc. Geol. Soc.*, **1664** (1971), 280.

[2] Demonstration of the . . . special 1:25000 sheet, Central Snowdonia, *Proc. Geol. Soc.*, **1663** (1970), 165.

Some rhyolitic magma was still available though thoroughly degassed and relatively immobile, so that it found it easier to spread laterally as sills, or to consolidate in minor vents, probably fault-located. Caradocian volcanicity in this part of Wales was brought to a close by the intrusion of a Basic Sill Complex of dolerites as sills and phacoliths.

(*b*) **The Intrusive Rocks.**—The Lower Palaeozoic rocks of **Wales** are penetrated by large numbers of intrusions. On the grounds of petrographical similarity it is clear that the hornblende-micro-diorities of Merionethshire represent the hypabyssal phase of the Rhobell-eruptions. Similarly the hypersthene-andesites intrusive into the rocks building Arenig Mountain must have been nearly contemporaneous with the eruptions from that centre. In both cases there is often difficulty in determining whether a particular rock is extrusive or intrusive. But in addition to these local types, there are dolerite sills of regional distribution and uncertain date. Although there were no Carboniferous or Tertiary volcanoes in Wales, intrusions of both these ages occur in Anglesey, Carnarvonshire and possibly farther south. Intense folding also took place in Wales during the Caledonian orogeny, and it is possible that these earth-movements were accompanied by intrusive phenomena. Nevertheless, most of the intrusive rocks, other than those of Pre-Cambrian age, must be regarded as probably belonging to the Ordovician cycle. The influence of the local centres is shown even in the regional sills: those in the Arenig district show affinity with the lavas by containing hypersthene, while the feldspar is stated to be less basic than in the dolerites—andesine in place of labradorite. In those areas where the facies of the extrusive phase was spilitic, the dolerites are usually albitized, while the spilitic dolerites (which were intruded into soft mud on the sea-floor) reproduce many of the characters of the lavas. Although in eastern Carnarvonshire the dolerites are often fresh, it is rare to find any original minerals among those of Snowdonia, where they are albitized and chloritized. In Merionethshire the dolerite intrusions are especially developed in the neighbourhood of Dolgellau occurring chiefly as sills, particularly in the Basement Group of the Ordovician. In Snowdonia, the doleritic intrusions are dykes in the Cambrian rocks, but sills and small phacoliths in the Ordovician volcanics.

Numerous concordant intrusions of albite-dolerite occur in the Builth-Llandrindod Ordovician inlier in Mid-Wales.[1] They seem to demonstrate the influence of depth of intrusion upon the form of the rock-bodies, which vary from sills in the Llanvirnian to small, widely

[1] Jones, O. T., and Pugh, W. J., A multi-layered dolerite complex . . ., *Quart. J. Geol. Soc.*, **104** (1948), 43; and *ibid.*, p. 71.

scattered bun-shaped masses in the Llandeilian. Some appear to be laccoliths, with feeders, of the conventional type.

Large numbers of intrusions (often composite) occur in the south-west portion of the Lleyn Peninsula, and are intrusive into the *D. bifidus* and lower beds. Petrographically these are linked with the Palaeozoic sills, dykes and laccoliths of Anglesey. Distinctive features are the occurrence of hornblende in even the most Basic types, and the large amount of olivine sometimes present. The rock-types represented include minverite (hornblende-dolerite), olivine-dolerite and hornblende-picrite. In mineral composition these rocks appear to be distinct from those occurring farther south, and Greenly claims them to be of late-Caledonian age.

The Acid intrusions of Snowdonia followed the eruptive phase, and include representatives of all the commoner Acid minor intrusive rocks. Phenocrysts of quartz, microperthite and orthoclase are frequently plentiful, and are embedded in a micro- to cryptocrystalline groundmass of soda-orthoclase and quartz. These minor intrusions are petrographically allied to the rhyolite flows. The granophyres (graphic microgranites) are characterized by the presence of augite, while the microgranites contain both this mineral and biotite. In two localities, Mynydd Mawr and Bwlch-y-Cywion, the place of these minerals is taken by riebeckite (Fig. 86). Mynydd Mawr is probably the plug of a volcanic vent of Bala age. A similar riebeckite-microgranite with a flow-banded rhyolitic marginal facies occurs near Llanbedrog, in the Lleyn Peninsula.[1]

Acid intrusive rocks (granophyres) in southern Merionethshire form two massive columnar sills in the neighbourhood of Dolgellau, one building the precipitous scarp of Cader Idris, and the other occurring in the foothills to the north.[2] Both very closely resemble certain of the volcanic rocks nearby, and doubtless represent rhyolitic magma which failed to reach the surface and consolidated under a sedimentary cover.

Few of the larger intrusions in Caernarvonshire have been studied in detail: those that have, indicate their composite nature, due in some cases to successive injections and in others to differentiation in place. An example of the latter is Penmaenmawr Mountain between Conway and Bangor. The intrusion becomes progressively more basic from the summit downwards. The upper part is a graphic microdiorite approaching markfieldite, while the rock building the lowest visible portion is nearly doleritic in composition.

Shropshire and the Welsh Borders.—Ordovician andesites occur

[1] Matley, C. A., *Quart. J. Geol. Soc.*, **94** (1938), 596 and 603.
[2] Cox, A. H., and Wells, A. K., The Lower Palaeozoic rocks of the Arthog district (Merionethshire), *Quart. J. Geol. Soc.*, **76** (1921), 283–305. See also Davies, R. G., *op. cit.*, 437.

in the hills of the Shelve district, near the boundary between Shropshire and Montgomeryshire (Fig. 140). Here lavas of lower Llanvirn age form the Stapeley Hills, attaining their greatest thickness and extent on Todleth. These lavas, tuffs and associated intrusive andesites contain phenocrysts of plagioclase, augite, and hypersthene

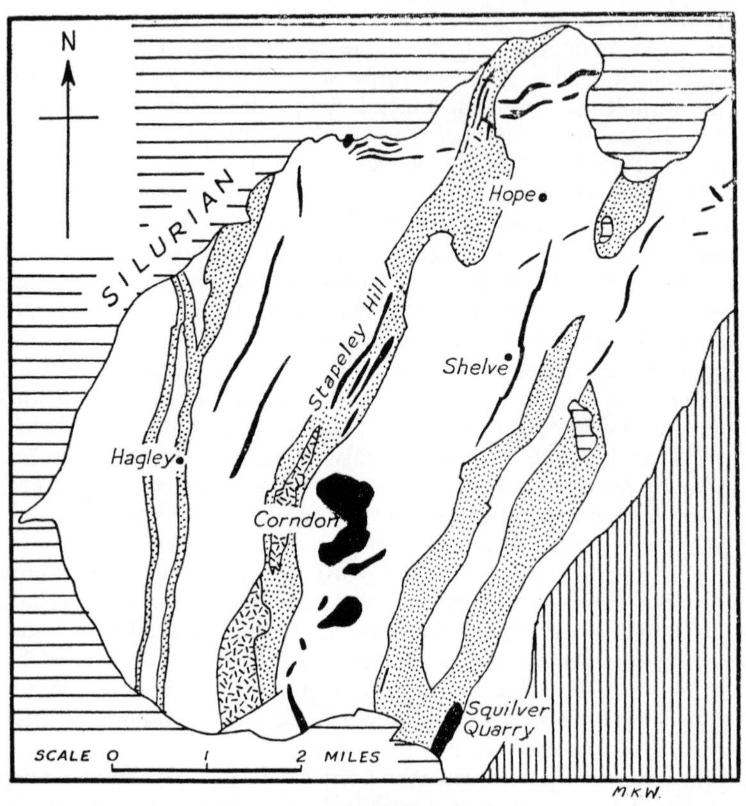

FIG. 140

Sketch-map of the Ordovician inlier of the Shelve area, Shropshire, showing the outcrops of the lavas and associated intrusions. *After W. W. Watts and F. G. Blyth.* Volcanic rocks, stippled; basic-ultrabasic, black; intrusive andesites, small dashes; Ordovician sediments, blank.

in a fine microlitic groundmass. In the same area thick sheets of similar composition are intercalated in the Bala rocks.

Precisely similar hypersthene-andesites occur in the Breidden Hills and near Builth, in the range of the Carneddau (Herefordshire and Radnorshire).

[1] Whittard, W. F., *Proc. Geol. Assoc.*, **42** (1931), 322.

The comparative uniformity in the petrographical character of the lavas is reflected in the minor intrusives, of which two main types occur. The first of these are intrusive andesites, found in the area of the Stapeley Volcanic Series on Stapeley Hill, Llanfawr, Roundtain, and Todleth.[1] In the Breidden Hills they form the greater part of Moel-y-Golfa. From their lithological characters some were formerly thought to be lavas. The second type is coarse-grained dolerite, resembling the gabbroid dolerites of North Wales. The Ordovician rocks of the Shelve area are pierced by many dykes and sills of dolerite.[2] At the Corndon and at Pitchfolds the phacolithic character of these intrusions is apparent. At the former locality the phacolith is intrusive into the *bifidus*-shales, a little below the overlying Stapeley Volcanic Series. In the Breidden Hills, hypersthene-dolerites containing both hypersthene and augite, and often with ophitic texture, are similar in form to the Corndon phacolith. In addition to the basic intrusives small outcrops of augite-picrite occur in the Shelve district.

The dolerites of this area are not all of the same age: cognate xenoliths of dolerite are found in the intrusive andesites referred to above. The latter, on Moel-y-Golfa, were definitely emplaced and exposed by denudation prior to the deposition of the unconformable Llandovery (Silurian) rocks. The dolerites of the Carneddau range, near Builth in south-west Radnorshire, cut Ordovician (Llandeilo) rocks, but not the Llandovery beds. Consequently some, at least, of the dolerites must be of Ordovician age. Blyth has shown that the Squilver gabbro (Fig. 140) formed a 'feature' on the early-Silurian coastline, as pebbles of this rock occur in the Upper Valentian beach-deposits forming the local base of the Silurian. As this gabbro metamorphoses Llanvirnian shales, the age of these intrusives is proved.

The Lake District.[3]—The vast accumulation of volcanic material known as the Borrowdale Volcanic Series is mainly made up of andesitic lavas and tuffs; but, as in the case of the contemporaneous cycle in North Wales, a change in the character of the magma took place towards the close of the volcanic episode, the andesites and andesitic pyroclasts being succeeded by rhyolites including welded tuffs.[4]

The great thickness originally assigned to the Borrowdale Volcanic

[1] Watts, W. W., The Geology of South Shropshire, *Proc. Geol. Assoc.*, **36** (1925), 359.

[2] Blyth, F. G. H., Intrusive rocks of the Shelve area, *Quart. J. Geol. Soc.*, **99** (1943), 169.

[3] Green, J. F. N., Vulcanicity of the Lake District, *Proc. Geol. Assoc.*, **30** (1919), 153; also Mitchell, G. H., Borrowdale volcanic series, etc., *Quart. J. Geol. Soc.*, **90** (1934), 418.

[4] Oliver, R. L., Welded tuffs in the Borrowdale volcanic series . . ., *Geol. Mag.*, **91** (1954), 473–83.

Series has been proved to be considerably over-estimated; but they probably exceed 10,000 feet. During the Caledonian revolution they were subjected to very complex folding and profound faulting.

The succession commences with true explosion tuffs, formed of material ejected during the first paroxysmal outbursts. Above these, pyroxene-andesites are followed by tuffs, and these in turn by andesites and rhyolites. The lavas are frequently autobrecciated, this being accentuated on weathered surfaces, and causing a deceptive resemblance to coarse agglomerates. Petrographically the lavas are very variable, both in texture and in the proportions of the component minerals, and rock types other than the dominant andesites may be present.

The Eycott lavas that occur at Eycott Hill, one mile from Troutbeck Station, near Keswick, and in the Cross Fell range, constitute a more basic type, probably more correctly classified as basalts. Some of the members of this series are strongly porphyritic, containing fine large crystals of plagioclase (bytownite to anorthite), also idiomorphic hypersthene altered to bastite.

Rhyolites form the highest part of the Borrowdale Volcanic Series. They are felsitic rocks, presenting strong flow structure, and occasionally containing garnets (as at Illgill Head). They occur in Langdale, on Crinkle Crags and Great Gable (Sty Head), in Long Sleddale and near Great Yarlside. The basal flow is a typical nodular rhyolite, which has been traced over a considerable area, but several of the rhyolites are intrusive, and welded tuffs occur.

The lowest lavas in the main Borrowdale Series, which appear to succeed the Lower Llanvirn conformably, must have been submarine, and poured out in deep water. Shallowing of the water is indicated by the strongly marked false-bedding at higher levels in certain of the tuffs.[1]

In addition to the main outcrops, the Borrowdale volcanics are represented in the Cross Fell inlier by a small thickness of andesitic lava and tuff (the Milburn Group) in the *bifidus* zone,[2] as well as by the overthrust masses referred to above. In the Sedbergh inlier rhyolitic flows occur in the Ashgillian.

The Lake District has its plutonic rocks in addition to a host of minor intrusives. Several of the intrusions occur at the junction of the Skiddaw Slates with the overlying Borrowdale Volcanic Series. The Eskdale Granite (45 square miles in area), the largest intrusion in the district,[3] the Buttermere and Ennerdale granophyre, the St John's Vale microgranite and the gabbro-granophyre sheet-complex

[1] Hartley, J. J., *Proc. Geol., Assoc.*, **36** (1925), 203.
[2] Shotton, F. W., The Cross Fell inlier, *Quart. J. Geol. Soc.*, **91** (1935), 149.
[3] Simpson, B., *Proc. Geol. Assoc.*, **45** (1934), 17.

of Carrock Fell all occur at this horizon. The Shap Granite (strictly an adamellite) is intruded into the Borrowdales adjacent to the Coniston Limestone Series, shown in black in Fig. 141.

The Skiddaw granite[1] outcrops in three small inliers, but its true size is indicated by the very extensive metamorphic aureole within which the slates have suffered not only thermal metamorphism but also metasomatism. The granite itself has been modified profoundly

FIG. 141

Sketch-map of the English Lake District, showing the major rock-groups and the more important intrusives. Ordovician, *horizontal ruling*; Silurian, *vertical ruling*; Borrowdale Volcanic Group, V.

by late-stage solutions which accumulated near the roof of the intrusion and have converted the northernmost outcrop of granite into the largest mass of greisen in Britain—the Grainsgill Greisen. The picrite of Great Cockup and many dolerite sills also occur in the Skiddaw Slate.

On general grounds it was formerly considered possible that the Lake District plutonic intrusions, the Shap, Skiddaw and Eskdale

[1] Hollingworth, S. E., and Dunham, K. C., Geology of the country around Cockermouth and Caldbeck, *Mem. Geol. Surv.*, **23** (1968), 114–29.

granites, might be Ordovician, related to the Borrowdale Volcanic Series; but these have now been dated and prove to be Caledonian, *e.g.* Skiddaw, 399m.y. (=late-Silurian–early Devonian).[1] Similarly, the age assigned to the Carrock Fell Complex varies according to different authorities from pre-Bala to Tertiary. Much depends

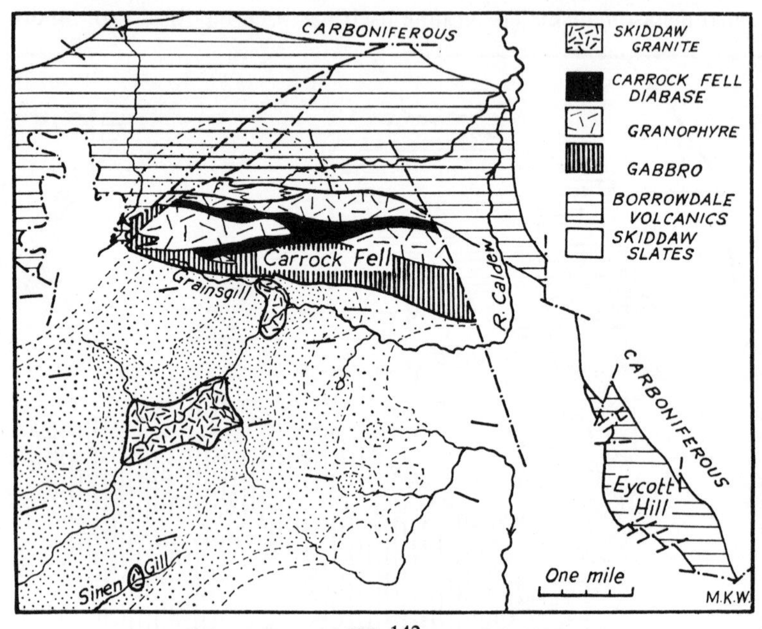

FIG. 142

Carrock Fell Complex and the Skiddaw Granite, Cumberland. (*After S. E. Hollingworth* (1938.))

Three zones of the metamorphic aureole of the granite are shown by stippling in the region of Skiddaw Slates. Differences of metamorphic grade cannot be distinguished in the Carrock Fell Complex which is therefore shown without stipple. Fold-axes in the Skiddaw Slates are indicated.

upon the identification of certain rock-fragments occurring in the basal conglomerates of the Coniston Limestone Series.

Petrographically the Lake District intrusives cover a very wide range of composition—from very acid granophyres and granites to ultrabasic picrites. Acid and basic types again predominate: there are few of intermediate composition. As in North Wales, the granophyres pass, towards their margins, into more basic modifications approaching markfieldite in composition (Ennerdale).

[1] Brown, P. E., Miller, J. A., and Soper, N. J., Age of the principal intrusions of the Lake District, *Proc. Yorks. Geol. Soc.*, **34** (1964), 331–42.

In the Carrock Fell Complex[1] (Fig. 142), north-east of Keswick the most interesting features are the steeply dipping sheeted form of the the complex, the close association of gabbro and granophyre; the presence of hybrid zones involving the two main types; and the differentiation of the gabbro. The latter is mostly quartz-bearing and hornblendic, while it is abnormally rich in ilmenite which increases towards the margins until ultimately it forms a considerable proportion of the rock.

Of the many minor intrusions, the following may be mentioned. The Armboth dyke is a spherulitic, or granophyric quartz-porphyry, composed of bright-red feldspar (orthoclase) and dark bipyramidal quartz crystals scattered evenly through a dun-coloured groundmass containing garnets. Porphyritic microgranites occur as bosses and dykes: thus the two laccoliths of St John's Vale, one of which is quarried near Threlkeld Station, are of this type, containing phenocrysts of orthoclase and small garnets; and dykes of a similar rock, with or without porphyritic crystals of quartz and feldspar, occur, for instance, at a number of localities. In the Wastwater district there are, connected with the Eskdale and the Ennerdale masses, innumerable dykes and sills, some of which contain feldspar phenocrysts up to two inches in length. A remarkable porphyritic microgranite containing, besides phenocrysts of a red orthoclase, plagioclase and quartz, large plates of muscovite and small flakes of biotite occurs at Dufton Pike in the Cross Fell inlier, and is known locally as the 'Dufton Granite'. A dyke of spherulitic felsite traverses the rocks of High Fell in Cumberland.

In the areas considered above, the intrusions are closely associated with lavas of essentially the same petrographic type; but in other tracts where Lower Palaeozoic rocks occur this is not the case. For example, in the Cambrian inlier lying west of the Malvern Hills[2] several sills and small bosses occur. They are not accompanied by lavas, and direct proof of their age is wanting. It is significant, however, that no intrusions occur in the adjacent Silurian strata, so that an Ordovician age is implied. This supposition is strongly supported by the facies of the rocks, which are spilitic. The types represented are spilitic andesites, vesicular spilites and spilitic olivine-dolerites.

The Loch Borolan Complex.[3]—An interesting intrusive complex

[1] Hollingworth, S. E., Carrock Fell and adjoining areas, *Proc. Yorks. Geol. Soc.*, **23** (1938), 208–18; Hollingworth, S. E., and Dunham, K. C., Geology of the country around Cockermouth and Caldbeck, *Mem. Geol. Surv.*, **23** (1968), 78–113.

[2] Blyth, F. G. H., On the intrusive rocks in the Cambrian inlier near Malvern, *Quart. J. Geol. Soc.*, **91** (1935), 463.

[3] Shand, S. J., Loch Borolan laccolith, North-West Scotland, *J. Geol.*, **47** (1939), 408.

occurring in the Assynt district of Sutherlandshire cuts and metamorphoses Cambrian rocks. It is affected by the Caledonian thrusts, so was believed to be of Ordovician age. In facies it differs from other intrusive rocks of this period, and, indeed, contains rock types unique in this country and uncommon elsewhere.

The complex has the form of a stratified laccolith, the members of which grade one into another, and hence were derived by differentiation in place. The highest rock exposed is a quartz-bearing soda-syenite (nordmarkite). Beneath this, megascopic quartz is absent, and with its disappearance nepheline and its alteration products come in, while the proportion of coloured minerals to perthitic feldspars gradually increases. The central zone of the laccolith is occupied by a melanite-nepheline-syenite, while the lowest portion exposed is a melanite-nepheline-gabbro. It is probable that the base of the intrusion (which is hidden) consists of a garnet-rich pyroxenite.

The minor intrusions connected with the Loch Borolan Complex are not less interesting and include nepheline-syenite-pegmatites, aegirine-aplites, pseudoleucite-porphyry and the problematical rocks, borolanites. These rocks were believed to contain pseudomorphs after leucite, since they resemble very closely undoubted pseudoleucites of other (American) localities, but Shand has thrown doubt upon this identification.[1]

At two other localities in northern Scotland syenitic complexes occur which, from their petrographic characters, are obviously comagmatic with the Loch Borolan Complex. Indeed one of them, the Loch Ailsh Complex, is only two miles distant from the latter, and like it, is intrusive into the Cambro-Ordovician rocks outcropping to the west of the great overthrust faults which bound the North-West Highlands. The Ben Loyal alkali-complex, on the other hand, builds a picturesque mountain group lying ten miles east of the overthrust zone, in the heart of the Moine Schists of uncertain age.

The **Loch Ailsh** mass[2] has the form of a sheeted complex or stratiform laccolith, consisting essentially of sodi-potassic syenites. The highest member contains quartz, and approximates closely to nordmarkite. Downwards, quartz fails, as in the Loch Borolan Complex, and the dominant type is pulaskite, with colour index of 10 or over, the mafic components including aegirine and melanite garnet, with riebeckite in one variety. Still lower, the proportion of coloured minerals increases, reaching about 70 per cent in a shon-

[1] On borolanite and its associates in Assynt, *Trans. Edin. Geol. Soc.*, **9** (1909–10), 202–15 and 376–416. *See also:* Tilley, C. E., Some new chemical assemblages of the Assynt alkali suite, *Trans. Edin. Geol. Soc.*, **17** (1958), 156–64.

[2] Phemister, J., in Geology of Strath Oykell and Lower Loch Shin, *Mem. Geol. Surv.* (1926), 22.

kinite zone, which in turn passes down into an ultrabasic layer including biotite-pyroxenite, biotite-hornblende-pyroxenite and hornblendite. This major portion of the complex is cut by a later intrusion of ultra-feldspathic syenite, termed *perthosite* by J. Phemister, on account of its extraordinarily high content of anti-perthitic feldspars.

The **Ben Loyal Complex** has been shown by H. H. Read[1] to consist of rock types closely comparable with the Loch Ailsh syenites. Pulaskite makes up the greater part of the twelve square miles exposed, but nordmarkite is also well represented.[2]

The three complexes, together with the associated minor intrusions, comprise an alkali-province of unique type so far as the British Isles are concerned, and about 400 m.y. old (Loch Borolan 388 and Ben Loyal, 403 m.y.) *i.e.* late-Silurian-early Devonian. It appears, therefore, that these complexes are much younger than was originally thought from the geological evidence.

SILURIAN

Igneous activity in the Silurian period was restricted to feeble outbursts in one Irish (Clogher Head), one Welsh and two English localities. In the **Tortworth Inlier,** lying north of the Bristol Coalfield, two bands of igneous rock occur in Silurian strata. The higher of these is considered to be a lava-flow, since it is accompanied by tuffs, while the lower is probably intrusive. The lava is an enstatite-andesite, containing many corroded xenocrysts of quartz which, as usual, are surrounded by reaction rims of pyroxene. The rocks of the lower band are more basic, and appear to be intrusive albitized olivine-basalts.[3]

The only other English locality where lavas of Silurian age are known to occur is near Shepton Mallet in the eastern parts of the **Mendip Hills.** The rocks, extensively quarried for road-metal, are purple and green andesites and andesitic tuffs. They contain conspicuous phenocrysts of plagioclase and bastite pseudomorphs after orthopyroxene.

Silurian volcanic rocks occur in West Pembrokeshire at Marloes, in Upper Valentian strata. They are olivine-basalt flows showing pillow-structure.

The lavas of **Clogher Head** in County Kerry differ from the other Silurian examples by being Acid in composition and by including nodular and banded rhyolites. The associated sediments are of Llandovery and Wenlock age.

[1] Geology of Central Sutherland, *Mem. Geol. Surv.* (1931), 174.

[2] King, B. C., The Cnoc nan Cuilean area of the Ben Loyal Complex, *Quart. J. Geol. Soc.* **98** (1942), 147.

[3] Reynolds, S. H., *Quart. J. Geol. Soc.*, **80** (1924), 106–11.

Mention may be made of the 'green streak' in the zone of *Monograptus argenteus*, in the Skelgill Beds of the Llandoverian of north-western England and central Wales. This interesting band is believed to represent a distant volcanic eruption, and consists of the finest wind-blown volcanic dust.

For the rest the Silurian was a period of tranquility: it was the calm before the storm, however, for with the close of the period came the main phase of the Caledonian, which is considered in relation to contemporary igneous activity in the following pages.

At this point we have to return to the main theatre of Caledonian igneous activity—Scotland, to examine the nature of the evidence in the deeply eroded roots of the Caledonides, *i.e.* the Grampian and Northern Highlands. Much progress has been made in recent years in unravelling the tangled skein resulting from several espiodes of folding, multiple metamorphism and metasomatism. The rocks involved comprise three major formations: the Lewisian, restricted on the eastern side of the Moine thrust to so-called inliers; the Moinian, essentially granulitic; and the more variable Dalradian. The inter-relationships are still obscure; but on general grounds it is widely believed that the Moinian is metamorphosed Torridonian, therefore Pre-Cambrian, but entirely reconstituted during the Caledonian orogeny. The Dalradian appears to span the time interval late Pre-Cambrian to early Palaeozoic.

It is possible to group the igneous rocks in these areas as (1) pre-metamorphic, (2) synmetamorphic and (3) post-metamorphic.

The Caledonian intrusives of pre-metamorphism date fall into two compositional groups: firstly, Basic to Ultrabasic, essentially gabbroic and doleritic rocks, now with concordant boundaries and especially forming massive sills, but also dykes and small bosses. The best-known is the **Portsoy Sill Complex,** dominantly gabbroic, but including also serpentinites and anorthosite. A second similar group of sills occurs in the South-West Highlands in Knapdale and in the Hebridean islands of Raasay (off Skye) and Islay. All these rocks were involved in the regional metamorphism; and although in parts of the more massive intrusions the igneous structures and textures have survived, the original minerals have been completely replaced by an assemblage stable under the then-existing conditions: andesine and common hornblende are characteristic; the rocks are termed amphibolites.

Secondly, pre-metamorphism granites also occur, the best known being the **Carn Chuinneag** and **Inchbae** coarse-grained granite gneisses, both very handsome rocks with prominent augen of pink alkali-feldspar. Both are intrusive into Moinian rocks (probably metamorphosed Torridonian) in which a zone of thermal metamorphism up to a mile wide was produced and within which original sedimentary structures were preserved. Isotopic dating suggests an age of the Carn Chuinneag granite of about 530 m.y.—that is, Lower Palaeozoic, and late-Cambrian on Holmes' time scale.

We pass now to what we may broadly refer to as the syntectonic features of the Caledonian orogeny, which are unique in British geological history and in which one process was particularly significant—migmatization. Reference to the processes involved was made in the discussion on the origin of granites. The migmatization was most intense in certain 'hot spots' where the regional metamorphic effects were at a maximum in both the Dalradian and Moine regions of the Highlands. In large areas in the Northern and Grampian Highlands the originally sedimentary Moine and Dalradian up to Middle Cambrian in age were thoroughly soaked in granitic ichors in the manner referred to in the chapter on the origin of granites, forming permeation gneisses. At a somewhat later stage of evolution a higher degree of mobility was achieved, and thin veins, sheets and larger bodies of more definitely granitic aspect invaded the country rocks forming injection gneisses. These bodies are the 'Older Granites' of earlier accounts. They are not the oldest Caledonian granites (see Carn Chuinneag and Inchbae above) and it is important to note that they are *not* Pre-Cambrian, although in some cases material of Pre-Cambrian age may have been involved in their formation.

THE NEWER CALEDONIAN INTRUSIONS

A small-scale geological map of Scotland shows a large number of major intrusions, some of very large size. These have been termed the 'Newer Granites' to distinguish them from the 'Older Granites', the nature and origin of which we have discussed above. A better name would be 'Newer Intrusives', for many rocks in addition to granites are included. It will be observed that the intrusions are aligned roughly parallel to the great faults which divide the country into regions and have the direction of the Caledonian mountain chains (the Caledonides of Suess). Doubtless the granites worked their way up into the cores of the mountains and have been exposed by subsequent profound denudation. As the intrusions came into existence at the same time as the mountain chains, it is best to refer to them as the Caledonian intrusions. They are not restricted to Scotland and England: others are found in the prolongation of the Caledonian chains in Ireland (the Newry granite, for example)[1] and Scandinavia.

With regard to the exact age of the intrusions a difficulty arises

[1] Richey, J. E., and Thomas, H. H. *Quart. J. Geol. Soc.*, **88** (1932), 787.

through lack of direct evidence. Isotopic data suggest that, broadly speaking, they have ages of the order of 400 m.y. The Galloway granodiorites and adamellites in the southern Uplands of Scotland are intrusive into Ordovician and Silurian rocks, so is the well-known Shap adamellite (two square miles) in the English Lake District; while the largest single intrusion in the British Isles—the Leinster granite—is definitely Post-Silurian and Pre-Carboniferous in age. The small intrusions in the Manx Slates in the Isle of Man, comprising the Dhoon[1] porphyritic microgranite, and Foxdale granite, are also probably of Caledonian age. Now, had the granites been in place before the culmination of the Caledonian earth-movements they would undoubtedly show the effect of the intense pressures to which they would have been subjected. They do not do so, and hence must be younger than the acme of the orogeny. In one or two cases the marginal portions of the granites show some foliation, and were therefore intruded before the earth-movements had entirely ceased. An upper limit to their age is fixed by the occurrence of boulders of the Kincardineshire granodiorite in the local base of the Lower Old Red Sandstone.[2] Not all of the intrusions, however, were of this age, as some, *e.g.* the Ben Cruachan and Ben Nevis granodiorites cut, and are therefore younger than, the Lower Old Red Sandstone lavas. The 'Newer Intrusives' thus include two age-groups: an earlier Caledonian, and a later (Post-Lower, Pre-Middle Old Red Sandstone) series. Among others whose age is not definitely proven are the Mountsorrel Granites[3] and associated dioritic and gabbroic intrusions. In their petrographic characters they are distinctly Caledonian.

When the petrology of the intrusions is examined it is found that there are no marked differences between the two series, such as might be used as a criterion of relative age, and in many cases it is impossible to date the intrusions precisely, on stratigraphical evidence.

Although the intrusions are dominantly Acid, many other kinds of plutonic rock are associated with the great granite bosses either as marginal facies or as satellitic intrusions situated near to the main masses. Chief among these are diorites and monzonites; though basic types, including gabbro and norite, are not uncommon, and ultrabasic rocks are sometimes found.

The 'Newer Intrusives' of Scotland are all either adamellites or granodiorites, with still more Basic peripheral modifications in some cases (tonalites and diorites).

The **South East** and **Central Highlands** are remarkable not only

[1] Nockolds, S. R., The Dhoon (Isle of Man) granite, *Min. Mag.*, **22** (1931), 494.
[2] Summary of progress, *Geol. Surv.* (1901), 111.
[3] Taylor, J. H., *Geol. Mag.*, **71** (1934), 1.

for the number of great intrusions found in these parts of Scotland, but also the striking difference in their mode of occurrence: the irregularity and complexity of the forms assumed by the 'Older Granites' contrasts strongly with the comparative simplicity of form and uniformity of the 'Newer Intrusives'. Of these several masses, the Kincardineshire adamellite is the least variable; it contains the most plagioclase and is interesting on account of the apophyses thrown out on its south-western margin. These rapidly pass into porphyritic microgranites.

Some of the Caledonian granites increase in alkalinity towards the north or north-west, indicating intrusion at a time when the country was still under the influence of the Caledonian earth-movements, and suggesting the partial separation of the products of earlier from those of later crystallization, with the migration of the latter into regions of less stress. The same principle is illustrated by the occurrence of small intrusions of intermediate, basic and ultrabasic rocks near to the major acid intrusions. These are of earlier formation than the latter, and in several cases occur at the south-eastern side of the main masses. This is well-illustrated by the Glen Tilt complex which comprises augite-diorite, tonalite, hornblende-adamellite, biotite-granite, muscovite-granite, and granite-aplite. The complex at Garabal Hill,[1] 400 m.y., near the head of Loch Lomond, includes augite- and mica-diorites, norites and rocks bordering on hornblende-peridotites, now largely serpentinized, as well as tonalite and granite. This complex occurs along the south-eastern margin of the porphyritic granite of Glen Fyne.

It would be unprofitable to discuss the petrology of each of these intrusions, and only the more important points can be here considered. A constant feature is the occurrence of abundant plagioclase in addition to orthoclase in the more acid intrusions: most of the 'granites' belonging to both age-groups are adamellites or granodiorites. This is illustrated by the so-called **Galloway** granites, which are distinguished as the Criffel-Dalbeattie, Cairnsmore of Fleet and Loch Dee *massifs* (Fig. 143).

The largest of the 'Galloway Granites' is the Criffel-Dalbeattie Complex[2] which consists of an early quartz-diorite forming a separate intrusion adjacent to the main one. The latter is granodiorite including an earlier aphyric member, which is marginal to the later, central, porphyritic granodiorite. The Galloway granodiorites are associated with a well-developed dyke-swarm,[3] some

[1] Nockolds, S. R., Garabal Hill–Glen Fyne igneous complex, *Quart. J. Geol. Soc.*, **96** (1940), 451.

[2] Phillips, W. J., The Criffell–Dalbeattie igneous complex, *Quart. J. Geol. Soc.*, **112** (1956), 221.

[3] Phillips, W. J., The minor intrusive suite associated with the Criffell–Dalbeattie granodiorite complex, *Proc. Geol. Assoc.*, **67** (1956), 103.

members of which trend in the same direction as the strike of the Silurian sedimentary rocks, while others are at right angles to this direction (N.W. to S.E. and N.E. to S.W.). The dyke rocks fall into three groups: a very small proportion (5 per cent) are lamprophyres, chiefly hornblendic types (spessartites), but a few are kersantites. The chief members of the swarm, to the extent of 95 per cent,

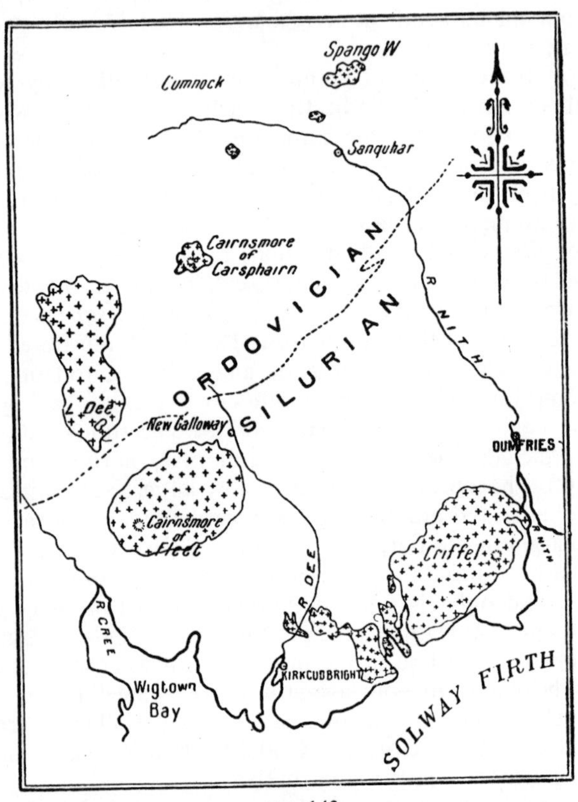

FIG. 143

Map of the Galloway district, showing the distribution of the granite complexes. Scale: 1 inch = 13½ miles.

broadly compare with the plutonic types, though the majority have been recorded as 'porphyrites' (Intermediate medium-grained dyke-rocks of dioritic composition), but this is really a field term only. However, the porphyritic microdiorites are associated with, and grade into, microtonalites and ultimately micrograpodiorites, both containing quartz, which in the latter is accompanied by subordinate potassic feldspar.

The Loch Dee Complex comprises a central outcrop of pale coloured biotite-granite, surrounded by biotite- and hornblende-tonalite. These tonalites are crowded with sedimentary xenoliths and pass into a strongly biotitic facies where they penetrate into the wall rock. Still more Basic rocks are quartz-norites and norites which are marginal in position. A zone of hybrid rocks is believed to have resulted from admixture of the sub-magmas which gave rise to the tonalite and norite respectively.[1]

The rock of the Cairnsmore of Fleet *massif* (390 m.y.) is less calcic, being typically a biotite-adamellite, grading into biotite-muscovite-granite.

In addition to the *massifs* mentioned above, smaller intrusions of grandodiorite associated with tonalite and quartz-norite, and evidently of the same age and origin, are found in the Galloway district at the Mull of Galloway, on the east side of the Cree, south of Creetown, at Cairnsmore of Carsphairn,[2] Spango Water and south of New Cumnock (Fig. 143). There are also many dykes of Caledonoid trend of which the chief are porphyritic microdiorites, accompanied by a variety of Acid and lamprophyric types.

The main structural and stratigraphical units established in Scotland extend across into Ireland where again Caledonian granites form important features in the Dalradian and Lower Palaeozoic tracts. The largest is the Leinster Granite; but the one which has been most exhaustively studied is the Donegal Granite illustrated in Fig. 97 and referred to in the discussion on the origin of granites. Similar granites occur also in the adjacent counties of Galway and Mayo.

The most alkaline of the Caledonian granites forms part of the **Leinster** Granite (625 square miles) in south-eastern Ireland. The main intrusion is an alkali-granite with an excess of potash over soda; but in some of the subsidiary intrusions, which are doubtless comagmatic with the main mass, soda predominates. Thus a sodic granite occurs at Aughrim, while another forms the summit of Croghan Kinshela and consists chiefly of a brilliant white sodic feldspar and grey quartz. In the north-eastern part of the complex three granites are distinguishable: the main type (central) is muscovite-granite, marginal to which is a porphyritic microcline-granite which occurs also in the roof zone of the former. Finally the outer zone is finer-grained and granodioritic while the adjacent rocks external to the complex proper are migmatites. A noteworthy feature is a striking fluxional arrangement of dark schlieren and,

[1] Gardiner, C. J., and Reynolds, S. H., The Loch Doon granite area, Galloway, *Quart. J. Geol. Soc.*, **88** (1932), 1.

[2] Deer, W. A., The Cairnsmore of Carsphairn igneous complex, *Quart. J. Geol. Soc.*, **91** (1935), 47.

particularly in the microcline-granite, of the porphyritic feldspars. Pegmatites and associated aplites occur within the complex.[1]

In most cases minor intrusions in the form of dykes and sills are closely associated with the 'Newer Intrusives' and are considered below (p. 503). Mica-lamprophyres are less widely distributed, but occur in Galloway and in the Lake District, where they are connected with the Shap adamellite.

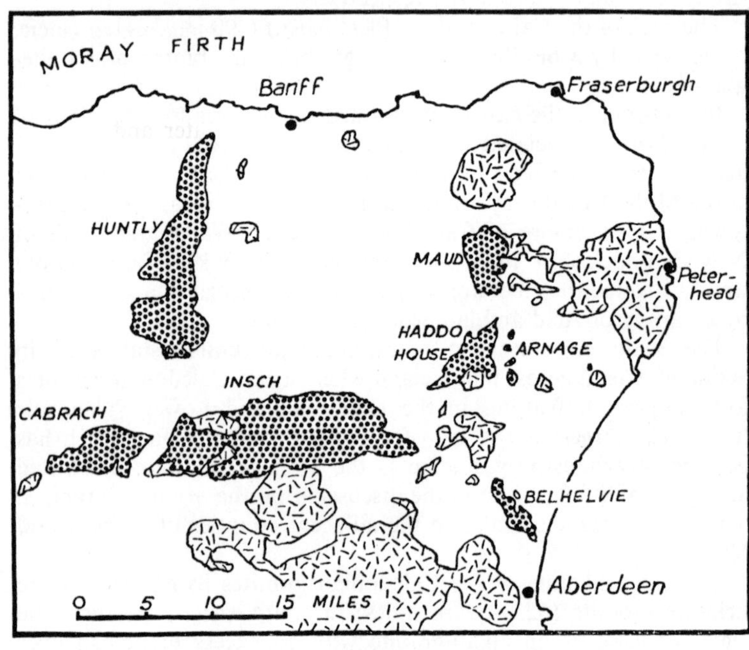

FIG. 144

Map of part of north-eastern Scotland, showing the distribution of the granitic and Basic igneous complexes. (*Based largely on the work of H. H. Read.*)

A second great group of 'Newer Intrusives' includes the complexes of Basic rock occurring in Aberdeenshire at Huntly, Haddo, Arnage, and Insch (Fig. 144). Although the various outcrops are separated by a mantle of Dalradian rocks, they were probably connected originally and are the exposed portions of a once continuous sheet.[2] The igneous rocks are essentially norites. Between the norite and the country-rock is a zone of '*contaminated norite*', differing markedly from both intrusive and invaded rock. The contaminated rock

[1] Brindley, J. C., The geology of the northern end of the Leinster Granite, Part I . . ., *Proc. Roy. Irish Acad.*, **56b** (1954), 159–90.

[2] Read, H. H., and Farquhar, O. C., The geology of the Arnage District: a re-interpretation, *Quart. J. Geol.*, **107** (1951), 423.

is crowded with zenoliths in all stages of absorption and is usually rich in minerals, such as garnet, cordierite (Fig. 53) and spinel, not normally found in igneous rocks. It is noteworthy, however, that some of the slides of the contaminated rock do not appear to differ essentially from normal igneous rocks. Thus one type from Kinharrachie, which consists of hornblende, plagioclase and interstitial quartz, is for all practical purposes a quartz-hornblende-diorite.

The Huntly mass[1] is a sheeted complex some 50 square miles in area, within which a wonderful variety of rocks of special petrological interest are exposed. They include peridotites, olivine-gabbros, troctolites, norites, as well as small granite bosses. Some of the Basic rocks exhibit a striking layering simulating stratification, rivalling the famous banded gabbros in Skye.

The Belhelvie Complex (Fig. 144) is essentially similar as regards its rock-types, and is best known for the troctolites (Fig. 117) which are represented in most teaching collections.[2] The age of the Complex is 456 m.y., *i.e.* Ordovician on Holmes' time scale.

DEVONIAN

Igneous rocks of Devonian age are important in **South Devon** and in **Cornwall,** and comprise basic and acid lavas (the latter being very subordinate to the former), together with tuffs and minor intrusions (Fig. 149 and 150). The extrusive rocks cover a large area in the neighbourhood of Totnes, particularly around the village of Ashprington, and are hence often referred to as the Ashprington Volcanic Series. From this locality they range southwards through Modbury, Saltash, Liskeard, and thence, swinging north of the Bodmin Moor Granite, reach the north coast of Cornwall near Padstow and Port Isaac. Within this tract the igneous rocks possess peculiarities indicating community of origin: the area is, in fact, a good illustration of a petrographic province, or rather, a portion of such, since the same series of Devonian lavas, tuffs and intrusions ranges across central Europe, through the Vosges and Harz Mountains, into Moravia and Nassau.

The original character of the rocks is best preserved in the northern part of the area, but towards the south the rocks become more and more affected by the Armorican earth-movements; indeed, in the Plymouth district they are so altered that their recognition as of igneous origin is difficult, while it is usually quite impossible to distinguish between lavas and intrusions.

[1] Read, H. H., *Geol. Mag.*, **61** (1924), 433, and Geology of Banff, Huntly and Turriff, *Mem. Geol. Surv.* (1923).

[2] Stewart, F., The gabbroic complex of Belhelvie, Aberdeenshire, *Quart. J. Geol. Soc.*, **102** (1947), 465.

In south-eastern Devonshire the eruptions commenced in early Middle Devonian times. The sites of the volcanoes have not been located, but it is clear that they must have been situated at a considerable distance from the shore-line of the Devonian Continent. Consequently the lavas are submarine, and are interbedded with limestone and other normal marine sediments. The basic lavas are spilites showing pillow-structure and the high degree of vesicularity characteristic of these rocks. These features are particularly well exhibited at Chipley, where a thickness of 70 feet of pillow-lava is exposed.[1]

In the Plymouth-Liskeard district the lavas are much decomposed, especially those belonging to the middle Devonian. Those interbedded with the Upper Devonian sedimentary rocks are slightly better preserved, and are seen to be chiefly spilites. Farther to the north, in the neighbourhood of Tavistock and Launceston, similar rocks occur in the Upper Devonian-Lower Culm.

On the north coast of Cornwall no igneous rocks are found in the Lower and Middle Devonian, but in the Frasnian division of the Upper Devonian they attain to their maximum development: at Pentire Point, near Padstow, the thickness of pillow-lava, well exposed in the sea-cliffs, exceeds 250 feet.[2]

In addition to the abundant basic flows, others of acid composition occasionally occur in the neighbourhood of Newton Abbot and Ivybridge. Like the spilites with which they are associated, a high soda-content is characteristic, and the rocks may be referred to the sodic rhyolites (quartz-keratophyres of some authors). They are much brecciated, but it is difficult to decide whether this is the result of flow-movements in an extremely viscous lava, or of explosions in the vents.

The minor intrusions range in composition from basic to ultrabasic, and comprise three chief types; dolerites, minverites and picrites.

The dolerites are coarse-grained, non-vesicular rocks forming sills, in some of which large quarries have been opened for road metal, as at Trusham in the Newton Abbot district.[3] The augite is a normal brown variety, which in some specimens merges into titanaugite. No basic plagioclase is found in these rocks, but albite, some of which is primary, usually makes up nearly the whole of the feldspar present. An aplitic modification of the albite-dolerites is not uncommon. It occurs in veins and segregations, and consists essentially of microperthitic feldspars.

[1] Geology of Newton Abbot, *Mem. Geol. Surv.* (1913), 54–6.
[2] Dewey, H., *Proc. Geol. Assoc.* **25** (1914), 165–73.
[3] Newton Abbot, *Mem. Geol. Surv.* (1913), 59–63.

The minverites are feebly developed in Devonshire, but are the dominant intrusive types in Cornwall, both in the neighbourhood of Plymouth and in the Padstow district. At the latter locality they form sills up to 70 feet in thickness. The type-specimens come from the Rock Quarry, St Minver, on the Camel Estuary. In these rocks TiO_2-rich amphibole is the dominant coloured mineral, though it is sometimes accompanied by olivine, titanaugite and bronze-coloured biotite. As in the lavas, the feldspar is chiefly albite, but some anorthoclase occurs in addition. The texture varies considerably: some types are coarsely ophitic; in others there is a tendency towards idiomorphism of the dark minerals; while others again are compact, fine-grained and resemble some camptonites.

No fresh olivine has yet been found in the basic intrusions, but serpentinous pseudomorphs, embedded in the pyroxene, are not uncommon in the dolerites. By increase in the amount of olivine, at the expense of the feldspar, there is a gradation to picrites, a few of which occur in this cycle. Teall has described the augite-picrite of Menheniot (Clicker Tor). Typical hornblende-picrite is found at Molenick in the Plymouth district, while augite-picrite occurs at Highweek in the Newton Abbot district.

Each of these intrusive types strongly resembles the contemporaneous rocks in the continuation of the province in central Europe.

THE OLD RED SANDSTONE LAVAS[1]

Igneous activity during the Devonian Period was by no means restricted to the submarine eruptions that occurred in the area now occupied by Devon and Cornwall: in northern Britain and north-western Ireland contemporary volcanoes poured out an enormous quantity of lava and ash of quite different types from those described above, although of the same age. The Old Red Sandstone cycle is complete, comprising an extrusive phase, the dominant type being andesite; a phase of major (plutonic) intrusion, during which some of the largest granite masses in this country came into being; and a phase of minor intrusion, when porphyritic microdiorites were injected as dykes.

The lavas were poured out from volcanoes of the central type located along lines of instability connected with the Caledonian earth-movements. One series of volcanoes bordered the Midland Valley of Scotland on the north, and extended across into Ireland.

[1] This volcanic cycle is of Devonian age, but the extrusives are interstratified with rocks of continental facies—the Old Red Sandstone. These contrast strikingly with the marine facies of the Devonian, typically developed in Devon and Cornwall in the extreme south-west of England.

Another series bordered it on the south (Fig. 145). Relics of the outpourings from the former occur in the **Ochil** and **Sidlaw Hills,** and the Lorne district of Argyllshire, including **Ben Nevis** and **Glencoe;** while the latter are represented by the lavas, tuffs and intrusions of the **Pentland Hills,**[1] the **Braid Hills,** the **Cheviot Hills** (Fig. 147), and a small area in North Ayrshire.[2]

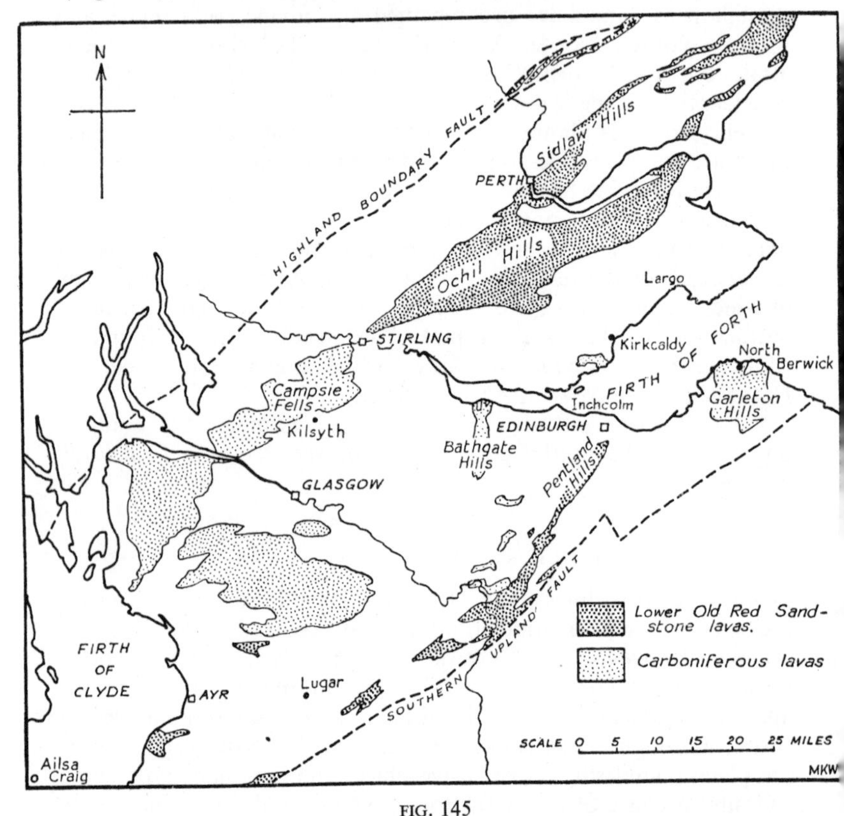

FIG. 145

Map of the Old Red Sandstone and Carboniferous lavas of the Midland Valley. of Scotland. (*Based on maps of the Geological Survey.*)

The volcanic rocks rest upon an uneven eroded land surface consisting of schists in the South-West Highlands, and folded Silurian rocks in the Cheviots and central Lowlands. Locally basal conglomerates and breccias are intercalated, and are of interest inasmuch as they contain pebbles of lava, thus proving the existence of

1 Mykura, W., The Lower Old Sandstone igneous rocks of the Pentland Hills, *Bull. Geol. Surv. G. B.*, **16** (1960), 135–55.

2 MacGregor, A. G., in Geology of N. Ayrshire, *Mem. Geol. Surv.* (1930), 28.

even earlier flows. The basal sediments have yielded fragmentary plant remains (*Psilophyton* and *Pachytheca devonica*) in Lorne and Glencoe, which prove the age of the overlying volcanic rocks as Lower Old Red Sandstone. Further, they are succeeded unconformably by Upper Old Red Sandstone in some localities.

The lavas themselves are frequently brecciated (block-lavas) and in some cases can only with difficulty be distinguished from interbedded agglomerates. The proportion of pyroclasts to lavas is small, as the eruptions were not of a violently explosive type. It is probable that the lavas were in part erupted under subaerial conditions: the upper portions of the flows are frequently reddened, due to atmospheric weathering between one eruption and the next. Occasional shales and sandstones occur interbedded with the lavas, and sand-filled crevices and 'sandstone dykes' are characteristic.

Although the succession has been the subject of careful study at many localities, it is possible to correlate individual flows over restricted areas only, because the lavas erupted from neighbouring centres differ in composition, although the volcanoes were so close to one another that their products interdigitate. The dominant type of lava is a basic andesite (verging on basalt), and this is interbedded with rhyolite and basalt.

The more Basic lavas contain small red pseudomorphs after olivine and bear a close resemblance to mugearites. In others olivine is less abundant and is accompanied by pale green augite. In the typical augite-andesites only the latter occurs as phenocrysts. Basalts and basic andesites with phenocrysts of feldspar are uncommon in the west (Glencoe and Ben Nevis), but a very distinctive type occurs near the top of the succession in the Pentlands. This rock is distinguished at the 'Carnethy porphyry', and is particularly rich in fluxionally arranged, platy phenocrysts of plagioclase. One of the basic lavas in Glencoe contains the rare red variety of epidote, withamite, occurring as blood-red crystals in druses.

The less Basic lavas include enstatite-, hornblende- and mica-andesites. In addition to phenocrysts of coloured minerals, these andesites are rich in porphyritic feldspars. The Cheviot lavas[1] total some 1,200 feet in thickness, and include dominant augite-hypersthene-andesites, glassy andesites, trachyandesites and curious types termed by the Survey Officers 'oligoclase-trachytes'. These we would classify as oligoclase leuco-andesites.

The Old Red Sandstone lavas which build the Sidlaw Hills in eastern Scotland are fresher than those in most other areas where volcanic rocks of this age occur. They include basalts, basaltic

[1] Carruthers, R. G., *et. al.*, Geology of the Cheviot Hills, *Mem. Geol. Surv.* (1932), 8.

andesites and true andesites in variety, together with rare trachyandesites and comagmatic minor intrusions.[1]

In chemical composition the lavas noted above are evidently closely related to the intrusive augite-diorites, kentallenites and monzonites of the plutonic phase. Some of the lavas are rich in orthoclase, particularly certain flows from Lorne, and consequently are to be regarded as trachybasalts or trachyandesites, according to their degree of basicity.

The dacites connect the Acid lavas with the hornblende- and mica-andesites, from which they differ chiefly in the occurrence of phenocrysts of quartz. They are so closely similar in appearance to many of the rhyolites, that in some districts the two types of rock have not been differentiated in the field.

The rhyolites are chemically related to the most Acid granites and microgranites of the succeeding phases, but vary considerably in texture among themselves. Some show beautiful flow-structure, others are cryptocrystalline through devitrification of an originally glassy rock, while many are spherulitic. The phenocrysts include albite, orthoclase, biotite and quartz. It is usually the case that the more acid andesites grade into rhyolites, but in the Pentland Hills they are associated with potassic trachytes.

One of the most interesting igneous rocks of this age forms a narrow outcrop (the rock is only 40 m thick) traceable for 80 km from Dunkeld in the Highland Border area. Its identification has caused considerable difficulty to petrologists who have studied it. First it was regarded as an intrusive 'quartz-porphyry' (= porphyritic microgranite). Later it was thought to be a dacitic lava; but now it is stated to be an ignimbritic ash-flow, in part welded.[2] Interest centres on its original extent which appears to have been comparable with some of the most extensive in New Zealand and North America, referred to in the account of ignimbrites.

In addition to the above, two massive ignimbrites have been discovered among the Old Red Sandstone volcanics in Glencoe.

THE OLD RED SANDSTONE MAJOR INTRUSIONS

Reference has already been made to the general characters of the post-Lower Old Red Sandstone major intrusives. They range in composition from Acid granites to Basic rocks rich in olivine, while locally feldspar-free Ultrabasic types occur. Outcrops of Acid rocks are far more extensive than those of Intermediate and Basic com-

[1] Harry, W. T., The Old Red Sandstone lavas of the western Sidlaw Hills, Perthshire, *Geol. Mag.*, **93** (1956), 43–56 and Old Red Sandstone lavas of the eastern Sidlaws, *Trans. Edin. Geol. Soc.*, **17** (1958), 105–12.

[2] Paterson, I. B., and Harris, A. L., Lower Old Red Sandstone ignimbrites from Dunkeld, Perthshire, *Proc. Geol. Soc.*, **1664** (1971), 282.

position, and in volume the former are in large excess over the latter, *i.e.* the magma, as represented by the plutonic rocks, was essentially Acid. Direct evidence of the age of these intrusions is sometimes forthcoming. Thus on Ben Nevis, in Glencoe and the Cheviot Hills, the granites cut the Old Red Sandstone lavas, and must therefore be younger than the latter. In other cases, however, such direct evidence is wanting, and the age of many of the granite masses in the Highlands is therefore largely a matter of speculation. It will be appreciated that these intrusions and the associated O.R.S. volcanic rocks represent the youngest phase of igneous activity in the Caledonian orogenic cycle.

With regard to the form of the major intrusions, it is clear that many are true bosses: the outcrops are in many cases roughly circular, the contacts are steeply inclined, while the manner in which they cut across the bedding and foliation planes of the country-rock is strikingly shown on the geological maps. On account of the marked similarity of composition between rocks of this age over very extended areas, it may be suspected that they are connected underground, and that the bosses are cupolas rising from the tops of batholiths.

The granites of Glencoe form part of one of the greatest intrusive masses in Scotland, known as the Etive Complex[1] (Fig. 146). The earliest and main portion of this complex is the Moor of Rannoch Granite; this is cut by the Ben Cruachan Granite, and the latter again by the Starav boss. The evidence seems clear that the dyke-phase, referred to below, supervened between the injection of the Cruachan and Starav masses. We thus have an interesting case of one granite boss penetrating another of somewhat earlier date. It is believed that the second injection was consequent upon the down-faulting of a cylindrical plug of country-rock (in this case the Cruachan Granite). It has already been shown that cylindrical or conical fractures may be produced above the top of an advancing plug of magma. In at least three cases in the area under consideration, such fracturing was followed by collapse or foundering into the magma underlying the tract surrounded by the fault. Such phenomena have been termed cauldron subsidences. In Glencoe and on Ben Nevis the lavas owe their preservation to such 'piston-faulting' (Fig. 66). In the latter locality, the igneous *massif* consists of three approximately concentric zones: the outer granite partly quartz-diorite, partly adamellite, the inner trondhjemitic granodiorite and, in the centre, the roughly circular outcrop of lavas resting on a floor of schist. As in the Etive Complex the dyke-phase intervened between the injection of the outer and the inner granites. In the

[1] Bailey, E. B., and Maufe, H. B., Geology of Ben Nevis and Glencoe, *Mem. Geol. Surv.* (1960), 2nd edn.

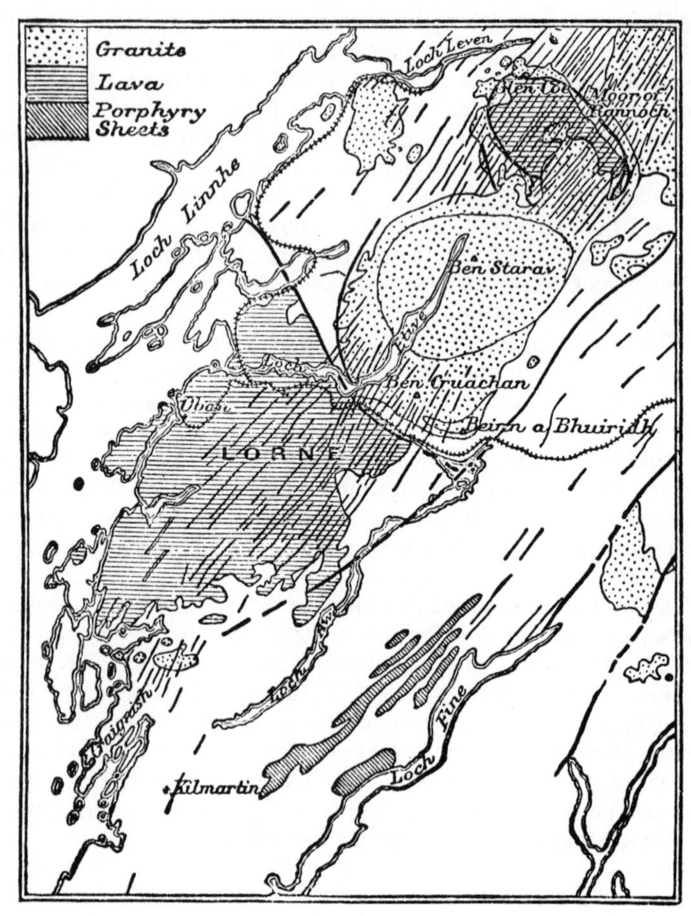

FIG. 146

Sketch-map of the volcanic district of northern Argyllshire (Glencoe), showing the distribution of the lavas of Old Red Sandstone age and of the dykes (which are chiefly porphyritic microdiorites), in relation to the Etive Granite Complex. The Highland schists are left white. Faults are shown by heavy black lines. (*After Clough, Maufe, and Bailey* (1909).

light of the clear evidence obtained at Glencoe, it is probable that cauldron subsidence operated on three occasions at Ben Nevis, while the outer granite itself consists of three fractions forming arcuate outcrops.

With regard to the petrography of the major intrusions the dominant types are adamellites and granodiorites,[1] *i.e.* a large

[1] The student should note that in the Geology of Ben Nevis and Glencoe, *Mem. Geol. Surv.* (1960), the rock-names are not used in exactly the same sense as in this textbook.

proportion of the feldspar consists of acid plagioclase, typically oligoclase. Locally the rocks grade into tonalite and quartz-monzonite. The larger intrusions are too numerous to be described in detail, but the Etive Complex may be regarded as typical. The earliest member is the Quarry Diorite: the Moor of Rannoch rock is essentially granodiorite, consisting of oligoclase, microperthite, quartz, hornblende and biotite, while sphene is a prominent acces-

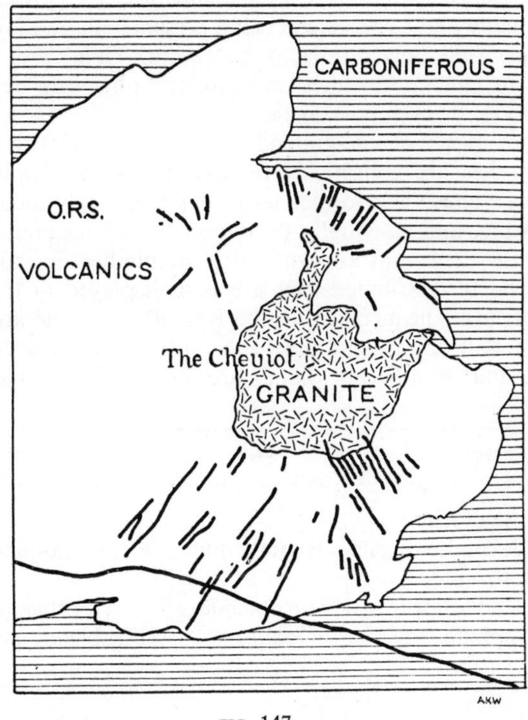

FIG. 147

Sketch-map to illustrate the Old Red Sandstone igneous cycle in the Cheviot Hills. Dykes are shown by thick lines.

sory. The marginal zone is more Acid, free from hornblende, and is porphyritic. The Ben Cruachan granite consists of a northern lobe of adamellite which has invaded the cauldron subsidence of Glencoe, and a southern lobe of tonalite. The youngest member of the Complex is the Starav granite the intrusion of which clearly involved the cauldron subsidence mechanism. It also is adamellite, largely aphyric, but the marginal facies is strongly porphyritic containing phenocrysts of orthoclase and oligoclase.

The Cheviot granite[1] (Fig. 147) may be partly unroofed laccolith, the area being 22 square miles. It is noteworthy on account of the modification it has undergone through assimilation of the country rock. The normal rock is a pink micrographic granite deficient in coloured minerals; but in places, particularly where the magma has penetrated *lit-par-lit* into the surrounding lavas, it is highly contaminated, of dioritic aspect and locally contains augite.

In the granophyric type biotite is the only coloured silicate; but in the marginal facies biotite is accompanied by pyroxene, both diopsidic augite and hypersthene being present; quartz may amount to a few per cent only, and may be absent. Thus these marginal rocks are, in fact, monzonites and diorites, obviously derived, by contamination, from granitic magma.

The more Basic coarse-grained intrusives associated with the Newer Granites are collectively referred to as the Appinitic and Kentallenite Suites. The former term seems to imply merely hornblende-bearing, and essentially products of 'wet' magma fractions. In this sense they contrast with 'dry' augite-bearing rock-types. They are widely distributed, but are best displayed in the South-West Highlands. The members of the Kentallenite Suite are centred around the Ballachulish granite. The types of rocks occurring in Argyllshire and the adjacent islands are shown in the Table.

Intermediate	Basic	Ultrabasic
Augite-diorite Hornblende-augite-diorite Monzonite	Kentallenite	Augite-picrite
Hornblende-diorite Appinite	Hornblende-gabbro	Hornblendite Hornblende-peridotite

The dominant rock-types are those in which pale-green augite (diopside or malacolite) is the characteristic coloured component: normal hornblende-syenites and diorites are rare among the intermediate rocks: but many of the meladiorites are rich in hornblende and have been termed 'appinites'.[2] With decreasing feldspar these rocks grade into hornblendite, and with the incoming of olivine into hornblende-peridotite of the Cortlandt type. In Colonsay, one of the Hebrides adjacent to the mainland, a similar series of intrusions

[1] Carruthers, R. G., *et. al.*, Geology of the Cheviot Hills, *Mem. Geol. Surv.* (1932), 87. Jhingran, A. G., The Cheviot Granite, etc., *Quart. J. Geol. Soc.*, **98** (1943), 241.

[2] Ben Nevis and Glencoe, *Mem. Geol. Surv.* (1916), 168.

has been described by the Survey.[1] Direct evidence of age is wanting, but on the ground of petrographic similarity they are referred to the Old Red Sandstone. Since, in some cases, they are intrusive into breccia or agglomerate it has been suggested that they mark the site of explosion-vents and are themselves vent-intrusions.

THE LOWER OLD RED SANDSTONE DYKE-PHASE

One of the most striking features of this igneous cycle is the well-developed dyke-phase which in general followed, though to some extent it overlapped, the plutonic phase. The dyke-phase reaches its maximum development in western Argyllshire, where the uprise of magma into a series of closely spaced parallel fractures formed the **Etive swarm** of N.E.–S.W. dykes (Fig. 146). The dykes are restricted to a comparatively narrow belt of country very much elongated in the direction of the dykes themselves. There can be little doubt that this tract was located above the body of magma, the comparatively thin roof of which constituted a belt of weakness in the crust. The magma must have been closely similar in composition to the lavas poured out during the extrusive phase, since the dominant types among the dyke-rocks are porphyritic and aphyric microdiorites.[2] In the former rocks idiomorphic crystals of hornblende and biotite, together with plagioclase ranging from oligoclase to andesine, occur as phenocrysts. The same pale green augite which characterizes the more Basic plutonic intrusions is occasionally found in the dykes, and is sometimes accompanied by orthorhombic pyroxene. In addition to the microdiorites more Acid, as well as more Basic types are found. The former include porphyritic microgranites, with phenocrysts of quartz and alkali-feldspar, set in a variable groundmass. The proportion of coloured minerals in these rocks is low, and they are evidently closely related to the leucogranites. Acid dykes without phenocrysts and deficient in dark minerals are not uncommon; it seems reasonable to regard them as aplites complementary to the lamprophyres which also occur. The latter are of two different ages: some occurring as horizontal sheets are probably the earliest intrusions in the cycle, while others are clearly members of the Etive and Ben Nevis swarms. As might be expected, the dominant types are rich in hornblende (hornblende-lamprophyres), sometimes with olivine in addition (olivine-hornblende-lamprophyres), while mica-lamprophyres with augite and olivine are but feebly represented.

The dyke-phase in the Cheviot Hills includes two swarms differing slightly in age, but widely in trend. They freely cut the lavas, but

[1] Geology of Colonsay and Oronsay, *Mem. Geol. Surv.* (1911), 28–37.
[2] Termed 'malchite' in the Ben Nevis and Glencoe Memoir.

few can be traced into the granite, which appears to be later than the majority of the dykes. The chief type is almost identical with the dominant type of lava, being augite-hypersthene-microdiorite or andesite, according to grain-size.

In conclusion, the igneous rocks of Lower Red Old Sandstone age in Scotland and the Border Country may be regarded as an excellent example of an igneous cycle. The magma was essentially calc-alkaline in facies and, although the three phases were distinct, the products are so closely related in composition as to leave no room for doubt as to their common origin. The contrast between the Devonian igneous rocks of South-West England and the Old Red Sandstone igneous rocks of Scotland and the Borders is to be accounted for in part by differences in the conditions of outpouring (the former being submarine and the latter subaerial), and in part by their different tectonic setting.

The main igneous cycle ceased in Lower Old Sandstone times, but there were feeble revivals of activity in Middle and Upper Old Red Sandstone times. The andesitic lava and associated tuffs at Rhynie in Aberdeenshire are interbedded with Middle Old Red Sandstone rocks. The tuffs are unique as they contain silicified plant remains which have proved invaluable to palaeobotanists. In the Orkney Islands lavas and pyroclasts occur both in the Middle and Upper Old Red Sandstone.

In the region of the Orkneys, north of the Scottish mainland, an explosive eruption of great violence brought to a close the period of folding and erosion that immediately preceded Upper Old Red Sandstone times. The products of the explosion were spread as a thick tuff over a wide area, and this is succeeded by a flow of olivine-basalt in the island of Hoy. Later still, the rocks, up to and including the Upper Old Red Sandstone, were cut by numerous dykes which 'form a petrological group of remarkably interesting characters' that cannot be exactly matched anywhere else in Britain.[1] The dyke-rocks include leucocratic types, but the majority are melanocratic and thoroughly Basic. The most interesting of the former are microsyenites that are very typical examples of their kind: normally highly feldspathic, one of them contains the highest percentage (nearly 11 per cent) of potash of all British analysed rocks, and consists almost exclusively of orthoclase. The melanocratic group includes three chief types: camptonites, three out of every four dykes falling in this category; monchiquites, to the extent of nearly a quarter of the whole number; and a few intrusive olivine-basalts. The camptonites may be regarded as the central type. They

[1] Flett, Sir John, in Geology of the Orkneys, *Mem. Geol. Surv. Scotland*, (1935), 173.

contain small olivines; zoned augite with green (perhaps chrome-diopside) cores surrounded by mantles of titanaugite, is characteristic. A third mafic component is brown amphibole. With the incoming of porphyritic plagioclase and the elimination of the brown amphibole the camptonites grade into basalts; while in the opposite direction, with the elimination of feldspar, they pass into monchiquites. In these the two essential constituents are olivine and augite, the latter making up two-thirds of the rock, embedded in brown glass, not in analcite, which, however, occurs in steam cavities and ocelli. Nepheline may occur as small crystals in the groundmass or as micropoikilitic patches. Finally, some varieties are rich in megascopic biotite and approach closely to alnöite. Comparison of the available analyses shows a fairly close correspondence between these monchiquites of the Orkneys and 'nephelinites', suggesting that the former are the dyke-equivalents of the latter; but the monchiquites are rather poorer in silica, alumina and alkalies, thus stressing their melanocratic character and lamprophyric facies.

The age of these dykes is uncertain: they differ in type from the Old Red Sandstone dykes of other parts of Scotland, and may be either Carboniferous, Permian or Tertiary. In this connection it is significant that both camptonites and monchiquites are recorded from among the 'Permian' dyke-rocks of Ayrshire, while one of the Permian vent-agglomerates in the same area has yielded blocks of monchiquite containing xenocrysts of anorthoclase.[1] At least one of the Orkney monchiquites also contains these xenocrysts, a fact which, taking into account the extreme rarity of such rocks, affords strong evidence that both occurrences are of the same age, *i.e.* Permian.

In retrospect it may be well to emphasize the close similarity, in all respects but scale, between the products of Old Red Sandstone igneous activity in northern Britain, and those occurring in the western mountain ranges of North and South America. Both cycles are of the Orogenic Continental type, and there can be little doubt that the parental magma was the same in both cases, and essentially andesitic (see discussion on origin of andesites).

[1] Eyles, V. A., in Geology of North Ayrshire, *Mem. Geol. Surv. Scotland*, (1930), 288.

CHAPTER II

CARBONIFEROUS TO TERTIARY IGNEOUS ACTIVITY

CARBONIFEROUS

DURING the whole of the Carboniferous Period Scotland was the scene of widespread igneous activity which reached a climax in Dinantian times, 350–60 m.y. ago—probably in late-Tournaisian times. In England contemporaneous eruptions led to the accumulation of tuffs and lavas in Derbyshire and the Isle of Man; while in Devon and Cornwall, the period of activity which commenced in Middle Devonian times persisted into the early part of the Carboniferous. In each of these localities the cycle was limited to the extrusive phase and the phase of minor intrusions. There are no plutonic rocks comparable in size with the Caledonian granites.

SCOTLAND

It is significant that the unstable area of the **Midland Valley of Scotland** was the site of most of the volcanoes of Lower Carboniferous age. The earliest eruptions occurred in the east, in the neighbourhood of North Berwick, and, as is usually the case, the lowest volcanic rock is a typical explosion-tuff. This is succeeded by a considerable thickness of basalt-flows, which, as a result of subsequent movement, are inclined towards the west, in which direction the lavas become progressively more Acid, indicating an increase in the acidity of the magma as time went on. Thus, in the Garleton Hills the basalts are succeeded by thick sheets of trachyte. Farther to the west, in Midlothian, lavas of the same age (Calciferous Sandstone) occur as outliers, as at Arthur's Seat, in Edinburgh, where they rest on the Cementstone Group; and on the north bank of the Firth of Forth between Burntisland and Kirkcaldy. In the latter localities the eruptions were slightly later in date, belonging to the higher part of the Calciferous Sandstone, and extending up to the base of the Carboniferous Limestone.[1] Still farther west, extensive

[1] The term 'Carboniferous Limestone' does not have the same age-significance in Scottish and in English stratigraphy: in Scotland the term covers only the limestone-bearing part of the Lower Carboniferous below which come the Oil Shale Group and the Cementstone Group. In England the term covers the *whole* of the Lower Carboniferous (the Dinantian).

outcrops of Carboniferous igneous rocks occur which are relics of a widespread plateau of basalt, the Clyde Plateau of A. Geikie, and, judging from existing outcrops, it must have once stretched continuously from beyond Arran in the west to Stirling in the east, and from the Highland Border southwards well into Ayrshire. A moderate estimate of the original area of the Clyde basalt-plateau is 2500 square miles. In the Glasgow district the lavas form the terraced scarps of the Campsie Fells in the north, the escarpments of the Kilpatrick Hills in the north-west, and the Cathkin Hills in the south. Petrographically these basalts are of the same types as those of the more easterly areas. Several volcanic necks, some of considerable size, have been located. It is probable that the stacks of North Berwick Law and the Bass Rock mark the sites of vents; while Arthur's Seat, in Edinburgh, is a composite vent, built partly of agglomerate and partly of vent-intrusions, which are lithologically identical with the associated flows[1] (Figs. 59 and 60). Such is the well-known Lion's Haunch (intrusive) basalt. Vents are also of common occurrence in the Clyde Plateau. Some are choked with basalt, while others are filled with agglomerate. The vent known as Meikle Bin in the Glasgow district is important, since the associated intrusions include phonolitic trachytes.

In the eastern part of the Midland Valley the volcanoes soon became extinct; but those occurring farther to the west persisted for a much longer period, and new vents were established, indicating a progressive migration of the centres of eruption to the west. Thus in the Bathgate Hills, in the neighbourhood of Linlithgow and Bo'ness, eruptions, which commenced late in Calciferous Sandstone times, persisted into the Lower Limestone Group of the so-called Carboniferous Limestone of Scotland; while at Bo'ness other flows are found in the Upper Limestone Group. This is also the case in the Saline Hills north of the Forth. Still later in date, in the Limestone Coal Group and Upper Limestone Group tuffs are interstratified with normal sediments at Dalry, North Ayrshire. Highly decomposed olivine-basalts, chiefly of Dalmeny type, occur in the Millstone Grit of the West of Scotland. In Ayrshire[2] these volcanic rocks reach a maximum thickness of some 500 feet at Troon; they extend northwards to Stranraer, westwards into Arran, and possibly across into Ireland in the Ballycastle coalfield. Their highly decomposed condition is due to contemporaneous weathering effected largely by acid water formed by rotting of the luxurious vegetation

[1] Clark, R. H., Petrological study of the Arthur's Seat volcano, *Trans. Roy. Soc. Edin.*, **63** (1958), 37–70.
[2] Wilson, G. V., and MacGregor, A. G., in Geology of North Ayrshire, *Mem. Geol. Surv.* (1930), 206 and 221.

which clothed the area at that time. The basalts in some cases have been converted into bauxitic clays.

An outlier of the volcanic region of the Midland Valley of Scotland occurs to the south in the Southern Uplands extending for some fifty miles from the Kirkcudbright coast to the Langholm area lying north of the Canonbie coalfield. The lavas are basalts of various types, closely matching those of the main outcrop to the north. They rest directly and unconformably on the Old Red Sandstone red marls and reach a maximum thickness of 250 feet. Some 1500 feet higher, near the top of the Cementstone Group and again above the latter in the lower part of the Oil Shale Group, basaltic volcanic rocks occur and doubtless originally linked up with their equivalents in the Midland Valley.[1]

The only other important tract in Scotland where igneous rocks of Carboniferous age occur is near the Anglo-Scottish border, in the **Tweed Valley.** In this case, also, the present outcrop is a mere fragment of a basalt plateau, which has disappeared under the influence of erosion. To the west of the present outcrop numerous plugs of basalt and agglomerate probably also belong to the Calciferous Sandstone cycle. This is the case with the interesting intrusions in the Eildon Hills, near Melrose, referred to below.

The Basic Lavas

The Carboniferous lavas of Scotland have been exhaustively studied by Hatch, Watts, Tyrrell,[2] A. G. MacGregor,[3] and S. I. Tomkeieff,[4] and detailed classifications have been evolved, based chiefly upon mineralogical composition. The characteristic minerals of the Lower Carboniferous basalts are plagioclase, ranging in composition from oligoclase to bytownite, but labradorite is typical; a pale brown augite, occasionally with a tinge of mauve, though not with the full mauve tint of titanaugite; and olivine partly altered into serpentine or iddingsite. The accessories include orthoclase, analcite, biotite, hornblende, ilmenite and apatite. The three essential minerals are present as phenocrysts in widely varying proportions.

The most distinctive megaphyric basalts bear type-locality names including the Markle type, with prominent plagioclase phenocrysts, the Craiglockhart type containing phenocrysts of pyroxene and olivine, and the Dunsapie type with phenocrysts of all three chief

[1] Elliott, R. B., Carboniferous volcanic rocks of the Langholm district, *Proc. Geol. Assoc.*, **71** (1960), 1.

[2] Tyrrell, G. W., *Geol. Mag.* (1912), 69–80 and 120–31.

[3] Classification of the Scottish Carboniferous olivine-basalts, etc., *Trans. Geol. Soc., Glasgow*, **18** (1928), 324; and Problems of Carboniferous-Permian volcanicity in Scotland, *Quart. J. Geol. Soc.*, **104** (1948), 133.

[4] Tomkeieff, S. I., Petro-chemistry of the Scottish Carboniferous-Permian igneous rocks, *Bull. Volc.*, *série* 2, **1**, 61–87.

constituent minerals. Corresponding microphyric types bear the names Jedburgh, Hillhouse and Dalmeny; but there are intermediate varieties, and the names cover ranges of compositions in a widely variable series of basalts. These, together with the associated mugearites and Intermediate trachytic-phonolitic types invite comparison with the modern oceanic island suite, the parent magma being alkali olivine-basalt.

It should be noted that these type-names are not restricted to lava-flows, but are applied also to small intrusive masses forming volcanic necks and sheets.

Most of these types are widely distributed in the Midland Valley, but there appears generally to be a greater development of the more feldspathic types in the west than in the east. Thus, on the Little Cumbrae, the lavas are Markle basalts and mugearites, while varieties richer in phenocrysts of coloured minerals are restricted to the numerous basaltic plugs and north-east or east-north-east dykes. In South Bute, however, the Markle, Jedburgh, Dunsapie and Craiglockhart types as well as mugearites are represented among the flows.[1]

The youngest Carboniferous lavas in Scotland occur in the **Mauchline** area in Ayrshire and were formerly thought to be Permian as the volcanics are succeeded conformably by bright red aeolian sandstones which also occur interstratified with the higher lava-flows. These must therefore have been erupted under subaerial conditions. The Upper Coal Measure (Westphalian) age of the Mauchline lavas is proved, however, by the occurrence of diagnostic plant remains in shales in the lower part of the volcanic succession. The lavas tend to be more Basic and more strongly alkaline than the Scottish Dinantian lavas as they include not only microphyric basalts but also nepheline- and analcite-basanites and feldspar-free monchiquites. The last-named occur also in vent intrusions and dykes and are noteworthy as containing xenocrysts and peridotite xenoliths. These lavas were erupted from some sixty small vents, some of which are choked with agglomerate.

The Intermediate Lavas

The less Basic lavas, which are restricted to the main Dinantian igneous cycle, are trachytes of various kinds. Some containing accessory nepheline are phonolitic trachytes. The intrusions associated with these flows are of the same composition and texture as the latter, and in many cases it is impossible to state definitely whether the rock is a flow or an intrusion.

The phonolitic trachytes consist essentially of sanidine, with,

[1] Tyrrell, G. W., Igneous geology of the Cumbrae islands, *Trans. Geol. Soc. Glasgow*, **16,** ii (1916–17), 244–74.

which is usually associated a little plagioclase, and a green and slightly pleochroic soda-augite, together with some nepheline. Such rocks form the laccolith of Traprain Law, the sill of Hairy Craig, and the stocks of North Berwick Law and the Bass Rock. The Traprain Law rock contains a little sodalite in addition to analcite and nepheline, while the microsyenite of the Bass Rock carries fayalite and nepheline.[1] In one case nepheline is sufficiently abundant to justify the use of the term phonolite. There is one exposure of this rock near Fintry, in the Campsie Fells.

A. G. MacGregor has described the Calciferous Sandstone lavas and associated intrusions from the fringe of the Clyde Plateau in North Ayrshire, and remarks that the occurrence of phonolitic trachytes here, as in the Garleton Hills farther east, emphasizes the alkaline affinites of the whole province. In this region also there is a unique nepheline-basanite in the same series, while the normal suite ranging from basalt to trachyte goes farther, including rhyolites.[2]

Trachytes are not uncommon in the central part of the Campsie Fells, round the Meikle Bin vent, where they form small elongated plugs or short dykes.[3] Their state of preservation does not allow of detailed description, but it is claimed that there is here even a greater variety of types than is to be found among the trachytes of the Garleton Hills, East Lothian. Some contain 'moss-like' pseudomorphs, which may well have been riebeckite originally. The latter mineral is found in the trachytes that occur as intrusions in the Eildon Hills. In some of the trachytes there is accessory quartz. In others large phenocrysts of sanidine are prominent, as at Peppercraig: these are perhaps best developed in certain rocks that are evidently links connecting the basalts with the true trachytes, *i.e.* they are trachybasalts so-called banakites closely resembling the Yellowstone Park type-rock, in that plagioclase as basic as labradorite is associated with much sanidine. The latter occurs as a peripheral zone round the former in the groundmass, and locally as phenocrysts two inches in diameter, enclosing smaller plagioclase crystals.

The Intrusions

As noted above, there are no plutonic intrusions belonging to this cycle, but the occurrence of ultrabasic plutonic rocks among the constituents of the agglomerates filling the vents and as numerous xenoliths in the basic lavas, points to the deep-seated presence of such rocks as dunite, augite-peridotite, pyroxenite and biotitite, none of which has as yet been exposed by denudation.

[1] Campbell, R., *Trans. Edin. Geol. Soc.*, **13** (1933), 126.
[2] Geology of North Ayrshire, *Mem. Geol. Surv.* (1930), 89.
[3] Bailey, E. B., in Geology of Glasgow district, 2nd edn., *Mem. Geol. Surv.* (1925), 185.

Minor intrusions are, however, numerous and varied in composition and habit. As the main eruptions were followed by revivals of activity at three later periods: in the Carboniferous Limestone, in the Millstone Grit, and into the Permian, it is difficult to assign a definite age to any particular intrusion. There is no clearly defined dyke-phase restricted to a relatively short interval of time; on the contrary the injection of the minor intrusions was spread over the whole of the cycle: the sills that are petrologically similar to the lavas with which they are interbedded were probably contemporaneous with these flows—the vent-intrusions and plugs certainly were. There can be little doubt that the trachytic plugs and other intrusions in the Midland Valley and in the Eildon Hills were contemporaneous with the upper lavas, of Cementstone age, in the Garleton Hills. Although the minor intrusions in the eastern part of the Midland Valley do not occur at higher horizons than the Carboniferous Limestone, in the western part of this tract they range as high as the red Sandstones in Ayrshire. In mineral composition these intrusions are clearly related to the Mauchline lavas, and were injected at a period somewhat later than the extrusion of the lavas. They occur in the form of sills reaching 400 ft in thickness; and although they have a moderately wide range of composition, they are evidently comagmatic and consist of different proportions of the following minerals: olivine, titanaugite, 'barkevikite,' labradorite, ilmenite and analcite, with nepheline and other minerals as accessories. The occurrence of primary analcite in many of these intrusions has been satisfactorily established. Its components in the magma undoubtedly decreased the viscosity, and hence allowed an unusual degree of differentiation, which is demonstrated by the stratiform nature of many of the larger intrusions. No intrusion in this cycle shows this more clearly than the Lugar Sill in Ayrshire,[1] which includes some half-dozen distinct rock-types, *i.e.* teschenite, theralite, picrite, peridotite and lugarite.

Two series have been distinguished among these minor intrusions, one characterized by analcite and the other by nepheline. The former series includes analcite-syenite, a typical example occurring at Howford Bridge, Ayrshire, teschenite and crinanite. The most distinctive members of the second group are theralite and lugarite. There are intermediate types, containing both analcite and nepheline, such as the well-known essexites of Crawfordjohn and Lennoxtown. Localized intrusions of basic lamprophyric rocks including camptonites and monchiquites occur in or near the Ayrshire vents.[2]

The most typical and the most widespread of these rock-types is teschenite, which is common in the Glasgow district, in Ayrshire, in

[1] Tyrrell, G. W., *Trans. Geol. Soc., Edin.*, **15** (1952).
[2] *Ibid., Glasgow*, **28** ii (1927–8), 281.

East Lothian and Fifeshire. Augite-picrite is frequently associated with the teschenites, of which it is an ultrabasic differentiate. This is the case, for example, in the Inchcolm sill which forms an island in the Firth of Forth. Typical teschenite forms the Salisbury Craigs Sill on the outskirts of Edinburgh, and part of Gullane Hill on the Forth. The most distinctive of these rocks, so far as appearance in the hand-specimen is concerned, are the essexites, by reason of the large number of euhedral, porphyritic titanaugites they contain. An olivine-rich variety, which is correspondingly poor in feldspar and thus intermediate between the essexites and picrites, occurs at Benbeoch and elsewhere in Ayrshire, and has been termed 'kylite'.

The latest episode in Scottish Carboniferous volcanicity marks a dramatic change in magma-type and in the forms of the intrusions. The magma-type was now tholeitic, typically oversaturated, the rock types involved varying from quartz-dolerite, containing a mesostasis of intergrown alkali-feldspar and quartz, to tholeiite with interstitial glass of the same composition. Both types are well represented in a regional swarm of dykes, unrelated to the Carboniferous volcanic centres, but widespread, extending northwards into the Grampian Highlands and southwards into northern England (Durham). In two areas the dykes are closely associated with quartz-dolerite sills: in northern England the **Great Whin Sill**[1] (295 m.y.) is probably the best-known named sill in the country (see map, Fig. 148) while its counterpart in Scotland underlies a large part of the Midland Valley. Formerly thought to be a suite of sills, the latter is now known to be essentially one sheet which sometimes splits and changes its horizon along steeply inclined dyke-like risers. In a maximum thickness of about 200m it displays a full range of grain-size variation from very coarse pegmatitic facies, through medium grained quartz-dolerites to tholeiitic chilled margins.

Summarily, the petrographic characters of the igneous rocks of Carboniferous age in the south of Scotland clearly indicate that they constitute an 'Alkali' suite: the dominant lavas are alkali olivine-basalts though the alkaline affinites of some are indicated by the occurrence of interstitial analcite. True andesites are unrepresented; but mugearites are interstratified with the basalts in several localities; while, as is commonly the case, trachytes, often nepheline-bearing, are among the late members of the lava succession. The intrusive phase is more obviously alkaline, with widespread teschenites and alkali-picrites carrying accessory analcite.

The regional east–west dyke-swarms which form such a striking feature of geological maps of the Midland Valley of Scotland are

[1] Holmes, A., and Harwood, H. F., Age and composition of the Whin Sill, *Min. Mag.*, (1928), 493–542.

quite different: they represent a tholeiitic magma-type. The dykes consist of either tholeiites (*i.e.* oversaturated basalts), or quartz-dolerites, of the same composition but coarser grain. It is interesting to speculate on the causes of the change of magma-type, from 'oceanic' to 'continental', at a time when the area was being warped and fractured by the Armorican earth movements. The only certain fact is that it was a regional change and clearly indicates the advent of magma drawn from a new and different source.

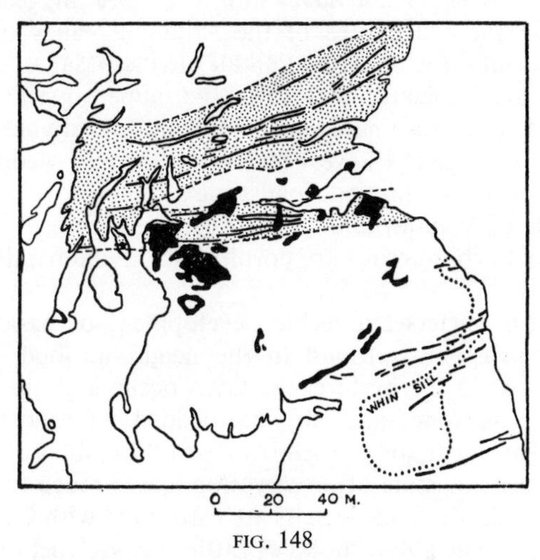

FIG. 148

Sketch-map showing Carboniferous volcanic rocks (black) and quartz-dolerite dykes in Scotland and North England. The dykes occur in the two areas dotted in Scotland; only a few of the dykes are shown. (*Based on maps by F. Walker and A. Holmes.*)

ENGLAND

The most extensive tracts of Carboniferous volcanic rocks in England occur in **Devon** and **Cornwall,** where the rocks form a direct continuation of the Devonian cycle, so that no further description is necessary. Emphasis may, however, be laid upon the fact that there is very little in common between the lavas and intrusions of this and the Scottish region: both series are essentially Basic in composition; but in the case of the Scottish rocks primary differentiation had not removed the composition of the magma far from that of average basalt, although the tendency to migrate towards the 'alkali pole' becomes noticeable towards the close of the cycle. On the other hand, the magma that gave rise to the spilites of the Devonian-Carboniferous cycle of South-West England was much more definitely alkaline (sodic) at the commencement of the cycle, and this character persisted throughout the cycle, with little appreciable alteration in the composition of the magma.

In **Derbyshire** igneous rocks of Lower Carboniferous age occur in the *Dibunophyllum* zone in the neighbourhood of Matlock, Miller's Dale and Tissington. The lavas include basalts of two different kinds. Sheets of normal olivine-basalt occur in multiple flows with more alkaline types. In the latter the dominant feldspar is oligoclase, together with some orthoclase, while augite is usually subordinate to olivine. These lavas are therefore mugearites. They occur, as do the mugearites of the Midland Valley of Scotland, in close association with normal alkali olivine-basalts.

The lavas are associated with tuffs and numerous sills of olivine-dolerite, and were presumably erupted from central volcanoes, since agglomerate-filled necks have been found in their vicinity. Locally the igneous rocks are termed 'toadstone'.

In the **Isle of Man** around Castletown the highest rocks referred to the Lower Carboniferous are porphyritic olivine-basalts, tuffs and agglomerates.

In northern **Somerset** a feeble development of basic lavas and associated intrusions is found in the neighbourhood of Weston-super-Mare. As in Derbyshire, the lavas occur high up in the Carboniferous Limestone, and are interbedded with normal marine sediments. They bear some resemblance to the spilites of Devon and Cornwall, but their state of preservation leaves much to be desired.

In the English Midlands basalts are associated with Carboniferous Limestone in the neighbourhood of Little Wenlock near the Wrekin, Shropshire, and with the Upper Carboniferous at Rowley Regis, Barrow Hill,[1] Pouk Hill, Kinlet, and the Clee Hills. Petrologically these rocks are closely similar, often extremely fine-grained analcite-bearing olivine-basalts, that is, analcite-basanites. Although some are definitely intrusive, others are just as clearly extrusive.[2] Thus the Titterstone Clee basalt is a sill; the Rowley Regis basalt is laccolithic; the Barrow Hill mass is intrusive. On the other hand the Little Wenlock basalt, which is finely exposed in quarries at Doseley, has been converted in its upper parts into a typical red bole, due to contemporaneous subaerial weathering. Although much of the Etruria Marl in the Upper Coal Measures in the northern Midlands superficially resembles red bole, the suggestion that it was formed of weathered basaltic material has not been generally accepted,[3] but

[1] Marshall, C. E., The Barrow Hill Intrusion, S. Staffs., *Quart. J. Geol. Soc.*, **101** (1946), 177.

[2] Pocock, R. W., The age of the Midland basalts, *Quart. J. Geol. Soc*, **87** (1931), 1; but compare Marshall, C. E., Field relations of the Basic igneous rocks associated with the Carboniferous strata, *Quart. J. Geol. Soc.*, **97** (1942), 1.

[3] Robertson, T., The origin of the Etruria marl, *Quart. J. Geol. Soc.*, **87** (1931), 13.

in certain fine-grained breccias in the Etruria Marl (the so-called espley beds) abundant shards of volcanic material may well indicate the reality of eruptions at this time.

THE ARMORICAN INTRUSIONS

Just as the formation of the 'Caledonides' was accompanied by the uprise of the 'Newer Granites' and the intrusions associated with them, so the birth of the Armorican or Hercynian chains coincided with the emplacement of the granite masses of South-West England and the deeply denuded mountain belt stretching through 'Armorica' into Spain. Five bosses of large diameter and several smaller satellitic intrusions occur in Devon and Cornwall: the former comprise the granites of Dartmoor (240 square miles), Bodmin Moor or Brown Willy 75 square miles), St Austell (33 square miles), Falmouth (or Carn Menellis) (50 square miles), and Land's End (75 square miles); while the Scilly Isles represent the highest points of a sixth large mass. The smaller intrusions include those of St Michael's Mount, Godolphin, Carn Brea and Carn Marth, Belovely Beacon, Kit Down and Hingston Down. These apparently isolated outcrops are connected underground and there is no doubt that the intervening stretches of killas are in some cases merely roof-pendants, and that further unroofing will extend the area of granite and diminish that of the country-rock. The granites occupy a belt of country having a Caledonoid trend; but they are individually associated with Armorican axes of uplift, and were injected into the crust near a line of weakness bordering the ocean.[1] This weak tract not only determined the positions of the Armorican intrusions, but also the location of the Devonian-Carboniferous igneous rocks, and probably also of the Permian Exeter lavas. The Armorican granites of this country are important not only from a petrological, but also from an economic point of view, because their intrusion was accompanied by a period of mineral emanation, when lodes were formed containing ores of tin, copper, lead, zinc, iron, molybdenum and arsenic. To the same period belongs the intense local alteration of the granites which resulted in the very valuable deposits of china-stone and china-clay rock for which parts of Cornwall are famous. The Armorican complexes have been subjected to detailed examination, and Dr A. Brammall's papers on the Dartmoor granite convey the results of years of intensive study.[2] Their importance in the study of petrogenesis has been referred to above.

[1] Dewey, H., *Proc. Geol. Assoc.*, **36** (1925), 109; Edmonds E. A., *et al.*, British Regional Geology: South-West England, 3rd Edn. (1969).

[2] See particularly The Dartmoor granite, *Quart. J. Geol. Soc.*, **88** (1932), 171, (with Harwood, H. F.).

It is evident that, like the Caledonian granites, those of Armorican age are composite intrusions. The Dartmoor granite consists of three chief members which were intruded in the order of decreasing basicity. The earliest intrusion was basic in composition, and is

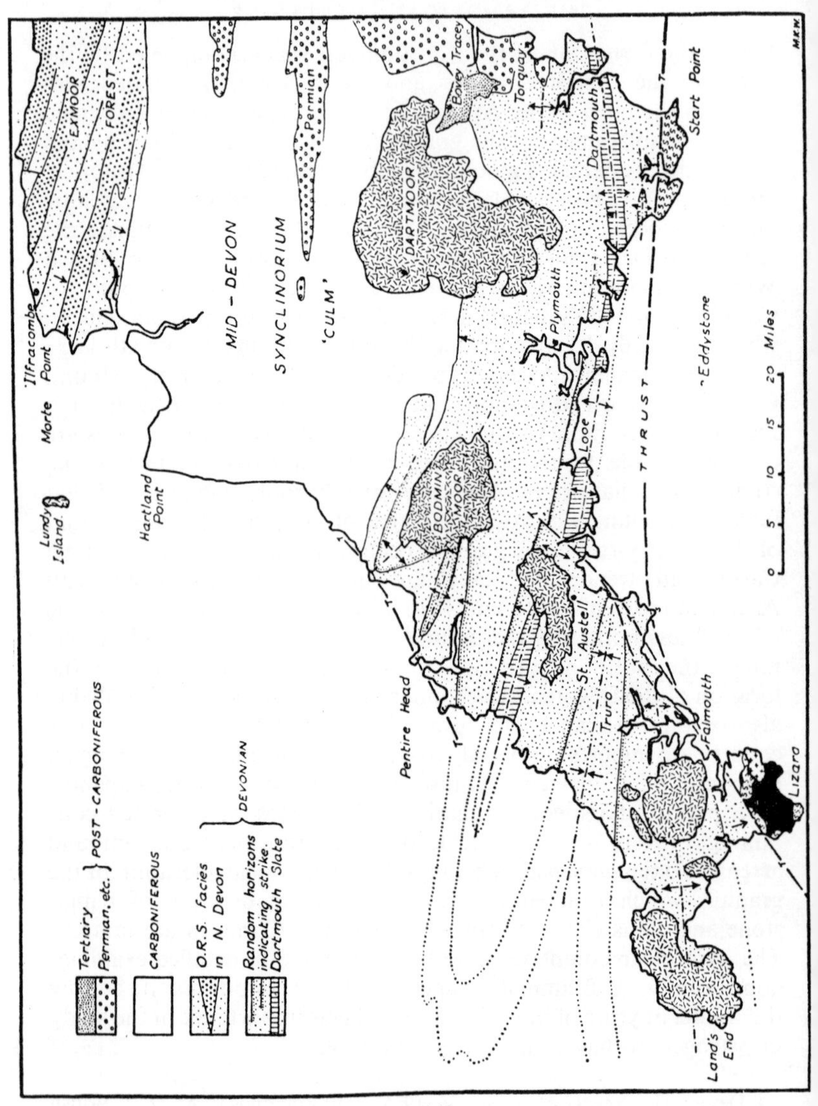

FIG. 149

Structural map of Devon and Cornwall showing the Armorican granitic complexes.

represented by numerous inclusions, some of large size, in the later intrusions. Next, the so-called 'giant granite' was injected under a roof of killas, Basic lavas, etc., and was later split into two or more sheets by a somewhat more acid 'blue granite', which is a valuable building-stone. Minor intrusions in the form of narrow dykes, veins and thicker sills are common in both of the main intrusions. They are of granitic composition, but in some cases show wide variation from the main type.

The so-called 'giant granite' is very coarse in texture, and rich in exceptionally large feldspar phenocrysts, which may measure 7 by 5 inches and consist of coarse microperthite. In the groundmass, microperthitic orthoclase is associated with subhedral plagioclase ranging from albite to oligoclase. The dominant coloured mineral is biotite, with which is associated a subordinate amount of muscovite. Accessories are very variable, and in addition to those of normal occurrence in granites, such as zircon, magnetite and apatite, include others of pneumatolytic origin, namely, tourmaline, topaz, anatase and brookite, as well as minerals resulting from assimilation of country rock by the granite. Among these are abundant almandine garnet together with cordierite, andalusite, sillimanite, corundum and spinel. The 'blue granite', which was intruded beneath the 'giant granite', and was thereby protected from contamination by assimilation, is practically free from these highly aluminous accessories.

The minor intrusions connected with the granites include basic as well as aplitic and pegmatitic 'differentiates' of the main types. One of these, the Bittleford pegmatite on Dartmoor, contains a small amount of gold and silver of pyrogenetic origin. Of equally wide distribution are the many dykes of microgranite (some of them porphyritic), known to the miners as 'elvans'. Generally the aplites are more leucocratic than the main granites and more subject to pneumatolytic alteration, with tourmaline in place of biotite. The best-known aplite, which is quarried at Meldon, north of Dartmoor, is of this type, and contains a notable amount of topaz, wolfram, etc.

The dykes in Cornwall are stated to be arranged radially to the granite bosses, but locally there is a pronounced parallelism in their trend. Very few are found in the granites themselves: most occur in the synclinal tracts between the anticlines into which the granites were intruded; but there is reason to believe that the plutonic rock lies at no great distance beneath the surface in the areas where the dykes are concentrated.

The most widely distributed type of dyke-rock, well exemplified by the Prah elvan, is a strongly porphyritic variety containing numerous feldspar phenocrysts up to one inch in length, together with bipyramidal quartzes and hexagonal biotites. In the elvans associated with the St Austell granite, porphyritic muscovite is not uncommon, while topaz, although widely distributed as an accessory, is especially characteristic of the intrusions near the Land's End mass. The dykes have suffered the same pneumatolytic modifications as the parent intrusions: they are often found to have been tourmalinized and kaolinized—in some cases tourmaline is the dominant coloured constituent.

The St Austell granite is economically the most important of the Armorican granites. Having suffered more severely from the effects of pneumatolysis than any of the others, its deposits of china-clay rock and china-stone are more extensive, although they are not absent from the other masses; while veins of greisen and schorl-rock are common. The active agents which effected the alteration of the granite rose along vertical or highly inclined joints in the granite.

The metalliferous lodes occupy a broad belt of country embracing the northern half of the Land's End mass, the Redruth-Truro district north of the Carn Menellis mass, the St Austell granite, the southern part of the Bodmin Moor intrusion, and terminates against the Dartmoor granite in the neighbourhood of Tavistock. Within this belt the dominant direction of the lodes is north-east to south-west, but south of Bodmin Moor is north–south, and in the extreme east of the belt nearly east–west. The only important tin-mining area at the present time is that lying on the north-west side of the Carn Brea granite. From the petrological point of view, some of the most interesting of the metalliferous deposits are the wolfram-pegmatites, which are of very coarse grain, and are typically developed near Buttern Hill to the north of the Bodmin Moor granite. Here microcline crystals up to three inches in length are intergrown with wolfram and quartz, while other constituents are apatite and tourmaline. These pegmatites often pass into veins of wolfram and quartz, and finally into pure quartz-rock. A detailed consideration of the genesis of the ores is outside the scope of this book, but it is interesting to note that a zonary distribution of the several ores, dependent on the temperature of their formation, has been demonstrated.

In the Channel Islands granites are magnificently exposed in cliff sections in Jersey, Guernsey and Sark. In Jersey[1] the plutonic rocks include a wide range of types, red-weathering alkali-granite forming the north-western, south-western and south-eastern bastions. The north-western granite has stoped its way into earlier basic intrusions and lavas, and is consequently crowded with xenoliths in all stages of dissolution. Because of assimilation of some of this basic material the red granite, which is normally deficient in coloured minerals and of leucogranitic type, is converted into a grey porphyritic rock,

[1] Wells, A. K., and Wooldridge, S. W., *Proc. Geol. Assoc.*, **42** (1931), 178.

with clots and schlieren of hornblende and biotite. The basic xenoliths are now of dioritic or monzonitic composition—consisting essentially of hornblende and plagioclase, with quartz, orthoclase and biotite. Normally the original structure has been destroyed; but rarely small patches are seen to have survived, and these consist of labradorite and titanaugite in ophitic relationship. Therefore some of these Jersey diorites are amphibolitized (olivine-) gabbros. But in addition, dioritic rocks are widely developed in the north and south-east of the island, and include the same types of xenoliths as the granites. These are *magmatic diorites*, though they may well have been produced from the same materials as the *metasomatic diorites*, though more extensively, at some deeper-seated source. They include types with brown and green hornblendes varying in size between needle-like crystals to stout prisms several inches in length, showing a preferred orientation and particularly striking in certain pegmatitic facies. Arrested stoping with widespread hybridization, and the development of local pegmatitic clots and aplitic veins are well displayed.

Dating the plutonic rocks of the Channel Islands is difficult in the absence of stratigraphical evidence. From their location between the granites of S.W. England and Brittany it has hitherto seemed reasonable to regard them as Armorican; but, as previously explained (p. 468) radiometric dating suggests that some, at least, may be as old as late Precambrian.

PERMIAN

Igneous rocks of Permian age are found in only one area in Britain—in Devonshire—though the nosean-phonolite which forms the Wolf Rock off the Cornish coast has a radiometric age of about 260 m.y. and must now be included amongst the Permian volcanic rocks.

In **Devonshire** the Permian lavas are restricted to the neighbourhood of Exeter, and are often referred to as the 'Exeter traps'. They lie at, or near, the base of the 'New Red Sandstone', either directly on the Culm, or separated from it by a thin stratum of red rocks. The lavas were erupted over a very irregular surface, from which rose prominent hills of Carboniferous (Culm) rocks. The main outcrops occur in the 'tongues' of New Red Sandstone that extend westwards from the neighbourhood of Killerton and Tiverton. In addition, small patches of lava associated with Permian conglomerates rest on Devonian slates to the south of Dartmoor near Kingsbridge, and also south-west of Plymouth. The actual centres of eruption have been located in some cases. In others the lavas are regarded as of puy type. It is possible that other vents are hidden beneath the Permian rocks, but some evidence has been adduced which indicates derivation from the high land in the neighbourhood of Dartmoor. It is possible that the Dartmoor granite itself occupies the site of a magma-basin, capped with a roof of Culm sediments, from the upper surface of which rose several volcanic cones. In the unroofing of the granite mass the cones have of course been destroyed; but it is significant that blocks of acid and intermediate lavas and

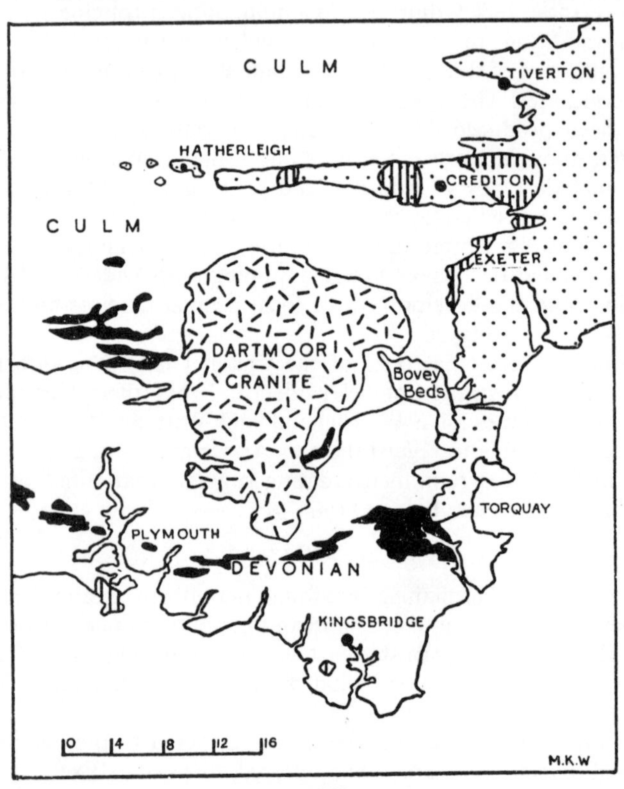

FIG. 150

Sketch-map showing distribution of the Exeter lavas and Devonian-Carboniferous volcanic rocks in Devonshire. Permo-Trias, spots; areas containing Permian lavas, ruled; Devonian and Carboniferous volcanic rocks, solid black.

strongly porphyritic microgranite occur in the Permian breccias, and decrease in size away from Dartmoor.

The systematic examination and identification of the rocks is rendered difficult by their extremely altered condition, the alteration dating from the Permian period. The lavas were the product of sub-aerial eruptions, and were extruded under desert conditions. Hence they suffered rapid disintegration, and their iron-bearing constituents were quickly oxidized, which gives the rocks a characteristic red colour. The upper portions of the lavas are often much fractured (block lavas), and the crevices filled with sand. Many of the flows are extremely vesicular, and in their general appearance are strongly reminiscent of the Old Red Sandstone lavas of the Pentland Hills, etc.

The Exeter lavas have been investigated by Tidmarsh,[1] who divided them into ten types grouped into three series. Each of the types exhibits several different facies. Among the lavas and associated intrusions are some to which the names of normal rock-types may be applied, such as minette, rhyolite, 'quartz-porphyry'. Most of the rocks, however, are of quite exceptional mineral and chemical composition. Among the features of special interest are: the occurrence of quartz-xenocryst-basalts; the lamprophyric facies of the lavas; and the occurrence at Loxbeare near Tiverton of a biotite-olivine-leucitite which is evidently an under-saturated minette. The leucites are small but typical, and show the regular groupings of minute inclusions so characteristic of the small leucites in chilled lavas. Apparently rapid chilling of the magma at a time when mica was in the process of formation gave rise to olivine plus leucite instead of more mica.

Although the range of composition of the Exeter lavas is small, it is significant that the Permian breccias contain abundant pebbles of strongly porphyritic microgranite rich in large phenocrysts of quartz and feldspar, also spherulitic rhyolites and andesites.

These Permian igneous rocks of South-West England form a small portion of a large petrographic province extending into Germany. On the Continent, however, the amount of igneous material is much larger and more varied in composition.

TERTIARY

Introduction

Throughout the whole of the Mesozoic Era there were no volcanic outbursts in the British area; but during the Tertiary Era igneous activity occurred over an enormous area known as the **Brito-Icelandic** or **Thulean** Province, which embraces North-West Britain, and extends northwards into Iceland, Spitzbergen, the Faeroe Islands and Jan Mayen, and westwards to Greenland. In Britain volcanic activity started in earliest Tertiary times—preliminary radiometric results suggest ages up to about 70 m.y. for some of the lavas—and may have continued until about 50 m.y. ago.

The distribution of the Tertiary volcanic rocks in Scotland and

[1] Tidmarsh, W. G., The Permian lavas of Devon, *Quart. J. Geol. Soc.*, **28** (1932), 712.

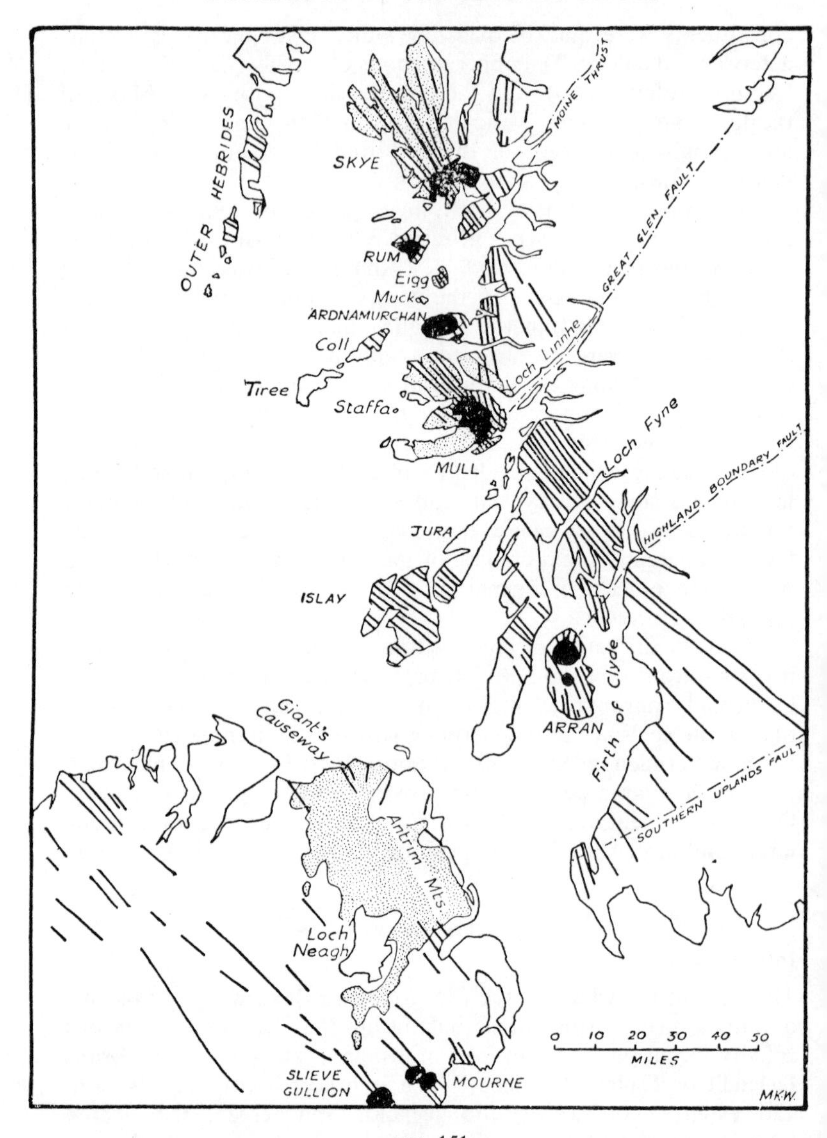

FIG. 151

Map of the Tertiary igneous rocks of Scotland and northern Ireland.
(*Based on maps by the officers of the Geological Survey.*)

Northern Ireland is shown in Fig. 151. Extensive tracts of plateau-building and flood-lavas are preserved in Antrim and in the Inner Hebrides, particularly in the northern part of Skye, and on the island of Mull and the adjacent mainland. In addition centres of localized igneous activity occur in Skye, Mull, the island of Rum (or Rhum, one of the Small Isles of Inverness-shire), on the Ardnamurchan peninsula, in central and northern Arran, and in three localities in Northern Ireland, namely the Mourne Mountains, Slieve Gullion and Carlingford. Since the end of the period of volcanic activity erosion has removed much of what must originally have been very extensive lava fields and has exposed the intrusive rocks of the volcanic centres listed above. Some, and perhaps all, of the latter represent the roots of volcanoes of central type, many of which had a long and complex history of activity. Because of fortunate circumstances of erosion and exposure, and the fact that it has been studied intensively by a number of distinguished geologists, it is probable that the structure and history of the Mull complex are better known than those of extinct volcanoes from any other part of the world.

The Tertiary igneous rocks have not suffered subsequent folding or metamorphism, and hence they are usually in an excellent state of preservation. The lavas can be examined in hundreds of miles of coastal sections, while the rocks of the various intrusion centres are generally well exposed in areas of considerable relief. It is not surprising, therefore, that the phenomena of Tertiary igneous activity have attracted much detailed research. The reader is advised to study the valuable summary provided by Dr Richey[1] and in addition reference should be made to the classic Survey Memoirs and accompanying 1-inch maps on Skye, Mull, Ardnamurchan and Arran.[2]

The Tertiary igneous cycle consists of the usual three phases: the eruptive (extrusive) phase, the phase of major intrusions, and that of the minor intrusions—the dyke phase.

The Eruptive Phase: the Lavas

Lavas of Tertiary age in Britain are overwhelmingly basaltic; those of intermediate composition are practically unpresented; but rhyolitic lavas and plugs occur rarely. It is a little difficult to deal with the basalts adequately on account of some confusion which has arisen in the use of the terms 'type', 'series', 'groups', 'modes of occurrence' and 'magma-types'. From the petrographic angle it appears that the basalts fall into two main types corresponding with

[1] Richey, J. E., Scotland: the Tertiary volcanic districts, *Br. Reg. Geol.*, (1948), 2nd edn.

[2] *Memoirs of the Geological Survey* relating to Tertiary igneous activity in Scotland include: Harker, A., The tertiary igneous rocks of Skye (1904); Bailey E. B., Clough, T. C., Wright, W. B., Richey, J. E., and Wilson, G. V., The Tertiary and post-Tertiary geology of Mull, Loch Aline and Oban (1924); Tyrrell, G. W., The geology of Arran (1928); Richey, J. E., and Thomas, H. H. The geology of Ardnamurchan, North-West Mull, and Coll (1930); Anderson, F. W., and Dunham, K. C., The Geology of Northern Skye (1966).

those recognized by Kennedy and Anderson, and discussed above under 'origin of basalt'. These are (1) over-saturated, olivine-poor, fine-grained basalts of tholeiite type, containing interstitial glass; and (2) alkali olivine-basalts, under-saturated as regards mafic constituents, containing some 20 per cent of olivine, though this figure may rise to up to 50 per cent in certain melabasalts of picrite-basalt type. Augite-rich melabasalts (ankaramites) also occur, but are much more important in the outlying parts of the region, for example Jan Mayen, where they dominate the basal group. Among the olivine-basalts aphyric and strongly porphyritic varieties are distinguished. Less common types of the latter contain very prominent phenocrystic plagioclases up to 3 inches in length and termed 'Big Feldspar Basalts' by the Survey. Other basalts are rich in microphenocrysts of plagioclase, while mugearites occur occasionally as flows, as well as minor intrusions (in Skye, the type locality).

The lavas were erupted under subaerial conditions and on account of the general absence of pyroclastic rocks it is inferred that there was little explosive volcanic activity, though several vents filled with agglomerate or even merely brecciated Chalk have been discovered in Antrim and evidently represent volcanoes of phreatic or steam-blast type.

In northern Antrim the lava succession[1] is interrupted at two levels by well-developed laterite horizons representing periods of quiescence during which the exposed basalts were deeply weathered. Elsewhere in North-East Ireland only one Interbasaltic Bed occurs. In addition individual basalts may have weathered reddened tops which in some cases are red lateritic clays or boles. Terrestrial conditions are further evidenced by very occasional plant-beds containing the impressions of leaves; by the preservation of trunks of trees overwhelmed by, and engulfed in basalt; and by very thin lignite or coal seams, including one associated with a few feet of Tertiary mudstone underlying the lavas of Mull, Morvern and Ardnamurchan. The leaf-beds, of which that occurring at Ardtun in South-West Mull is the best known, provide the only direct evidence of the age of the lavas, and of the climatic conditions at the time. Actually the plant-remains are not sufficiently distinctive to fix the age within the Tertiary Epoch precisely, but the palaeobotanists concerned are agreed that they probably represent an early Eocene flora. An early Tertiary age is indicated by the stratigraphical relationships: in Antrim the lavas rest directly on an eroded surface of Upper Chalk; in the Hebrides they overlap on

[1] Patterson, E. M., The Tertiary lava succession in the northern part of the Antrim Plateau, *Proc. Roy. Irish Acad.*, **57** (1955), 79. A useful summary of the Tertiary igneous rocks in north-east Ireland may be read in *Proc. Geol. Assoc.*, **71** (1960), 441.

to other Mesozoic and older strata. It is confirmed by radiometric dating.

With regard to general features, the lavas vary in thickness from a few feet in the case of certain ropy (pahoehoe) lavas to 150 feet, the average being perhaps 40 feet. Although the central parts of the flows are massive (and indistinguishable in this respect from basaltic minor intrusions) the upper parts are often slaggy and extremely vesicular, so that they weather readily, giving rise to a terraced type of scenery. Mapping is greatly assisted by this terracing, and it has proved possible to trace individual flows and to prove that they maintain their thicknesses in many instances for long distances.

Some of the Tertiary basalts are world famous for the magnificent columnar structure which they display, particularly at the Giant's Causeway in Co. Antrim,[1] and at Fingal's Cave in the Hebridean island of Staffa. On a point of mineralogical detail it may be noted that at certain localities (including the Giant's Causeway) the lavas are strongly vesicular and provide well-crystallized specimens of many varieties of zeolites and other minerals for collectors.

Some difficulty has been experienced by surveyors in distinguishing between the more massive lavas and sills of the same petrographic type. This was notably the case in Skye where a group of Basic 'sills' were proved later to be extrusive. Genuine Basic sills do occur, however, sometimes underneath the volcanic pile, sometimes intruded into it. Noteworthy examples are the 300-foot sill (which is of sufficiently coarse grain to be classed as gabbro), forming Fair Head in North-East Ireland, and the 100-foot Portrush Sill.

An additional complication is the occasional occurrence of composite flows with sharply defined and sometimes pseudo-intrusive contacts between the lower and upper parts of the lava. Such basalts have been described from Skye[2] and Northern Ireland.[3]

The Interbasaltic Horizons automatically divide the lavas of northern Antrim into three series of which the Upper and Lower Series are petrographically alike, consisting of olivine-basalts; but the Middle Series consists of tholeiites. In parts of Scotland the lavas are divided on a different basis into (1) the regional basalts and (2) the products of local volcanoes of 'central' types, of which the most important is the Mull Volcano. The regional basalts of the whole province are comparable in status with the flood-basalts of the Parana Basin or the Deccan. In the topographical sense they are plateau-building and constitute the Plateau Group of the authors of the Survey Memoirs to which attention was drawn above. The

[1] Tomkeieff, S. I., The basaltic lavas of the Giant's Causeway District of Northern Ireland, *Bull. Volc. Naples*, *série* ii, **6** (1940), 89.
[2] Kennedy, W. Q., On composite lava flows, *Geol. Mag.*, **68** (1931), 166.
[3] Walker, G. P. L., Some observations on the Antrim basalts and associated dolerite intrusions, *Proc. Geol. Assoc.*, **70** (1959), 179.

lavas erupted by the Mull Volcano or any other volcano of the same type constitute localized Central Groups. Only relics of the original basaltic plateau have survived; but even so they occupy approximately 2000 square miles, of which some 1500 square miles occur in North-East Ireland where they are magnificently exposed on the Antrim coast. The thicknesses which have survived vary widely from a maximum of 6000 feet in Mull (equally divided between the regional and the local groups), 2000 feet in Skye, but only 300 feet in Ardnamurchan. The lavas of the Mull Central Group were erupted from a volcano which changed in character during its long history. At one important period it was a great caldera the dissected remains of which occupy the south-eastern parts of the island. Within the caldera some of the lavas are unique among the British Tertiary basalts in exhibiting pillow structure which proves that they were erupted into a crater lake at that time occupying the caldera depression.

During its early active life the Mull Volcano must have been in many ways similar to the existing volcano of Hawaii: a broad shield built up mainly of basaltic lavas and with a summit caldera, from the rim of which the main effusions of lava took place. There must have been many other volcanoes scattered over the lava plateau in early Tertiary times, the sites of some of the more deeply dissected being now represented by the intrusive complexes which form the subject of the next part of this account.

It seems probable that in the reservoir underlying and feeding the Mull Volcano, basalt magma was locally differentiated to a greater extent than was normally the case for the plateau region as a whole. Various genetic relationships have been suggested between the 'plateau' and 'central' types of magma involving, amongst other factors, fractional crystallization with the formation of some magma fractions depleted in crystals and others with crystals added.

Similar local differentiation has produced lavas of Intermediate and Acid composition near to the intrusion centre of the Cuillin Hills of Skye. This centre must have been the site of a volcano similar to that of Mull at the time of the lava eruptions, and gave rise to up to 2000 feet of trachytes and overlying rhyolitic tuff, agglomerate and lava.

In the Central Complex of Arran, the preservation of blocks of Mesozoic sedimentary rocks (which are otherwise unrepresented in the island) suggests the former presence of a collapsed caldera. Some of the blocks may have been down-faulted about 3000 feet. Within the caldera, volcanic cones were built up of andesite, dacite and rhyolite and their pyroclastic equivalents.[1]

[1] King, B. C., The Ard Bheinn area of the Central Igneous Complex of Arran, *Quart. J. Geol. Soc.*, **110** (1954), 323–55.

There is no doubt that lavas of the so-called Central Group were erupted from the Mull Central Volcano, and the same mode of origin is probable for other lavas of restricted distribution around other plutonic centres (*e.g.* the Cuillins of Skye); but much more uncertainty surrounds the source and manner of eruption of the more extensive flows of the so-called Plateau Group. Sir A. Geikie long ago suggested that the latter might have been erupted from fissures; but until recently no proof of a direct connection between feeding dyke and a lava flow had been forthcoming. Furthermore the members of the Tertiary dyke swarms which would provide the obvious feeding channels for the lavas are in the great majority of cases younger than the lavas which have survived denudation. On general grounds, however, it cannot be doubted that many of these dykes must have reached ground level and therefore acted as feeders during fissure eruptions. It is inconceivable that parallel, presumably tensional, fissures could be restricted to the *lower* levels of the basaltic plateau. The chances of seeing a direct link between any one of these hundreds of dykes and a particular basaltic lava must be extremely slight; yet it has proved possible to demonstrate that fissure eruptions did actually occur in North-East Ireland when the basaltic plateau was in process of formation.[1] About thirty dolerite plugs occur within the area of the Antrim Plateau, and these are thought to have formed by enlargement at points along fissures through which basaltic magma was passing. Fissure eruptions have occurred within historic times in Iceland, from the Laki Fissure for example.

Major Intrusions of the Central Complexes

The chief interest of the British Tertiary Igneous Cycle lies undoubtedly in the unique variety of ring-dykes and cone-sheets which collectively make up the central complexes. The ring-dykes and all the larger bodies such as bosses and stocks associated with them are described for convenience as 'major' intrusions, although in fact they vary greatly in size and in the coarseness of texture of their component rocks. All these intrusions constitute the basal structures of volcanoes of central type exposed by the removal of the cover of ejected material—surface lavas and tuffs—under which they consolidated. Evidently the rocks of the intrusive phase rose into the base of the volcanic piles which had accumulated during the eruptive phase. It is impossible to estimate the thickness of the cover; but it should be remembered that the *total* thickness of lavas locally reached 6000 feet. Frequently the intrusive rocks are

[1] Patterson, E. M., 'Evidence of fissure eruption in the Tertiary lava plateau of North-East Ireland,' *Geol. Mag.*, **87** (1950), 45–52. See also Walker, G. P. L., *op. cit. supra.*

now exposed at the same levels as the lavas which must have flanked the original volcanoes. None of the intrusions is 'plutonic' in the sense of being of really deep-seated character, though most of the rocks involved fall in the coarse-grain category.

In the main the rocks forming the major intrusions are either thoroughly Basic or thoroughly Acid—intermediate types are virtually absent. The close association of contrasted Acid and Basic rocks is displayed in various ways in each of the complexes. In Skye the largest of all the Basic masses—comprising the 'banded gabbros' of the Cuillin Hills—forms sombre and rugged topography which

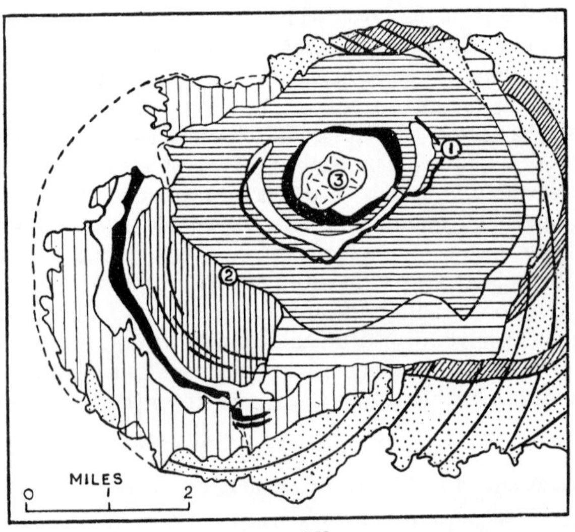

FIG. 152

Simplified sketch-map of central intrusion complex, Ardnamurchan, Scotland. Ring-dykes of Centre 1, oblique ruling; of Centre 2, vertical ruling, and black and white; of Centre 3, horizontal ruling, and black and white. Tonalite of Centre 3, small dashes. Sediments, lavas, etc., spots. A few cone-sheets round Centres 1 and 2 are shown. (*Based on maps in Ardnamurchan Memoir*, 1930.)

contrasts strongly with the adjacent Red Hills formed of granophyres that have given rise to scree-covered, rounded mountains.

In Mull the emplacement of gabbro and granophyre tended to alternate in time, to form a complex series of mainly arcuate intrusions which encircle the principal caldera and to a lesser extent enclose a later intrusion-centre located to the north-west of the caldera. Ardnamurchan witnessed a similar alternation of gabbroic and granophyric intrusions. In this case the intrusions form concentric arcs or rings about three well-defined centres (Fig. 152) which shifted with time along an E.-W. axis. Other aspects of the close association of Acid and Basic rocks are referred to below.

The order of intrusion can often be deduced from the evidence of cross-cutting relationships, marginal chilling and the occurrence of xenolithic inclusions. Some of the contact-relationships in the Tertiary province, however, are very misleading, particularly between intrusions of contrasting Acid and Basic composition. Veins from the Western Granophyre of Rhum (Fig. 156), for instance, penetrate the adjoining mass of Basic and Ultrabasic rocks occupying the centre of the island, suggesting that the granophyre was intruded later than the Basic rocks. However, other evidence makes it reasonable to suggest that the granophyre is actually older:[1] the intrusion of very hot Basic magma caused localized melting and mobilization of the adjacent granophyre. Basic magmas in general crystallize at higher temperatures than Acid ones, so that at an appropriate stage of cooling, melted granophyre could be injected into somewhat consolidated and jointed gabbroic rocks by a process which is appropriately termed 'back-veining'.

The proportions of rock-types vary considerably in the different centres. Thus the Mourne Mountains[2] complex in Northern Ireland (Fig. 154) is composed entirely of granite forming five successive intrusions. Most of the northern half of Arran is occupied by a complex of granites of both coarse- and fine-grained types. In Skye the area covered by granophyres almost equals that of gabbros; but in Rhum, Mull and Ardnamurchan, Basic and Ultrabasic rocks are dominant.

Since it is impossible to describe the intrusion centres systematically in the space available, certain aspects of interest are selected in relation to intrusions which cover the full composition range from Ultrabasic to Acid.

The largest of the Ultrabasic complexes, in Rhum (Fig. 156), provides a convenient starting point for this survey. The rocks are composed dominantly of anorthite-rich plagioclase (An_{85-90}) and Mg-rich olivine in varying proportions, with chromite a notable accessory. The most conspicuous and interesting feature of this complex is the occurrence of well-defined layered units of the order of a hundred feet in thickness. Within each such unit the rock-type varies from 'peridotite' at the base to bytownite-troctolite ('allivalite' of Harker) above, with the development of some bands of pure anorthosite at the top. Because of the contrasts in colour and weathering properties of the alternating olivine-rich and

[1] Dunham, A. C., and Emeleus, C. H., 'The Tertiary geology of Rhum, Inner Hebrides', *Proc. Geol. Assoc.*, **78** (1967), 391–418.

[2] Richey, J. E., 'Structural Relations of the Mourne Granites', *Quart. J. Geol. Soc.*, **83** (1928), 658; but *cf.* Emeleus, C. H., in 'Geology of North-East Ireland', *Proc. Geol. Assoc.*, **71** (1960), 446.

plagioclase-rich layers, these form a very striking feature of the twin mountains of Allival and Askival which dominate the centre of the island. Harker, in his original description of the Complex,[1] believed that the layers were separately and successively intruded; but this has been shown to be inconsistent with the contact relations between the layers.[2] The complex has now been re-described in great detail.[3] Each layered unit resulted from the bottom accumulation of olivine crystals that settled out of the magma body as a result of their early crystallization and high density. The factors controlling the rhythmic precipitation of crystals are not well understood. It is believed, however, that the chromite, magnesian olivine and calcic plagioclase represent concentrates of the earliest precipitated minerals formed at the highest temperatures in a body of basaltic magma which may have been repeatedly replenished from below while the liquid products of differentiation were separated from the cumulates by periodic surface eruptions.

The largest of the Basic and Ultrabasic complexes form the Cuillin Hills in Skye.[4] The complex, originally considered as a single body of 'banded gabbros', comprises several distinctive units with a concentric pattern characteristic of ring intrusions. One of the outer rings is of the same kinds of Ultrabasic cumulates that are found in Rhum: dunites, peridotites and allivalites. The central part of the complex consists largely of olivine-rich eucrites (*i.e.* gabbros containing bytownite), much of it showing excellent banding, and some evidence of cryptic layering.

Layering is of special interest for the evidence it affords concerning the form and mechanism of emplacement of the gabbroic intrusions. Where it is gently dipping and shows evidence of grading it can be ascribed largely to gravity stratification on the Skaergaard pattern, even though evidence of cryptic layering is very slight (as in Skye), or missing altogether. In a number of cases, however, the layering dips quite steeply, suggesting either that it originated by a mechanism other than bottom-accumulation, or that it has been deformed and tilted subsequent to its original formation. The hypersthene-gabbro[5] of Centre 2, Ardnamurchan, was the first of the Tertiary ring in-

[1] 'The Geology of the Small Isles of Inverness-shire', *Mem. Geol. Surv.* (1908), 68–92.

[2] Tomkeieff, S. I., 'On the petrology of the ultrabasic and basic rocks of the Isle of Rhum', *Min. Mag.*, **27** (1945), 127.

[3] Brown, G. M., 'The layered ultrabasic rocks of Rhum, Inner Hebrides', *Phil. Trans. Roy. Soc.*, B, **240** (1956), 1–53; Wadsworth, W. J., 'The layered Ultrabasic rocks of South-West Rhum, Inner Hebrids', *Phil. Trans. Roy. Soc.*, B, **244** (1961), 21–64.

[4] Wager and Brown include a comprehensive account in *Layered Igneous Rocks*, (1968), 408–24, based largely on the work of Carr, J. M. (1952), Zinovieff, D. S. (1958) and Weedon, P., *Trans. Geol. Soc. Glasgow*, **24** (1961), 190–212.

[5] Wells, M. K., The structure and petrology of the Hypersthene-gabbro intrusion, Ardnamurchan, Argyllshire, *Quart. J. Geol. Soc.*, **109** (1953), 367.

trusions to be studied with these possibilities in mind, and it was concluded that the layering in this case was due largely to differential flow movements in crystallizing magma. It was suggested that the magma was forcefully injected, developing a shape which was similar to that of a funnel or inverted cone in depth (roughly parallel to the layering structures) and causing an up-doming of the overlying rocks. The mechanism of ring-faulting postulated for the origin of ring-dykes outlined on p. 154 seems inadequate to account for the

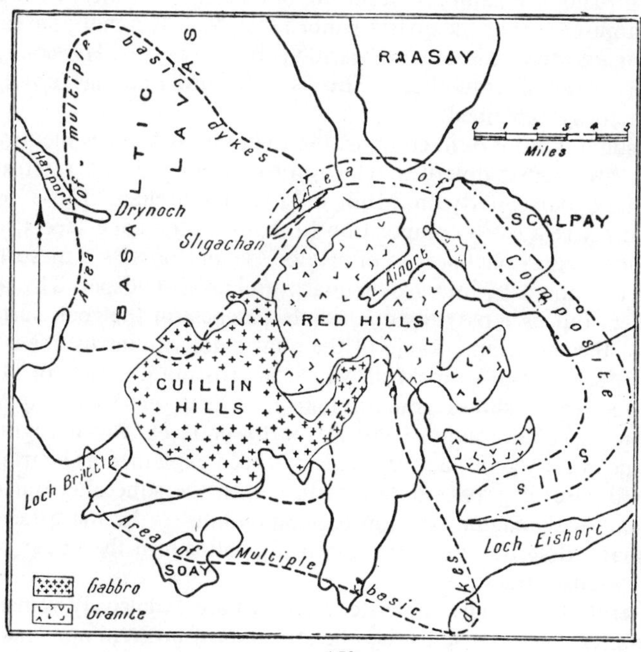

FIG. 153

Map of a portion of the Isle of Skye showing the distribution of the gabbro and granite, and of the multiple basic dykes and composite sills. (*After A. Harker.*)

considerable width of the outcrop of the Hypersthene-gabbro intrusion, and the same is true of a number of other gabbroic ring-intrusions in the Tertiary province, *e.g.* the impressive Great Eucrite of Centre 3, Ardnamurchan (Fig. 152) in which the concentric inner and outer margins are in places over a mile apart. It is important to note that in both cases an unknown amount of the original body of gabbro has been cut away during the emplacement of later intrusions. It has been suggested[1] that the ring faulting and subsidence

[1] Skelhorn, R. R., and Elwell, R. W. D., Central subsidence in the layered hypersthene-gabbro of Centre II, Ardnamurchan, Argyllshire, *J. Geol. Soc.*, **127** (1971), 536–51.

associated with such later intrusions may in fact be the agency by which earlier-formed layering—perhaps originally flat-lying—could become tilted. The process would in effect be one of fault-drag. One extra line of evidence is available in the case of the Hypersthene-gabbro: palaeomagnetic data[1] suggest that however it evolved, the layering acquired its present structure while the rocks were very hot.

Some of the 'eucrites' and olivine-gabbros are closely associated with quartz-gabbros, which may form either separate contiguous ring-intrusions (examples occur in both Centre 1 and Centre 2, Ardnamurchan); or the quartz-gabbro may be only a marginal facies grading into the main olivine-bearing gabbros (*e.g.* the Hypersthene-Gabbro of Ardnamurchan). This is a feature that has not been satisfactorily explained.

Of slightly more Acid composition than the rock-types mentioned above are quartz-dolerites which form a number of interesting intrusions particularly in Mull and Ardnamurchan. These rocks are closely related in composition to many of the cone-sheets; they commonly carry phenocrysts of plagioclase and possess a mesostasis of micrographically intergrown quartz and alkali-feldspar. The latter weathers out as mossy-textured white patches on the rock surface. One of the most remarkable of the quartz-dolerite intrusions is the Glen More ring-dyke in Mull, which in crossing a region of about 1500 feet relief shows gradation in a vertical sense from quartz-dolerite (specific gravity 3·06) at the lower levels to microgranite (specific gravity 2·50) above.[2] This has been explained as a straightforward case of gravity differentiation with the squeezing out of a residual Acid magma fraction (corresponding to the mesostasis of the quartz-dolerite), and injection of the latter into the upper parts of the arcuate fracture.

Several of the quartz-dolerite intrusions are extensively veined by granophyre, giving rise to the phenomenon of net-veining. This is a form of composite intrusion and demonstrates very clearly the extremely intimate association of Acid and Basic magmas which is so characteristic of the Tertiary province. There can be no doubt in many cases that the dolerite and granophyre magmas were intruded together, as shown by the way they share the same intrusion cavities. This is clearly demonstrated in the extensively veined Quartz-Dolerite ring-dyke of Centre 2, Ardnamurchan[3] which has a number

[1] Wells, M. K., and McRae, D. G., Palaeomagnetism of the hypersthene-gabbro intrusion, Ardnamurchan, Scotland, *Nature*, **223** (1969), 608–9.

[2] Tertiary and Post-Tertiary Rocks of Mull, *Mem. Geol. Surv.* (1925), 306.

[3] Wells, M. K., The structure of the granophyric quartz-dolerite intrusion of Centre 2, Ardnamurchan, and the problem of net-veining, *Geol. Mag.*, **91** (1954), 293; Skelhorn, R. R., and Elwell, R. W. D., The structure and form of the granophyric quartz-dolerite intrusion, Centre 2, Ardnamurchan, Scotland, *Trans. Roy. Soc., Edin*, **66** (1966), 285.

of sills projecting from its outer margin. The vein complexes are scattered throughout all parts of this complex intrusion. Despite their simultaneous injection, the Acid and Basic magma fractions have only occasionally mixed to give gradational intermediate rock-types. Generally the granophyre veins cut and brecciate the dolerite quite cleanly, due presumably to the lower temperature of crystallization of the granophyre compared with the dolerite. It is interesting to note that heat has locally been transferred from the Basic to the Acid magma so that at the granophyre/dolerite contacts the latter is sometimes chilled. In the Slieve Gullion complex granophyre occurs as vertical pipes in dolerite. Again the latter is chilled against the former.[1]

Major intrusions of intermediate composition are extremely rare in the Tertiary province. Hybrids of very heterogenous character, collectively termed diorites by Tyrrell, have been formed by the acidification of Basic rocks in the Central Complex of Arran, and to a limited extent such rocks occur also in the other centres, notably in Skye, as exemplified by Harker's 'marscoite'. Almost the only well-defined intrusions in this range of composition, however, occur as bosses at the focus of Centre 3 in Ardnamurchan. These are of tonalitic and quartz-monzonitic composition: the latter was intruded as the youngest member of the complex about 55 m.y. ago.

The major intrusions of granite have been listed above. Although some possess typical granitic texture (*e.g.* the coarse member of the North Arran (Goatfell) intrusion) the majority are granophyric to some extent. Small drusy cavities lined with well-formed crystals of smoky quartz, alkali-feldspar, mica and occasionally rarer minerals such as beryl, are characteristic, particularly of the Mourne Mountains and Lundy granites. As explained on p. 230, the granophyric texture is widely accepted as a feature resulting from simultaneous crystallization of quartz and alkali-feldspar from a melt. A high-temperature magmatic origin has been confirmed for one of the Skye granites from the evidence of the quartz and feldspar, the latter occurring as the high temperature form of alkali-feldspar in the chilled marginal facies and changing to perthite nearer the centre of the intrusion.

Tertiary granophyres scarcely need this kind of evidence to confirm their magmatic origin: however, the intensive study to which these rocks have been subjected has now extended our understanding to the point where we can assess the probable origins of the magma. It appears likely that, regardless of the part played by differentiation of Basic magmas, a substantial contribution has been made by crustal rocks (*e.g.* possibly Torridonian arkoses and more probably

[1] Elwell, R. W. D., Granophyre and hybrid pipes in a dolerite layer of Slieve Gullion, *J. Geol.*, **66** (1958), 57.

Lewisian gneisses and kindred formations deeper in the crust) which have been partially or completely melted as a result of invasion of the crust by large volumes of very hot Basic magma rising from the mantle. Part of the evidence for this lies in trace-

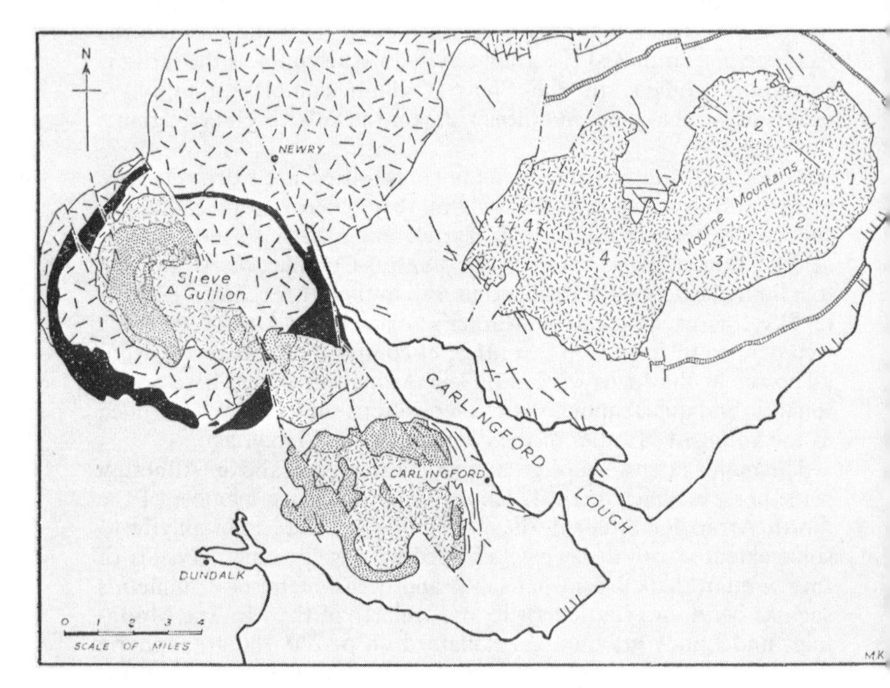

FIG. 154

Map of part of north-eastern Ireland showing the igneous complexes of the Mourne Mountains, Carlingford and Slieve Gullion.

Newry granodiorite (Caledonian), *large dashes*; Mourne Mountains granite, *small dashes*; Tertiary gabbros, *stippled*; granophyres, *dot-dash ornament*; the acid ring-dyke of Slieve Gullion, *black*. One cone sheet is shown round the Mourne granites, which are numbered in order of intrusion.

element geochemistry, particularly the ratio of strontium isotopes[1]; part is derived from experimental data on the melting of the granites[2]; and part is from the field relationships at the contacts between Acid and Basic intrusions. The significance of back-veining

[1] Moorbath, S., and Bell, J. D., Strontium isotope abundance studies and rubidium-strontium age determination on Tertiary igneous rocks from the Isle of Skye . . ., *J. Petrol.*, **6** (1965), 37–66.

[2] Brown, G. M., Melting relations of Tertiary granitic rocks in Skye and Rhum, *Min. Mag.*, **33** (1963), 533–62.

in showing that rocks of appropriate composition can be melted locally to give rise to Acid magma, has been discussed above.

Some of the Tertiary acid ring-dykes were affected by volcanic gases which caused internal brecciation and led to the development of intrusive rocks of agglomeratic appearance. The nature and modes of origin of such intrusive tuffs were discussed in the chapter on Pyroclastic Rocks.

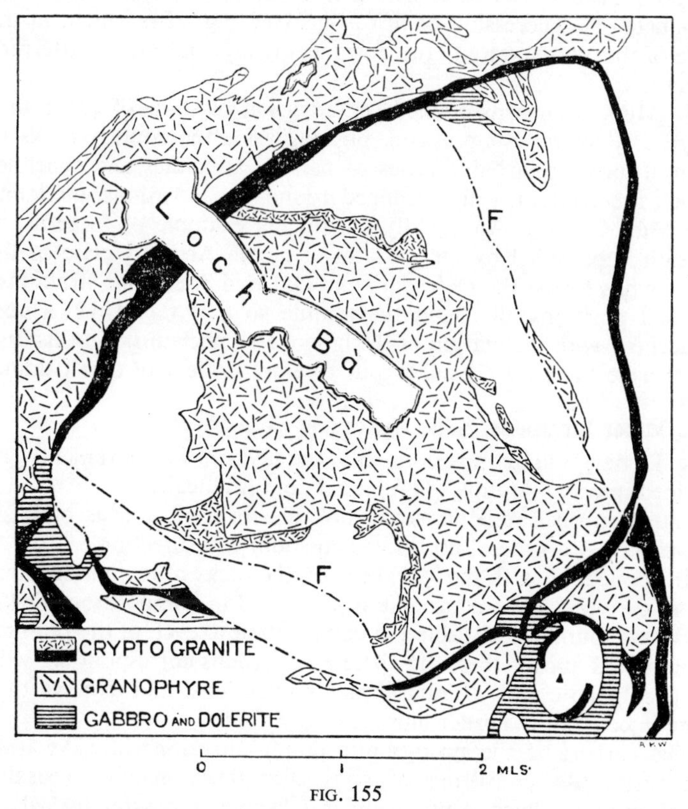

FIG. 155

A Tertiary ring-dyke and associated intrusions, Loch Bà, Mull. (*After E. B. Bailey and others*, in Tertiary and Post-Tertiary Geology of Mull, *Mem. Geol, Surv.*, (1924).) For 'cryptogranite' read 'felsite'.

It is reasonable to assume that the surface expression of ring-dykes of this kind would be the development of agglomerate-filled vents. A good example is provided by an arcuate series of such vents which partially encircle Centre 1 in Ardnamurchan.

The acid intrusions vary greatly in their form and mechanism of emplacement. In the case of a ring-dyke such as the Loch Bà felsite of Mull (Fig. 155), the mode of intrusion is reasonably certain. This is one of the few cases where subsidence of a central block bounded by a ring fault can be shown to have occurred, the displacement amounting to about 3000 feet. Intrusion took place after the faulting. It is possible that cauldron subsidence may have occurred to produce some of the larger granitic and granophyric masses such as those of the Mourne Mountains, and the Red Hills. In both cases it has been shown that the granophyric bodies are products of a succession of intrusions (see Fig. 154) which, in the case of the Western Red Hills form a roughly concentric pattern of outcrops.[1]

In Mull, room was created for intrusion of some of the granophyres by lateral compression, pushing aside the country-rocks to form a most remarkable series of parallel anticlines and synclines which are concentrically arranged round the intrusions encircling the Mull Caldera. An equally convincing example of forceful intrusion is provided by the granites of North Arran. Uprise of the granite has forced the Dalradian and Old Red Sandstone strata into vertical positions all round the granite so that the strike is now concentric with the margin of the latter. The mechanism in this case must have been closely analogous to emplacement of a salt dome.

The Minor Intrusions: Sills, Dykes and Cone-Sheets

The Tertiary minor intrusions, like the lavas, may conveniently be divided into two groups, (1) regional, and (2) local.

Intrusion on a regional scale took the form of swarms of Basic dykes having a general north-west to south-east direction.

In connection with the local centres, the rocks of the minor intrusions were much more variable in form and in composition. Dyke-rocks are found closely allied in composition to each of the plutonic types noted above: there are three chief groups corresponding with the Acid, Basic and Ultrabasic major intrusions, and in Skye they were injected in the order named.

The earliest of the post-granite minor intrusions in Skye were composite sills consisting of an earlier Basic member (basalt), followed by a later Acid member (graphic microgranite) which usually split the former down the centre. They are restricted to an arcuate belt lying north-east of the granite of the Red Hills (Fig. 153).

The local minor intrusions of Acid composition are clearly related to the great granite intrusions, but are slightly more Acid than the latter. In texture they vary considerably. Some are typical porphyritic microgranites with large phenocrysts of quartz and feldspar set in a groundmass which varies from microcrystalline to crypto-

[1] Wager, L. R., *et al.*, Marscoite and related rocks of the Western Red Hills Complex, Isle of Skye, *Phil. Trans. Roy. Soc.*, A, **257** (1965), 275–307.

crystalline. Some of the visible offshoots from the granites are indistinguishable from rhyolites. In a parallel series the quartz and feldspar of the groundmass are intergrown, yielding micrographic, cryptographic and beautifully spherulitic rocks. Rarely these rocks become less Acid, and by failure of the quartz pass into microsyenites, which in Skye and Arran closely approach trachytes in general appearance. These are the 'bostonites' and 'orthophyres' of Harker.

In areas near to the granites, intrusions of a late date include tholeiites and pitchstones, often associated together in composite intrusions, though both do occur alone. The pitchstones in particular are interesting on account of their rarity. First recognized by Jameson about 1700, they were more fully described by Judd, but new features, such as the occurrence of fayalite, are still being brought to light by careful examination. The best known are the pitchstones of Arran and of Eigg. In the former island they occur as sills, but more commonly as composite dykes, such as those found on the foreshore at Tormore, associated with strongly porphyritic microgranites and tholeiites.

Local minor intrusions of Basic composition are in like manner related to the great gabbro and eucrite intrusions: they occur in large numbers in the neighbourhood of the gabbro of the Cuillin Hills in Skye, round the eucrites of Ardnamurchan, and in Mull, Rhum and Arran. Among them are the thick sills of crinanite in South-East Arran, and sills of crinanite and olivine-dolerite in northern Skye. Included in this category also are certain radially disposed dykes of basalt and dolerite, others that are tangential to the boundaries of the major intrusions, and the remarkable inclined sheets of conical form—the **cone-sheets** of Bailey. These were described by Harker from the Cuillin Hills, where they occur as segments of cones inclined inwards towards the centre of the gabbro intrusion at angles up to 45°. It has since been discovered that cone-sheets are developed on an even more spectacular scale in Mull and Ardnamurchan. In the former locality they are concentrically disposed around two centres of eruption. Their injection was not restricted to one short episode as in Skye, neither are they all of Basic composition: some are Acid, and the intrusion of cone-sheets, which started quite early, persisted in Mull intermittently until the close of the cycle. The early cone-sheets are innumerable. They are chiefly olivine-dolerites from 30 to 40 feet in thickness, and are somewhat irregular in their attitude as they repeatedly cut one another. The majority are inclined inwards at angles of about 45°. Somewhat later in date came another series of less basic, finer-grained cone-sheets, the rocks of which include variolites, tachylytes, basalts of the tholeiite type, and quartz-dolerites. These later sheets are so numerous that in a measured section the ratio of igneous rock to 'screens' of country-rock is approximately 2:1.

Passing now to the consideration of the **regional dykes** we meet one of the most striking features of the Tertiary cycle: the dykes are remarkable not only on account of their vast numbers, but also by reason of their regularity of trend over a very extensive area. The regional direction differs little from north-west to south-east, indicating a tension acting at right angles thereto.

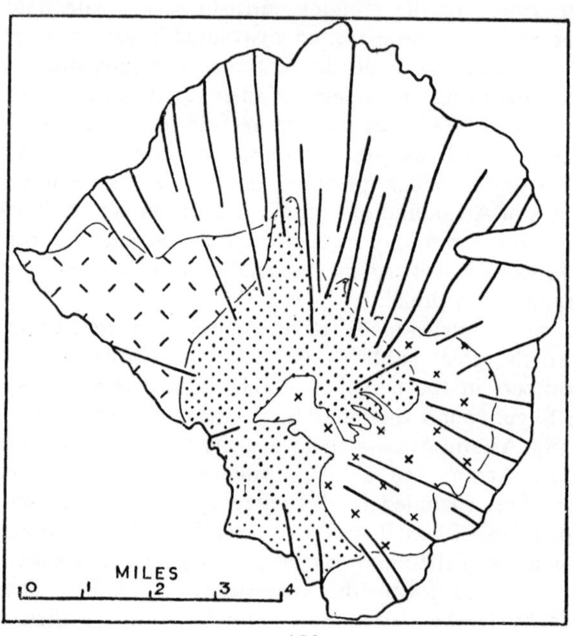

FIG. 156

Sketch-map showing relation between radial dyke-swarm and plutonic intrusions in Rhum. Ultrabasic, *spotted*; basic, *crosses*; granite, *dashes*; Torridonian, *blank*.

Localization of swarms of north-west dykes was no doubt due to the presence of wedges of basic magma lying at no great distance beneath the crust and causing belts of weakness. There is a marked tendency for the dykes in the neighbourhood of the plutonic centres to swing away from the regional direction and to crowd in towards the centres, while few occur in the intervening belts: the fractures occupied by the dykes crowd in towards the weak spots in the crust, rather than maintain their direction through adjacent stronger parts. Thus the dykes are concentrated into 'swarms': these comprise the Skye, Rhum, Mull, Islay, Arran, Mourne and Carlingford-Slieve Gullion swarms.

Although chiefly concentrated in the Hebrides, in the adjacent parts of the mainland of Scotland and in northern Ireland, Tertiary dykes are also found in northern England, for example, the Cleveland dyke; occasionally in the Midlands, such as the nepheline-olivine-dolerite of Butterton in North Staffordshire; while in North Wales they have been recorded from Anglesey[1] and various localities in Caernarvonshire.[2] Those of North England have been shown to converge upon the Mull swarm,[3] while the Welsh dykes clearly belong to the Antrim swarm.

The dykes were formed from Basic magma which, rising rapidly, suffered little contamination or differentiation. Essentially the dyke-rocks are dolerites or basalts according to their grain-size. Those of the thinner dykes are indistinguishable, under the microscope, from basaltic flows, and in a representative series from this suite, all stages from basalt-glass (tachylyte) to quite coarse ophitic dolerite may be found. In addition to normal olivine-bearing and olivine-free types, others bear evidence of a tendency on the part of the magma to migrate towards the alkali pole: common augite is replaced by deep purple-brown titanaugite, while there is a moderate amount of analcite, or other zeolite, in the interstices between the labradorite and other earlier formed minerals. These analcite-dolerites constitute the Crinan type of Flett, and it should be noted that no means of distinguishing between the 'crinanites' in the north-west dykes (presumably of Tertiary age) and identical rocks found in close association with 'Permian' intrusions of the Midland Valley of Scotland has yet been discovered. Among the intrusive basalts, two types may be specially mentioned. One, first described from the Cumbrae Islands between Arran and the mainland, is a porphyritic basalt with large pale-coloured phenocrysts of anorthite set in a groundmass of laths of labradorite, enstatite and augite, with an abundant glassy base.[4] The second type is an andesitic basalt with intersertal texture (tholeiitic), the type-occurrence being the tholeiite of the Brunton dyke in Northumberland. These are of essentially the same composition as the rather coarser quartz-dolerites. In contrast to the other dyke-swarms, that of Mourne consists largely of Intermediate (andesitic) rocks.[5]

Among the latest minor intrusions are some of Ultrabasic composition, which occur chiefly in radiating dykes in or near to the

[1] Greenly, E., Geology of Anglesey, *Mem. Geol. Surv.*, (1919), 684–90.
[2] Matley, C. A., *Quart. J. Geol. Soc.*, **69** (1913), 525.
[3] Holmes, A., and Harwood, H. F., The tholeiite dykes of the North of England, *Min. Mag.*, **22** (1929), 1.
[4] Tyrrell, G. W., *Geol. Mag.*, (1917), 305–15, 350–6.
[5] Tomkeieff, S. I., and Marshall, C. E., The Mourne dyke swarm, *Quart. J. Geol. Soc.*, **91** (1935), 251. See also Tomkeieff, S. I., and Marshall, C. E., The Killough-Ardglass dyke swarm, *Quart. J. Geol. Soc.*, **96** (1940), 321.

Basic plutonic complexes. Petrographically they resemble the peridotites which were intruded at the commencement of the plutonic phase, and include picrites (with anorthite), augite-periodotites and dunites.

We have finally to notice a few scattered intrusions of rare rock-types occurring on the outskirts of the Tertiary Province. These are of more strongly sodic character than the majority of the normal rocks of Tertiary age in this country, and include the sodic granites of the island of Rockall,[1] the riebeckite-microgranite of Ailsa Craig, and the riebeckite-trachyte of Holy Isle, Arran.

It remains to mention two other complexes which may be appropriately considered here.

Lundy Island[2] lies off the Somerset coast, well within range of the Armorican granites of south-western England. The island consists almost entirely of granite, cut by a varied assemblage of dykes. Petrologically the granite is much like the West of England types, the earlier intrusions being sodi-potassic, with perthitic orthoclase the dominant feldspar; but there is a larger proportion of albite-oligoclase in the later types. In certain points of detail the granites recall the Tertiary intrusives of the Mourne Mountains, particularly in textural similarities: the occurrence of miarolitic texture, with small beryls in the cavities, among the normal minerals. Of the dykes over nine-tenths are basalts and dolerites—some are olivine-bearing, some quartz-bearing, some are classed as tholeiites, while another link with the Tertiary Province is afforded by the pitchstones and occasional tachylytes like those of the Hebrides. Less acid dykes occur, some being trachytic and others are micro-syenitic. The isotopic age, 52 m.y., proves the Lundy granite to be early Tertiary.

Secondly, on maps showing the Tertiary dykes, a swarm unrelated to any of the visible volcanic centres crosses Islay and Jura. The inference is that a concealed complex lies a little distance from the coast of Jura.

With the close of the Eocene cycle, igneous activity ceased in Britain. Looking back over this brief account, one is struck by the variety of the rock-types occurring among the British rocks, by the comparative perfection of the record so far as some of the cycles are concerned, and on the other hand, by the many problems connected with the genesis, age and relationship of these rocks, which still await solution.

[1] Sabine, P. A., The geology of Rockall, North Atlantic, *Bull. Geol. Surv. G. B.*, **16** (1960), 156 and 178.

[2] Dollar, A. T. J., The Lundy Complex: its petrology and tectonics, *Quart. J. Geol. Soc.*, **97** (1941), 39.

INDEX

Bold numerals refer to maps or figures

Index